ferro·cement

design, techniques and application

ferro·cement

design, techniques and application

bruce bingham photographs and illustrations by the author

B. Bingham

cornell maritime press, inc. — cambridge, maryland

Library of Congress Cataloging in Publication Data

Bingham, Bruce,   1940-
    Ferro-cement: design, techniques, and application.

    Bibliography: p.
    1. Concrete boats.   2. Yacht-building.   I. Title.
VM323.B56   623.82'07'34   74-4255
ISBN 0—87033—178—7

Printed and Bound in the United States of America

To

SABRINA

who waited long, listened much, loved more,
suffered most, asked for little but gained least.

contents

section one

the builder's decisions

section two   patterns

section three   setting up

acknowledgments

It would be virtually impossible for me to extend individual thanks to all those persons and organizations who have contributed so unselfishly toward the production of this volume. I have met with overwhelming and continual cooperation which has made my work much more gratifying than it otherwise would have been. I have sorted through hundreds of pounds of catalogs, technical reports, photographs, letters and tape recordings from builders, professional and amateur alike, and their comments have been both valuable and necessary toward the development of ferro-cement excellence. I most particularly extend my deepest gratitude to the following contributors.

Jay R. Benford and Associates, naval architects; Samson Marine Design Ltd., naval architects; Peter Ibold, naval architect; William Preston, naval architect; Edward S. Brewer and Associates, naval architects and consultants; Steven R. Seaton, naval architect; Joseph Hartog (Holland Yacht Design); Helmut Kanzler (Ferro-Crete Marine Company); Continental Corporation, yacht builders; Marine Crafts International, yacht builders; Cleary Marine Enterprises, yacht builders; Bett's Marine, yacht builders; Ferro-Boat Builders, Maryland; Platt Monford (Aladdin Products, Inc.); A-1 Marine Surveyors, San Diego; Great Lakes Ferro-Cement Corp., yacht builders; Adhesives Engineering Company; International Paint Company, Inc.; Howard-Riffel-Sims Inc.; Morse Controls Divisions of North American Rockwell; Wm. E. Hough Company (Wagner Hydraulic Steering); The Edson Company; Buck Algonquin Company, Inc.; Richmond Ring Company; Alson's Products Corporation; Department of the Environment, Industrial Development Branch, Fisheries Service, Ottawa, Canada; Division of General Research, Bureau of Reclamation, Denver, Colorado; Radiation Division Department of Applied Science, Brookhaven National Laboratory; Department of Navy, Office of Naval Research, Arlington, Virginia; Professor Robert Brady Williamson, Department of Civil Engineering, University of California; Kenneth A. Christensen, University of California; Tra-Con Incorporated, Industrial Polymers; Mallas de Acero para la Construccion, S.A., Madrid, Spain; Lt. j.g. Paul G. Snyder, USCG; American Association of Corrosion Engineers; International Lead Zinc Research Organizations, Inc.; Perkins Marine Lamp and Hardware Corporation; Adler and Barbour, Marine Hardware; Bernd Blanke, Associate Editor, *Kazi Magazine,* Japan; Clark and Sheri Mason; Mr. and Mrs. Jack Wardell; Mr. and Mrs. Wayne Heyerly; Don Terry; Tom and Jean Carroll; Norman Bonnenburger; Brad Wallingford; Tom Mingoia; Wesley Storer and Terry Bragg; Mr. and Mrs. Howard Kiefer; Jack R. Whitener; Richard Munezer, North American Ferro-Cement Marine Association; John and Steve Daniels, master plasterers; Frank Beumer, master marine electrician; James Crowther; Gerry Acord; Richard Reynolds.

To my own staff, past and present (not necessarily in order of importance): Susan Neidenbach (general office); Jan Jackson (stenographer/typist); Charles Steven Wray (draftsman); Cheryl Ann Stayton (stenographer/typist); Patricia Evans (stenographer/typist); Christine Elston (stenographer/typist); Heidi Baldwin (darkroom); John Coudray (darkroom); Paul Kotzebue (draftsman); Debby Bingham (proofreader); Lisa Steele (general office); Becky Bush (general office); Fred Bingham (author's father, critic and proofreader *par excellence*); Helen Gustafson (manuscript typist); Sharon Jepson (proofreader and manuscript typist); and especially Jessica Bush, my loyal and patient secretary, stenographer, typist, friend, and confidante, whose fortitude and continuous encouragement never faltered throughout this incredibly grueling project.

I must also extend my gratitude to the invaluable outside services which aided in the production of illustrations for this manuscript: Santa Barbara Photoengraving, Santa Barbara, California; Dailey Engraving, Walnut Creek, California; Camera-Graphics, Lafayette, California.

My acknowledgments would not be complete without a mention of thanks to the real unsung heroes—the hundreds of individual amateur and professional boatbuilders with whom I had the opportunity to speak and question (names, unfortunately, have been lost in their numbers) and to the well-intentioned persons who have become subjects of my (constructively) critical photographs and comments.

Bruce Bingham

Neither the field of ferro-cement nor this volume on its marine application could have been conceived without the contributions of Pier Luigi Nervi, one of the greatest structural engineers of our time. Born in Sondrio in 1891, he received his engineering degree at Bologna and was awarded his professorship in technology and construction techniques at the University of Rome. As a modern master of prestressed concrete, his most widely influencial works include thin-shelled airplane hangars, warehouses and the Turin Exhibition Halls. Throughout his lifetime he experimented with ferro-cement in marine applications, having designed and built the 165-ton motor yacht, **Irene.** The crowning glory of Nervi's work, seen here, was the Olympic Sports Palace in Rome, which stands today as a tribute to his work in ferro-cement.

foreword

In the days of wooden boats the amateur builder was always advised to build a dinghy before trying his skills on a more complex cruiser as in this way he would gain valuable practice in handling tools and materials. Besides, the dinghy would prove useful as a tender for the larger craft when it was launched. Indeed, the builder was told not to attempt too large a cruiser as a starter and to work his way up through a succession of craft until, finally, someday he might hope to build a 40-footer. It was also stressed that none but the most patient and skilled worker should ever tackle anything as large as 40 feet without an ample supply of outside labor.

With the advent of ferro-cement, things have changed drastically and not for the better. Ferro-cement has come to be thought of as a miracle material by far too many amateur builders. Men who should know better have jumped on the bandwagon and touted the material for super strength, simplicity of construction and low cost. As a result, too many hapless amateurs without the skill to build that first dinghy have attempted projects far beyond their ability or their financial means. The outcome has been a larger number of hopelessly inadequate craft, many of them never to be completed and eventually to be abandoned as empty hulls sitting in backyards or city dumps. Unfortunately each one of them represents a considerable waste of labor and money and each one is also some would-be sailor's broken dream.

Ferro-cement is not a miracle material and though good boats can be built of it there is just as much patience, skill and hard labor required as with any other type of construction. There is almost as much money required as well since even 50 percent savings on the cost of the hull material works out to only about a 15 percent savings on the overall cost of the completed craft.

Bruce Bingham, being a competent and realistic young designer, has not espoused the wild claims about ferro-cement and has put a great deal of study into the true merits and the defects of the material as a boatbuilding medium. This book reflects his honest and intelligent approach to the subject and will provide both the amateur and the professional builder with complete and up-to-date information on the latest techniques in ferro-cement construction and its recent offshoot, **FER-A-LITE.**

Possibly the most useful part of the book as far as the amateur is concerned are the chapters on designers and boat design in general. Too many books have been written about boat construction, in ferro-cement and other materials, which fail to discuss the factors that go into a good design. There is little point in telling an amateur how to build a boat if, at the same time, you do not tell him how to select a suitable design to start with. All the patience, skill and quality materials in the world cannot be combined into a fine boat unless there is a well planned design as a foundation for the project.

Bruce has presented the selection of a design thoroughly, discussing in detail the advantages and disadvantages of various features and rigs. While I do not always agree with his conclusions (that, of course, is the reason my designs do not look exactly like Bruce's), I feel that in attempting to educate the builder about design, material costs and the other vital aspects of the overall job of building a boat, he has done a great service. It should prevent many amateurs from selecting an unsuitable design or one beyond their means or ability.

One point where Bruce and I disagree is on the need for lofting. He believes the amateur is best advised to work from full-size patterns while my opinion is that the builder will learn more about his future vessel if he prepares his own lofting. Regardless of our feelings in this regard, the book contains an excellent section on lofting that should dispel any terrors the amateur might have about this relatively simple process.

In the long run though, no one can learn how to build a boat by reading a book anymore than he can learn to ski by reading a book. Boatbuilding can only be learned by getting your hands dirty and sweating a lot while working on an actual vessel. A book is simply another tool to be used in this boatbuilding process and like any other tool, if it is a good one, the work will go easier. *FERRO-CEMENT: Design, Techniques and Application* is a very good tool indeed and will provide the builder with the knowledge he needs to do a workmanlike job.

Ted Brewer
Edward S. Brewer and Associates
Naval Architects, Marine Surveyors
Director, Yacht Design Institute
Brooklin, Maine

section one

the builder's decisions

ferro-cement, the dream builder

Every now and then there comes along a material claiming to fill a gaping void in the endless needs of common man. Revolutionary developments conjure up visions of dwindling housework, maintenance-free finishes, overwhelming costs savings and eternal strength.

To add to the list of recent miracles such as plastics, teflon and synthetic fabrics, we now boast of cement boats so lasting, so inexpensive that any American schoolteacher may trek through the gold-plated yacht club gates and romp with the yachtsmen elite.

Promises, promises! Technological advancement, history tells us, is never such a one-sided matter. There goes with it many drawbacks, pitfalls, failures and slipshod applications, not to mention a great deal of responsibility for research and leadership by its developers.

We must, therefore, look at ferro-cement yacht building from many directions in order to form a truly objective opinion. The rosy pictures painted by some of the promoters of ferro-cement construction as being so cost-cutting, laborsaving and quality-producing, require a long, healthy scrutiny in order to put things into their proper perspective.

While from this I may seem somewhat skeptical about concrete, I must point out that, as a designer, I am both convinced of its potential as well as committed to its growth by virture of years of study and personal application. I am of the firm belief that a day will come when concrete as a boatbuilding material will find its rightful place of respect within the American yachting community . . . but it has a long way to go! It will take a great deal of delicate grooming, a lot of polish and many more fine finished examples than presently exist to sway the opinion of the established boating fraternity.

Ferro-Cement in a Nutshell

The term "ferro" refers to a steel armature of wire mesh and rod which is impregnated with a carefully formulated concrete to form a thin but amazingly strong hull shell. The armature is usually preformed over a temporary wood mold (which may be either upright or inverted); this is the element that dictates the quality of the shape of the finished boat. (Plates 1 & 2)

While relatively little actual expense is involved in building this mold, a tremendous effort must be expended to ensure

perfect alignment and fairness or the builder may produce a mass of bumps and hollows, the norm rather than the exception in concrete boats.

The armature, stapled or tied together with wire, becomes extremely rigid and capable of maintaining its own shape with a minimum of inside support and may be stripped of the mold for accessibility prior to cementing.

The concrete itself may be initially pumped into the substructure, applied by trowel pump or Gunite. One hundred percent permeation is absolutely essential for strength and finishing must be done by hand. (Plates 3 & 4)

Plate 1. Wood mold construction for a Samson Marine Design 63' **Sea Witch**.

Plate 2. The steel-wire armature for a Bingham 46' **Andromeda.**

Once the hull has been completely covered with concrete and troweled smooth, the curing process begins: twenty-eight days of continual wetting and high humidity around the structure, accomplished by covering the boat or scaffolding with thin plastic sheets to prevent fresh air from drying the hull. (Plate 5) Curing may also be accomplished with steam in a much shorter time.

Once cured, the mold is removed unless an open or web system is employed. (see Chapter 8) Internal structural components are added to the hull and the builder finds himself well on his way toward becoming the owner of a new boat. (Plate 6) Very few critical skills are required of the amateur and most of these can be acquired through practice. They are fully explained in many other fine books on the market, from joinery to engine installations, and may be obtainable at your local library.

Acceptability

The greatest deterrent to the future acceptance and growth of ferro-cement within the yachting community lies not in the construction but rather in the design aesthetics currently employed. The increasing impression among knowledgeable boatmen is that ferro-cement vessels are less than good-looking, an attitude caused by the fact that the greater majority of cement boats under construction are in the hands of those with little or no sailing or cruising experience whatsoever. Most designs have been chosen because of some superficial element rather than the practical and are being perpetrated by "overnight experts" basing their opinions on hearsay from other unreliable sources. The evidence can be clearly seen in the resurgence of obsolete design theory, of revised "old-timers," of rigs and deck plans utterly impractical for long-range cruising.

This lack of experience and marine exposure is seen in low standards of acceptability, especially in fairness and finish. (Plates 7 & 8) Many boats appear on the scene with crooked waterlines and bumps and hollows on their topsides. These are the norms rather than the exceptions. These flaws can be attributed only to the untuned eye of the amateur and in all but a few rare cases these errors in construction could have been corrected or avoided through the exercise of patience and objective personal observation, not to mention a little more guidance and leadership from the designers.

Strength, Maintenance and Repair

Several well-known designers and universities have recently published the results of independent testing of ferro-cement using various formulas of both steel substructures and concrete mixtures. How many times have you been told that ferro-cement is "super" strong? Unfortunately, "strength" is a comparative term and derives its value as such. You must ask yourself the question, therefore, to what shall we compare ferro-cement in order to establish its relative strength. Is ferro-cement as strong as a soda cracker? Considerably more so! Is ferro-cement stronger than steel? Certainly not! Let us, therefore, establish the position of ferro-cement on scales for tensiles, compression, elasticity, and impact resistance to penetration.

Figures 1-4 hardly justify the claims for strength being perpetrated upon the unknowing public. Perhaps the comparisons are somewhat out of proportion, however, in that the figures represent cubic inches of the various materials. We certainly know, for instance, that we would not expect to build a 40' vessel with 1" steel plate, as it would be overly strong and heavy. At the same time, it would be absurd to

Plate 3. John Daniel, veteran of over 150 plastering jobs, pumps concrete onto another new Bingham-designed **Flicka** hull.

attempt construction of a 1" Douglas Fir-planked 40' vessel without relying upon a complicated frame structure. Let's compare the various materials once again, using the appropriate thicknesses normally found on a boat of this size. (Figs. 5 & 6) Even when ferro-cement is compared in this manner, it still falls to the low end of the scale in almost every respect.

The ferro-cement figures to which I refer are those derived from the most common use of ¼" mild steel rod and ½" hexagonal mesh armature. The strength figures of ferro-cement can be increased significantly through the use of high-tensile reinforcing rod, square welded mesh, as well as impregnation of the mortar with polymer or vinyl additives. Their use, however, is still restricted through lack of knowledge and availability, and increased cost.

Concern for chipping can be disregarded if certain precautions are taken during construction such as rounding the sharp edges and adhering to recommended cement mixtures and curing procedures. If, however, such small damage is detected, application of epoxy grout or any commercial polyester filler will permanently correct any such scars.

Major damage from impact requires the chipping away of broken concrete, reshaping or replacement of the steel armature and the application of fresh cement. This procedure is far less complicated and considerably less expensive than repairs in other materials, rarely requiring professional services or lengthy overhauls. This aspect of ferro-cement construction is one of its finest attributes and can be of tremendous benefit to the cruising man who plans on sailing into areas where boatyards may be scarce or nonexistent. In many cases, small underwater punctures can be closed immediately by applying bedding compound or other nonsoluble paste to the fracture. Vaseline may even work!

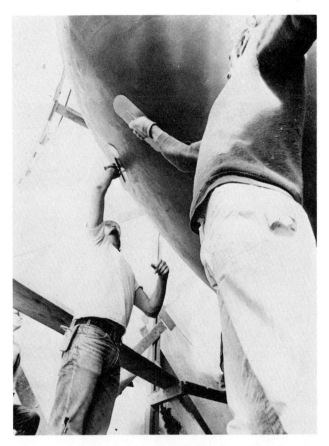

Plate 4. With the grace of ballerinas, John and his son, Steve, stroke a hull surface with steel trowels for the final finish on the Bingham-designed **Andromeda** stern.

Plate 5. The "hothouse," a temporary covering used to control humidity during curing.

Plate 6. An **Andromeda** ketch fully cured and braced. The stains on the topsides are caused by mineral deposits left by running water. A Bingham design. (Malloy Photo)

Plate 7. Only one of hundreds of abandoned concrete hulks lying idle in boatyards throughout the country. The builder thought he knew a "better way" of doing it but lacked experience, guidance and patience. He is now deeply in debt with an unsalvable vessel.

STRUCTURAL STEEL	60,000
ALUMINUM	42,000
FIBERGLASS ROVING	35,000
DOUGLAS FIR	2,150
FERRO-CEMENT (TO FIRST CRACK)	1,600
PHILIPPINE MAHOGANY	~1,200

Fig. 1. Average tensile strengths (p.s.i.) of common boatbuilding materials.

STRUCTURAL STEEL	60,000
ALUMINUM	32,000
FIBERGLASS ROVING	17,000
FERRO-CEMENT (AVERAGE FULL CURE)	10,000
PHILIPPINE MAHOGANY	6,750
DOUGLAS FIR	2,000

Fig. 2. Average compression strengths (p.s.i.) of common boatbuilding materials.

STRUCTURAL STEEL	28.9
ALUMINUM	10.0
DOUGLAS FIR	1.6
PHILIPPINE MAHOGANY	1.5
FIBERGLASS ROVING	1.45
FERRO-CEMENT	1.3

Fig. 3. Modulus of elasticity (x 10^6) of common boatbuilding materials.

STRUCTURAL STEEL	490
FERRO-CEMENT	168
ALUMINUM	166
FIBERGLASS	96
DOUGLAS FIR (12% MOISTURE)	34
PHILIPPINE MAHOGANY (12% MOISTURE)	33

Fig. 4. Weights (lbs./ft.^{3}) of common boatbuilding materials.

MATERIAL	NORMAL THICKNESS	WEIGHT lbs./ft.²	SHELL WEIGHT
STRUCTURAL STEEL	7/32"	11.0 (FRAMES)	12,100
FERRO-CEMENT	3/4"	11.0	12,100
PHILIPPINE MAHOGANY	1 3/8"	6.8 (FRAMES)	7,400
DOUGLAS FIR	1 3/4"	6.5 (FRAMES)	7,150
FIBERGLASS	3/4"	6.0	6,600
ALUMINUM	1/4"	5.5 (FRAMES)	6,050

Fig. 5. Thicknesses and weights of common boatbuilding materials as would be used on an average 40,000-pound displacement auxiliary yielding approximately 1,100 ft.² hull surface.

MATERIAL	NORMAL THICKNESS	MODULUS of ELASTICITY ADJUSTED (x 10^6 p.s.i.)/in.
STRUCTURAL STEEL	7/32"	6.3
DOUGLAS FIR	1 3/4"	2.8
ALUMINUM	1/4"	2.5
PHIL. MAHOGANY	1 3/8"	2.06
FIBERGLASS	3/4"	1.09
FERRO-CEMENT	3/4"	.98

Fig. 6. The adjusted modulus of elasticity of various materials as would be used on an average 40,000-pound displacement auxiliary.

Appearance

When properly faired, a cement hull can produce a finish comparable to aluminum or steel but sometimes requires a great deal of grinding and sanding if the builder has not been careful. The outcome depends entirely upon the degree of the builder's patience, fortitude and availability of man-hours. (Plates 9 & 10) Epoxy sanding sealers and finish coats are usually employed with excellent results requiring no more upkeep than any other material. There is little tendency of the steel to bleed through, providing the substructure has been completely covered with concrete, while zinc anodes grounded to the steelwork alleviate any concern over electrolysis. Cement hulls do have the inherent drawback of sweating (much like fiberglass), but this can be easily solved with the application of cork, foam, acoustical tile or other insulations.

Weight

Until just recently ferro-cement yachts were considered to be unreasonably heavy vessels. (Fig. 5) The weight for the normal 7/8" shell has been generally about 13 pounds per square foot regardless of the size of the boat. Thus, the greater the displacement, the lighter the relative weight of the shell. Assuming the same hull thickness, a boat of 38' on the waterline will be of a very normal cruising weight, but as this length decreases her relative displacement will increase, being accompanied by a smaller ballast capacity.

Technology is now advancing so rapidly that shell thicknesses of only ½" and less are being built into boats under 20' in length (claimed "impossible" three years ago) and concrete deck weights are approaching those of wood structures. (Plate 11) Some designs calling for cement decks and bulkheads lower the ballast capacity, risking stability while sacrificing sail-carrying power to the detriment of performance. The shape of the hull can make up for some tenderness but this is not always done in some plans being distributed, so the prospective builder should be fully aware of drawbacks of constructing this cheaper approach. It is wise to note at the same time that cement deck structures raise the yacht's center of gravity which, in most circumstances, cuts severely into potential sail power. Cement deck construction at the same time as the hull also complicates the installation of bulkheads and interior components, while creating severe lighting and ventilation problems during building. In a sense, it is like putting your trousers on before your underwear. Any professional shipwright will agree.

Plate 8. The builder of this vessel continually said, "It won't make any difference!" But, if the exterior finish is any indication of how she's built, the difference could be life or death.

Plate 9. **Gray Gull,** from the drawing board of Steve Seaton, is one of the finest examples of ferro-cement ever seen. Concentration on aesthetic detail and finish yields a vessel that would stand up to any comparison.

Plate 10. The incomparable **Limmershin**, Steve Seaton designed, is as proud, proper and beautiful as any of her fiberglass, wood or aluminum counterparts. (Featherstone Marine Photo)

Plate 11. Upon completion of plastering, this 20' **Flicka** sloop will have a hull shell of only ½" thickness. Its tremendous strength is derived through the use of square-welded mesh and spring-steel reinforcing rod. A Bingham design.

Plate 12. She may not be the best looking boat in the fleet, but financially the **Ferro-Maid** has paid her way many times over. Initial cost and low maintenance are the keys to competitive commercial fishing.

7

Plywood deck structures over wood beams are quite common and very sensible in ferro-cement construction, lending greatly to a more professional finish of the completed boat. Interiors tend to be less sterile in appearance and very strong but admittedly more complicated to build and quite expensive; and great care must be taken in creating a firm, watertight seal at the sheer. Internal components can be integrally attached to the hull by bolting to reinforced cement frames or with epoxy grout, as is now fairly common in the commercial boatbuilding industry.

Future of Ferro-Cement

In spite of the many failures caused by anxiety, lack of building and cruising experience, gross underfinancing, misleading promotions, and a frequently lackadaisical leadership, ferro-cement has a great deal going for it as a material tuned to amateur ability. Few other methods can promise as large a hull for the low cost and ease of construction and repair while yielding a product worthy of pride and confidence. As long as the builder is aware that creating a boat of any material is no mean accomplishment; that spare time may be translated into better than a year of knuckle-busting, back-breaking effort, the results depend more on a good eye, ingenuity, patience and pocketbook than on the material. It is safe to predict that ferro-cement will win its rightful place of respect and admiration in the world of American boating. Its application in workboats, fishing vessels and tugs cannot be paralleled for performance or upkeep and all that ferro-cement requires now is polishing and grooming to bring it up to the finest yachting standards we have become accustomed to in the past. (Plates 12 & 13)

Plate 13. Because a power yacht is more hull than equipment, the inverse proportion to sailboats, a greater material cost savings can be anticipated with ferro-cement when compared to other materials. This Hartog-designed 86-footer would have cost thousands more in wood or aluminum.

the naval architect

Who is he?

Obviously the yacht designer doesn't just happen simply because it's what he wants to be, nor is it the result of intensive study alone. It is much deeper than this. It is a combination of many elements that only time and experience can mix into the proper proportions.

The extremely delicate subtleties of marine aesthetics are so critical that only years of yachting exposure can instill the finest sense of visual proprieties. What may be perfectly acceptable to a less experienced boatman may be quickly felt (if not actually seen) by the lifetime sailors around him. These subtleties may appear to the trained eye as almost imperceptible angles and curves, awkward relationships or as "rink-a-dink" mismatches of sizes and shapes. As most designers have been literally reared amidst boatyards and marinas, their visual awareness is extremely acute, knowing the impact the imperfection may eventually have on a vessel's value.

No reputable architect ever made the grade without first having thousands of miles of offshore sailing as the breeding ground for his theories. Only at sea can the designer find faults in his dimensions and feel the problems of livability under way. He quickly finds that a perfectly workable galley at dockside becomes a nightmare at a 20° heel and that his comfortable little stateroom turns into a dungeon. It is in the crashing swells of the open ocean and the violent pitching of the ship under him that he discovers doors are too small, berths too narrow, spaces confining and stowage lacking. Because even the largest yachts are still very small in comparison to most homes, miniaturization and cramming of interior units become much more apparent.

Even the laziest cruising designer has maintained an active hand at racing, not necessarily because of his competitive urges but because it keeps him in tune with the details and requirements of performance which can (and should) be transmitted into even the heaviest schooner. Granted, you may never race your own boat, but the conscientious architect is largely responsible for getting you to your destination safely with the least amount of effort, fuel and resources.

Of course, the architect has spent countless hours on study but it doesn't end when he opens his office for the first time, lest he become static and out of touch. His walls are already jammed with texts and reports but he will continue to add shelves as technology and theory change. Many hours per week must be given to research on materials and methods as well as updating catalogs and design files, reviewing new equipment and personally observing construction to find better ways of doing things. The more the

designer learns, the more he realizes how little he knows and how much opportunity for improvement lies before him. He cannot become complacent for one instant.

The normal day of the designer is not one of cranking out drawings at the board or puffing leisurely on his pipe while dreaming up new creations. It is one of painstaking calculations (hardly a romantic duty), answering phones and letters, supervising his staff (such as it is), meetings with clients or prospective ones, "punching" hardware catalogs and compiling specifications.

Most of my colleagues work late into the night and on weekends to make up for time mysteriously lost but it is the cost of any successful operation in an increasingly demanding field. In a word, the architect eats and sleeps boats day in, day out. I have had the experience many times of finding in the hours before dawn the solution to a problem I had wrestled with for weeks. More than once I've jumped to my drawing board out of a sound sleep to put a hot idea on paper. A regular occurrence now is to work 'round the clock to meet a pressing deadline and both my dog and I have gone without dinner out of my reluctance to turn out the light on an exciting project. The girls in the office always know when there's something new in the wind as evidenced by the full ashtrays and empty coffee cups in the morning.

The Responsibility of the Architect.

Based on his experience and training, the architect must not only weigh the criteria of the individual owner or the general public, he must impart hundreds of recommendations concerning every detail of construction and rigging. Each new boat necessitates careful evaluation of all that affects the vessel's seaworthiness, convenience, price and finished appearance. These translations are incorporated into as many as 50 engineering drawings and dozens of calculation pages requiring weeks to accomplish. In developing coefficients to assess performance and construction, the designer continually refers to the competitive design files for comparisons upon which to judge his new creation. Every line drawn, each specification, is the result of all boats which have gone before and a new creation, even before it is built, should never be thought of as "unproven." The characteristics and idiosyncrasies of all predecessors are thoroughly sorted and modified (as necessary) before releasing the offspring for construction.

The role of the architect does not end with his signing his approval on the last engineering drawing. Feeling that each new boat is the direct reflection of his own philosophy, experience and sense of quality, he is deeply concerned with the actual construction when the new vessel begins to emerge from the wire mesh, yards of fiberglass or oak chips. During the development of a prototype for a new stock design, the architect will spend as much time at the construction scene as the lay-up men and carpenters.

Stock Designs

The architect's decision as to what type of boat he will release to the public is not based on his personal taste alone; if this were the case he would be designing boats for himself until the end of time with only an occasional sale. Because he is in business for profit, he must objectively respect his market by reviewing the requests of his correspondents, by keeping abreast of boat popularity, by hopefully outguessing trends in purchasing, and boatmen's tastes.

His choice of design is not always correct and occasionally he may develop a boat pleasing to himself but uninviting to his critics. The result is the loss of his valuable time, his investment in material, overhead and promotional costs. It may take years for this tube of drawings to pay for its own keeping.

In all stock designs it requires thousands of dollars for a new release to be issued. As a result, the architect must sell the package **without modification** to many builders before he makes his money and it is only through volume that he can earn his living. The builder should realize this fully before asking the designer to make changes ... it is like buying canned stew in a supermarket and telling the counter clerk to leave out the potatoes.

If alterations of a stock package are desired, discuss them with the designer, as he may be aware of many factors which would prohibit (or make very costly) the simplest-looking change. The alterations would be conducted on an hourly basis and can quickly mount up out of proportion for extensive overhauls.

"Simple Changes"

Let's look at a few typical examples of change requests.

"Can I build your boat as a sloop instead of a ketch?" On the surface this sounds very simple, but ... the vessel would have to be rebalanced for directional stability, a new sail plan would have to be calculated and produced; rigging and mast loads recalculated; new mast, boom, spreader, club drawings produced; new running rigging details produced; new stock hardware sizes established; deck plan redrawn; deck construction redrawn, new tanks, chain plates, spreader brackets, mast head, mast step weldments and hull steps detail executed; trimming weights for possible modification rechecked; possibly redrawn tank installations to make way for new mast steps; and alteration made in accommodations plan to reflect new mast positions. As you can see, the original, seemingly simple request may, in fact, be extremely complicated, requiring over 100 hours of design and engineering time. A project such as this could reach as high as $2,000.

"I'd like to move the galley aft and install a second head with shower!" Sounds pretty easy, huh? Well, here we go again! Modify accommodations plan; modify general construction plan; draw new joiner sections; draw new thru-hull sections and modify thru-hull profile; modify plumbing schematic; modify electrical schematic; draw new shower sump details and specify required hardware. This little job could rack up 30—40 design hours at a cost between $450—$600. There's no simple way if it's to be accomplished properly.

Extensive Changes

There is a point of diminishing returns when it comes to modifying a stock plan package. This occurs when the extent and cost of alterations exceeds that which would be incurred if a totally new design was generated, specifically tailored to your very own requirements.

An example of such a modification request: "I would like the draft reduced and a centerboard added to your stock design," or "I would like a long keel version of your fin keel design." These requests demand that a completely new lines drawing be executed; table of offsets relofting; tracing for reproduction; complete recalculation for stability, displacement, weight, rigging loads; balancing for directional stability; new construction drawings and assorted details; new sail plan (if such is indicated by calculations), etc. As you can see, these alterations have evolved into an endless chain reaction which cannot justify the attempt to change the original vessel. It would be far better to start from scratch.

Be sure you know how complete the drawing inventory is before labeling an architect as expensive. I have seen stock designs range from $15 to $1,000 and from only three blueprints to sixty. Do not expect a thousand dollars worth of detail and consultation from the fifty-dollar package.

A typical stock plan package inventory should include the following drawings or details if the builder is to avoid guesswork:

Sail Plan or Outboard Profile
Accommodations Plan and Profile
Lines and Offsets (or Loftings)
Scaffold or Skid
Mold Profile and Typical Section
Armature Detail
    Center Member
    Bulkhead Web Detail
    Deadwood and Shafting Installation
    Engine Beds and Mount Weldments
    Mast Steps and Step Weldments
    Stemhead Installations
Thru-Hulls and Ports
Thru-Hull Piping Sections
Tank Detail
Major Interior Construction Plan and Profile
General Construction Section (scantlings and specifications)
Sheer and Bulwark
Bulkheads or Joiner Sections
Cabin Construction
Engine Installation and Exhaust System
Plumbing Schematic
Electrical Schematic
Standing and Running Rigging Detail
Masts and Spars
Rigging Weldments
Chain Plates
Steering Installation
    Rudder and Stock Construction
    Rudder Heel Bearing
Deck and Profile Layout
Miscellaneous Details
    Hatches
    Doors
    Rails
    Drawers
    Moldings, Coamings & Fiddles
    Tables
    Dorades
    Portholes and Windows
    Ladders

Some designers supply lofted mold patterns with stock designs, and I feel it is a good bet. Considering that the professional loftsman has spent years learning his trade properly with its overwhelming complications (see Lofting), I hardly think the amateur should risk failure of his vessel on the difference of a hundred dollars or so for these patterns. Some builders may be concerned over shrinkage and distortion of the printed loftings, but they are drawn with these tolerances in mind. I have never seen a paper station pattern off by more than ¼ inch in 20 feet.

Warning! Warning!

Never try to outguess your designer by making structural or layout changes without talking it over with him first. Most often he will give you his immediate concurrence but he knows the factors which may ultimately be affected. "Make it stronger" is the most common hail of the amateur builder. One of my own clients substituted steel bulkheads for 1" ply, while another added almost 4,000 pounds to his boat by simply modifying the mesh recommendations. The result of this "overbuilding" in the name of strength was disastrous failure. Stability, trim, and strength can be easily upset by instant experts and unqualified engineers.

Every architect automatically builds into structural elements as much as a five-times safety factor over what is required under the most demanding circumstances. Trust him or talk it over before you barge ahead. (Plate 14)

Special Equipment or Hardware

Many prospective builders are extremely satisfied with the general layout, rig or construction as seen in a stock design . . . but would rather install a Chrysler engine than a Perkins diesel. This often creates many problems, because the

engine sizes may be different, as well as the whole requirements, shaft alignment, mount positions or controls.

Many builders prefer to substitute instrumentation, install autopilot, roller reefing or furling, steering vanes, cockpit dodgers, etc. Such items are rarely within the normal realm of the architect's responsibility.

Impractical Alterations

We have received hundreds of letters and drawings from prospective builders or owners asking that bowsprits be added to vessels not designed as such, or that great cabins be allowed on boats which cannot accommodate them. Requests for changes in headroom, engine or tank positions, cabin extensions, modification of the vessel's normal dimensions or rescaling are quite common. What must be remembered is that all vessels are carefully calculated for proper flotation, balance and performance. Such changes, therefore, would seriously upset these performance characteristics or construction.

Custom Design

While the vast majority of drawings issued are for unmodified stock designs, each owner necessarily is searching to fulfill dozens of very specific requirements which may not be available from the shelf. The case in point is the shopper who not only dislikes potatoes in his stew but dislikes the can it comes in, as well.

Usually, in very specialized cases custom design will be required, such as in the development of high performance racing machines, very large boats or single-purpose working vessels. These designs are impossible to sell to the general public on a large enough scale to pay for themselves, so you alone will bear the total cost of development. Don't expect

Plate 14. The builder of this popular Samson **Sea Mist** followed his plans to the letter while referring to books on marine joinery and checking well-executed existing boats for detail. The obvious result is a seaworthy, beautifully finished yacht, worth many times the material investment.

the architect to put your custom design into his stock catalog while charging you only a stock fee. Occasionally the architect may be able to sell custom drawings to another builder, in which case he may give you a partial rebate on your fee, but it is not standard practice, so don't expect this to happen.

Now it boils down to just how badly you dislike potatoes, as well as the can! How intent you are at owning a particular boat must be translated into dollars and cents. It's as basic as that. The old supply and demand story!

Of course, it does cost money. An extensively detailed 40' vessel, for instance, can involve as much as 300 hours of engineering and drafting, 20 hours of calculations, 20 hours of material, parts and equipment research. We assume, here, that the builder has very little construction experience and requires guidance and detail far beyond that needed by the professional builder. For an extremely proficient and experienced sailor and builder, the design hours may be reduced significantly. In the first instance, the creation of the design for the complete amateur may cost as high as $3,400, while only $2,500 might be required of the professional. This cost difference is not to be considered as a professional discount, but rather strictly one of design involvement.

In addition to this possible design cost, the difference may be the charges for lofting. Most professional builders are capable of (and prefer) lofting the custom vessel within their own facility. The amateur, on the other hand, may lack the experience and equipment to accurately execute such a massive and critical undertaking. Most amateur builders, therefore, request that the architect conduct the lofting, as well as tracing patterns for reproduction. As lofting time is usually proportionate to the length of the hull, changes are based on the L.O.A. and are added to the custom design fee. Custom lofting of our 40' example, for instance, would run about $600. Design and lofting, then, could cost as much as $4,000. Now you can see why a stock yacht design (without modification) is a relative bargain if the owner is willing to sacrifice some of his personal requirements. Unfortunately, it necessitates the sale of dozens of the same design in order for the architect to break even, hence the reluctance by him to allow countless modifications.

Preliminary Custom Design

Many critical decisions have to be made before getting deeply caught in (possibly) the wrong groove. Your construction budget must be established, your building capability evaluated, your layout and rig requirements considered and the appearance of the proposed vessel discussed. This is what the preliminary drawings are all about. They are your assurance that you and your architect are on the same wave length.

Before the preliminary drawings are started, you should inform the designer of every possible detail you desire, including the service to which the yacht will be subject, special equipment and hardware, displacement, L.O.A., L.W.L., beam, draft, sleeping accommodations, rig, special sails, the general appearance you would like, headroom, power range, etc. If you have made rough sketches, they will help immeasurably, or, even better, clippings from magazines of boats which suit your fancy.

From the accumulated information, the architect will produce a tentative Sail Plan or Outboard Profile, Basic Line Drawing, and Interior Layout or Accommodations Plan.

These drawings will measure from 30" to 48" in length and will usually indicate all requirements asked for. In addition to the drawings, a displacement calculation will be conducted to help establish construction cost, as well as rough ballast, water, fuel and hull surface estimates. A hull lay-up system will often be recommended in addition to other primary construction specifications.

The preliminary work must always be paid for in advance. If it were otherwise, we would spend the rest of our time doing preliminaries without ever receiving a nickel from well-intended prospective customers. Once the preliminary fee has been received, it usually takes only about two weeks to finish the work.

Final Custom Design

After you have received your preliminary prints and figures, take plenty of time to review them carefully. Feel free to mark them up, indicating changes you would desire in the finished work. At the same time, carefully consider the amount of detail you will require for successful construction. When you feel you are ready to commission the final design, return the marked drawings to the architect along with your check for his retainer to begin. The actual fee must be discussed with him personally, of course.

Design Fees

It is common knowledge that the best-known architects have created names for themselves through racing, significant achievements or promotion. Because of their recognition, their services are much more in demand, and consequently they can charge more for their services. If the owner's prime requirement is total confidence in achieving construction quality or the ultimate in performance, he then must turn to an architect who has proven himself in that given field. This is not to say that a lesser known architect is less qualified.

If your prime requirement is one of aesthetics, interior layout, or a particular building procedure, most designers are eminently qualified to handle your requests. In this case the fees will range somewhat less than the "big gun," but in either instance will be guided by a universally accepted code of ethics which has stabilized most charges.

A rule of thumb is that the designer will receive from 5½% to 7½% of the value of the commissioned vessel (based upon a qualified survey). This fee will be at the high end of the scale for racing boats, as they require many special calculations, extra details and often tank testing.

Other designers find it fair to base their fee on displacement, as it is a better guide to a yacht's complexity. This fee system runs from 7.5¢ to 10¢ per pound and may vary according to rig and extra details required according to the amateur's capability.

Lofting Fees for Custom Design

It has never been the general practice of naval architects to loft their own designs, because it has been the standard procedure for the professional builder to undertake this function. Lofting by your designer, therefore, is most generally charged separately from the execution of the working drawings. Because lofting is more closely associated with the mileage accumulated on one's knees rather than the displacement of the vessel, then the rates charged for lofting are usually based on L.O.A. The charges may range from $15 to $25 per foot. It should be assumed that

Plate 15. The price paid by this amateur designer is wasted time and money. This vessel's windward performance will be highly questionable; her motion in a seaway will be intolerable, her rigging totally inadequate under any circumstances; and she is destined to be a waterfront "eyesore." Upon commissioning, her Volkswagen engine failed and this event may have been a blessing in disguise. Needless to say, her resale value and insurability will suffer markedly. Makeshift abominations such as this impair the respectability of ferrocement and degrade the fine art of yacht design.

Plate 16. This little sloop, while nicely built, was converted from an existing wood design. Once cured, she was launched and ballasted (hull stains indicate two waterlines) prior to installing the interior and rig. The builder made no calculations to predetermine stability or sail area and the finishing will be accomplished on a "by guess and by God" basis. Plagiarism runs rampant in the ferro-cement field and "instant architects" are cropping up like mushrooms, thus breaking down the credibility of the **bona-fide** marine engineer. Those who attempt such conversions without the originator's permission run the risk of an ill-performing yacht as well as a lawsuit.

the lofting fees will include tracing onto mylar or linen for reproduction.

Amateur Design

During my tours of ferro-cement construction sites, I have run into scores of vessels which caught my eye by virtue of their unusual or unacceptable proportions as well as their lack of visual balance and coordination. Upon investigation it was discovered that many of these builders had decided to design their own boats for the reason that "stock vessels available did not meet their requirements" (most notably, they did not have large enough accommodations—in these instances the owner should have been looking at a larger vessel). The basis for their architecture was the reading of one or two books on the subject and the recommendations of other amateur builders. The price for this kind of self-gratification is the production of an obviously amateurish spectacle having little or no resale value with the probability of a nonfunctional interior layout. Performance and stress coefficients are so critical that one error in calculations by the do-it-yourself designer could seriously endanger life and limb.

There is no justification whatsoever for an inadequately trained layman to attempt such a design project beyond producing his own preliminary drawings. He could easily be sacrificing thousands of hours and the investment of a tremendous amount of money. (Plate 15)

The Conversion of Wood Designs

Thousands of lovely and practical wooden vessels have been produced over the years that, indeed, would make excellent ferro-cement yachts and their conversion is both reasonable and desirable. This conversion is much more than simply adjusting the shell weight and ballast factor and it can be just as complicated in its development as the origination of a new cement design. For the reasons that I have discouraged the amateur designer, I must also warn against the home builder attempting his own alteration for the sake of saving a design fee. The gamble will never pay against the overwhelming odds of disastrous results. (Plate 16)

Take advantage of the professional designer's versatility, training and experience. It will be the best dollar you could spend for the guarantee of being satisfied by your building efforts.

Designer/Professional

Under no circumstances should a prospective builder expect the architect to give away his services. He is not in the business of distributing free boatbuilding lessons or answers available through many well-known books. Those interested in ferro-cement should first read a little about it before using the designer's valuable time.

A simple request is usually all that is needed to receive spec sheets on the designer's inventory. Remember that naval architects are primarily engineers and, with few exceptions, have limited promotional budgets and rarely the aid of a marketing firm. They have attempted to impart all necessary information through their spec sheets at considerable expense and numerous unusual or complicated questions should be answered by the purchase of the designer's time or plan package. He should receive the same courtesy given to any other professional. (Plate 17)

Plate 17. Accurate scale models often precede lofting for the purpose of helping the designer "feel" the vessel and to detect any unpleasant shapes which may not be apparent in the lines drawing. The 42' **Doreana**, constructed by the author, is 31½" long and used as a visual study. Tank testing may also be called for, but it is normally restricted to high performance racers and expensive custom designs and is always paid for by the prospective owner.

3

the cruising hull

Long-range cruising has a lot in common with sky diving—they both flirt with danger, they both demand physical and mental stamina, they both require training and practice, and . . . neither is a joke in the slightest sense. Sailing, like sky diving, may seem exciting and peaceful at the same time, with the resulting exhilaration over the control of one's own destiny. But there is a lot to lose if it's not done properly.

It is just as important to your own safety as it is for the protection of your investment of time and money to start off with the best equipment for the purpose intended . . . the boat itself (or parachute, as the case may be). While many experienced sailors and cruising people have a sound basis for their requirements in a boat, I am directing this chapter toward those who have never actually spent great

periods of time at sea or living dockside. It will be impossible for me not to reject many "great" ideas, but my task here is one of instilling a sense of realism, acceptable proportion and practicability.

Size

While we are all in the habit of referring to a boat by her length (L.O.A.), this really does not tell us much more than what to expect to pay for dockage. A fifty-foot boat, for instance, is not simply twice the size of a twenty-five-footer; it will actually be more than three times the volume because we have also added to beam, freeboard and draft. There are great variances in overhang, waterline and section shapes among boats of the same length, with tremendous effects on the usable volume. It is at displacement, there-

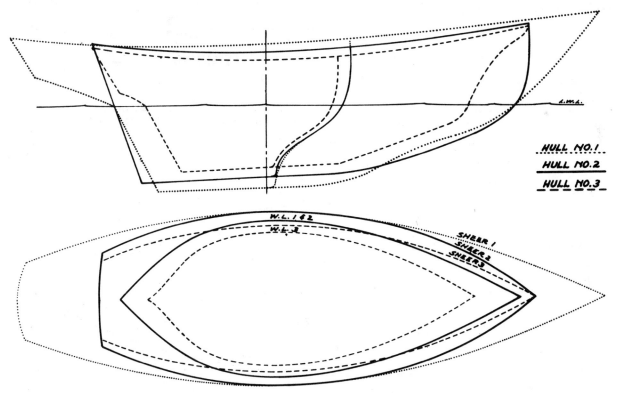

Fig. 7. As you can see, length and size are not synonymous. You can clearly see that hull No. 2 (20' L.O.A.) is as long on the waterline as hull No. 1 (28' L.O.A.). Both hull No. 1 and hull No. 2 share the same waterline shape as seen in the plan view while their sections are almost identical. On a displacement basis, the 20-footer would be only a shade lighter than the 28-footer and, hence, these two boats may be considered to be almost the same usable size. Hull No. 3, while being the same 20' L.O.A. as hull No. 2, is obviously a lot smaller.

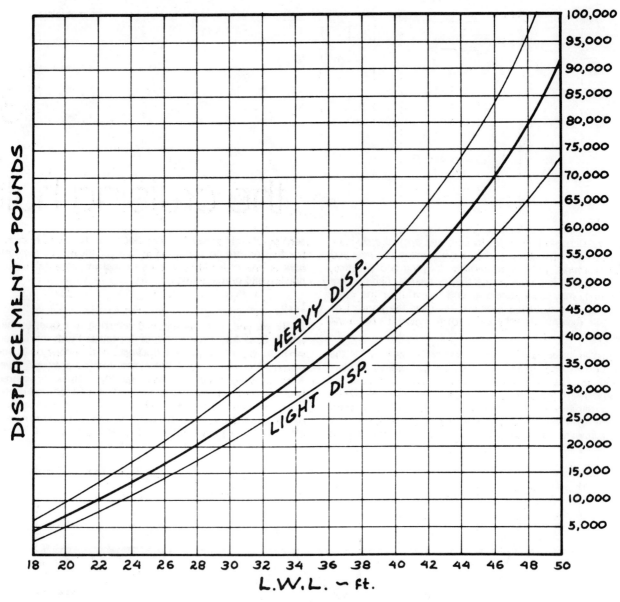

Fig. 8. L.W.L./Displacement Table.

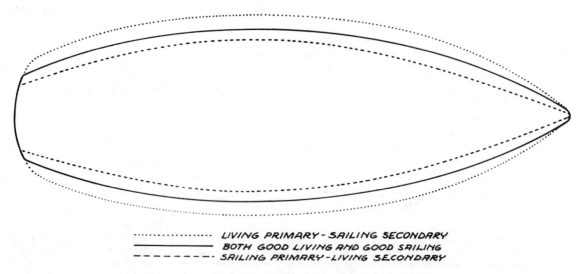

·············· LIVING PRIMARY - SAILING SECONDARY
————————— BOTH GOOD LIVING AND GOOD SAILING
– – – – – – – SAILING PRIMARY - LIVING SECONDARY

Fig. 9. Beam variations.

16

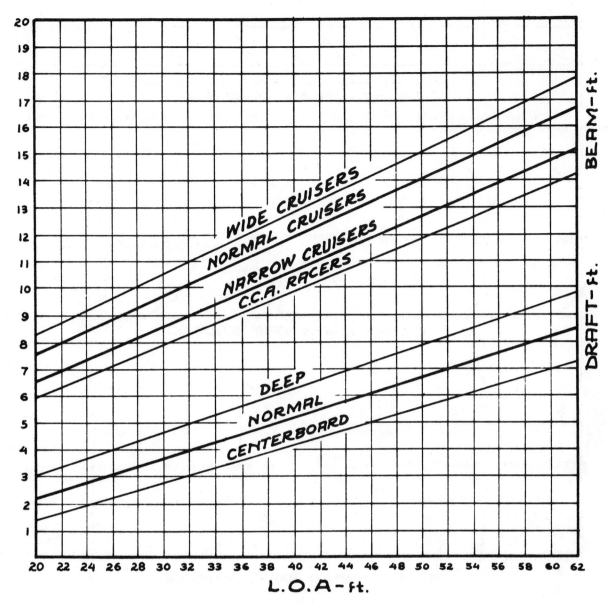

Fig. 10. Draft and Beam vs. L.O.A.

fore, that the cruising man should look to judge a prospective vessel. Displacement is an indication of a yacht's roominess. A hull that is 36' L.W.L. and displacing only 28,000 pounds will be so cramped inside and so fine in shape that she will be of little use other than for racing. At the same L.W.L., the boat weighing 48,000 pounds will be great for living but will be such a pig that she will be extremely slow in moderate air and will be hard to handle. Of course, the criteria, speed and roominess, is up to the owner, but he should be aware of the extremes as well as norms upon which to judge in order to avoid ending up with a totally useless boat. (Fig. 7)

Using the table (Fig. 8), boats which lie above the normal line will make the best long-range cruisers and live-aboards, while those below the middle line should be considered for bay sailing and coastal cruising.

Speed

Because the potential boat speed it is the direct product of the wave length she creates, a simple formula may be used to determine this potential quite accurately:

1.25 $\sqrt{\text{L.W.L.}}$ All boats can reach their hull speeds under the proper conditions but because the length-to-displacement ratio affects resistance, the heavier vessel will require more wind or power. Putting in a larger engine or increasing sail area will not change this hull speed; it will only cause her to reach it sooner. These requirements have been carefully calculated by your designer and it is extremely dangerous to attempt to alter them.

Proportions

Certain ratios have been found most practical for certain types of sailing as they affect livability and performance. To change these proportions to a marked extent is to be very disappointed in her characteristics.

In the accompanying table, I have drawn lines to illustrate norms for beam and draft-to-length, which may be confidently used as a guide. Remember that a 55' L.O.A. with only a 12' beam will have an interior like a long hallway lined with components, yielding very impractical walking or working space. The vessel of the opposite extreme may not sail to windward worth a hoot. (Figs. 9 & 10)

Fig. 11. Bow variations. Moderate overhangs lend themselves most satisfactorily for speed, space, comfort and safety. Extremely long or short ends may compromise your cruising requirements.

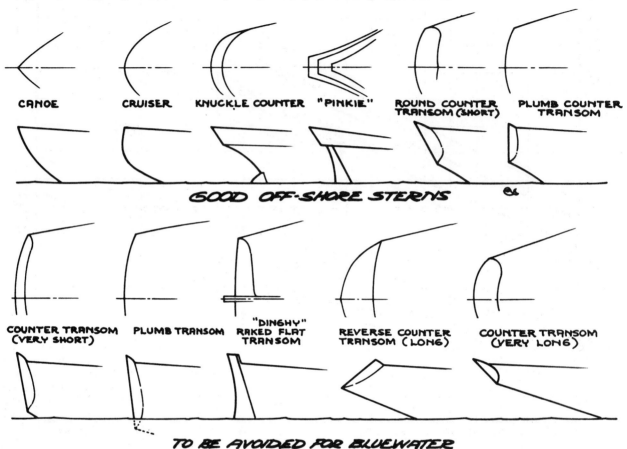

Fig. 12. Stern variations.

Bow Types

Another important factor to cruising people is overhang, which affects space as well as the motion of a hull in a seaway. Let us first look at bow profiles:

The long overhang lends to great reserve buoyancy. If built to an extreme, this type of bow will pound like hell to weather but will be extremely safe downwind, as it will prevent the digging action. The space afforded by extreme overhang is almost useless, as it is like crawling into a narrow tunnel.

The short overhanging plumb bow with a narrow "V" section indicates very little reserve buoyancy to react to an oncoming wave so that as the swell passes, the boat will respond very slowly. This action yields an extremely comfortable ride, but she will be very wet on deck because of the tendency of the hull to plow right through a wave. Downwind the short overhang and deep forefoot will dig in, magnifying the broaching characteristic. The benefit of the plumb bow is that it yields the maximum usable space and greater potential speed, a product of maximum waterline length.

The angles of the long and short overhangs can be designed in several combinations by the architect to create a specific type of ride most desirable under varying conditions. The spoon bow with the low angle of entry will attack a wave aggressively, then soften her motion as the water rises up the stem.

It is a shape conducive to speed in heavy going and allows for moderate space usage—it is most usually seen in bay sailers and coastal cruisers.

The clipper bow, with its high angle of entry, gives a very soft ride to windward because of its initial plowing action, but as it deepens into the face of a swell, it gently reacts upward. The clipper bow is spacious but is somewhat slow when it begins to blow up.

Stern Types

Stern types are just as important as bow profiles and are not simply decorative. Large boats have been driven under because of an improper transom in heavy seas and particular attention should be paid to this factor when choosing a design. The good sea boat should have her reserve buoy-

LONG, SHALLOW UNDERBODY

EXCELLENT DIRECTIONAL STABILITY AND STEERS A FINE STRAIGHT LINE BUT MAY BE DIFFICULT TO MANEUVER IN LIGHT AIR, DOCKING, OR INTO HEAD SEAS. SACRIFICE WINDWARD PERFORMANCE WITHOUT CENTERBOARD

NORMAL, MODERATE UNDERBODY

GOOD DIRECTIONAL STABILITY WITH MANEUVERABILITY. GOOD REACHING AND WEATHER PERFORMANCE. GOOD SPACE USAGE.

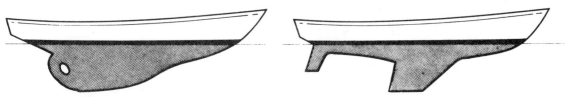

DEEP, CONCENTRATED UNDERBODY

SUPER MANEUVERABILITY AND WINDWARD PERFORMANCE. REQUIRES CONSTANT HELM ATTENTION. TYPICAL OF RACERS.

Fig. 13. Underbody variations.

ancy well above the waterline so as not to be driven wildly forward at the onrush of a following sea. Sections and waterlines should be well pointed to allow a gentle lifting over each wave; canoe and cruiser sterns are ideal in this respect and are extremely safe in broaching situations.

Long overhangs and wide stern bilges must be avoided at all costs if long-range cruising is intended as these sterns can push the boat dangerously into the wave trough, causing the bow to dig treacherously, resulting in broaching or pitchpoling.

Just as unwieldy as the long overhang is the short, broad transom which has been created to satisfy buyers wanting the romance of the "great aft cabin." This abnormal distortion to develop usable stern space not only lacks necessary reserve buoyancy but also poses a huge target for the push of following waves. In this instance we combine two detrimental effects in one. What is saddest about this arrangement is that "great cabins" are always cramped and clumsy living spaces on all but 30-ton frigates (on which they are only hideous imitations). I shall talk more about this later.

Reverse transom counters serve absolutely no purpose other than giving the illusion of a very modern profile while using up deck and lazarette space and destroying reserve buoyancy. Their development was the result of a loophole in a racing rule and, consequently, they have become popular in mass production issues. (Fig. 12)

The Underbody

Since 1965, certain loopholes and modifications of the existing racing rules allowed some competitive advantage in

the use of the fin keel and spade rudder. More properly I should say that these racing rules did not take into account the reduction of wetted surface and increase in overall performance resulting in the obvious winning trend of these types of boats. Since the introduction of the fin and spade, they have become deeper (high aspect) and more radically separated, causing some vessels to lose directional stability. The consequence of the deep separate fin and rudder is that these boats need continual tending of the helm, particularly sailing downwind in heavy seas; but this is the cost that the racing skipper has been willing to pay for the advantages of speed and pointing ability.

In what I feel to be an overreaction to the nasty temperament of some of the fin/spade rudder hulls, many cruising skippers swing to the opposite end of the spectrum. Hundreds of home builders now feel that a very long underbody is necessary for self-tending and control, but I must warn that this philosophy may also be erroneous. While the hull with a lengthy underwater profile will track a straight line beautifully in a seaway, she will be slow in reacting to intermittent wind variations and will sacrifice windward performance accordingly. The long-keeled hull develops markedly less windward lift than a fin-keel variation and will never do as well when pointing. What could be a dangerous characteristic is that they are reluctant to come around when tacking in light or moderate air as well as being almost impossible to maneuver when docking. (Fig. 13)

Rudders

Because many builders have been taught to believe that ferro-cement is the sole solution for every construction

19

situation, thousands of boats are now being built with ferro-cement rudders. In some cases amateurs are using steel plate. In some instances, these rudders could weigh several hundred pounds and have severe negative buoyancy; i.e., they will sink. This brings up an extremely important and dangerous fact.

As long as there have been sailboats, skippers have known that a proper boat will tend to veer into the wind on all points of sail, especially when hit by a knockdown (gust of wind). This inherent characteristic not only improves the performance of a vessel to windward but also eases the heeling forces when taking such a knockdown.

If a rudder has negative buoyancy (as in the case of ferro-cement or steel plate), the additional weight will cause it to react exactly opposite to that desired when heeling or rolling, thus causing the boat to continually track downwind. The heavy rudder will also place undue strain on the steering gear and heel bearing while creating a deceptive and treacherous feel on the steering wheel. If the prospective builder regards his own well-being and the safety of his vessel, he must reject this type of construction without further consideration.

One of the best-known, time-proven and simplest types of rudder construction is that of constructing a marine laminate which sandwiches the rudder stock. After it is shaped to the proper foil, it is fiberglass-covered and installed. Some yacht builders will mortice a small lead block into the laminate, as well, in order to ensure that the rudder is of neutral buoyancy, thus creating the most faithful and delicate steering control.

Summary

There is a happy medium that a long-range sailor should attempt to secure in his design in order to most suitably meet the acceptable requirements of control with speed and windward ability.

4

stability

I can't think of anything that perturbs me more than to be browsing through the design section of a boating publication only to find that the editors have omitted some of a vessel's vital statistics. In some cases we find beam, length and draft but not the displacement, while in other instances the displacement is listed without the ballast. The most irritating of all, however, is the notation of ballast capacity alone, as this tells me absolutely nothing . . . even less (if that can be) when a boat's total weight is absent.

It is distressing to me that the sailing public places such a tremendous emphasis on ballast capacity or the ballast-displacement ratio. This is not to say that these factors are unimportant, only that even the most experienced yachtsmen have been misled into favoring or disliking a boat on this basis. It amazes me how little is understood about stability and the many varying factors which affect it. In the following illustrations I will attempt to show that ballast plays only a small part in regard to sail-carrying ability and resulting performance.

The Righting Arm

This term refers to: (1) The portion of a vessel's underbody which is being pushed upward by the water forces; i.e., buoyancy on the lee side of the hull, (2) all weights aboard the vessel pushing downward on the windward side of the

vessel. These two factors will increase and decrease at varying rates according to the degree of heel—as the hull increases its heel angle, the center of buoyancy moves further outboard while her center of weights may move inboard. The vertical position of the center of gravity as well as the shape of the hull have a tremendous effect on the lateral shift of the righting arm.

Referring to the vertical position of the center of gravity: If this center is above the heeled waterline, it will not only decrease the righting arm but will increase the heeling arm. It takes a few inches up or down to make the difference, whether the position of the center of gravity will be a benefit or hindrance as the hull increases its heeled angle.

Heeling Arm

This term refers to: (1) All weights pushing downward on the lee side of the vessel, (2) that part of a vessel's buoyancy pushing upward on the windward side of the hull, (3) the amount and position of wind pressure on the sails, (4) the underwater side forces pushing against the hull as a result of a vessel's tendency to sail angularly to her course.

Again, it is important to note that if a vessel's center of gravity is above the heeled waterline, the C.G. will be a heeling force.

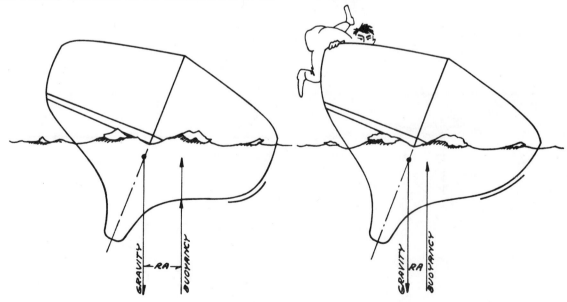

Fig. 14. Effect of chine or bilge shape on righting arm.

It is the plight of the designer to develop combinations of hull shape, construction, accommodations and sail plan that will ultimately result in the longest possible righting arm against the shortest heeling arm. The greatest challenge to the architect is to lower the center of gravity as much as possible so as to be able to increase the yacht's sail area and ultimate performance.

Let's see how this is done:

1. Hard bilges increase the righting arm by adding relatively more buoyancy to the lee side of the hull, thus shifting the center of buoyancy to leeward. (Fig. 14)

2. Without respect to the shape of the bilges, a wide beam also shifts the C.B. farther to leeward as the hull increases its angle of heel, again adding to the effect of the righting arm. (Fig. 15)

3. Hollow garboards (the inward curve of the hull at the keel) raise the C.B. while full garboards, then, help to increase the righting arm. (Fig. 16)

4. The lower the center of the lateral plane, the greater will be the heeling forces due to the crabbing (sailing angularly) of the vessel along its course through the water. (Fig. 17)

5. The higher and greater the center of the sail area, the greater will be the heeling force. (Fig. 18)

6. Comparing one boat against another, vessel "A" carries 500 gallons of fuel and water located directly over the ballast—cement or steel tanks. The engine is located below the waterline and she may be using plywood bulkheads and deck as well as lower interior positions. She carries a light-weight aluminum spar and her construction weight considerations have allowed 12,000 pounds of ballast. Vessel "B" has the same fuel and water but located at the sides of the hull, the engine is on the waterline and she has been built with cement decks and bulkheads. She also has a high salon cabin as well as carrying heavier wooden spars. Vessel "A's" center of gravity is 14" below L.W.L., allowing her to carry a thousand square feet of sail and be stiff in a blow, while vessel "B" can carry only 880 square feet under the same conditions. (Fig. 19)

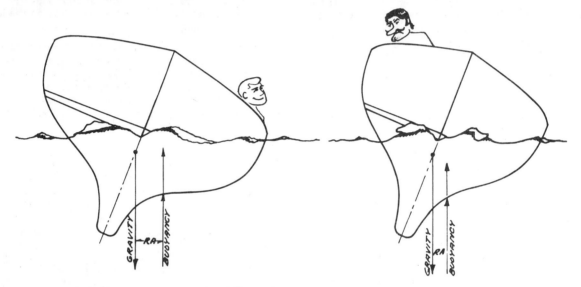

Fig. 15. Effect of beam on righting arm.

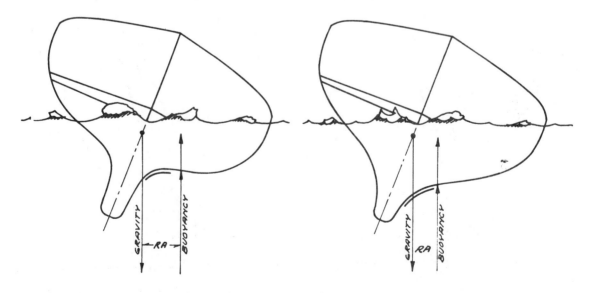

Fig. 16. Effect of shape of the garboard on righting arm.

Conclusion—the lower the C.G., the less will be the heeling forces and the greater the righting arm.

The foregoing graphic examples have been somewhat exaggerated and simplified, of course, but have been meant to illustrate the hundreds of very slight differences which can quickly add up to thousands of pounds of forces which make a yacht stiff or tender. My comments and drawings have not been intended to be specific recommendations to the prospective buyer but, rather, guides to the necessary considerations which must be taken into account when passing judgment upon a design. In short . . . do not discount the yacht with a low ballast-displacement ratio; she may have many other factors "heavily" in her favor and may be superior in terms of stability. Take everything into account when choosing a boat on this basis.

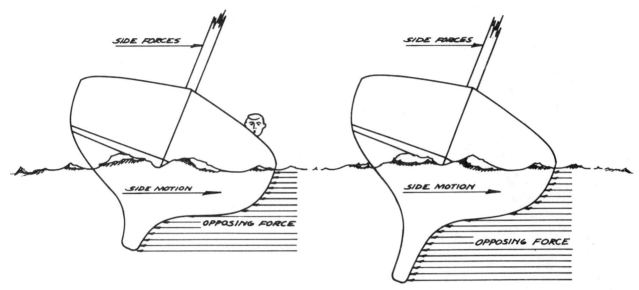

Fig. 17. Effect of lateral motion as a healing force.

Fig. 18. The more stable the vessel, the more sail-carrying potential; hence, superior performance.

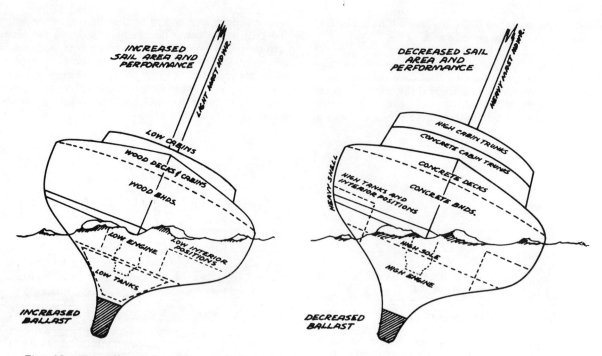

Fig. 19. The effect of weights and their locations on center of gravity and ballast capacity resulting in stability variations.

the cruising layout

Reality and Practicality

One of the greatest problems that the naval architect has to face is not the actual development of the hull or its calculations, but rather of instilling in his client a sense of reality and practicality. It is inherent in all of us to want more for our money, more spacious layouts, and more functional elements within our boats. But it is when we lose our sense of reality and ask too much of a vessel's given displacement that we begin to miniaturize and compact beyond acceptable limits. For decades designers have been using standard dimensions for the lengths and widths of berths, counter tops, lockers and interior elements; these units are based on known minimums for the greatest use of the least amount

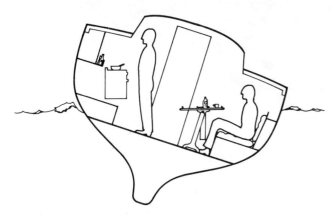

Fig. 20. What may look like a large boat on paper may become extremely small regarding human elements—even smaller during actual sailing conditions.

of space. For a prospective builder to demand more of a vessel's size than that which it would normally yield is to run the risk of investing one's time and money in a cramped and uncomfortable hull with little or no resale value.

I was once asked by a fiberglass yacht manufacturer to design a 30' boat that would be in direct sales competition with at least a dozen other well-known stock boats. In order to create "one better," I was asked to include five berths; a galley with gimbaled range, sink, and refrigerator; head with shower; chain locker; salon table; hanging locker; and 6½' cockpit. The hull of this proposed vessel was to have a long overhanging bow (looks fast), a reverse counter transom (looks even faster), and to be of a light displacement, fin-keel type. I told my client that it would be virtually impossible to incorporate all of these features without increasing the beam requirements and L.W.L. His comments were, "Why don't you shorten the berths an inch or two,

reduce the hanging locker and head widths and move her entire layout a foot forward to accommodate the cockpit? All you have to do is move this or that around a little and you've got it." At a glance it may all sound very simple, but considering that the original hanging locker was only 20" wide to begin with, moving the layout forward would have raised the fo'c'sle berths almost 9". I felt I could go no further to comply with his wishes. Quite obviously, all of these changes would have had a disastrous effect on comfort and the feeling of openness.

I highly recommend that anyone anticipating the purchase of a design for a boat intended specifically for cruising or living aboard make a concerted effort to spend some time on existing boats of the same general size and nature, carefully measuring every element and personally evaluating the space relationship in respect to his own requirements. Most specifically, attempt to establish in your own mind what you consider to be too large or too small; i.e., lockers, drawers, shower floor, galley counters, cupboards, shelves, etc. Remember at all times that a good part of your sailing will be done at night (which causes a vessel to close in on itself) and at heeled angles up to 30 degrees. (Fig. 20) When taking these measurements, think primarily in terms of the number of people who will actually be using the boat under normal circumstances, the quantity and type of food you will be carrying, personal effects and clothing requirements, adequate unobstructed volume for workability and elbow space. Upon establishing minimum or optimum sizes and relationships, lay these elements out for yourself in a plan and profile sketch. In doing so, do not allow L.O.A. (length overall) to cloud your perspective, as it should be only the net result of the displacement necessary to incorporate your criteria. In other words, don't simply start off with a predetermined hull length, stuffing it with an interior, as it should be the other way around; i.e., an interior wrapped up in a hull. This one sketch, in itself, will be the greatest help in choosing a design as well as aiding you in understanding some of your architect's problems.

Size Relationships

For the purpose of helping you render this sketch more realistically, I have listed the following minimum recommendations that should not be reduced under any circumstances:

Bunk width: 2' at the shoulder, 18" at the foot
Bunk length: 6'3" on center or at least 4" longer than a
    given person's height
Bunk height: Minimum 15", maximum
    (lower berth only) 3'
Settee width: 20"

Settee height: 16"
Sitting headroom: 3'
Standing headroom: 6'2"
Walk-thru door width: 24"
Hanging locker depth: 2'9" (from sheer)
Hanging locker width: 18"
Galley counter width: (flush space) 20"
Galley counter height: 33"
Vanity width: 18"
Vanity height: 30"
Cabinet door width: 14"
Sliding hatch width: 2'
Cockpit seat height: (average) 17"
Cockpit width: (at working end) 30"
Cockpit coamings: (average) 7"
Toe-rail height: 2"
Fiddle height: (counter and table edges) 1¼"
Drawer depth: 5"
Scuttle width: 2'
Bookshelf width: (at widest point) 9"
Over-counter clearance: 14"
Table width: 18"
Galley, ranges and stoves always face the center line.
Refrigerators may face inboard only if well fiddled.

cause it is difficult for the layman to sense the third dimension or volume. The interior elements of a boat are always miniaturizations to some degree in comparison to units we would normally find in a house; i.e., berths narrower than beds, lockers shallower than shoreside closets, overheads lower than ceilings, etc. When looking at a yacht's layout, try to keep these factors in mind, as space is always at a premium. (Fig. 21)

One exercise that may be of great help to the prospective builder when reviewing study plans is to carefully scale, then sketch a normal human figure on a piece of tracing paper, then superimpose it over any portion of the vessel. By doing this, you will quickly discern inadequacies in sizes and shapes. I believe some designers have purposely broken down yacht interiors into many hopelessly small segments in order to make a yacht appear much larger. The "tissue man" will uncover these discrepancies in proportions. If more builders had been aware of this simple trick, they could have saved many gruelling hours and huge investments in vessels which when completed were totally unaccommodating. (Fig. 22)

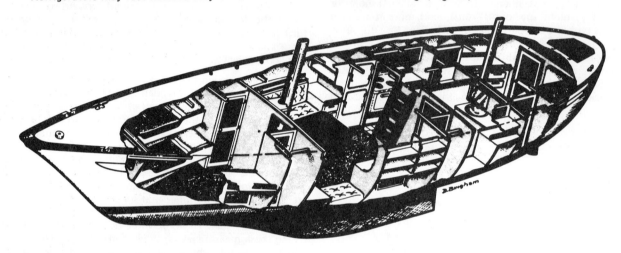

Fig. 21. By comparison with most boats her size, the 48' **Andromeda** is spacious and practical. Considered as huge by most knowledgeable sailors, she is still much smaller than an average one-bedroom apartment. A Bingham design.

Toilets should always face the center line (otherwise they are very difficult to stay on).

Berths should parallel the center line (otherwise you could wake up on your head with your shorts around your neck).

Galley or head sinks should be well inboard and above the waterline. (If positioned at the sides of the hull, they may overflow under heeled conditions.)

Water and fuel tanks should be as nearly amidships as possible so as not to alter trim when full or empty.

Remember: The above figures represent the absolute minimums as they probably would be applied to, say, a 25' minicruiser; for permanent livability and long-range cruising, they should be increased accordingly.

Reviewing Plans

Looking at the designer's accommodation plan can be very deceiving to someone who has not actually drawn one, be-

Continuous vs. Stepped Sole

When laying out your proposed vessel or selecting existing designs, it is wise to avoid variations in the plane of the cabin sole that require steps or ladders, as they take up valuable floor space as well as increasing the chances for injury when the boat is in a seaway. This is especially true at night or when the interior of the boat has become wet (unavoidable during heavy weather). The necessity for additional stairways is particularly true in the case of dual or aft cabin arrangements, as they create a necessity for traversing the engine or tanks compartment in all but the largest boats. (Figs. 23 & 24)

Aft or Great Cabins

Several dimensional planes are affected simultaneously as we follow the lines of a hull toward its stern: The profile increases its degree of rise, the waterlines converge more

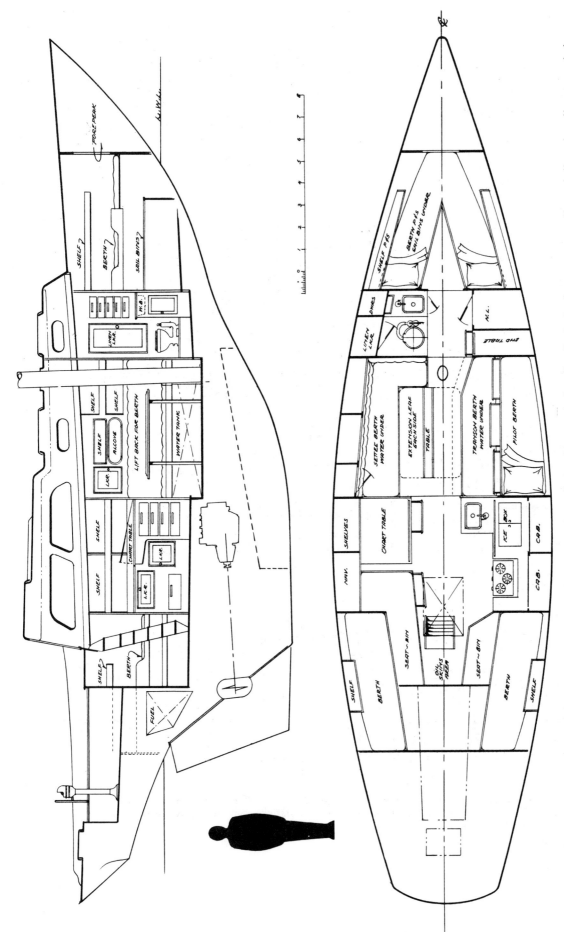

Fig. 22. At a glance, this popular stock "cruiser" seems to have all the comforts of home—but notice the lack of passage space through the salon, the somewhat lacking privacy of the head, the duplication of tables within a foot of each other, the amount of valuable living space allotted to sleeping, and the absence of "elbow room." If this boat looks big to you, cut out the silhouette of the average man and superimpose him onto the accommodations plan.

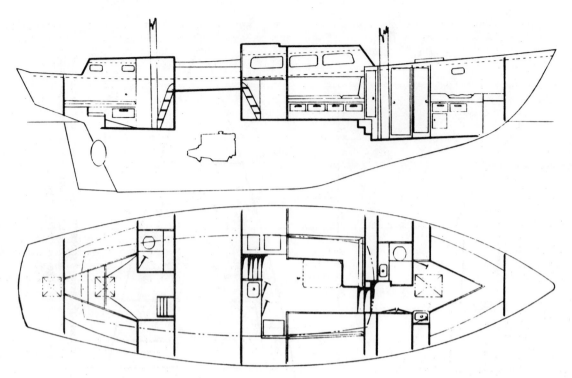

Fig. 23. "Ups 'n downs" can waste a great deal of valuable space while complicating travel through the hull. In this layout, the most livable portion of the hull has been unquestionably turned over to the engine, tanks, sail, lockers, and cockpit. Otherwise, this vessel's layout has many attributes.

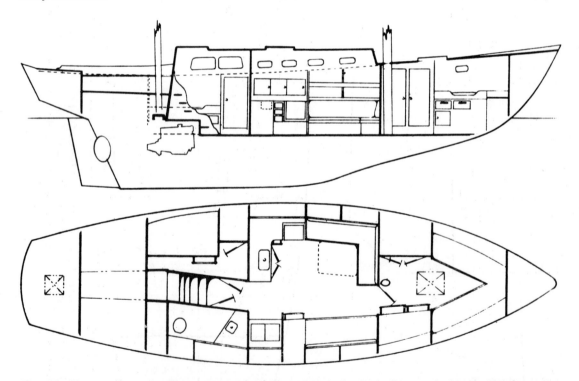

Fig. 24. The continuous sole arrangement of the same hull yields the greatest potential for using space realistically. It is far easier to construct and more variations can be evolved than would be possible in the previous example.

drastically, the sheer travels inboard and at a much lower level than amidships. These diminishing planes result in shallower and more noticeably triangulated sections than those farther forward.

Obviously, these smaller stern sections mean less usable volume, being embraced within strangely shaped proportions. Discounting the difficulty of constructing joinery in the stern area, we are faced with the even more important problem of the angles at the side of the hull and the decreased headroom. The net product is almost always a "delta" shaped compartment necessitating the installation of "V" or "U" layouts of settees, berths, etc.

These are difficult forms to live with in terms of mobility and elbow space, usually compounded by a deficient sole. The most normal arrangements call for the installation of either quarter berths (in a pseudo-stateroom) or a salon revolving around a table. Occasionally one even sees a double berth of sorts shoved into the counter (it's like sleeping in a shoe). (Fig. 25)

have found most generally that those few persons asking for this arrangement have spent little or no time with it and simply like it for its appearance. On this point I must concede that a dual-cabin profile can be extremely handsome.

There has been some warranted concern about the following sea breaking into the aft-cockpit position and thus endangering the helmsman, but as the highest percentage of boats built today are of aft-cockpit type, it is significant that we hear so very few reports of this occurrence. If the cruising hull is designed or chosen properly to begin with, we can eliminate this possibility as readily as that of a boat's rolling over.

The Flush Deck

In the 1920's the appearance of a flush-deck vessel was quite common. The topside working space was flat and expansive with few obstructions on which to trip or snag lines and the mobility of crewmen handling canvas was considered an extreme pleasure. The massive deck plan was

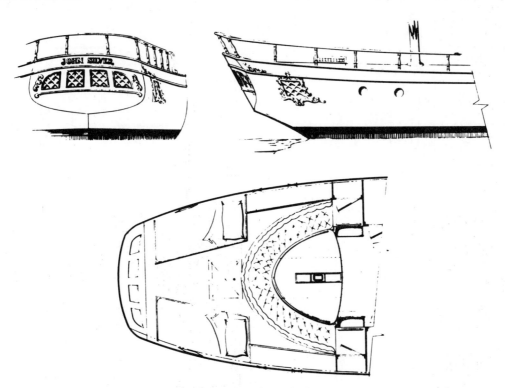

Fig. 25. Even the great cabin on this 65-footer is a hoax which hardly justifies its own nomenclature. It is a far cry from the captain's cabin we see so often in movies.

Don't get me wrong; I'm not against aft cabins . . . at least not always. I personally feel that they are cramped and inadequate on most boats under 35' L.W.L. and of little benefit for the price that must be paid. In smaller vessels it requires that the normal aft-cockpit sail-stowage position be moved forward into the choicest part of the hull while almost always requiring the parties to go topside to get from the main to the aft cabin. In larger yachts such as the **Andromeda,** we are able to find both the headroom and floor space necessary to install a passageway along the starboard side which also incorporates the galley; but it should be noted that her L.W.L. is 37.5 feet. Even in this length, it took a considerable amount of juggling to accomplish. (Fig. 26) I have designed smaller aft-cabin type cruisers, but only for the purpose of meeting the curious demand for it. I

interrupted only by the occasional hatch or skylight which contributed to a cleaner and more efficient appearance.

Down below, the flush-decker was like a ballroom with her interior components well outboard and comparatively wide soles. Because of the lack of a deck shoulder (the notch in the deck section at the edge of a cabin trunk), the additional space added not only to the feeling of interior volume, but also was used to incorporate overhead cabinets and shelves (normally absent on a cabin-trunked vessel). (Fig. 27)

It is well to remember that during this period yachts commonly were much larger than the average cruiser of today and the popularity of the flush deck was not hampered by a lack of standing headroom in most sailing yachts. As we

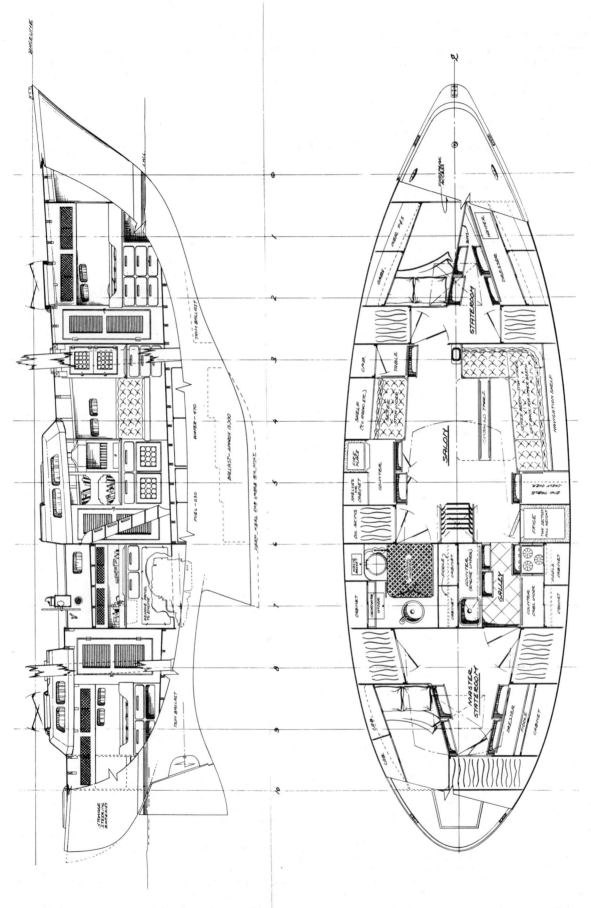

Fig. 26. Andromeda's interior fully maximizes every inch of her 48'. One of her most notable features is the passageway from the forward to the aft portions of the vessel. The cockpit does not infringe on the interior but her aft stateroom is still massive. Also notice that the basic functions of the vessel have been separated from each other. A Bingham design.

approached the 40's, however, the tight monetary situation forced owners to purchase smaller boats and, as a consequence, the impressive flush-decker disappeared from the design pages. Almost all of the yachts produced between 1930 and 1960 were of the "normal" cabin-trunk version (Fig. 28), with a few notable exceptions such as the big racing craft.

But alas! In the late 50's came the **Medalist**, a stock fiberglass sloop of only 38' L.O.A., boasting the return of the "good old days." This design was considered to be short-ended, of slightly higher freeboard than most boats of its day and it bore a radically cambered flush deck with a small "bubble" cabin trunk. Her topsides were studded with two fixed deadlights and her 6'3" standing headroom ran the entire length of the cabin sole. At the outset, the **Medalist** was poorly received because of its progressive profile and wide beam; but the **Rabbit** (the most notable of them all) became widely known for its practicality and livability. Since this introduction, the flush deck has become extremely popular and desirable for cruisers and racers alike.

The design of a flush-decker generally requires a higher sheer and a slightly lower sole line. The latter may intrude on water and fuel capacity when the tanks are located in the bilge, but this can be overcome with proper considerations. The aesthetics involved need not be overly "modern" (as has been the general rule), as more traditional cabin trunks and transoms may be incorporated. Ventilation below is created by hatches and ventilators, while lighting problems are solved with clear acrylic hatch tops and fixed side ports. Within the myriad of vessel types available, I consider none to be superior to the flush-decker. (Fig. 29)

The Pilothouse

Within the last two years I have had many requests from prospective clients to convert my existing designs or to generate new ones that would incorporate an inside steering station or "full-blown" pilothouse. I have no personal objections to the latter, as motor sailers do have their place as excellent coastal cruisers, but this is as far as I can bend. For blue-water sailing, the large doghouse windows usually required of the motor sailer can be extremely dangerous in a crashing sea, while the high cabin position decreases

stability and multiplies wind resistance. I must grant that it can be both romantic and comfortable to be able to "cocktail" while the view passes by, but this is an image straight out of Hollywood that rarely has a practical application except at dockside or in protected waterways.

There are several other factors that should be considered in regard to the raised pilothouse:

1. As the sole of the pilothouse becomes higher, the deck becomes lower relative to it, thus decreasing the area of the standing headroom.
2. When the distance between the sole and the underside of the deck is diminished to 4'7", then all settees must be moved inboard to accommodate sitting headroom within the confines of the cabin sides rather than under the side decks. In this circumstance, what realistic use can be made of the space behind the settees?
3. By raising the sole, we increase the void under it. A portion of this is undoubtedly given to the engine, generator, pumps, etc. What do we do with the space left over? If we install fuel tanks, it is to the detriment of stability, as their positions must be raised very near or above the waterline. I can think of no sadder waste of space than total emptiness in a situation where space is at such a premium. (Fig. 30)

Inboard Steering

I have turned away sizeable design fees by refusing to incorporate inside control stations aboard blue-water cruisers. While this fad has grown in popularity, I believe it is found among those with the least practical experience. As any well-found sailor knows, handling a vessel is much more than simply looking where it is going. A seasoned helmsman steers not only by sight but by feel, hearing, smell and a sixth sense which I cannot describe. To sail otherwise is unhealthy, if not foolish.

The premise of inside steering is that it affords comfort to the helmsman during the chill of night or in heavy weather. It's a nice thought! It is precisely during these times that the skipper is needed topside. The prudent sailor must not only see his course but his sails and rigging, as well; he must be in a position to react instantly to any emergency situation. He must be able to see 360 degrees at all times.

(Text continues on page 36.)

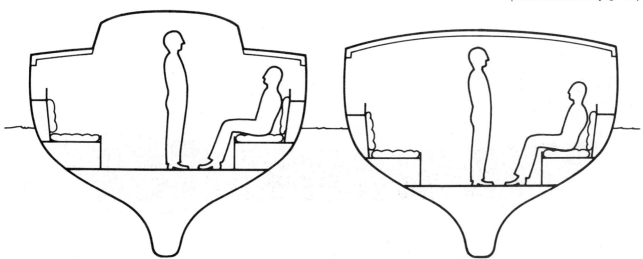

Fig. 27. While requiring slightly more freeboard than the conventional cabin model, the interior of a flush-deck vessel is usually far more spacious.

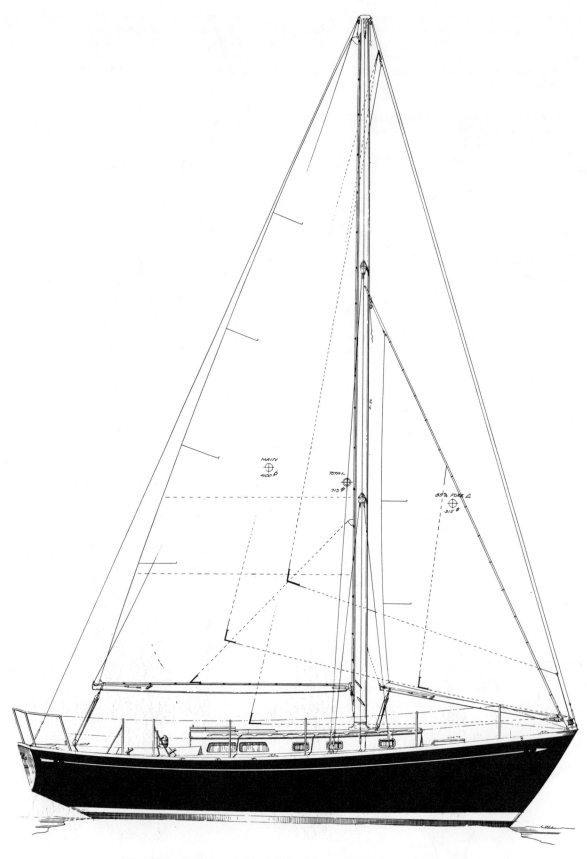

Fig. 28A. **Regina,** a typical full-cabin yacht. A Bingham design.

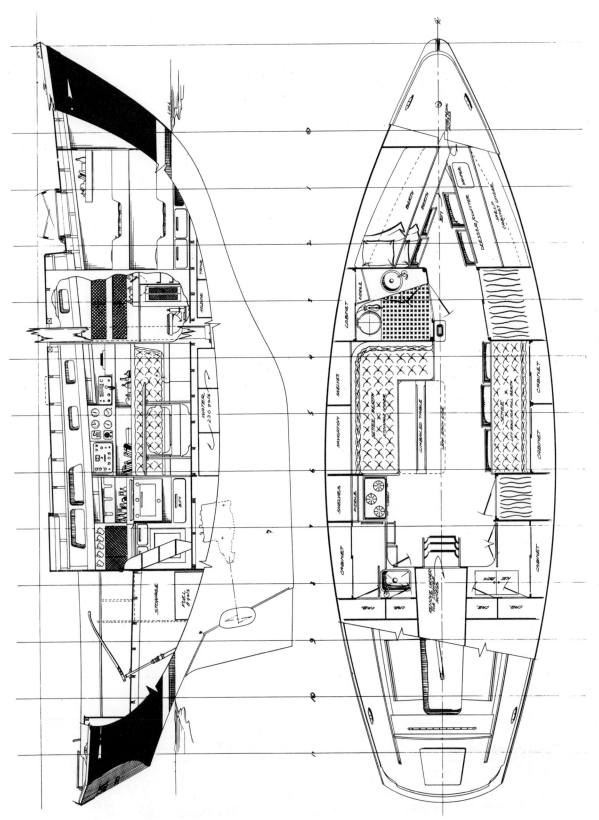

Fig. 28B. Interior view of **Regina**.

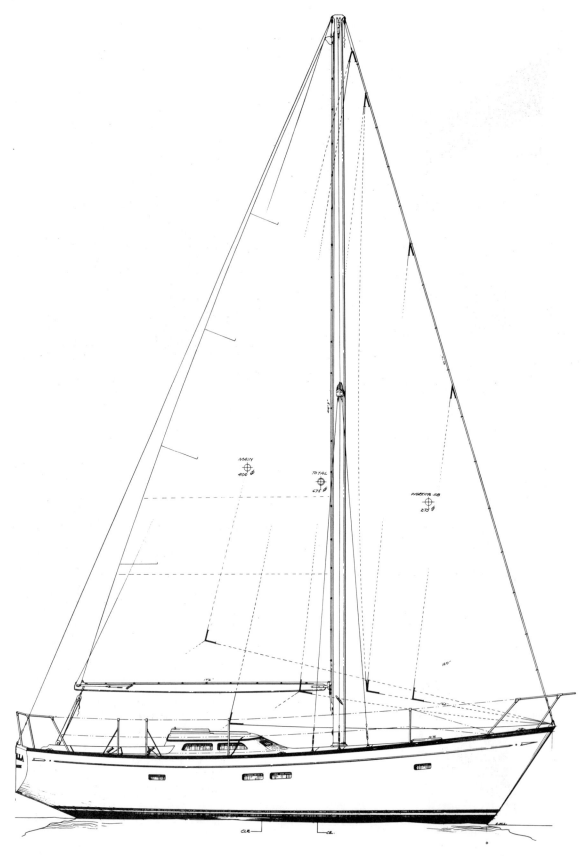

Fig. 29A. **Panella,** virtually the same hull as **Regina,** requires slightly more freeboard in order to create the flush-deck arrangement. A Bingham design.

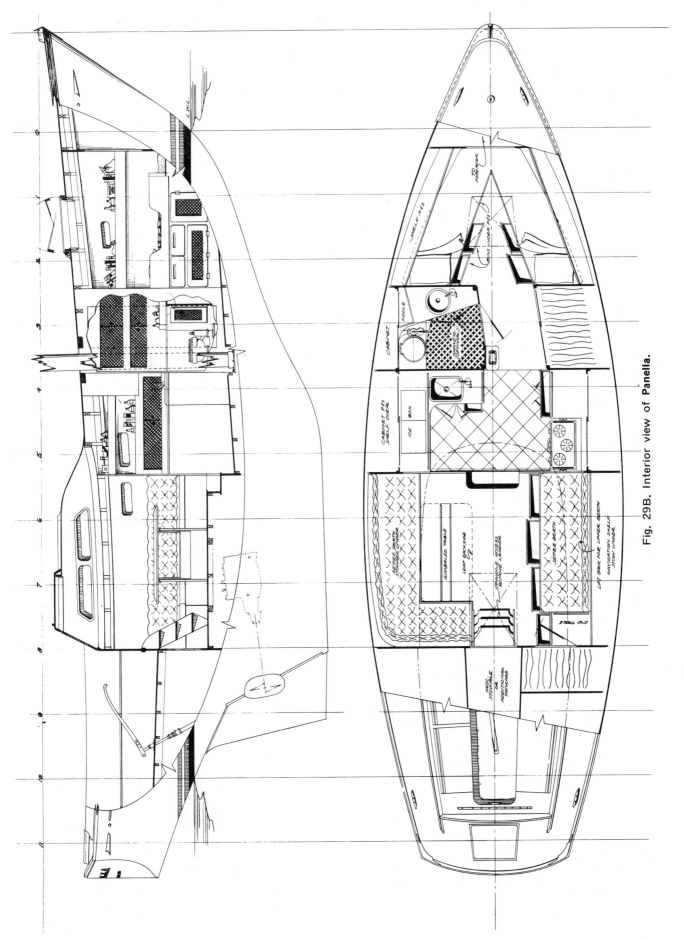

Fig. 29B. Interior view of **Panella**.

35

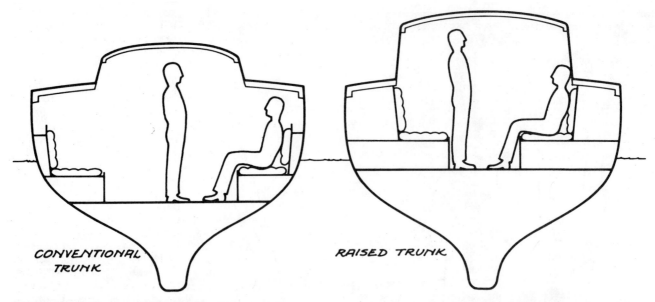

CONVENTIONAL TRUNK

RAISED TRUNK

Fig. 30. It is obvious from these sections that the raised salon or pilothouse model has sacrificed a major portion of livable space for the questionable advantage of installing larger windows. Her higher center of gravity, compared to the more conventional model, will surely be a detriment to her performance.

(Continued from page 31.)

The Dodger

This is a canvas "convertible" top that can be raised from the cabin trunk to cover the forward end of the cockpit. It is designed in such a way as to afford total visibility through clear acrylic panels in the front, sides and top. When not in use, it is simply unbuttoned from the cockpit coamings and folded forward. When raised, it gives the helmsman warmth and protection from sea spray. It can be made so that the aft end can be closed off to hold in the cabin heat while still allowing the skipper immediate accessibility to everything on deck. The only requirement for the use of the dodger is that the wheel position be far enough forward so that the shelter completely encloses the steering position. They are manufactured by Island Nautical, Inc., City Island, New York, or they can be custom-made by any sailmaker. Why use valuable interior space when the solution is so easily at hand? Even a short dodger as shown in Fig. 31 offers full protection from wind and spray and the heat from the cabin below is a great boon to the lonely helmsman.

Fig. 31. A well designed, collapsible dodger provides 360° visibility while still affording full protection from the weather. It may be designed to completely enclose the cockpit in order to maintain a cozy environment for the helmsman during inclement weather and conveniently removed or folded away on nice days.

Duplicating Components

A 50' schooner or ketch is one hell of a big boat, but foot for foot would not equal even the usable space of a one-bedroom cottage. Many builders, however, would expect to cram into this boat the accommodations of a three-bedroom ranch.

Take sleeping requirements, for example: How often do you invite six or eight people into your home to spend a week or two? I suggest that the builder include no more permanent berths than he will actually require with **emergency** provisions only for the occasional overload. I do not believe that all settees need be 6'3" in length in order to be convertible bunks when they can be shortened to 4'6", thus making room for an additional locker, end table, fireplace or longer galley counter. In small boats, of course, this may not be possible.

The question concerning the number of people who will **normally** be on board also raises the question as to whether we **really** need two heads aboard a vessel of less than 50' L.O.A. The occasions when your guests will be in line to use the toilet will be very rare and the additional distance one has to walk should not even be considered, as most single heads are conveniently centralized. I believe it is better to have one large head with toilet, sink, cabinets, hamper and elbow room than to split the accommodations into two smaller units.

At sea, your navigation duties will require only about 1½ hours a day (unless you are out of practice) or may not even be used at all when sailing long rhumb lines. Of course, when you're coastal piloting, this time may be increased to possibly five minutes out of the hour. You can see, then, that even under the most demanding circumstances a chart table will be active only 12% of the time when under way and completely idle dockside. The question is, then: Is a separate navigation area really necessary when the salon table, properly designed, could adequately serve the same purpose? On very large boats the salon may be too distinct, so a chart table may be warranted, but even under this circumstance it should serve dual purposes, such as a dresser top, or it could be a folding type to save space. In any event, there must be generous shelving for instruments and books.

You can see from the previous example that the duplication of specialized components can be very expensive in terms of the loss of stowage or the expansion of compartments; too much of one thing always means the loss of something else.

Integrated Components

As already mentioned, settees can be designed so as to be used for berthing. Two bunks can evolve from the same unit if designed with a lift-back or transom-extension seat. While the space underneath can hold tanks, canned goods or tools, the same settee may also incorporate a back-shelf.

"L" corners of counters and cabinets, which normally would be useless for stowage, become great places for the installation of water pumps, heaters and items that do not require constant attention. I have also seen "L" corners occupied by the foot end of berths that project from an adjoining compartment. As the "L" corner usually occurs in the galley or head cabinetry, the builder might consider installing a covered wastebasket, as there is usually no other practical position for one.

Not only can cockpit seats be used to great advantage as sail stowage or to enclose the foot end of quarter berths, but they may also double as hanging lockers, even in the smallest vessel, when accessible from the interior.

Under some circumstances, the engine covering may become welcome counter space. Overhead and cabin trunk corners that do not lend to standing headroom (over end tables, etc.) may be used to suspend navigation instruments or even a small television, thus reducing the necessity for occupying overlays or shelf space for these items. The possibilities for developing dual-purpose combinations are endless and the examples mentioned should be enough to stimulate the builder's ingenuity and imagination. The desired effect is always to squeeze the greatest potential out of even the smallest bit of volume.

Space and Lighting

Many prospective owners looking at an accommodation plan feel that an interior is being wasted unless every inch of a vessel's floor plan is utilized. While I believe in building into a boat as many usable components as possible, I am also of the philosophy that an interior should have "air." What I mean by this specifically is enough room to allow a person to walk through the interior without having to squeeze himself past lockers, through doorways, around tables or other people. It is my firm opinion that the entire sole should yield standing headroom so that the occupants are not required to bend over before sitting down or to duck under a deck carlin while using the head sink. A flush-deck type vessel is excellent in this regard, although generally requiring slightly higher freeboard and camber. When a yacht is heeled, it should not require the occupants to lean against cabinets or lockers in order to stand upright and there should never be so little space that when two people are in the same compartment, it feels like a dozen. The latter aspect becomes acute during night watch changes, when several people may be dressing or removing their clothing at the same time.

The prospective owner should always keep in mind that he may be sailing for weeks at a time and finding himself confined to his vessel for endless months. A vessel which has not allowed for the feeling of expanse can quickly become a dismal cell in which the exhausted skipper and crew must live. Coziness is one thing, but entombment is another.

In the same regard we should consider colors, lighting and open wall space, as they can also add to or detract from the sense of space. One of the greatest tricks is the use of mirrors, which can be applied to a bulkhead (self-adhesive mirrored panels are available from most interior decorating or large department stores), and I have even seen mirror kits with nautical murals etched into them. They are extremely handsome and add markedly to lighting, and can make a cabin look twice as long as it actually is. Care should be taken, however, not to mount mirrors in areas where they will be bumped into by people or equipment. The use of mahogany or teak is certainly very warm and pleasant within an interior, but I believe its use should be confined to trim such as drawer fronts, handles, fiddles, hatch coamings, carlins, etc. On a boat that is intended for living aboard, an overabundance of dark wood can create a very drab interior requiring additional lighting to achieve the same effect as if we had used light colors on bulkheads and overheads. We should always consider the reflective properties of the interior colors. Certain hatch covers may be of clear, frosted or tinted acrylic which, of course, will add greatly to the lighting below and one should also consider the use of flush deadlights in the deck, as well. These are simply acrylic plates or prisms inserted in a rabbeted cutout in the deck covering. When properly constructed, they are perfectly watertight and do not clutter the deck in any way. Remember that windows can be too big, as large expanses of acrylic can be broken by green water.

Interior Positions

There are characteristics in the motion as well as in the shape of the hull which dictate the usability of the hull space. The designer cannot simply place units where he finds ample room. He must consider the pitch and roll of the boat in a seaway at the same time. Because the longitudinal center of motion of the hull is well aft, the forward end of the vessel becomes a huge teeterboard which attempts to throw its occupants and contents into the air at the crest of every wave while driving them through the sole on the downstroke. The sensations felt by the occupant are weightlessness followed by heaviness, then weightlessness again, which can create nausea and difficult sleeping. Anyone who has ever had to change his socks in the forepeak knows how difficult these conditions can be. It is primarily for this reason that the bow of the boat should be used for sail stowage and emergency berthing only. The triangular sections of the bow are also very restricting in terms of lockers, drawers, etc.

The portion of the vessel most conducive to good sleeping and all-around stability is the quarters which revolves more closely around the pitch center and should be chosen, if at all possible, for the master berths. This same center should also be the first choice for the location of the galley because it is the position of the least motion. However, the galley should never be located in an area subject to continuous traffic, while at the same time it should be near enough to a hatchway to allow for plenty of ventilation when cooking. Remember that head and galley sinks should be well inboard so as not to be submerged when heeled, thus avoiding the possibility of overflowing. Positioning sinks against the hull creates complicated drainage problems. Because of its very nature, the salon should be in the largest part of the hull as it will be the most used.

I have never felt that eating spaces should be combined with sleeping quarters or that the galley should be exposed to a berthing compartment. It has always been my belief that there should be very distinct dividing lines between compartments of different functions so as to afford privacy as well as maximum function of specialized space; for instance: 1) Galley and eating section; 2) Eating section and salon; 3) Berthing with head facilities.

To set up a boat otherwise can mean that a crewman's slumber can be interrupted by a round of poker or that he may wake up in his skivvies some morning at the same time the ship's lady is preparing breakfast in the same room. The skipper should not have to crawl over the salon table to turn in for the night, nor should the crew be required to traverse the master stateroom to use the head. Each crewman should have means to get away from the others on board; this is imperative for compatibility under long-range conditions.

thoughts on rigs

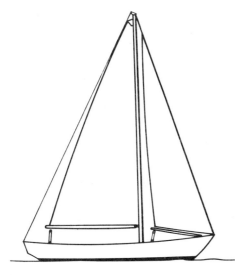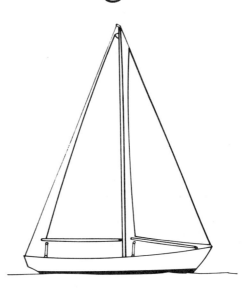

Fig. 32. The sloop (l.) and cutter (r.), both rigged with self-tending jibs.

It is not my intention in this section to get into a basic lesson on sailing, but only to point out some of the factors that could play an important role in the efficiency or ease of handling so important to the cruising man.

The Sloop and Cutter

The sloop (or cutter) is a vessel carrying only one mast and flying a mains'l and heads'ls. There has always seemed to be a controversy over the differences between a sloop and a cutter. I will not attempt to settle this dispute and will, therefore, consider them to be one and the same for the purpose of this dissertation, anyway. It has always been my contention that the cutter's mast is stepped slightly farther aft than that of a sloop and in some cases is even found amidships. The sloop or cutter may fly one or two heads'ls (Figs. 32 & 33) and this arrangement is quite often interchangeable on the same design.

The mains'l on a sloop or cutter is always much larger than normally found on a ketch or schooner of the same L.W.L. Because of this, most cruising skippers have avoided this rig. In heavy air the large main may be difficult for one person to handle and always requires reefing (usually a two-man job) under storm conditions. Should the sloop or cutter lose her spar at the partners, she has lost everything and has no alternatives for setting emergency sails. The concentrations of stresses on a sloop or cutter rig require very hefty spar and boom sections, more massive hardware, as well as

longer sheets and halyards. Although the rig is simpler than the ketch or schooner, it takes a highly tuned and experienced crew to handle it under deep sea conditions. The prime advantages of these sail plans are the simplicity of rig and superior windward ability.

A vessel that boasts two masts, the forward mast (the main) being taller than the after (the mizzen) and the mizzen being positioned forward of the intersection of the hull and the waterline, qualifies as a ketch. (Fig. 34) However complicated in definition, it is by far the most practical and most widely used rig for cruising. The benefits of this rig are manifold.

Because the sail area is distributed in smaller sails on two masts, the main and mizzen require less individual manpower than the main alone, as found on a sloop. The main and mizzen can be sheeted conveniently to the cockpit position and need not be tended simultaneously. Should the ketch lose either mast, she is able to fly a sail from her remaining spar. The ketch's greatest merit, by far, is her wealth of possible sail combinations. Rather than attempt to explain them all, I refer you to Figs. 35 & 36. While the ketch is not quite as efficient to windward as the sloop and cutter, she makes up for this lack with her reaching capability by flying a mizzen stays'l. Shortening sail can be accomplished by reefing in extremely heavy weather, but usually it is done by raising and lowering the sails in balanced combinations.

Self-steering vanes aboard the ketch are somewhat impractical because of the interference of the mizzen boom. The ketch rig, however, has the inherent capability of maintaining a given course by sheeting dihedrally. What this means is that the fores'ls are sheeted somewhat closer than normal, while the mizzen is sheeted full. In doing so, it will take several adjustments of the sheets to settle the boat down in the required direction; but once this has been accomplished, her directional stability will be as efficient as most other self-steering systems. "Heaving to" is done in

The choice of a schooner is generally made on the basis of aesthetics and salty appearance, as it is truly the most regal of yachts. The schooner's one advantage over all other types of rigs is that its reaching ability cannot be surpassed, but I believe that this is offset by the sheer immensity and complexity of the rig itself. The schooner almost always requires a bowsprit, and either a boomkin or running backstays on the main, while her staying may be twice that of a ketch. I have found that the position of the masts creates difficulties in laying out a practical interior and deck plan.

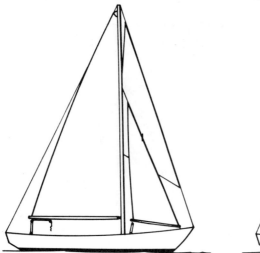 

Fig. 33. The sloop (l.) and cutter (r.), both rigged with double heads'ls. Because this jib arrangement is more normally found on the cutter, sloops rigged in this fashion are often confused with cutters.

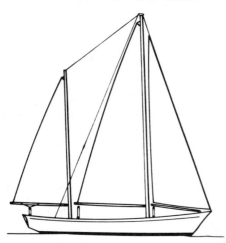 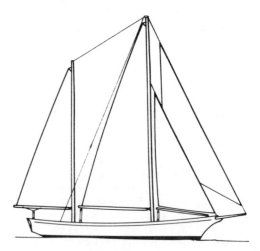

Fig. 34. The ketch as it is most commonly rigged.

basically the same manner, but the dihedral sheeting angles are accentuated by backing the jib hard to windward.

The Schooner

In essence, one might consider that a schooner is simply a ketch rig in reverse; that is, her largest sail (main) is aft of the smaller (usually fores'l) while carrying jibs or heads'ls. There are many variations of the schooner rig ranging from two to five masts (or more), some being gaff or baldheaded (Fig. 37) and others Marconi (Fig. 38) and stays'l rigged (Fig. 39). I shall not attempt to explain each but refer you to the illustrations.

If the schooner is Marconi-rigged, the size of the mains'l may be larger than the cruising man is willing to handle, also causing reefing problems at the far end of the main boom. If gaff-rigged, the schooner's hardware and halyard requirements are so increased as to demand a larger crew. As picturesque as the schooner may be, I cannot recommend it over the ketch rig as an ideal deep-sea setup.

Exotic Sail Plans

So far, I have spoken only of the most common rigs used for long-range cruising. There are many other arrangements that may be acceptable, while others should be completely discounted for the blue-water sailor.

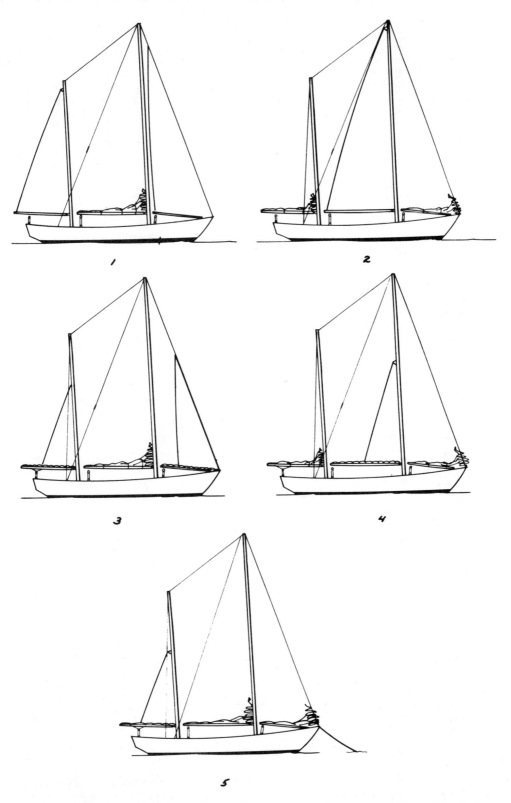

Fig. 35 (1—5). The ketch under various conditions of shortened sail, the key to her flexibility. In each case the vessel's helm can be perfectly balanced until she finally puts out the hook.

Fig. 36. Until recently, the ketch had always been considered as strictly a cruising man's rig; but now under I.O.R. (the racing rule for all-out speed freaks), there is a resurgence in the ketch's popularity. Few other rigs can spread as much canvas with as much flexibility. Here she has set her mizzen stays'l to improve her reaching capability.

Fig. 39. The stays'l schooner.

Fig. 37. The baldheaded schooner.

Fig. 40. The stays'l or wishbone ketch.

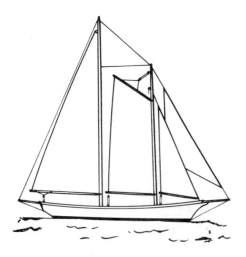

Fig. 38. The Marconi schooner.

Fig. 41. The hermaphrodite brig, the most common type of square-rigger among yachts.

At a glance, the yawl may appear similar to the ketch, but a yawl's mizzen is always stepped aft of the intersection of the waterline. For the purpose of discussion, we can consider that a yawl has the windward advantages of the sloop with the drawback of carrying a very large main and jib; but the dividend is that she can fly a mizzen stays'l while reaching. A yawl's mizzen is not a driving sail, but one used solely for balancing, and its development is the result of a loophole in the racing rules (not so with cruising yawls back in the 1880's). So, as you can see, a yawl may be considered as somewhat of a half-breed sloop-ketch, and it makes an excellent cruiser.

A variation of the ketch evolved in the late 1930's for the purpose of increasing sail area while lowering the center of effort, and in many respects it resembles the stays'l schooner. This type of rig bears the logical name of Stays'l or Wishbone ketch. (Fig. 40) This arrangement utilizes the normal mizzen and heads'ls, but in the place of the mains'l is flown a club mizzen stays'l and the wishbone sail (named after the split boom necessary on this rig) sheeted to the mizzen masthead. The stays'l ketch is an extremely efficient reaching and downwind vessel and a superb performer in light and moderate air, but it requires more sheets and halyards than most schooners and should not be considered unless extra crew members are invited along permanently.

One often sees in boating magazines articles on feather and junk rigs, as well as sail plans which rotate, fold up for easy stowage, bend, extend, retract, or otherwise claim remarkable breakthroughs in efficiency. While I am not against innovation, I strongly discourage your considering trusting your life and limb to the questionable attributes of unknown contraptions or to those that require complicated rigging or delicate handling but lack the known efficiency and safety factors of the more common sail plans.

Brigs, Barks, and Other Square Rigs

There is nothing more stirring to me than the sight of a square-rigged vessel entering a harbor surrounded by skiffs and dinghies crowded with curious onlookers. It conjures up visions of the clipper days (Fig. 41), the very core of our nautical heritage. Who cannot truly love figureheads, dolphin strikers and fife rails? The epitome of a deep-sea sailor's dreams is that of commanding a crew out in the yardarm while strolling the poop deck with a long glass under arm. As serenely beautiful as all this may seem, the cruising man's consideration must return to the practical and realistic for both safety and economy.

All square-riggers of whatever breed require complicated hardware, miles of line and rigging and a large crew tuned to an exhausting degree. The miniature toy ships that we see on rare occasions are the results of spending huge sums of money for the least efficiency. They will not sail to weather worth a damn; they make an incredible racket in a seaway by virtue of the banging and clattering about of blocks and shackles; and they can be extremely dangerous in situations of dire emergencies. I therefore discourage the choice of the square-rigger without futher ado.

Single vs. Double Heads'ls

As I read through the yachting magazines and some of the design brochures, I am aghast at the way some of the basic rudiments of sailing have been ignored. The most notable is the sudden resurgence of what has become commonly termed the "cutter" or "double heads'ls" rig. It must be pointed out that the use of two heads'ls instead of one requires twice the hardware, extra sheets and more hands, and increases the possibility of failure with numbers as well as complicating trimming, tacking and jibbing. While one of these sails may be self-trimming on a club or boom, the forward or flying jib always requires tending when working to weather. As you can see, this drawback imposes difficulties on the single handler or the skipper with a minimum crew. It is also well known that the power efficiency and pointing ability is somewhat less than that of the sloop or single-head-rigged vessel.

The objection to the single heads'l has been that the one large jib, by virtue of the increased sail area, requires more power to handle. While this may be true on a 50' vessel, in the smaller ranges I believe this to be a fallacy. Let's first consider what is commonly called the "working" jib. This sail is cut so that the clew is slightly forward of the spar and shrouds. Its shape allows for the attachment of a jib boom that, once trimmed, need not be readjusted after each tack. If the forward end of the jib boom (or club) is attached to the deck a short distance aft of the forestay, the draft of the sail will automatically increase desirably as the vessel is sailed off the wind. (Fig. 42) If the club gooseneck is attached on the forestay turnbuckle (Fig. 43), however, this advantage is lost. From time to time it may be advantageous to set the working jib in the "genoa" fashion, in which case the clew is simply detached from the club outhaul. The only drawback that I have ever found with the self-trimming jib is that when broad-reaching or running, the jib club tends to swing inboard, particularly in light air. Under these circumstances the jib boom may be held in a forward position with a short length of shock cord attached to a forward cleat, but this is possible only if the jib club is deck-mounted. One of the greatest dividends of the cutter-headed or double arrangement is that it simplifies the reduction of sail in heavy weather, as it rarely necessitates sail changes, but rather only the dropping of one of the jibs (usually the flyer). The answer to this disadvantage to the single-headed rig is roller furling (Fig. 44), and there are many other benefits, as well.

Roller Reefing

This is a system by which booms may be rotated for the purpose of reducing sail area. This is normally accomplished with a crank attached to a gearbox at the gooseneck and was originally conceived as a quick and efficient substitute for the cumbersome reef-point system. (Fig. 45) It is most popular aboard both cruisers and racers, but it still has several disadvantages. The cutback of the roller gooseneck always places severe localized stress at the tack of the mains'l as well as creating uncontrollable wrinkles in this area of the sail. If used often, roller reefing can eventually ruin a good piece of canvas. It also precludes the possibility of attaching blocks or other hardware to the boom in positions other than at the aftermost end (unless employing specially designed yolks). As a consequence, the boom section must be of slightly heavier construction with a recessed sail track to prevent lumping and tearing of the sail when reefed. The use of a boom vang is always complicated or eliminated by this, especially on large vessels. Rather, reefing always flattens the shape of the foot of the sail, resulting in the loss of draft control. Reefing with this system usually requires three people; i.e., one man cranking the gooseneck while controlling the halyard, the second feeding and straightening the sail at the gooseneck, the third pulling out the leech of the sail to prevent gathering and wrinkling. As you can see, roller reefing is not as foolproof as most

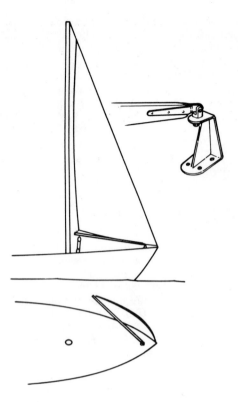

Fig. 42. A jibboom with gooseneck attached to deck pedestal. Note that the draft of the jib increases as its sheets are eased.

people are led to believe and it is extremely expensive; but from the standpoint of speed it is a definite improvement over more traditional systems.

Jiffy Reefing

By far the quickest and most inexpensive system yet developed for cutting sail is this method which employs a tack purchase (a line attached to one side of the spar, passed through the tack reef cringle and back to a cleat on the opposite end of the mast) and a clew purchase (a line attached on one side of the boom, passing through the clew reef cringle and returning to a cheek block or cleat on the opposite side of the boom). These reef cringles may be connected with a wire sewn directly into the sail, or may be simply interconnected through extra layers of sailcloth or reinforcement. At a glance, jiffy reefing appears to be quite similar to the reef-point system, because the ties are usually attached to the sail along the reefing line. (Fig. 46)

To use jiffy reefing, the sail is lowered to a predetermined point (marked on the halyard), the tack purchase is drawn up tight and cleated, followed by the tightening of the clew purchase. This will automatically put a strain on the internal reef while all the reinforcing will have the effect of controlling the foot of the sail, which may be adjusted to suit the conditions. Once the sail has been reefed in this manner, the excess fold of sailcloth along the boom is rolled up to the foot and tied out of the way. As you can see, this can be a one-man operation and does not require any sophisticated equipment whatsoever, while leaving the boom unobstructed for the use of a vang, block bails, cleats, etc.

Bowsprits

Most of the 1930-40 vintage ketches required bowsprits, but this is not necessary if a hull has been balanced accordingly. Bowsprits should always be avoided on the long-range cruiser, as they can be drowning poles in heavy weather and have caused more broken legs than I care to mention. Consider the likely circumstance of being called on deck out of a sound sleep to reduce sail some stormy night. The seas may be 10' to 20' high with gale winds, the decks awash with green water, the vessel snapping and pitching violently, while the bowsprit is shaking and vibrating high above the sea and plunging out of sight in the oncoming greybeards. These are the very conditions that have caused the mournful loss of many hearty seamen through the centuries; and who on this night will draw the short straw for the privilege of risking his life at the end of this precarious perch? I have seen bowsprits planked to a width allowing convenient walking space, plus the relative safety of a pulpit, but there remains the problem of the submergence of its rider into the face of an oncoming breaker. I cannot recommend bowsprits under any circumstance other than for coastal cruising. I admit that they can be picturesque, but they add to the dockage charges at the same time.

Backstays and Boomkins

I can think of no other element more dangerous on a yacht, whether cruising, coasting or racing, than running backstays. (Fig. 37) Although they do not always require setting up when tacking, it takes quick work and coordination to slack and haul simultaneously during a jibbing situation. They automatically require at least two additional hands on

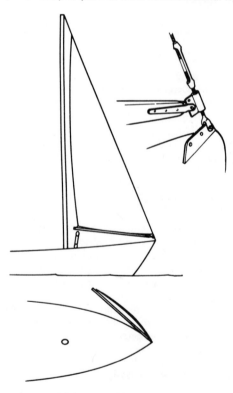

Fig. 43. A jibboom gooseneck attached to the forestay turnbuckle. The jib's draft is not changed as the sheets are taken up or eased.

deck and have long been considered a headache even under the most controlled conditions. If the worse were to happen, such as a Chinese or flying jibe when yawing downwind with only a man at the wheel, the risk is extremely high of ripping the whole damned rig right out of the hull. This could have the effect of annoying the owner, as it would not only upset his sleep but would require extended rowing to reach the desired destination. Toppling masts and snapping booms play havoc with the brightwork as well as causing immeasurable expense for replacement of torn sails and sprung chain plates.

Fortunately, running backstays are usually required on schooners only where the height of the mainm'st and overhang of the main boom create difficulty in rigging the standard single backstay. This is, in fact, absolutely impossible on the gaff-headed schooner, but may be installed occasionally on the Marconi with the addition of a boomkin. (Fig. 38) This is somewhat of a bowsprit (sternsprit) affair that may be single- or double-legged, reinforced by a bobstay in the same fashion as the bowsprit. I don't feel that they impair the appearance of a vessel and are certainly more desirable than the use of running backstays. Laughingly, some schooner owners have even boasted that they make excellent outdoor bathrooms. So, who's laughing? The marina owners!

Goosenecks

In order to set any sail properly, an equal amount of strain must be placed on the luff-bolt rope and the body canvas. In order to accomplish this, a tremendous strain is required that generally is built up with halyard winches or gun tackles, as previously described. On even the smallest boat these mechanical advantages may not be enough. Additional tightening of the sail luff may be aided through the use of a sliding gooseneck (Fig. 47), available in many styles. Once the sail has been hoisted, it will be found that through the use of a small purchase attached to the sliding gooseneck the boom may be lowered as much as two feet on larger vessels, thus creating the desired sail tension. The common fixed gooseneck should be avoided. (Fig. 48)

Another important advantage of the sliding gooseneck is that the height of the sail on the mast may be adjusted to suit the wind conditions; i.e., in heavy air a lower sail position is desired.

Turnbuckles and Deadeyes

Because there is an infinite variety of turnbuckles on the

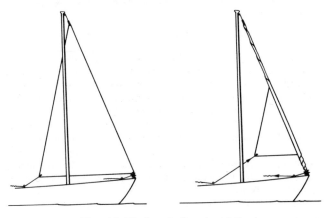

Fig. 44. Roller furling heads'l.

market, great care must be taken in their choice. Marine hardware manufacturers have taken great pains to develop metals and styles that will withstand shock and strains as well as the corrosive effects of salt water. Under no circumstances should the builder consider the use of common industrial grade merchandise. Admittedly, there may be a substantial price difference, but your life may depend upon these items and no attempt should be made to skirt this expense. Particularly dangerous is the hook-and-hook turnbuckle that is found in every corner hardware store; these will never do at all. But I have absolutely no objection to hot-dipped galvanized marine hardware as opposed to bronze or stainless steel, as it has been engineered to withstand the same calculated loads. Just be sure your turnbuckles have hot-dipped galvanized threads.

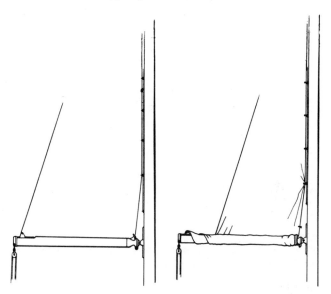

Fig. 45. Boom roller reefing. Note the wrinkles and localized strains which occur with this type of gear.

On the more traditional vessel, such as schooners and some ketches, deadeyes may be substituted for the turnbuckle in most positions. They may be set up to a remarkable tension. When executed properly, these may be homemade and at a fraction of the cost of large turnbuckles. They may measurably increase the aesthetic value of a yacht, provided they are in keeping with its character. Because the deadeye lanyards continue to stretch under the rigging load, they require continual adjustment and occasional replacement. They show considerable flexibility under sudden loads, a great safety feature.

Lazyjacks

This salty nomenclature refers to line bridles on each side of the sail running from an upper position on the mast to the boom. Their function is to allow the lowering of sails without dumping the canvas into the drink or cockpit and they are particularly useful on gaff-rigged vessels in that they prevent the wild swinging of the gaff and greatly aid furling.

Lazyjacks have always been common gear aboard schooners, but are rarely seen on ketches or sloops for some unknown reason. Their practicality cannot be disputed and they are a tremendous boon to the cruising skipper as they aid considerably when reefing or furling sails. (Fig. 49)

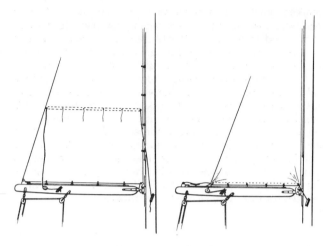

Fig. 46. Jiffy reefing. Just as quick as roller reefing but considered to be more efficient while only a fraction of the cost.

Sheet Leads

If possible, all heads'ls should be cut to share a common miter line, most particularly all jennies and the working jib. The reason for this simple measure is to eliminate unnecessary deck hardware, because all sheet leads then share the same block position. This necessitates a very high-cut working jib and may be somewhat less than desirable, but it certainly makes a great deal of sense from the standpoint of the gear inventory and economy.

Needless to say, all sheets should be within easy reach of the helm. The jib club sheet may be routed through carefully located deck blocks or "eye" fairleads. The number of purchases used should allow one man to handle sail easily in

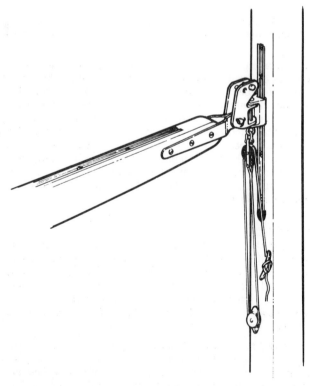

Fig. 47. A sliding boom gooseneck with downhaul purchase.

moderate air and there should be an allowance for attaching an additional purchase for heavier air. This requires that the sheets be extra-long to facilitate this alteration while under way.

Travelers are very nice things, I suppose, as they allow for relatively slight adjustments of the boom sheeting angles, but at the same time they have the knack of localizing the strain and shock of the boom load. I think this is a rather dubious attribute of the long-range cruiser. (Fig. 50) On my own racing boat I have a sliding traveler for the main boom that follows for adjustments of the lower block position; but I have found that one particular setting seems to serve to good advantage under most conditions. In my opinion a boom vang is more important to the sheeting angle than any other factor. In lieu of the traveler I always recommend the split-sheet purchase (Fig. 51), which is simply the separation of the lower double block into two singles.

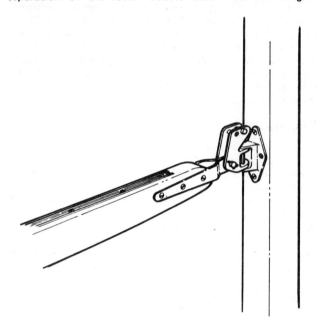

Fig. 48. A fixed boom gooseneck.

These must be located at specific angles in relation to the mast, and may offer the advantage of the sheet being handled from either side of the boat, at the owner's option.

If roller reefing is not used, at least two upper sheet blocks should be rigged on main and mizzen booms. This arrangement relieves the strain on the hardware fastenings as well as the boom itself.

Winches vs. Gun Tackles

The use of winches for heads'l sheets and halyards could mean the reduction of crew requirements from six to four on a 50' vessel and while the initial cost may be extremely high for this apparent luxury, it must be weighed against the expense involved in maintaining the extra crew. For instance, two additional people will not only be consuming foodstuffs that must be paid for but they will also increase costs in terms of refrigeration for preserving their required perishables, in water usage, electrical drain, berthing, locker space and beer. The factors of speed and safety through the use of winches should not be ignored as it could mean the difference in the time required to bring a yacht about and under way to avoid danger. They are not simply fancy items found aboard only the gold-platers and the most

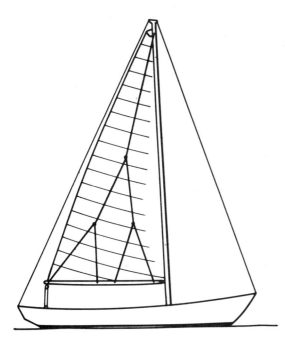 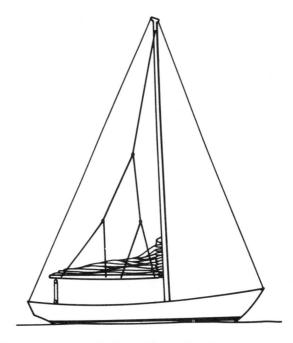

Fig. 49. Lazyjacks, most commonly found on schooners, can be both a laborsaving device and a tremendous safety factor when furling sails in rough weather aboard the sloop.

sophisticated racing boats; they are extremely vital to the prudent long-range skipper. If they cannot be installed upon commissioning the yacht, they deserve a place near top priority for additional gear, as the owner's budget permits.

The use of gun tackles (Figs. 52 & 53) has long been the norm aboard workboats and sailing fishermen, to provide adequate power for handling sail. In this instance, my use of the word "adequate" is rather loose, as block purchases are slow and in heavy air often require the luffing of the vessel in order to horse a large heads'l fully home. Gun tackles should never be considered as permanent substitutes for winches, but should be rigged only to allow the commissioning of a vessel at a slightly lower cost.

The gun tackle itself is simply an open snatch block with becket attached to the hauling end of the sheet or halyard. The initial work on the line brings the sail in as close as possible without a mechanical advantage, at which time the line is brought up under a cleat (or cheek block) into the snatch block. Hauling may commence once again, but this time with a two-to-one purchase. The use of the snatch block should always be preferred to the arrangement of a halyard block at the head of a sail, because this requires

twice the amount of line as well as adding undesirable weight aloft.

Fig. 51. Twin lower block boom sheeting.

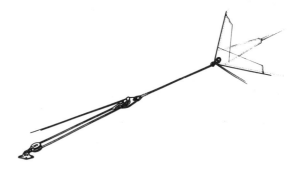

Fig. 52. A gun tackle with open snatch block set up on a jib sheet.

Vangs

Aboard the hot racing machines we continually hear references to vangs and kicking straps. The cruising skipper may immediately take the defensive position in saying that he only intends to loaf his way around and that such high-performance items have no place on his heavy-displacement ketch. I must protest this casual dismissal of what is actually a very important piece of safety gear that not only adds

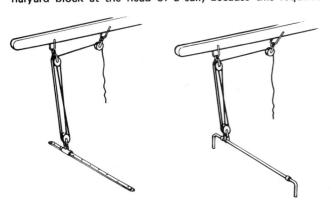

Fig. 50. Boom travelers.

47

to one's confidence under way but increases measurably the windward and reaching performance of any rig.

Looking through sailing magazines, my attention is occasionally drawn to a vessel with a severe twist in her mains'l. (Fig. 54A) Quite often this twisting causes luffing of the sail aloft as well as noticeable backwinding by the jib. These elements break down the power and drive of the sail, which can lower the pointing angle by 2° while cutting the speed potential under a given condition.

The boom vang is a tackle that is attached to the approximate center of the boom and led to a padeye on deck. (Fig. 54B) Because it controls both the twisting and draft, one can actually see the character of her sails change as strain is

Fig. 53. A gun tackle with open snatch block set up on a halyard.

applied. As the vang and mainsheet are used in conjunction with each other, the boom angle can be adjusted to a fine degree according to wind conditions. Of course, when the vessel is being tacked continually, as when beating through a narrow channel, the boom vang should not be used.

In very light air, when the boat may be rolling and sails slatting over an oily swell, the vang may be used to quiet the thrashing of the mains'l. When the craft is reaching in a moderate breeze, it eliminates the tendency of the boom to rise and fall undesirably. When the vessel is on a run, the vang prevents accidental jibbing, which is downright unhealthy in a high following sea.

As you can see, realistic evaluation is necessary when appraising a yacht's complete profile. It is much more than the decision of a ketch over sloop, but rather manpower and efficiency and their application under a variety of conditions. Moderation should be the key to choosing a sail plan, but there is little sense in short-cutting necessary and efficient equipment for the sake of its price tag. It can mean the difference between a quick and comfortable passage as opposed to one wrought with misery and blisters—and maybe tragedy.

The Tradewind Self-steering Rig

I have already mentioned the benefit of dihedral sheeting aboard the ketch rig for maintaining a given course. There is another self-steering method which has always been extremely popular, primarily because it involves a minimum of equipment and expense. This may be called the "twin jib" or "tradewind" rig and, as its name implies, it requires the use of two jibs set simultaneously, each clew being attached to a pole on opposing sides of the mast. (Fig. 55) The tacks of these jibs are not made fast to the stem, but rather some distance aft in order to produce a greater amount of sail exposure. The clew ends of the pole are sheeted to the stern of the vessel so that each pole forms approximately a 20° to 25° angle to the wind forward

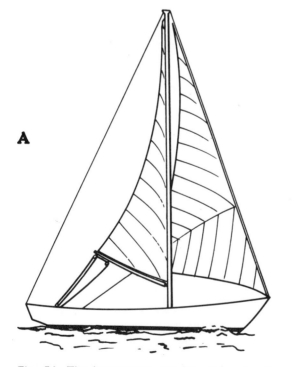
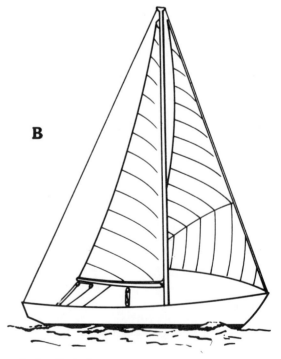

A   **B**

Fig. 54. The boom vang can be used not only to control sail draft for increased performance, but may also prevent accidental jibes when running dead before the wind.

from the spar. Reviewing the diagram, one can see that if the vessel changes or veers from the preset course, the windward jib will develop greater power than the other, thus forcing the vessel back to the desired heading. Literally thousands of vessels have crossed the Atlantic on the tradewinds using this arrangement without a man having to touch the wheel from the Canary Islands to Barbados. The only drawback to this system is that it can be applied only when running or broad reaching. For windward work some other self-steering method must be used. As most sail plans usually show a reacher or high-cut jenny, sail makers usually will use this as a basis for construction of twin jibs, simply duplicating the normal jib.

The tradewind rig can be used aboard any type of vessel, whether sloop, schooner, or what have you.

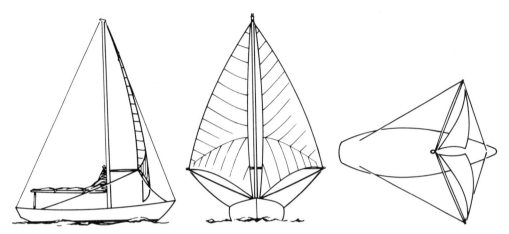

Fig. 55. A "tradewind" rig may be used as an effective self-steering device when running or broad reaching.

cost

Warning

I had a call one morning from a gentleman interested in commissioning a custom design. Upon my asking a number of leading questions, he related his requirements to me: A Marconi schooner 55' to 60' L.O.A. to cruise with three couples with additional berthing for two, flush-deck profile, clipper bow, diesel engine, and a water and fuel capacity for a 3,000-mile range. He wanted the hull constructed of ferro-cement, plywood bulkheads, teak deck and many shoreside conveniences such as deep freeze, etc. This information enabled me to refer to my own charts forthwith and, after establishing the basic displacement, I was able to determine roughly what the materials and equipment would cost. At the same time I asked him about his building budget and his astounding answer was only $10,000! Upon catching my breath, I became the bearer of sad tidings. His required vessel would have demanded financing in the range of $37,000 minimum!

Unfortunately, this is only one example of the many calls I receive weekly. What is so distressing is that these prospective builders, when dissatisfied with my building estimates, call scores of other designers until they are pacified by a lower figure. All this adds up to the obvious impression that ferro-cement is some kind of a miracle material which will enable the amateur to construct a vessel at only a fraction of the stock commercial price for a comparable yacht. How sad, how sad. As we have recently read so many books, advertisements and promotional materials on ferro-cement claiming 50% savings in hull cost over all other methods of construction, there is little wonder that the inexperienced person has been misguided. What rarely comes to the forefront is the fact that the bare shell comprises only about 10% of a finished yacht. While we may be able to build a 40' hull for $2,000, the finished boat will soar to approximately $21,000. This is still considerably less than we would expect to pay for a commercial counterpart.

Without a doubt, the greatest obstacles to quality are lack of experience, lack of patience and lack of money; the worst lack is money. Thousands of builders across the country have convinced themselves that they can beat the system by plunging headlong into projects far out of their reach. There is nothing more heartrending to me than to see the construction of a vessel halted only halfway toward completion. It would be far wiser for the prospective builder to evaluate his available funds objectively and to choose a vessel requiring considerably less than that figure. The result will be a finished boat well-executed and within his expectations. I suggest that enthusiastic builders complete a smaller yacht than hopefully desired in order to gain the feel and experience of the enormity and complexity of boatbuilding. This boat may be sold in due time to help finance a larger endeavor providing, of course, that the builder still maintains his optimistic outlook.

As I have toured so many building sites, I have seen countless abandoned boatbuilding projects, broken hearts, spirits, homes and bank accounts. I have observed hundreds of potentially fine vessels bastardized and jury-rigged for the sole purpose of cutting down on what may have already been an inadequate budget—material shortcuts, critical hardware home-built rather than purchased commercially, alterations in building specifications for the sake of stretching the dollar. It is for these reasons that I have attempted to take a realistic look at the cost analysis of cement construction on a retail material basis. It is my belief that the following tables will return the cost of this book a hundred times over. The reader has the option of ignoring the numbers before him and he is free to pass them by. But I must warn that the tables in this chapter are the results of months of calculations on actual construction projects, including not only my own designs but those of other major architectural firms.

Price Shopping

The builder must bear in mind that the cost tables included in this chapter are based on what I have discovered to be average prices for standard marine equipment and materials (listed retail, 1973). I have made no attempt to include shipping charges but have assumed crating for large items such as engine and galley range.

Checking and rechecking my catalog files made clear that a considerable price range existed between materials and in most cases these price differences were due to the quality of the finish of the item while not necessarily affecting strength or reliability. I strongly urge that the prospective builder write to the various engine equipment and hardware manufacturers for their most recent catalogs. Because some of the promotional material may not include prices, be specific in asking for this information. In some cases you will be referred to local dealers or distributors. It will also be necessary to contact several lumberyards for quotations on your basic requirements. Because many of the small neighborhood do-it-yourself outlets will not special-order unusual timber, it is best to deal with larger lumberyards. In some instances the big outfit may stock air-dried woods (a must!) and usually will have grade "A" spruce and fir. Exotic woods and large cuts should always be ordered well in advance of their actual need; otherwise you may have to pay a higher price to take whatever you can get at a moment's notice.

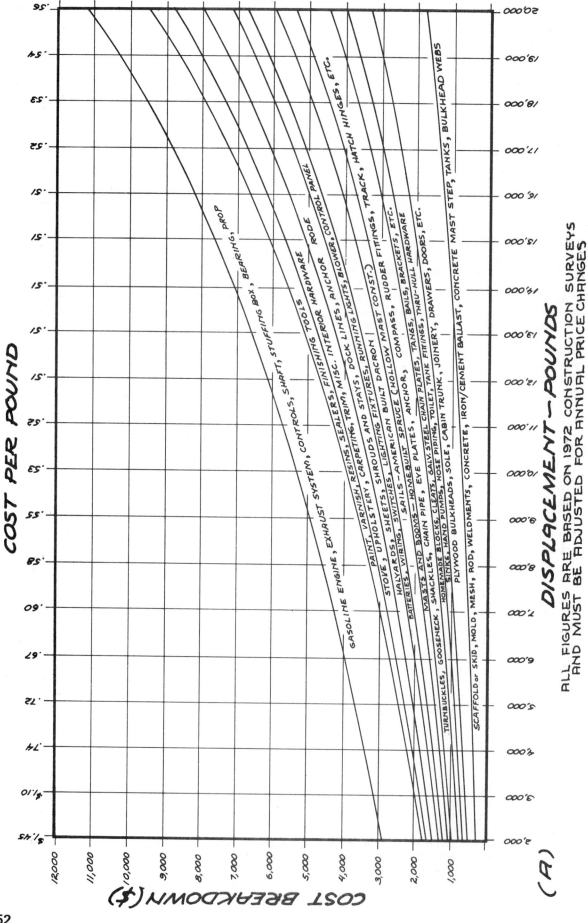

Fig. 56A. Basic amateur construction cost breakdown.

COST PER POUND

DISPLACEMENT—POUNDS

COST BREAKDOWN ($)

ALL FIGURES ARE BASED ON 1972 CONSTRUCTION SURVEYS AND MUST BE ADJUSTED FOR ANNUAL PRICE CHANGES

(A)

GASOLINE ENGINE, EXHAUST SYSTEM, CONTROLS, SHAFT, STUFFING BOX, BEARING, PROP

PAINT, VARNISH, RESINS, SEALERS, FINISHING TOOLS

UPHOLSTERY, CARPETING, TRIM, MISC. INTERIOR HARDWARE

STOVE, SHEETS, SHROUDS AND STAYS, DOCK LINES, ANCHOR, RODE

SWITCHES, LIGHTING FIXTURES, RUNNING LIGHTS, BLOWER, CONTROL PANEL

SAILS—AMERICAN BUILT DACRON

HALYARDS, WIRING, SPRUCE (HOLLOW MAST CONST.)

BATTERIES, COMPASS, RUDDER FITTINGS, TRACK, HATCH HINGES, ETC.

MASTS AND BOOMS—HOMEBUILT, ANCHOR, TANGS, BAILS, BRACKETS, ETC.

EYE PLATES, CHAIN PLATES, THRU-HULL HARDWARE

CHAIN PIPE, GALV. STEEL TANK FITTINGS, DRAWERS, DOORS, ETC.

MASTS AND BOOMS—HOMEBUILT, CLEATS, TOILET, TANK FITTINGS, JOINERY

TURNBUCKLES, GOOSENECK, SHACKLES, HOSE PIPING, SOLE, CABIN TRUNK,

HOMEMADE BLOCKS, HAND PUMPS, CONCRETE MAST STEP, TANKS, BULKHEAD WEBS

SINKS, PLYWOOD BULKHEADS, SOLE, CABIN TRUNK, IRON/CEMENT BALLAST,

CONCRETE, WELDMENTS, MESH, ROD, SCAFFOLD or SKID, MOLD,

52

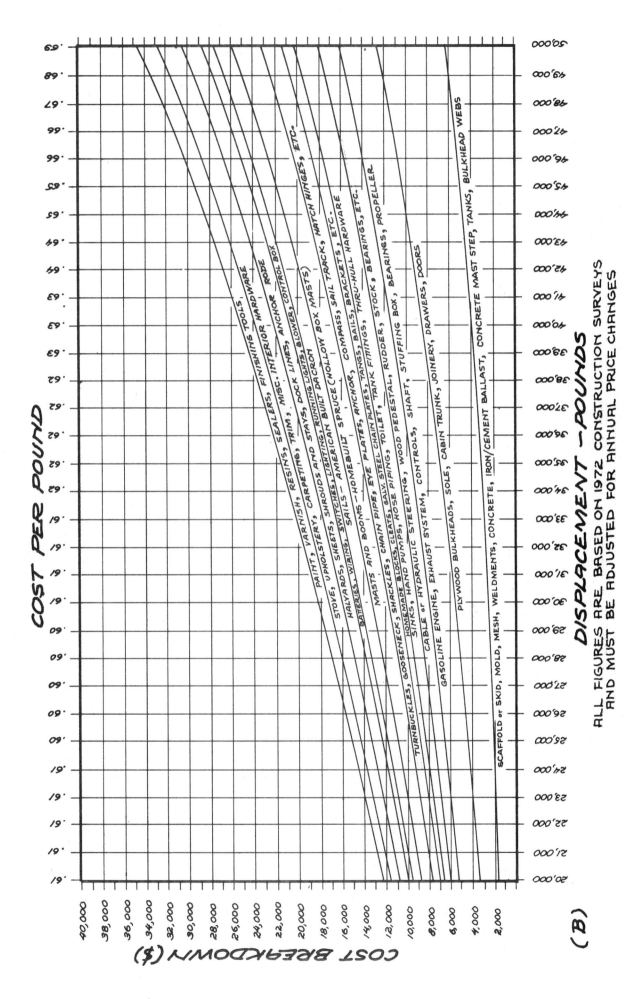

Fig. 56B. Basic amateur construction cost breakdown.

53

The lumber to be used for the scaffolding, skid, station molds, etc., need not be new lumber and, with patience, you may be able to locate a wrecking firm which can supply this material at a fraction of the normal cost. From time to time listings for used and rebuilt engines appear in the classified sections of the boating magazines and the local papers. This one factor alone may reduce the commissioning cost by as much as $1,000 but injects the question of reliability. A used engine may be a good bet as long as you assume that it may require replacement within several years. Anchors, winches, stoves and rigging hardware can also be found the same way. The cost for sails as listed in the first table is based on the price per square foot generally quoted by American makers. Because this item can easily equal the cost of the hull itself, you may consider purchasing the sail inventory from the Orient for approximately 25% less.

As you can see, price shopping can have a tremendous effect on the total cash outlay. The additional time spent, while tedious, could reduce the cost of a 45' boat by as much as $4,000. Unfortunately, tracking down materials cannot be done without having the complete plan package in hand, but the purchase of a set of drawings for this one purpose could easily save the builder from plunging into a project that he could not ultimately complete.

The General Estimate

By individually calculating the hundreds of items and materials necessary to the construction of scores of vessels, a proven system has evolved which may be used by designers and builders alike for determining the finances necessary for commissioning a yacht.

Naval architects have known for decades that most yachts of a given displacement ultimately require virtually the same basic materials and hardware regardless of the design of the yacht. For instance, a ketch will need more spar length than a sloop but the mast sections of the ketch may be smaller, thus offsetting the differences. The ketch will carry a little more sail area and hardware, but the gear aboard the sloop will be heavier and more sophisticated. Regarding the interior layout, the same general rule applies; i.e., the more complicated the accommodation plan, the less ballast will be carried, etc.

The accompanying table, therefore, has been based on displacement and may be used for estimating commissioning cost with surprising accuracy. You will notice that the finished yacht has been broken down into its basic elements as well as a completed figure. This may be read off as a dollar hull cost per pound. Using this table the building enthusiast may now financially evaluate designs that appear in magazines and study plans, and later maintain the proper cash flow pace during the actual construction of his boat. As for the budget, if the boat-builder discovers that he has spent more for a particular item than that illustrated by the table, it will be necessary for him to carefully price-shop the succeeding items in order to complete his yacht for a predetermined cost.

I must openly admit that builders can beat the system by bargain-hunting and scrounging but this is rarely done without significantly sacrificing finished quality, appearance or strength. I have recently read of a 42' ketch being completed for $6,000 and similar vessels for slightly more. Other designers in the ferro-cement field may openly criticize this cost estimate system but it should be remembered that one of their own designs may have been included

in my analysis. Remember . . . we are talking about averages and acceptable American marine standards.

The finish of the yachts, as shown on the accompanying charts, must be considered as plain and utilitarian. The price breakdowns do not include extensive uses of fancy bright woods, extensive or convenient hardware, chrome plating, etc. I am showing only minimums for Spartan construction.

Diesel Engine

This installation is the most requested alternate feature and with very good reason. While diesel engines are vastly more expensive in larger sizes and require more space and weight, they are noted for their reliability and longevity. Because gasoline fumes are heavier than air, there is always the risk of volatile gases collecting in the bilge to await an unsuspected spark and resulting explosion, but with a diesel installation this fear is eliminated. The safety factor of a diesel engine can reduce marine insurance rates considerably, thus helping to defray the additional cost of the installation. Another important factor is that diesel oil, not gasoline, is the most universally available fuel, being at the same time less expensive—each penny saved when filling up goes toward compensating for the higher initial cost of the engine.

In my own work, I specify engines largely upon serviceability. While less expensive "off-brand" equipment may cut down the vessel's commissioning cost, it may turn out to be a headache when you try to replace bearings in Katuka. Lesser-known machinery also has a tremendous impact on the yacht's resale value as it transmits the element of doubt to the new owner.

I always install an engine with an integral heat exchanger-cooling unit rather than having to match a proper remote unit at greater expense. Heat exchangers designed for a given power plant have proven extremely reliable under virtually all conditions. They are also more compact and do not require disrupting hull protrusions. To add the cost of an average diesel engine to the basic estimate, apply the price difference as shown in the accompanying table. (Fig. 57)

Wood and Planked Decking

Of all the builders presently involved in the construction of ferro-cement yachts, the vast majority have leaned toward vessels of character or of a traditional type. Most are planning on applying teak decking over concrete or plywood for the obvious reasons of accentuating the nautical and the fine-yacht atmosphere. Unfortunately, the general quality of the hulls that I have seen are no more deserving of teak than Mammy Yokum a mink; but as long as the amateur builder insists on this expense and assumes this critical task, I feel I should point out several factors that might help to create a more realistic perspective.

All lumber (except plywoods) should be air-dried, but it is becoming increasingly difficult to obtain properly seasoned lumber without paying a tremendous premium. Because the large deck in teak requires so much timber, air-dried material becomes a necessity to avoid severe shrinkage, the sheering of fastenings and the opening up of payed seams. Teak is one of the most difficult lumbers to procure, even kiln-dried, and the prices will stagger the most affluent amateur boatbuilder. Because deck planking rarely exceeds ¾" for sheer-laid decks and ½" for straight-laid decks, a large percentage of your investment will ultimately turn to dust when the planks are to the proper thickness, the milling charges alone adding an additional 20% to 30% over the cost of the wood.

Another problem that faces the builder when purchasing teak is buying the required quantity in a matching hue, as the color can range from buff to brown to sienna. Considering that all decking timber must be quarter sawn, this may take a tremendous amount of shopping around and most probably will require that the wood be ordered to meet these specifications. Anything less is totally unacceptable as well as a terrible waste of money and time.

The reason given for the use of teak decking is that it is "traditional," yet its use is only as recent as the twentieth century. Looking back into history we find that pine and fir species have always been the most widely used because of their availability, easy working qualities, longevity and footing. With rare exceptions they were the decking standard during the clipper era, the revolutionary period and before, without respect to continental or national boundaries. These woods, being much lighter in weight than teak, yield a tremendous advantage for the ferro-cement builder as they will add only about a pound per square foot over the normal deck weight (assuming that we can lighten the sub-deck structure).

Admittedly, teak is the epitome of yachtsmanlike quality but the builder must decide whether he wants a gold-plater or a functional, shiplike cruising boat. Once pine or fir has been oiled down, the finish is strikingly handsome, rugged and easy to maintain. It is for these reasons that I highly recommend them while discouraging teak and, as the accompanying table will show, there is an overwhelming cost difference.

As was explained in Chapter 4, ferro-cement decking can rob a vessel of much of its hard-earned performance. For my own boat, I would prefer a two-layer plywood laminate over spruce deck beams. The time factor compared to that of wiring up a steel armature cancels itself out while guaranteeing deck fairness rarely obtained in ferro-cement. The interior of a boat with deck beams is very striking, creating a warmth and personality unmatched by any other system. When fiberglassed and finished with a 2-part epoxy

paint, it is no more difficult to maintain than other materials. I have, therefore, included this alternative on the accompanying cost table. (Fig. 58)

Ventilators and Opening Ports

Ventilation is an absolute necessity aboard any ship. While the basic cost breakdown accounts for slide- or lift-hatches, I did not consider opening ports or cowl vents as an immediate necessity for commissioning and, therefore, they should be considered as accessory equipment to be installed during construction or any time thereafter.

Good ventilation can eliminate foul seagoing odors, increase the life of all woodwork and cool an interior measurably when sailing in torrid or humid regions. I recognize, however, that standard ventilators and opening ports are extremely expensive and I have shown some typical examples which the prospective builder may use as a general guide to cost. (Fig. 59) The required quantity of each porthole or vent must be added to the basic cost estimate. It is best to assume that each unit will run at least $50.

Pressure Water

Until recently, this feature has been considered an unnecessary luxury even aboard very expensive custom yachts. It is considered by many blue-water sailors to promote the indiscriminate and wasteful use of the ship's most valuable commodity. If the prospective builder believes the hand pump to be such an inconvenience as to require the household comfort of simply turning a faucet, he may not be ready for sea at all, since there are many more serious sacrifices that he will have to face. There is also the added factor of drainage of electrical power, which should always be conserved for compass, running lights, and emergency situations. Any use of electrical equipment must always be paid for in fuel.

"But I'd like to have a shower!" For decades cruisers have gotten along quite well with a deck-mounted gravity tank, usually painted black to create warm water even in cold climates. The gravity tank is quickly filled daily by the use of a hand rotary pump, then routed to the galley or head. As you can see, there is more than one way to skin the cat, and very efficiently and economically at that.

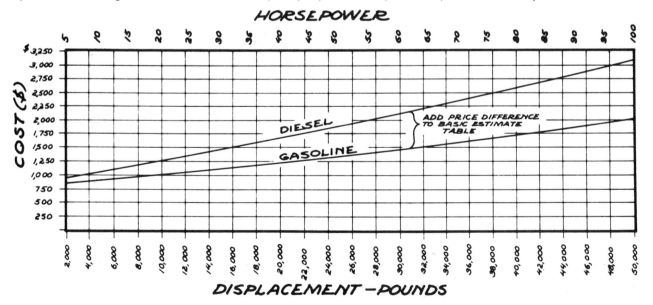

Fig. 57. Additional anticipated expense for diesel engine installation.

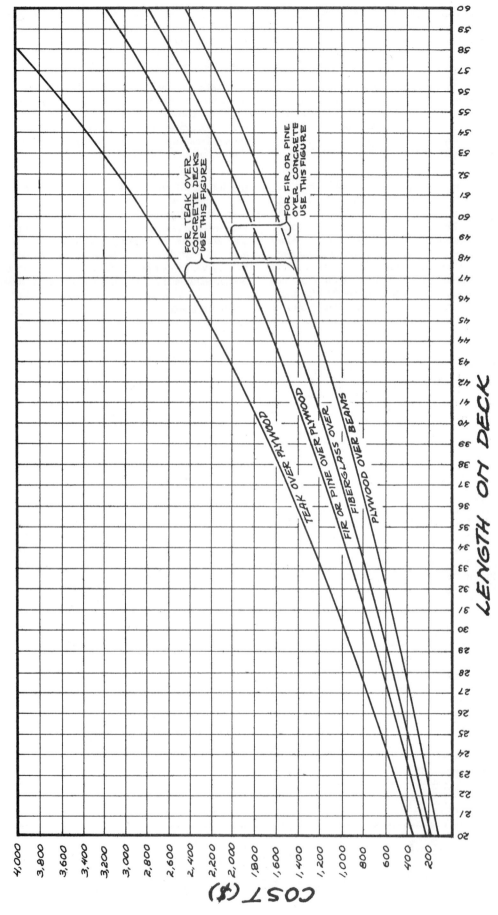

Fig. 58. Anticipated additional expense for wood deck systems.

LENGTH ON DECK

COST ($)

FOR TEAK OVER CONCRETE DECKS USE THIS FIGURE

FOR FIR OR PINE OVER CONCRETE USE THIS FIGURE

TEAK OVER PLYWOOD

FIR OR PINE OVER PLYWOOD

FIBERGLASS OVER BEAMS

PLYWOOD OVER BEAMS

56

There are, however, many excellent lightweight and inexpensive pressure units available that will fill your needs most adequately providing you pay very strict attention to the quantity of water being used on your long voyages. They may be switch- or faucet-actuated while requiring occasional charging and servicing. I have listed the most common pump types, including in their costs the required faucets and other equipment necessary with the pressure system. You need to add one system only to the basic cost estimate chart regardless of the size of the boat. (Fig. 60)

The Shower

Regardless of whether the builder uses an electric or gravity pressure system, the same requirements prevail as to the floor grading, sump and shower head. Without a doubt, the most practical and economical shower is what has become known as the "telephone" type. (Plate 18) This consists of a wall unit incorporating the hot and cold knobs and a flexible hose attached to the shower head, the latter resembling a push-button dictaphone. The terminology for this hardware is derived from the fact that it looks exactly like a telephone when hanging in the stored position. From the standpoint of water consumption, it uses considerably less than the more domestic overhead type because the user has the opportunity of localizing the coverage or turning off by releasing the button when not actually rinsing. For extended ocean passages or global cruising, owners will undoubtedly find themselves in the position of having to buy fresh water which may, then, be of questionable purity. If such voyaging is the intention of the prospective builder, he would be wise to consider the installation of an additional and completely separate pressure system for use with salt water. This can be used with great economy for washing dishes, clothing and showering, saving the fresh water for the final rinse only. You should assume an additional cost to the basic boat of approximately $150 for each shower installation. This will take into account the required sumps and discharge pumps.

Hot Water

One need only spend a brisk day in the fresh air to know the sticky feeling resulting from the penetration of the salty air. Imagine what 10, 20, or 30 days offshore can do to the sailor's hair and complexion. The salt air has been known to dry one's skin to the point of causing cracking and bleeding and, unless there is a way of washing off this film, it can irritate the eyes severely. Shipboard cleanliness and refreshment can also have a measurable impact on the crew's morale, especially after battling a gale or two when much sleep is lost and one's energy and spirit are at a low ebb.

There are many other obvious reasons for the use of hot water on shipboard and, generally speaking, there is no other convenience more welcome aboard a cruising boat. Unfortunately, it is a commodity that must be paid for with fuel and, gallon for gallon, can be overwhelmingly expensive on a transoceanic passage. There are three basic types of hot-water systems which may be considered aboard the cruising boat, each having its advantages and drawbacks.

The first type consists of a tank with an internal coil through which hot engine water passes. The greatest advantage of this system is that it requires no burners, no separate fuel or pilot light. It is extremely compact and may be installed in any remote position. Some models are available with provisions for inserting an auxiliary auto-

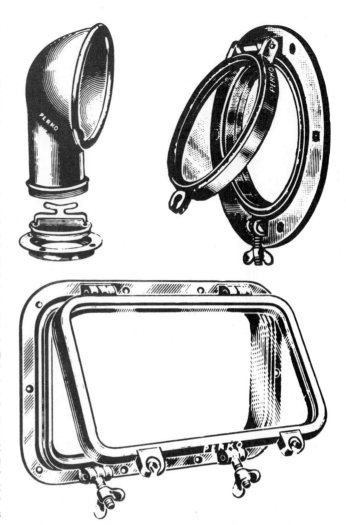

Fig. 59. Some typical ventilators. Courtesy, Perkins Marine Lamp and Hardware Co.

matic electric heating element for dockside use. The obvious drawback is that the system requires running the engine approximately 15 minutes to heat about 8 gallons to 160°. Of course, once the water has reached this temperature, it should stay hot for some time and, consequently, showers, laundry, dishes and the daily charging of batteries should be accomplished at the same time to reduce unnecessary engine operation. One source for these units is Sen-Dure Products, Inc., Bay Shore, New York.

	PUMP WITH PRESSURE CUT-OFF	FAUCETS ADDITIONAL TO ONE IN GALLEY AND HEAD
HEAVY DUTY	$80 to $150	$15
LIGHT DUTY	$45 to $75	$15

Fig. 60. Anticipated additional cost for various pressure water systems.

Diesel-fired hot-water heating boilers are available from several sources and have the convenience of not requiring engine operation. Their drawbacks are that they may be extremely heavy if built of cast iron and they may encourage the unnecessary use of fuel unless the crew makes it a definite point to shut down the burner when not actually

Plate 18. A "telephone" shower may be wall mounted or hand held. An on/off button permits water saving. (Alson's Products, Covina, Ca.)

being used. They also require a stack. A typical unit is manufactured by Shipmate Division, Richmond Ring Company, Suderton, Pa.

The last type to be considered is an extremely compact boiler fired with propane. (Plate 19) This is extremely fast heating, quick recovering and economical. It may be mounted anywhere in the vessel but most usually under cockpit seats, as it requires an air intake and exhaust vent. Its only disadvantage is that there is the possibility of small escapements of gas which, being heavier than air, may find their way to the bilge. This problem can be highly volatile as well as increasing insurance rates. The best I have seen are from Bowen Division, Atwood Vacuum Machine Co., Rockford, Illinois and Mobile Temp, Brighton, Michigan.

Plate 19. The LP gas water heater is compact and does not rely on engine operation. When properly installed, it is as safe as any comparable system. (Courtesy, Atwood Vacuum Machine Co., Rockford, Il.)

Cabin Heat

If the owner intends to spend a great deal of time aboard in colder climates, a heating system becomes an absolute "must!" If the heat source has not been specifically designed for this purpose (i.e., the galley stove), it is sure to cause an unacceptable waste of precious fuel. I do not generally recommend the use of an open fireplace because its principle is radiant with only a minimum of air convection and therefore should be considered only for its aesthetic value. A hot-water heater coupled to a pipe circulating throughout the interior is most reasonable and economical but requires a small rotary pump.

Many companies manufacture compact cast-iron heaters and pot-bellies which will serve quite adequately aboard vessels in the smaller sizes; while wall-mounted, open, cabin heaters that burn charcoal (Plate 20), coke, LP gas, etc., may suffice.

In my opinion, the best cabin heaters are the compact kerosene burners. (Plate 21) They rarely exceed the size of a portable television and, yet, deliver up to 11,000 BTU.

Plate 20. A cozy little charcoal wall heater. (Shipmate Div. of Richmond Ring Co.)

Their cost is ridiculously low; they are beautifully styled, and may be stored in the lazarette when not in use. I would suggest this type above all others and would recommend one small heater of approximately 2,500 BTU be located in each compartment. But, warning! They consume oxygen and create carbon monoxide, so the areas should be well vented wherever an "open" heater is used. They are available from Valour Engineering, Ltd., Bromford, Edington, Birmingham, England; Alladin Industries, Nashville, Tennessee; Greenford Products, Skokie, Illinois; Therm'X Corp., San Francisco, California.

Refrigeration

Only as recently as the past two years, I have noticed an ever increasing opinion that shipboard mechanical refrigeration or freezing is an absolute necessity in order to afford any degree of comfort at sea. I cannot dispute that it is a tremendous convenience but, as before, I must advise that it is a luxury requiring the expenditure of valuable fuel while under way. It is wise to remember, also, that virtually every type of foodstuff (i.e., meat, vegetable, etc.) is universally available in canned, dehydrated, pickled or smoked forms and that you will have little trouble in maintaining a delicious and balanced diet over long periods of time. It is also imperative that you consider installation costs of refrigeration units.

If your prime intention is to spend most of your time on board while tied to the dock, I would immediately recommend purchasing a 110V-AC refrigeration or freezing unit. These are most commonly of the condenser-evaporator type (Plate 22) and may be purchased specifically for marine use. They may also be dismantled from an old household refrigerator for installation in your ice chest. The compressor-condenser portion of the unit may be installed remotely. For use when at sea, the 110V-AC condenser-evaporator "frig" units may be operated by routing DC current through an inverter, but this will result in a slight increase over the normal amperage. Assume that you will be

Plate 22. Typical condenser/evaporator and coldbox units available for marine installations.

Plate 21. A portable, compact, high-output kerosene heater, ideal for marine use, by the Aladdin Division of Greenford Products, Skokie, Il.

refrigerating no more than 10 cubic feet. The cost of an AC condenser-evaporator unit will run approximately $200. To install the required AC-DC inverter, you should expect an additional outlay of approximately $350.

If you believe it more convenient to install a DC condenser-evaporator refrigerator unit, your initial cost will run at least $250 upward. A DC-AC converter for dockside use will cost about $200 additional.

While the condenser-evaporator refrigerator systems are by far the most common for shipboard, there are many other types available. The most notable and convenient of these is the "holdover plate" refrigeration system. (Plate 23) This requires the installation of a compressor, which may be run off the main engine or from a separate power unit. The "holdover" plate system will require approximately 30 to 40 minutes of engine running time to maintain adequate

refrigeration for 10 cubic feet of space throughout a 24-hour period. These systems are extremely efficient and require an outlay of approximately $650 for a 10-cubic-foot capacity.

Other refrigeration systems that might be considered are of the thermoelectric type, cold-water condenser type and the absorption type. The latter requires a small continuous flame of kerosene, propane, or other fuel. This may be objectionable from the safety standpoint as well as odors and cleaning problems.

As there are many variables which must be considered by both you and the manufacturer of your refrigeration system, it is far better to contact them directly with your requirements so that they may be afforded the opportunity to recommend the most efficient and economical system. The most common names in marine refrigeration systems are: Adler-Barbour Yacht Services, New Rochelle, New York; Birken Manufacturing Company, Venice, California; General Thermetic, Mt. Vernon, New York.

Plate 23. A typical holdover refrigeration installation requires periodic refreezing but is compact and mechanically simple. (Courtesy, Adler & Babour, New Rochelle, N.Y.)

Generators

When I look through my equipment file and run across the generator section, I continually see promotional photographs of people aboard boats watching television, running vacuum cleaners, making toast, shaving, broiling steaks, listening to hi-fi's, and otherwise engaging in all sorts of shoreside activities requiring 110V current. Rubbish! Every one of these items is available in either DC current or fired by the galley stove. When I am faced by clients who ask for generating equipment, I must always return with, "What ya gonna use it for?"

The need is usually for recharging batteries, but most engines are available with optional "heavy-duty" generators or alternators producing over 30 amps. If the owner is not wasteful with his electricity, one hour of running per day will easily keep up the average daily battery drain (assuming no fancy equipment or electric refrigeration is used). As pointed out earlier, this engine running time may also serve the dual purpose of heating the ship's water ration. The engine need not be operated at full power but may be idled at the generator cut-in speed with only a minimum of fuel consumption. If the owner still insists that he can't live without an electrical plant, I could never recommend anything beyond 10 kw; Honda produces one no larger than a breadbox that may be conveniently hand-carried anywhere aboard or ashore. The accompanying chart may be used for estimates to be added to the basic cost breakdown. (Fig. 61)

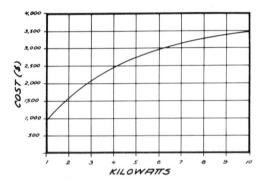

Fig. 61. Anticipated additional cost of various AC generating systems (includes power unit, muffler, auto-demand accelerator, shoreline transfer and exchange cooling).

Winches

In Chapter 6, I spoke of the timesaving and safety factor of winches. They should be first priority on the list of accessory equipment although they need not be aboard for commissioning. They are extremely expensive hardware and have a broad range in price from one manufacturer to another. Winches are available as "double" or "single" action types, which means that they may have one or two speed ratios by reversing the direction of the handle.

For the highly tuned racing machine, winches may be machined and polished to the accuracy of a fine watch and can cost as much as a small car; but I shall immediately dismiss this quality, as function is the ultimate requirement. Some of the most rugged hardware comes from England and Australia and at considerably lower cost than that of most American manufacturers. I can also recommend most highly the Merriman line out of Boston.

It is impossible for me to anticipate the requirements of each boat owner. So, for the purpose of the estimate chart, I have assumed the use of two halyard winches, two large and two small heads'l winches and a mainsheet winch. (Fig. 62)

Aluminum Spars

Nobody likes varnishing, especially from a bos'n's chair, and aluminum spars do have the advantage over wood of easy maintenance. Assuming the same strength characteristics, aluminum is somewhat lighter and may add considerably to the stability of the vessel.

While solid laminated masts are quite practical to construct, the hollow box spar is extremely difficult and should not be undertaken by the amateur builder without first ordering a precut kit from a professional spar builder. "Gluing up" is a procedure demanding careful planning and coordination of manpower; if it is not accomplished properly, the mast may be a total failure. Purchasing of "aircraft" grade spruce lumber is next to impossible these days, but "mast and spar" grade Sitka spruce and grade "A" Douglas fir are most acceptable.

You can easily see why the trend has been toward aluminum masts. Any reputable manufacturer may produce this custom item for your boat by simply converting your wooden mast and spar plan. They are experts at this and are called upon to do so every day.

In the accompanying chart I have assumed both mast and booms as well as that hardware which is normally supplied by the spar maker; i.e., tangs, bails, masthead, track, outhauls, gooseneck, spreader brackets and sheaves. (Fig. 63)

Roller Reefing

While I have already discussed the possible disadvantages of this reefing system, many builders may still prefer its use. Generally speaking, its cost is the same for either wood or aluminum booms. Add the cost difference factor to the basic estimate table.

Homemade vs. Stock Hardware

All of my designs include detailed drawings for tangs, bails, goosenecks, mastheads, travelers, cleats, blocks, etc. All items requiring plate may be made out of mild steel, galvanized after fabrication, or in stainless steel produced by a professional welder. If the builder chooses the additional expense, he may procure all of these items from marine manufacturers or spar builders. Apply the cost difference shown to the basic estimate. (Fig. 64A & B)

Steering Vane Gear

The cost of these units is extremely flexible but generally may run from $600 for use aboard a 30-footer to $800 for a 50-footer. The price range also depends upon the type of mounting required by the stern of the boat per rig and her displacement.

Autopilot

Here, again, many installation variables must be considered by the manufacturer, and there is no general rule of thumb which may be applied to cover all the possible situations. I can only say that generally autopilots run from $1,100 to $1,700. Here is where price shopping comes into play.

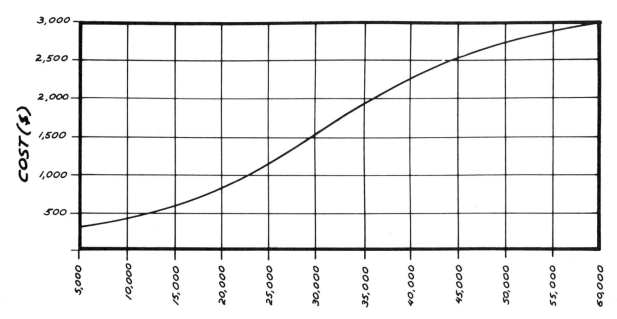

Fig. 62. Anticipated additional cost for basic cruising winch requirements.

APPLIES TO BOTH CHARTS

DISPLACEMENT (lbs.)

25   30   35   40   45   50   55   60

APPROXIMATE L.O.A.

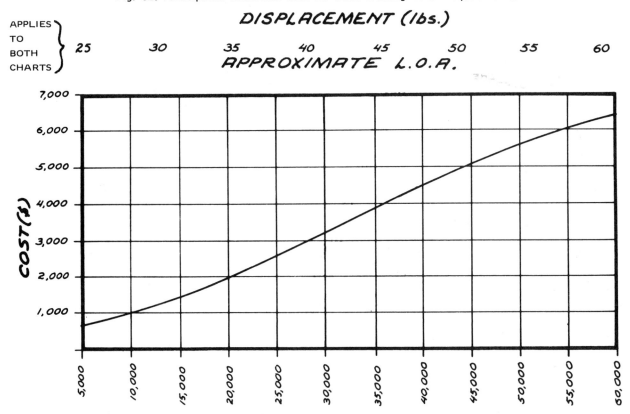

Fig. 63. Anticipated additional cost for aluminum booms and masts.

Navigation Systems

On my own boat I would consider RDF to be the most functional piece of gear and would expect to pay about $180 for a portable unit. Automatic direction finders (ADF), however, can easily run over $2,000 and should be considered only by the skipper who has everything.

The second most important navigation unit to consider would be LORAN, and these are available for as low as $1,000; but for greater reliability and accuracy, $1,500 would be far more reasonable to expect.

A nonrecording depth sounder with a 250-foot capacity should normally cost between $150 and $200.

Wind speed and direction indicators range between $250 and $500.

A recording speed indicator and log can easily flex between $350 and $700.

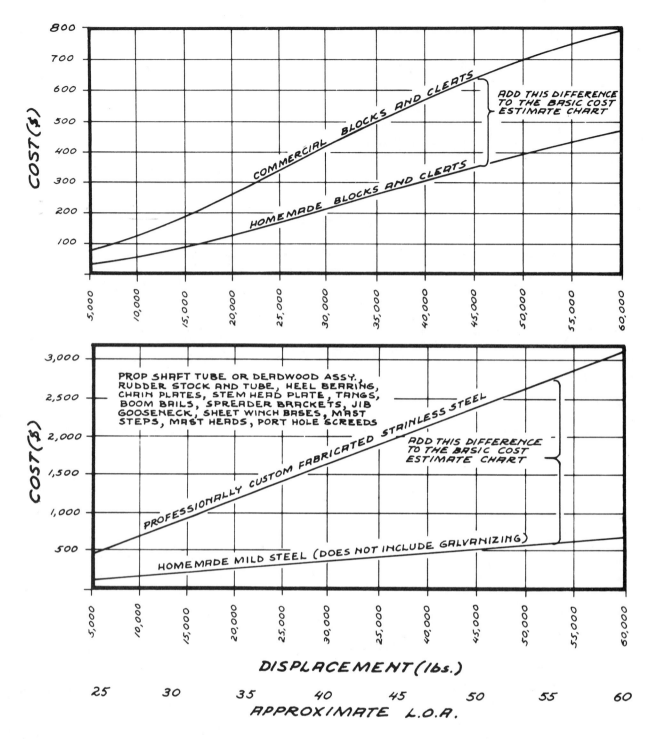

Fig. 64. Anticipated additional cost for commercially manufactured hardware vs. homemade.

Once again, this is an extremely delicate area in my trying to pin down your specific taste and long-range requirements and I would highly suggest contacting all available manufacturers for the purpose of weighing instrument quality and price.

Odds and Ends

As you may imagine, I could go on and on with a continuing list of optionals, extras and accessories, but I believe that those listed will give the prospective builder some firm groundwork for building a realistic budget.

Other items that should be considered before making the big leap: Space rental, sales tax, launching fees, transportation and shipping charges, equipment rentals, professional welding and plastering services, variations in interior decorating, electronics, navigation instruments and tables, initial fuel, surveying, insurance, initial dockage, sail covers, linens, galley and eating utensils, and lastly—the christening champagne and launching party.

8

construction methods

Choosing the System

For many years there have been available a variety of books and brochures on ferro-cement construction written by prominent marine architects and engineers. Each claims that his suggested system is the only method of construction of a ferro-cement boat while all other systems are more time-consuming, more costly, lesser in strength or generally more difficult. The amateur builders themselves have formed many conflicting opinions ranging from good to bad or indifferent, which only add to the existing confusion.

I received a letter recently from a prospective client saying, "If the designers can't make up their minds as to which system is best for constructing a boat, how the hell can the amateur builder be expected to make any kind of intelligent choice?" Unfortunately, I couldn't agree with this gentleman more. To insist, for instance, that the inverted cedar mold or the upright open-truss system is the one true construction method is like saying that there is only one way to iron a shirt! I personally feel that this one-sided, narrow-minded outlook is not only dangerous but totally absurd. Based on history, as well as on hundreds of test reports, I find all of the construction systems presently being employed have not only marked benefits but severe drawbacks at the same time and that the construction system chosen depends more upon the builder's capability and equipment than on any other factor. Ultimately, about the same strength, cost and time factor will be achieved regardless of the method chosen. In order to help the amateur builder solve this problem on an intelligent basis, I have described the available methods as well as the benefits and disadvantages of each.

Upright Open Pipe Frame

Description:

Pipe station frames are bent to accurate shapes on the lofted patterns allowing additional pipe to extend above the sheer line. This facilitates the attachment to overhead beams or scaffold at their prescribed intervals. Pipe deck beams may also be bent and welded to the pipe frames prior to their positioning, as well as the steel sole-beam brackets. Once the stations have been hung, the keel and sheer re-bar are sprung from station to station and weld-attached, followed by layers of reinforcing rod wire-tied to the outside of all frames. The required number of outer mesh layers are wire-tied to the armature and carefully darted and faired during application. Integral structures, such as tanks, bulkhead webs, mast steps and engine beds, may be formed of reinforcing rod prior to wiring in the inner hull mesh. Wooden thru-hull blanks are installed in the hull, followed by plastering. Most hull designs may be adapted for construction by this method.

Structural Attachments:

As mentioned above, re-rods for integral structures may be completely formed prior to plastering and completely meshed at the same time. The mast steps and engine beds are not poured until after the hull has cured in order to prevent sagging under this tremendous weight. Bulkheads or bulkhead web attachments may also be integral, but some builders prefer to bond in plywood using epoxy grout. In areas where hardware must be thru-bolted, prelocated waxed dowels should be inserted into the armature to eliminate troublesome drilling.

Advantages:

The structure is based on the use of common black iron pipe. This system does not require a complicated or expensive temporary wood mold, which is usually an expendable cost with other construction systems. Bracing or trussing of pipe frames to prevent distortion on larger hulls may be common pipe or reinforcing rods. A full scaffolding may not be required on smaller vessels providing that welded braces do not interfere with meshing and it is important to note that the open-pipe-frame system allows total accessibility to the inside of hull during meshing and plastering (ideal for building integral structural units). One of its greatest advantages is visual inspection and assurance of concrete penetration as well as permitting finishing the interior at the same time as the exterior. The complicated and dangerous operation of rolling over is eliminated with this system.

Disadvantages:

This construction method may require extensive scaffolding and bracing, particularly on large hulls. Because no mold is employed, wire tying should occur at each rod intersection to achieve proper fairness and rigidity. The pipe-frame system tends to distort under the weight of fresh concrete unless it is extensively suspended by wires or rods attached to the scaffolding. Unavoidable ridges stand out on the inside of the hull at all frame positions, which complicate the fitting of counter tops, berths, sheer clamp (if wood deck is used), etc. Because of extensive metal work, the builder should be an expert metalsmith and willing to invest in expensive equipment. Large pipe frames may be extremely difficult to handle during movement and erection because of their natural flexibility; precautions must be taken to prevent distortion, and some builders prefer to attach frames to temporary wooden molds to alleviate this fault. Because no ribbanding is used for fairness, only high-tensile, hard-drawn or spring steel should be used for the best results, wired (**not** welded) at each intersection.

to each frame. (Plate 25) Integral structures are formed of mild steel rod and inner mesh is applied. Outer mesh layers are now applied, taking care to wire-tie through all other layers at each intersection. Once the hull has been perfectly faired by "jogging" or "wedging," the hull may be plastered from either side. Tanks, bulkheads, etc., may be completed at the same time. Only vessels that have been designed for this system should use the open-truss-frame method, as it creates a space-loss problem which the architect may not have accounted for.

Structural Attachments:

Plywood bulkheads may be bolted directly to web frames if

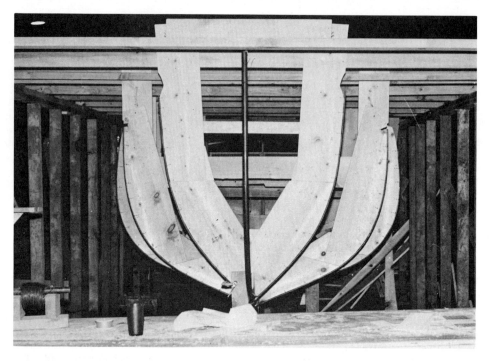

Plate 24. The popular 17' Benford catboat using the upright pipe frame system with temporary wood molds.

Variations:

Many builders have used the pipe-frame system as a substitute for the partial wood mold. With this variation, all mesh and rod is applied to the outside of the pipe, the pipes being removed after the hull has been plastered.

It is also possible to combine pipe construction with the upright wood-mold method. The wood molds serve only to stiffen the armature and, of course, are removed before plastering. (Plate 24)

Warning:

Under no circumstances should hull rods be welded to the pipe frames except in rare instances, as this creates shrinkage and severe corrugations in the hull while weakening the reinforcing rod markedly.

Upright Open Truss Frame

Description:

Mild-steel-rod station trusses or web frames are welded to shape on the lofting, then suspended from overhead beams or scaffold at their prescribed intervals. Keel and sheet rebars are weld-attached, followed by two layers of high-tensile, hard-drawn or spring-steel reinforcing rod wire-tied

their positions coincide or they may be of reinforced concrete. The engine beds, tanks and mast steps are usually designed directly into the station webs, while cabinets and joinery are attached with fiberglass or epoxy grout. The engine beds and mast steps, although preformed, should not be poured until the plastered hull has completely cured, to prevent sagging under this excessive weight. Bolt holes for hardware should be prelocated with waxed dowels to eliminate tedious drilling.

Advantages:

Because of the additional rigidity of the station trusses which remain in the finished hull, the hull shell may be of somewhat lighter construction than that of all other systems and, therefore, may be considered ideally suited for higher performance hulls or vessels which incorporate relatively flat surfaces, such as in most power boats. Construction is also well suited for ferro-cement decks, bulkheads and tanks because of the internal accessibility. The inherent lateral stiffness of truss frames alleviates the necessity for internal suspension and staying while leaving the hull wide open for working from both sides. Rolling over is not necessary. This system affords the visual assurance of complete concrete penetration.

Plate 25. Upright, open-truss frame system. Designer unidentified.

Plate 26. A Bingham-designed 42' **Doreana** ready for meshing using the partial wood-mold system.

Disadvantages:

This system may require extensive scaffolding, particularly on large hulls. It is imperative that the builder be an expert welder willing to invest in expensive equipment or rentals. Because the truss frames and bulkheads are extremely flexible, if not mounted on a wooden backing, they may be inadvertently bent or distorted. The excessively deep frame ridges on the inside of the hull create a severe loss of space, especially at berth positions, and they may obstruct walking areas if they intrude into the sole. The fitting of counter tops and other joinery is complicated by the internal frames and headroom is reduced if cement decking is employed. Because no mold is used, fairing may be extremely difficult to achieve unless the highest grade of spring-steel reinforcing rod is applied.

Variations:

The open-truss system may be combined with the pipe-frame method, using trusses only in the positions of tanks and bulkheads to reduce the space-loss problem in other areas.

Warning:

Under no circumstances should the hull shell reinforcing rod be of mild steel nor should these rods be welded to the station trusses, as severe weakening and distortion will surely occur.

Upright Partial Wood Mold

Description:

Wood station frames are carefully cut to shape, assembled and suspended from overhead beams or scaffolding at their prescribed intervals. Once aligned, all stations are interconnected horizontally with continuous wood ribbands spaced vertically apart. (Plate 26) The inner mesh layers are stapled and faired over the mold ribbands, followed by the required reinforcing rod, also stapled. If concrete tanks, bulkheads or attachments are required, "L"-shaped starter rods are inserted through the armature from the outside and wire-tied into position. The outer mesh layers may now cover the hull being wire-tied to the armature or stapled to the mold. Once the thru-hull blanks have been installed, the vessel may be plastered from the outside, taking great care to ensure penetration along the backside of the mold as it cannot be visually inspected. Most hull forms may be adapted for construction in this way.

Structural Attachments:

Epoxy grout may be used to connect all bulkheads, joinery and beds to hull or "L"-shaped starter rods (previously described) may be inserted through the armature and finished as webs. Bulkheads, beds, etc., may not be completely formed or plastered until after the hull has completely cured and the mold has been removed. The plastering of these units requires that all loose aggregate be thoroughly removed from the joint areas and concrete bonding agents must be used to create adhesion of the new concrete to the hull.

Advantages:

This system is ideally suited for the minimum weld and wire-tying requirement, resulting in rapid construction while permitting extreme accuracy in fairing. It also allows a large degree of visible concrete penetration as well as eliminating rolling over. Deck molds may be incorporated

into the hull stations, which may be wired up and plastered at the same time as the hull.

Disadvantages:

As with all other upright systems, an extensive scaffolding is required, especially on larger hulls. Because of the mold interference, integral members such as beds, bulkhead webs, tanks, etc., must be constructed in two stages, creating additional work to achieve a proper "cold" joint. The hull mold ribbands and stations also impede the troweling of the inside of the vessel while creating unsightly ridges in the concrete which must be ground and glazed with grout upon removal of the mold. The staples remaining in the concrete must be chiseled flush, ground slightly below the surface and glazed to prevent the possibility of bleeding.

Variations:

Some builders have recommended that the mold ribbands be left in the finished shell to serve as a hull ceiling as well as attachment base for bulkheads and joinery. I believe this variation to be pure folly, as moisture will cause the eventual deterioration of the staples holding the ribbands to the hull—dry rot is inevitable, especially in areas where the ribbands are enclosed behind cabinets, lockers, berths, etc.

In order to aid in plastering, some builders have successfully cut away the ribbands between the stage molds, thus exposing the armature. Because short sections of ribbands still remain along the station mold edges, the armature remains rigid and suspended.

Upright Open Mold

Description:

Wood stations are cut to exact shapes from the loftings and suspended from the overhead beams or scaffold. All stations are interconnected horizontally with continuous wood ribbands spaced vertically apart. The inner mesh is stapled to the mold, followed by re-rod. Re-rods for mast steps, engine beds, tanks and bulkhead webs are shaped to finished form, ensuring that their intersections at the hull are worked well into the hull armature. The outside layers of mesh are now applied by wire-tying through all others at each rod intersection. The hull is blocked up along the keel while the sheer and bilge are suspended from the scaffold with wire or re-rod at close intervals. Thru-hull blanks are installed into the armature, then the entire mold is completely removed, including ribbands, leaving the entire armature fully exposed. (Plate 27) The required mesh in applied to integral structurals, overlapping edges slightly onto the hull. The vessel may be plastered from both or either side and allowed to cure in the normal fashion. Once the hull has been braced securely, the suspension rods are cut loose at the hull, ground below flush and glazed over with grout to prevent bleeding.

Structural Attachments:

Ferro-cement bulkhead webs, as well as sole-beam brackets, etc., may be completely fashioned prior to removal of the mold, but mesh may not be applied until the hull has been stripped of all lumber. Prelocated, waxed bolt-hole plugs can be inserted prior to plastering—it may occur at the same time as the hull. If builders prefer, bulkheads and joinery may be bonded in with epoxy grout.

Advantages:

This system requires little welding and offers the ultimate

in penetration assurance. A very low hull weight results because of the absence of corrugations such as found in the partial-mold system caused by concrete built up between ribbands. Internal structures and bulkhead attachments are truly integral with no "cold" joints required. This method affords maximum accessibility for working on armature and plastering and allows the smooth finishing of the interior. As in all upright systems, it eliminates the necessity for rolling over.

Disadvantages:

A complicated scaffolding system is necessary and this method requires the most extensive and careful wire-tying, as the hull must be absolutely rigid to allow the removal of the mold. To accomplish this, all staples must be removed. Because the armature must stand fair without the benefit of interior structures, adequate suspension wires must be employed between scaffolding and hull. To prevent the possibility of distortion, only the highest grade of hard-drawn or spring-steel rod should be used.

Variations:

Because this system so closely resembles the pipe-frame and open-truss systems, once the mold has been removed, it is quite possible to completely fabricate ferro-cement bulkheads which may be plastered at the same time as the hull. Another possibility is in the removal of the station molds only, leaving the ribbands intact, which reduces much of the wire-tying. This, of course, will produce the same internal ridges as the normal partial-mold system. A third variation is to break away the ribbands while leaving the station molds in place.

Braced and Cantilever Molds

While these are not actually variations of the foregoing construction systems, they are methods for eliminating the necessity for erecting a complicated scaffolding. For brace-molding, each station is set upon blocking and diagonal supports are extended from the sheer to the ground (the mold must extend about a foot above the sheer to allow the diagonal brace attachment). To prevent fore-and-aft movement, additional diagonals must run from the sheer of each mold to the keel of the next mold. For cantilever construction, the center stations are supported in the preceding fashion but successive stations are positioned by means of longitudinal and diagonal braces from one station to the next toward the bow and stern. (Plate 28) Only occasional center shores and transverse diagonals are necessary.

Both the side-brace and cantilever systems may be used for pipe framing as well as open-truss and partial-wood mold, but should not be considered for the open-mold system.

Advantages:

Eliminates full scaffolding, particularly when overhead is limited. It is ideally suited for small boats where the support of the vessel may become a major cost factor. The lengths and sizes of timbers used are reduced significantly with this system.

Disadvantages:

The cantilever mold creates difficulties in effecting alignment and rigidity. Because the weight of the vessel is concentrated on fewer ground supports, adequate measures must be taken to eliminate sinking, especially if the ground becomes water-soaked during curing. Considering that bracing is minimal, caution must be exercised when working or walking on the mold as this could cause misalignment.

Inverted Open Pipe Frame

Description:

Pipe frames are bent to shape and supported, sheer down, on a timber strongback or skid. If the ground is fairly hard and level, they may be simply blocked to proper alignment. In large hulls additional timber supports must be placed under the bilge to prevent sagging as weights build up on the armature. The keel and sheer re-bar are weld-attached and the inner layers of mesh are wire-tied to the outside of the frames. The reinforcing rods are sprung into position, also wire-tied, at which time the integral structural rods may be inserted, followed by the outer mesh layers. The exposed pipe frames may remain in the hull and call for mesh strips to cover them. Allowance may be made for their removal, however, and this results in an interior free of ridges. Thru-hull blanks are installed in the normal manner and the plaster is applied from the outside while smoothed from within. Most hull forms may use this method.

Plate 27. A fisheye view of the upright open-mold system after removing stations and ribbands prior to plastering. Vessel is a converted Herreshoff **Marco Polo.**

Structural Attachments:

Bulkheads or bulkhead webs, as well as tanks, mast steps and beds, may be fully integral with their re-rods worked well into the armature. Plywood bulkheads may be epoxy-grouted into their respective positions using epoxy grout. Hardware attachments may be thru-bolted using prelocated waxed dowels to eliminate drilling.

Advantages:

This system eliminates the necessity for complicated and

expensive scaffolding and no internal mold is used. Beds, mast steps, webs, etc., employ an integral armature. Because both sides of the hull are completely exposed, it affords maximum accessibility for wiring and plastering. Bulkhead webs (or full bulkheads) and tanks may be cemented at the same time as the hull.

Disadvantages:

The builder must be a pretty good welder as well as willing to invest in or rent expensive metal-working equipment. Some experience is required for bending pipe frames fairly. Because the builder must fight gravity as well as work in confined quarters, it is very difficult to apply mesh to the inside of the pipe system. If the frames are removed, the procedure may be painstaking while necessitating grinding and glazing of the inner hull. Additional internal bracing may be required on larger hulls and it is imperative that a person's weight be kept off of the armature during con-

struction. Plastering must employ men working on the inside as well as the exterior of the hull and rolling over introduces the possibility of damage and injury.

Variations:

It is quite possible to combine the pipe-frame system with that of the open-truss. Some builders may also prefer to mount the pipe frames on a temporary wood backing to combat flexibility.

Inverted Open Truss Frame

Description:

Mild-steel-rod station trusses or web frames are welded to shape upon the lofting, then erected, sheer down, upon a skid or strongback. Keel and sheer re-bars are weld-attached, followed by the inner mesh layers. Hard-drawn or spring-steel reinforcing rod is wire-tied to each frame as well

Plate 28. Cantilevered station mold eliminates erecting a complicated scaffolding, but must be well braced for rigidity.

Plate 29. A Samson houseboat hull built to the inverted open-truss system.

as through the inner mesh to eliminate sags. (Plate 29) The integral structures are also formed of mild steel and built into the hull at this time and their meshing may be applied at the same time as the outer hull layers. The entire hull armature is wired completely, thus compacting all layers. The concrete is applied from the outside and back-plastered from within. Tanks, bulkheads, etc., may be completed at the same time. Only those vessels designed for this construction system should employ the open-truss-frame method, as it creates a space-loss problem that the architect or builder may not have accounted for.

Structural Attachments:

Plywood bulkheads may be bolted directly to web frames if their positions coincide or they may be of reinforced concrete. Engine beds, tanks, and mast steps are usually designed directly into the station webs, while cabinets and joinery are attached with epoxy grout. Bolt holes for hardware should be prelocated with waxed dowels.

Advantages:

Because of the additional rigidity of the station trusses which remain in the finished hull, the hull shell may be of somewhat lighter construction than that of other systems and, therefore, may be considered ideally suited for power boats and high-performance hulls when relatively flat sections are encountered. Ferro-cement bulkheads and tanks may be plastered at the same time as the hull and penetration is visually assured from both sides. A smooth interior finish is also possible because of interior troweling.

Disadvantages:

It is imperative that the builder be an expert welder, willing to invest in expensive equipment or rentals. Because the truss frames and bulkheads are extremely flexible, they may be inadvertently distorted during movement or erection. The excessive frame ridges on the inside of the hull create severe loss of space, especially at berth positions, and may obstruct walking areas if they intrude into the sole. It is imperative that the builder not walk directly on top of the armature as it can cause permanent distortion. The fitting of counter tops and other joinery is complicated by the internal frames and reduces headroom if cement decking is constructed. Because no mold is used, fairing may be extremely difficult to achieve unless the highest grade of spring-steel reinforcing rod is applied. Rolling over may be very complicated, costly and dangerous.

Variations:

The open-truss system may be combined with the pipe-frame method, using trusses only in the positions of tanks and bulkheads, thus reducing the space-loss problem in other areas.

Warning:

Under no circumstances should the hull-shell reinforcing rod be of mild steel nor should they be welded to the trusses or at intersections, as severe weakening and distortion will surely occur.

Inverted Partial Mold

Description:

Wood station frames are cut to shape and supported, sheer down, on a timber strongback or skid. All stations are interconnected horizontally with continuous wood ribbands which are spaced vertically apart. The mold is now covered

with common window screen and sometimes with an additional layer of thin vinyl. (Plate 30) Thru-hull blanks are waxed and attached to the mold, followed by the inner layers of mesh stapled in position. At the time the second layer of reinforcing rod is applied, starter rods for internal structures may be inserted through the mold and attached to the hull armature. When outer mesh is positioned, wire-tying may be used in areas that require additional compaction. The hull is plastered from the outside only but requires that a man be positioned inside to observe the penetration against the inner screen. Once cured, the vessel is righted and the mold removed.

Structural Attachments:

Engine beds, mast steps and bulkhead webs require that "L"-shaped starter rods be inserted through the mold while being attached to the armature. These structures may not be completed, however, until after the mold has been removed. Care must be taken to clean these joint areas of any loose aggregate and they must be treated with a bonding agent prior to plastering these units. The builder may prefer to use epoxy-grout fillets when installing bulkheads and joinery.

Advantages:

This system builds a very fair hull with minimum labor, welding and wire-tying. As an inverted system, no scaffolding is necessary. It allows the visual check of penetration and deck may be plastered at the same time as hull.

Disadvantages:

Requires skid or strongback while necessitating hull attachment webs to be constructed in two stages, creating "cold" joints. Excessive hull thickness may result due to the sagging of the screen and vinyl barrier between the ribbands under the weight of the fresh concrete. The internal mold complicates the construction of integral members, as they may be completed only upon the removal of the mold. The "cold" joint created at the integral structures and the hull must be cleared of all loose aggregate and treated with a bonding agent before their plastering proceeds. All mold staples remaining in the hull after the removal of the mold must be ground or chipped below flush and filled with grout to prevent bleeding. This system requires rolling over, which injects the possibility of damage.

Variations:

Some builders have recommended that the mold ribbands be left in the finished shell to serve as a hull ceiling as well as an attachment base for bulkheads and joinery. I believe this variation to be highly questionable as moisture will cause the eventual deterioration of the staples holding the ribbands to the hull—dry rot is inevitable, especially in areas where the ribbands are enclosed behind cabinets, lockers, berths, etc.

Inverted Closed or Cedar Mold

Description:

Wood station frames are cut to shape and erected sheer down on a timber strongback or skid. The entire mold is completely strip-planked (Plate 31) and covered with thin vinyl. Thru-hull blanks are waxed and attached to the mold, followed by the inner mesh, re-rod and outer mesh layers, all being staple-attached. The hull is plastered from the outside only, taking care to penetrate as fully as possible. Once cured, the hull is removed from the mold and inverted.

Plate 30. A Samson hull being built to the inverted partial wood-mold system. The mold shown here is one of the fairest the author has seen and should be considered as an excellent standard of quality.

Plate 31. The cedar-mold construction system, once popular, has lost favor because it creates mortar penetration problems. One hull has been produced from this mold.

Structural Attachments:

If ferro-cement web attachments and structures are used, the mold must be drilled for the insertion of "L"-shaped starter rods. The actual forming of these structures may not occur until after the hull has been removed from the mold, which will cause a "cold" joint when these units are plastered. If the builder prefers, all bulkheads and joinery may be attached with epoxy grout.

Advantages:

The cedar-mold method assures maximum fairness and easiest application of the rod and mesh layers. There is absolutely no wire-tying and a minimum of welding. While the mold itself may be somewhat more intricate than the other wood systems, it may be saved after production of the hull, allowing it to be used for succeeding hulls in rapid succession. Some designers have suggested that the decking be wired and plastered at the same time as the hull but this, of course, means that the mold would have to be destroyed for removal. Closed molding is suitable when more than one vessel is required of the design.

Disadvantages:

This system requires a substantial strongback or skid support system and also that the builder be an excellent woodcraftsman. If any wrinkles are created in the vinyl barrier, they are reproduced on the inside of the hull and must be glazed over. The greatest disadvantage of the closed-mold system is that there is no penetration assurance, as the visibility of the inner side of the hull is nil. As a consequence, vessels built by this method usually have massive voids requiring back-plastering after the hull has cured. While most builders consider back-plastering to be an easy solution, the strength of this method is highly questionable. Finishing of the interior requires that all hull staples be chiseled flush, ground below the surface and filled with grout to prevent bleeding. The installation of integral structures requires forming only after the removal of the mold and must be plastered as a separate operation, creating "cold" joints.

Variations:

Some builders have recommended that the mold planking be left in the finished shell to serve as a hull ceiling as well as an attachment base for bulkheads and joinery. I believe this variation to be highly questionable as moisture will cause the eventual deterioration of the staples holding the planking to the hull and dry rot is inevitable, especially in areas where the ribbands are enclosed behind settees, lockers, berths, etc.

Ferro-Cement Decks

There is no doubt in my mind as to the substantial cost savings of concrete decking plus the advantage of being integrally molded with the hull, since it completely solves the problem of joining two dissimilar materials. They are extremely strong, even when very light armatures are used, and can add substantially to standing headroom providing that beam webs or trusses are not used. Cement decks require little craftsmanship but great patience on the part of the builder and, as mentioned in the previous hull systems, may sometimes be plastered at the same time as the hull.

There are drawbacks which should be considered, however. Because of the relatively flat surfaces encountered, all flaws in fairness become visually accentuated; I have seen relatively few smooth ferro-cement decks. If the decking is wired up or plastered prior to installing the interior, it will create severe fitting problems as well as notably reducing ventilation, lighting and communications with those working above. I know of no professional builders who would consider decking a vessel before installing interior units. It is somewhat like putting on one's underwear after the trousers. The weight factor of ferro-cement decks is a severe detriment, as explained in Chapter 4. This is particularly true on sailboats, but not as critical on powered yachts. I firmly believe that it takes just as much time to wire up and fair a concrete deck as it does to build a wood structure.

Wooden Decks

Wooden decks are carefully engineered to produce a given strength which is always multiplied by substantial safety factors. So, any objections to this construction are fully unwarranted when comparing with ferro-cement. Wooden decks do, indeed, cost more to build than ferro-cement but the result is usually a much fairer appearance as well as being considerably lighter, adding greatly to the ballast capacity and stability of the vessel. Wooden decks are usually two layers of ply laid in separate stages over cut or laminated deck beams. This requires the greatest patience and skill, but other systems are available that eliminate this problem completely. Many fine custom yachts have been constructed with a slightly heavier deck laminate laid over temporary camber molds. Once the deck has been completed, the molds are removed, resulting in an unobstructed and smooth interior. This system increases headroom substantially while cutting down construction time by hundreds of hours. Most decks, although originally specified for a beam system, may be constructed by this method at the owner's option.

Cement Bulkheads

As with cement decks, concrete bulkheads are preferred by many builders for reasons of cost and strength. The same engineering premise holds true as that of decking in that their positions, thickness and attachment far exceed the stress and compression factors to which they may be subjected. In short, ferro-cement bulkheads are most generally overbuilt, extremely heavy and rarely good looking. Unless hundreds of prelocated bolt plugs are inserted, the attachment of joinery is complicated beyond belief. The heavy bulkheads have a considerable effect on stability.

Wooden Bulkheads

When a plywood bulkhead exceeds the size or shape possible from the normal 4' x 8' sheet, then construction becomes a laminate of thinner layers which produces one continuous piece. They result in a weight of less than half that of ferro-cement and their attachment to the hull may be by means of a ferro-cement web and bolts, or an epoxy-grout fillet. In terms of construction time, hundreds of hours may be saved, as the tedious job of "wiring up" is eliminated as well as simplifying screw, glue or nail joining of cabinets and other interior structures. Plywood does not shrink as solid timbers do and dry rot is rarely a problem. Their finish is most always smoother than that of ferro-cement and affords the possibility of easy alterations and joinery any time after the commissioning of the vessel. Fir ply when properly sealed takes a satisfactory painted finish, or vinyl may be applied. At slightly higher expense, a phenolic coated ply may be used, which yields by far the best looking painted work.

Engine Beds and Mast Steps

Because of the vibration, shock and pounding that these units must withstand, the merits of integral construction cannot be disputed. We do find an occasional cracking after long service, but this rarely affects the actual strength of the units providing that they have been liberally reinforced with a webbing of re-rod worked well into the hull armature. By virtue of the sheer weight of these structures, they should never be poured until after the hull has been cured, since sagging or distortion may result. The "cold" joint created between the beds and the hull is actually desirable, as any cracking that might occur terminates at the "cold" joint rather than extending into or through the hull.

Some builders may prefer to use a more conventional timber construction, in which case hardwood structures may be fiberglassed and epoxied to the concrete shell or fastened with a liberal epoxy-grout fillet. The main advantage of this system is one of saving weight, as concrete engine beds in a 40' hull can easily weigh several hundred pounds.

Tanks

Most designers in ferro-cement have been employing integral concrete tank structures using an armature somewhat lighter than that of the hull. Their obvious advantage is that they tend to add to hull rigidity and are only a fraction of the cost of separate steel tanks. Because the tank sides follow the surface of the hull, an increase in capacity results compared to other systems. Their disadvantage is one of greater weight and it is a messy job troweling the inside, as it requires a person to work in cramped quarters lacking ventilation and proper lighting. If the hull is stressed in collision or grounding, any resultant cracks or breaks in the hull also become openings in the tank. The only solution to this problem is that of lining the tanks with neoprene or thyocol, as this material is elastic enough to remain intact under most conditions.

Mild steel tanks may be custom-built by professional welders and have the obvious advantage of being separate from the hull, remaining relatively free from hull damage under impact. Water tanks should always be galvanized but fuel or gas tanks should be left untreated.

Fiberglass or fiberglassed ply-core tanks may be constructed by any amateur and may be built directly into the hull or separately, considering the advantages and disadvantages mentioned previously. These tanks are light, inexpensive and quickly constructed. They tend to inject a strange taste into the drinking water unless steam-cured and some polyesters are found to be incompatible with fuel oil and must be thyocol- or neoprene-lined.

Structural Metals

Needless to say, the higher on the galvanic scale the metals are, the freer will be the vessel from electrolysis as well as fatigue, wear and scoring of bearing surfaces. The drawbacks are essentially those of cost and difficulty of welding that usually requires professional outside services. A Monel propeller shaft or rudder post will easily double the cost over mild steel but is many times stronger and resistant to corrosion. The same premise also holds true for all other underwater parts and rigging weldments.

Rudders

Contrary to popular belief, steel plate and concrete rudders should never be considered for use on any sailboat under any circumstances whatsoever. Although this construction has already been discussed in Chapter 3, I cannot overstress the problems of steering control that result from their use. Timber or plywood laminate construction should be the only system considered by the amateur builder for these units.

Summary

As you may now see, many similarities exist between all available construction methods and their differences are primarily those of procedure but have relatively little to do with the finished product. Assuming all materials to be equal, one system rarely produces more strength than another, while the time and cost factors fluctuate only slightly.

Since the popularity of ferro-cement, naval architects are not only being asked to design boats but to offer step-by-step construction techniques as well. No designer could make a living if he had to offer descriptions of all the methods within one design package and, in a consequence, he must settle for one or another as an economical expedient, knowing at the same time that the builder may employ an infinite variety of construction combinations. So, then, the construction method chosen by most designers is not always as firm as many amateurs have come to believe, and with the application of the individual's ingenuity, he may be able to modify the recommended system to that which matches his capability. It is always suggested that the builder discuss his desires with the architect before proceeding headlong in a different direction, but usually the consequences are negligible. It is usually when a builder makes drastic changes in materials or design that he finds himself in trouble, such as altering wooden deck beams to steel, plywood bulkheads to ferro-cement, or stainless steel chain plates to mild steel. This kind of homespun architecture can drastically affect the strength or serviceability of the finished boat. Do not be too hasty in judging a design simply by its recommended construction system. It is most usually a point of departure from which to proceed.

section two

patterns

9

patterns

Although I will be fully describing the basic lofting procedures, it is an art which has taken even the best loftsman many years of practice, trial and error, and a deep understanding of lines development to become proficient in this exacting trade. It takes a sensitive eye to discern minute variances which are rarely apparent to an amateur builder, especially when laying down large hulls, as an error of a quarter inch or more may not always be as obvious as one might think. Considering that the loftsman is working on a two-dimensional plane and only interpreting for the third, it is often a difficult matter to "feel" the shape of a compound surface. Complicating this matter is the fact that the loftsman rarely has the opportunity to view his work from a distance other than his standing position, which can inject

a deceiving perspective. As I have not been able to clarify, within the scope of this work, all the complications of orthographic projection, it is impossible for me to forewarn the amateur builder as to the scores of problems that can arise. It is quite beyond me how the untrained person can be expected to execute this grueling process without injecting gross errors, simply through his lack of knowledge on the finer points of drafting.

My only suggestion, therefore, is to consider stock plans that include prelofted pattern sheets, as these will save the builder hundreds of hours with the assurance of proper fairness. Some designers have argued that these commercial patterns are inaccurate because the printing paper tends to

Plate 32. The author traces scrieve lines onto stable maylar to be used as reproduction masters for printed lofted patterns.

shrink and swell with changes in humidity. My experience in this area has shown a variable factor of approximately ¼" in 12', which is distributed proportionally in all directions. As you can see, the inherent error on a 12' beam would be less than ¼", which is well within the transverse tolerance of a 40' boat. Very few builders are capable of construction to this accuracy. We must also remember that molding will sag or distort under the weight of mesh and concrete and that fairing of the hull armature will vary from one builder to the next. When we consider the slight additional cost for the use of prelofted patterns, it is beyond me how anyone would rather attempt laying his own lines with a questionable degree of correctness. (Plate 32)

Assuming, then, that you have purchased a completely lofted body plan on paper panels, it is wise to clarify exactly what these lines represent; i.e., inside of the hull, outside of the hull, or inside of the mold ribbands. In all but the latter instance, it will be necessary to allow for these thicknesses when cutting the station molds. For instance, if the station lines are drawn to the outside of the finished hull, the thickness of the hull and ribbands must be deducted from the lofted line in order to establish the actual station mold perimeter. If the mold ribbands are let into the station molds, only the deduction of the hull thickness will be required. Your lofted patterns should include

the station lines, intermediate bulkhead webs (if truss construction is used), deck beams or crown pattern, expanded transom face and transom radius templates, stem molds or templates, and any other critical hull shapes such as the deadwood section.

After receiving your lofted paper panels, it is best to unroll them to their full length, weighting the corners to keep them flat. Do not disturb them for several days, as it will allow them to adjust to the local humidity as well as straighten their curl. The patterns should be taped (not stapled) to a wood flooring, as this will allow for the eventual driving of nails directly into the body plan for the purpose of laminating or transferring. Each panel will bear a series of registration marks or "bull's eyes" for proper alignment and the edges of each sheet should overlap slightly to cause their coincidence. (Fig. 65) With all but the open-truss construction systems, the lofting should be covered with a clear polyethylene sheet as this will prevent tearing and the sticking of glue, while facilitating cleaning. Once rolled out, the lofting should not be folded in any way as it could increase its wearing effect and distortion. Study the loftings as well as your mold drawings thoroughly before proceeding with your construction.

Picking Up Patterns (for Wood Molds)

The first question most generally asked by the amateur

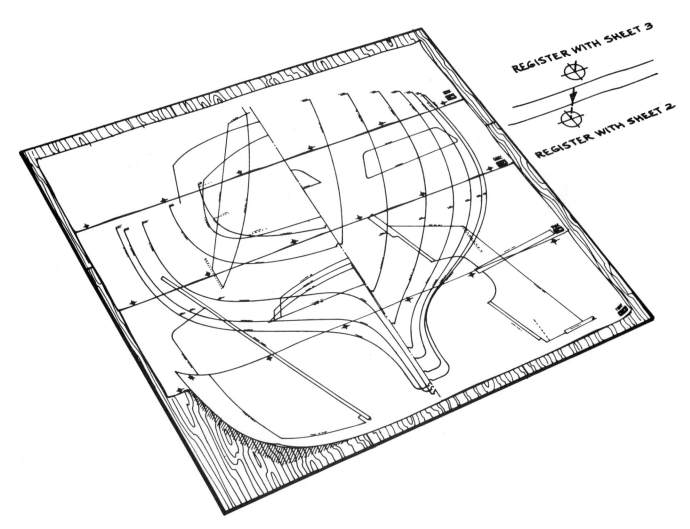

Fig. 65. Laying down printed patterns onto a firm surface. Note the method used for aligning individual panels.

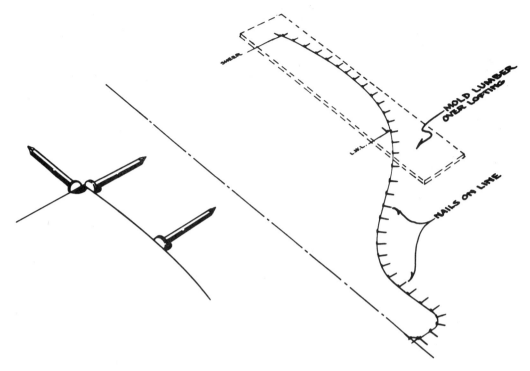

Fig. 66. The "nail transfer" method of picking up patterns.

builder is, "How do I transfer the lines?" There are many methods available which, in part, depend on whether you have lofted the boat yourself or have purchased stock patterns, as the latter instance will allow you to slip lumber under the printed sheets. In the following paragraphs I will describe briefly the most popular transfer systems; namely, "nail transfer," "the pickup grid," "the tailor's wheel" and "carbon transfer."

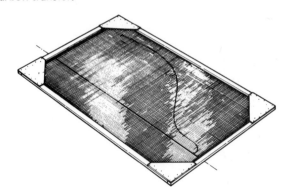

Fig. 67. The "pickup grid." A wood frame covered with window screen.

The Nail Transfer System

The nail transfer system requires that the builder lay down a series of nails along the line to be transferred so that their heads are directly on the line with their shanks being perpendicular. Each nail head is rapped with a hammer, which will imbed it into the lofting floor (or scrieve board). (Fig. 66) In the areas of tight curves, the nails should be spaced closely, while straight lines need only one at each end. Where corners or angles are to be displayed, two nail heads should be placed touching each other so that their intersections will be precise and clear. It is absolutely necessary that the waterline and sheer positions be included as well as

several nails on the center line at the keel to facilitate an accurate joint line. Once the necessary nails have been positioned on the lofting, the mold lumber is placed on top of the line so as to cover the greatest portion of the curve possible while maintaining an adequate width for strength. The upper side of the lumber is now hammered with a mallet to impress the nail heads into the underside. Upon removing the board, the desired line will be faintly visible and should be immediately redrawn with pencil for clarity. Once the lumber has been cut, it may be used as a pattern to duplicate the opposing side, as required for station molds. It should be noted that the basic station shape may be preassembled before transferring or assembled in sections as each line segment is cut. The nail transfer method may be used for all forms of construction, but in the case of "open truss" an actual transferring to mold lumber does not occur. It will, however, allow the builder to scribe lines into the welding floor or scrieve board if printed patterns are being used.

The Pickup Grid

The pickup grid is simply constructed by first forming a rectangular wooden frame of 1 x 2"s, having the corners diagonally braced with gussets to prevent distortion. Ordi-

Fig. 68. Tailor's wheel and chalk bag may be used for transferring patterns to lumber.

77

nary window screen is stretched across this frame to cover its entire area and staple-attached at close intervals. To use the pickup grid, lay it directly onto the lofting and trace the desired line onto the screen with a white or colored chalk pencil. As before, be sure to clearly indicate the center line, waterline and sheet. (Fig. 67) The mold lumber is now placed under the grid and the original chalk line retraced. Obviously, the screen will have an abrasive effect on the chalk which will cause a powder to fall through the screen onto the lumber, thus transferring the necessary lines. When using this system, it will be necessary to rotate or continually sharpen the chalk to allow for precision marking. The pickup grid may be used in most forms of construction but has little value with the open-truss method.

The Tailor's Wheel

The tailor's wheel is available at any local sewing shop. (Fig. 68) It is guided along the desired line of preprinted patterns or white butcher's paper (somewhat translucent for visibility) which will create thousands of small holes in the paper. The lumber is now positioned under the paper and the perforated line patted with a colored chalk bag (also from the sewing shop), thus transferring the line directly to the lumber. Do not fail to indicate the center line, waterline and sheer marks. This system is particularly useful in the

case of open-truss construction. The disadvantage, however, is that it may require the repeated lifting of the printed sheets. It is my recommendation that a butcher-paper transfer sheet be used, in any event, to eliminate this problem. Extreme care must be exercised so as not to disturb the chalk on the lumber until you have traced the lines with pencil.

Carbon Transfer

Carbon transfer is a variation of the tailor's-wheel method and requires lightly gluing or taping carbon paper to the underside of the preprinted lofted sheets so that all lines are completely covered. The station mold lumber must be carefully positioned under the print so as not to tear the carbon, which will now allow the builder to trace directly through with a ballpoint pen.

In the case of open-truss construction, lines must be transferred onto a welding floor or scrieve board. (Fig. 69) Caution must be exercised so as not to cause shifting of the pattern during tracing. This may be done by drawing dashes rather than a continuous line except in those areas of tight detail. To facilitate web-frame construction in continuous sections without having to flop them over, it is best to repeat the tracing of the complete opposing sides by replacing the carbon paper to the front of the printed pattern.

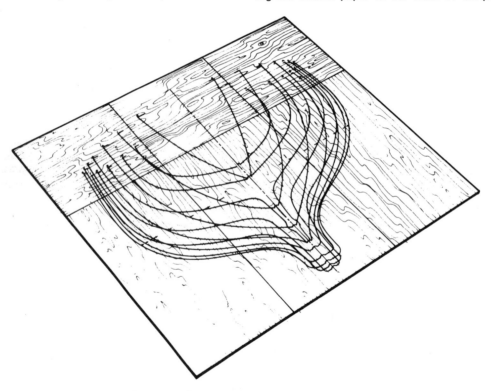

Fig. 69. The Body Plan transferred to a wood surface in preparation for welding truss frames. This is more commonly called a "scrieve board."

78

10

lofting

Purpose

When the lines of a vessel are drawn, the hull is usually represented as being only 30" or so, which may be 1/5 to 1/25 the size of the actual boat. Any errors drawn into the lines by the architect, then, are magnified on the finished vessel by the ratio of the linear scale. As accurate as the designer's efforts might be, it is virtually impossible for him to avoid minute imperfections in fairness; these imperfections must be detected and remedied prior to hull construction.

The reasons for errors in the scale drawings are numerous and often not within the architect's control. It usually takes many days to develop a Lines Drawing, in that the designer may be simultaneously conducting calculations to check his work. It is not unusual that his drafting paper may shrink or stretch as much as 1/8" during this period, which obviously creates ambiguities within the drawing itself as well as complicating dimensioning and offsets. It must also be noted that, when measuring from a small-scale drawing, the thickness of a pencil line may represent as much as 1/4" on the finished vessel, an error in hull fairing readily apparent to a trained eye. A remaining factor worth consideration is the accuracy sought by the designer himself. Many designers, knowing that their hull lines will be carefully faired full size by a qualified loftsman, do not attempt to achieve absolute precision. If a few line intersections do not coincide, he simply leaves them as they are so long as these points are within a "reasonable tolerance" for his calculations and subsequent drawings.

Lofting, or laying down, is the process of scaling the architect's Lines Drawing to full size. Throughout this procedure, ultimate accuracy must be sought and absolute fairness achieved (if such is really possible). The full-size lines also serve as the basis for tracing the hull mold patterns as well as integral structures, shaft alignments and even the shapes required for bulkheads and joinery. It must be realized that while lofting takes a great deal of time and physical energy, it is really a timesaving element and the "proving ground" for the construction of your hull. Lofting has the effect of developing the train of analytical thought by requiring consideration of building procedure and sequence. It also requires the careful superimposing of parts to be used for creating the hull form, which may avoid untold hours of haphazard cutting and fitting. Often, an additional hour on the lofting floor eliminates many days of backbreaking handwork on the vessel. For this reason alone, short-cutting this vital ritual is hardly worthwhile. (Fig. 70)

Lofting vs. Patterns

Some designers have suggested that the only way to construct a boat accurately is for the individual builder to loft his own vessel. I cannot fault this theory and it has been generally accepted by the industry as true. Many promoters of amateur construction, however, take the complexities of lofting very much for granted as they are normally encountered in their own daily work. What many architects seem to forget is that most amateur builders (particularly in ferro-cement) have little or no experience in yacht construction, drafting, body development, or even sailing and that the necessary "feel" is completely lacking in most aspiring candidates.

In my years in the design field, I have known very few really good loftsmen and, I dare say, they are few and far between. I, myself, have lofted many hulls under the guidance of designer Charles Morgan and I was continually amazed by his instinct and prowess when reviewing my work. On many occasions he would "sight" a spline that I thought I had laid "perfectly," only to detect flaws of 1/8" in a 50' line. His sense of fairness was absolutely uncanny, but even he freely admits that lofting is more of an art than a science. For this reason alone, I must question the judgment of the amateur builder who rejects the use of pre-printed patterns (when available) in the name of superior accuracy, which his lack of experience may preclude.

Printed lofted patterns do inject errors. There is no doubt about it! This is partly due to the shrinking and stretching of the paper as it is run through the machine as well as varying humidity conditions after the patterns have been delivered. Unless the patterns have been grossly abused or have gone without protective controls, the tolerances of error should never exceed 1/4" in 12', and most boatbuilders will agree that this is most acceptable for cruising hull construction. It should also be remembered that the fluctuations encountered with preprinted patterns occur in all directions and remain proportionate. In short, if the designer's own lofting has been conducted accurately and he has been diligent during tracing, you have little reason to expect anything less than a perfectly fair hull, well within the required precision.

In my own stock plans, I supply both patterns and Design Offset for those who would rather exercise this option.

Types of Lofting

In stock design work, many architects perform the lofting at their own facility or commission an independent builder

(Text continues on page 83.)

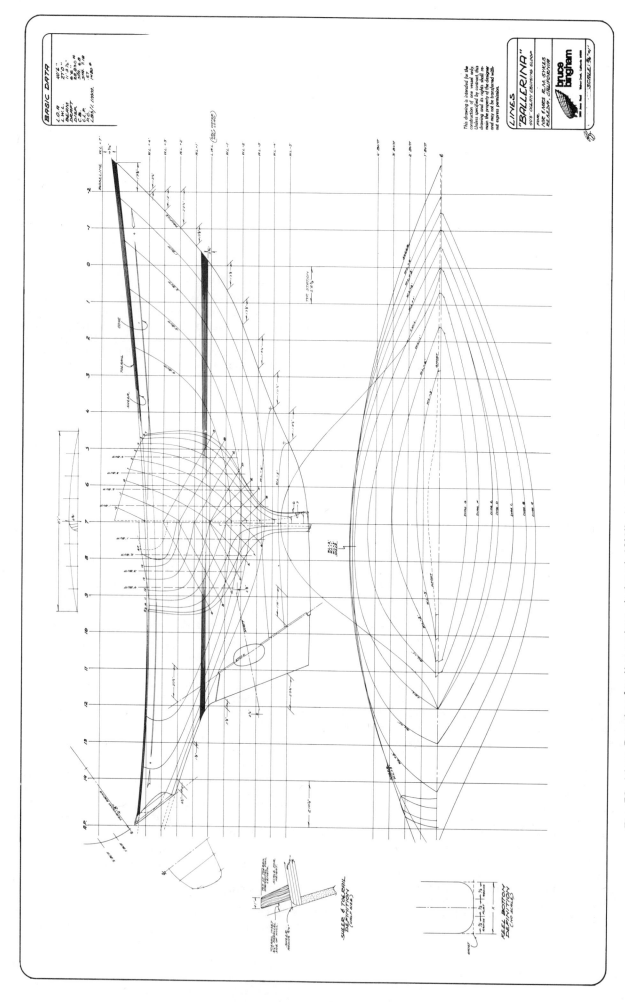

Fig. 70. Lines Drawing for the custom designed 40'6" sloop **Ballerina**. Original scale: 3/4" = 1'. A Bingham design.

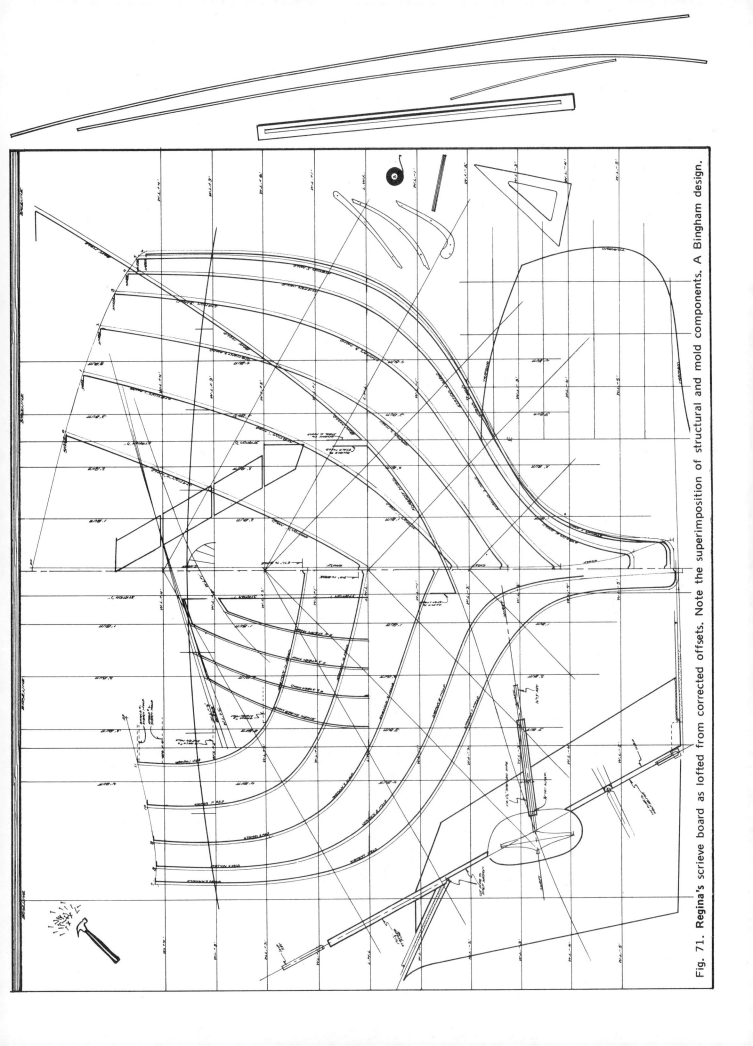

Fig. 71. Regina's scrieve board as lofted from corrected offsets. Note the superimposition of structural and mold components. A Bingham design.

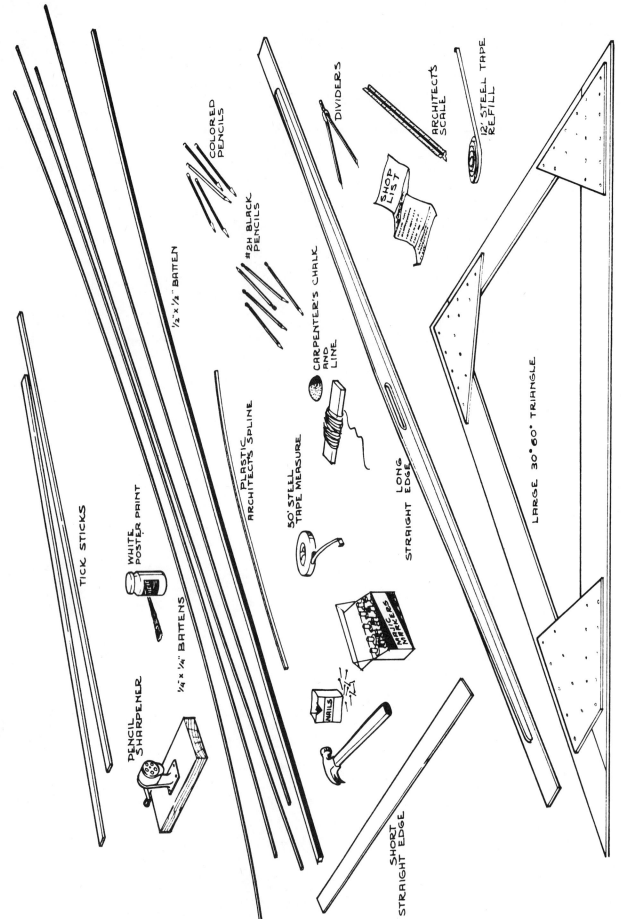

TICK STICKS

PENCIL SHARPENER

WHITE POSTER PAINT

¼" x ¼" BATTENS

½" x ½" BATTEN

COLORED PENCILS

#2H BLACK PENCILS

DIVIDERS

ARCHITECT'S SCALE

12' STEEL TAPE REFILL

SHOP LIST

PLASTIC ARCHITECTS SPLINE

50' STEEL TAPE MEASURE

CARPENTER'S CHALK AND LINE

LONG STRAIGHT EDGE

LARGE 30°-60° TRIANGLE

MAGIC MARKERS

NAILS

SHORT STRAIGHT EDGE

Fig. 72. Lofting tools.

82

(Continued from page 79.)

to do it. Once the lofting is completed, the lines are remeasured and corrections made in the Table of Offsets. These offsets are always designated as "Corrected" or "Lofted Offsets" and are the indication that the builder may trust their accuracy. When laying down lines of a vessel from Corrected Offsets for wood-mold hull construction, it is not absolutely necessary to reproduce all of the lines shown on the Lines Drawing, but only the stations which appear on the body plan (end view), the profile (side view), expanded transom and transom radius templates, and whatever other parts are necessary for the hull-mold construction. You do not have to fair the entire hull in all views when working from Corrected Offsets. (Fig. 71)

If, on the other hand, the lines of the vessel have not been prelofted prior to distributing the Offsets to builders, they may be designated as "Design Offsets." This is your indication that all lines must be reproduced full size in all views and that you bear the delicate responsibility for producing absolute fairness and accuracy. (Fig. 86) Lofting to this degree is lengthy, exhausting and expensive. It especially requires a thorough knowledge of body development, the idiosyncrasies of compound projection and a "feel" for the shape of the finished vessel. Unfortunately, the latter element cannot be taught and can be gained only through experience, imagination and insight.

Practice Makes Perfect

One of the best exercises that I can suggest for learning the lofting art is to literally redraw the Lines Drawing using ¾", 1" or 1½" to-the-foot scale. This is far less risky than simply forging ahead without having first encountered and solved the basic problems. Of course, this drawing would be recreated from the offsets supplied to you, and the basic instructions contained in this chapter should be followed through.

Because it is absolutely necessary that you fully understand the mechanics of the Lines Drawing, it is imperative that you do a considerable amount of reading beforehand. I would suggest thoroughly reviewing *Yacht Designing and Planning* by H.I. Chapelle, *Skene's Elements of Yacht Design* by Francis Kinney, and *Sailing Yacht Design* by Douglas Phillips-Birt.

The Tools

If you are lofting from corrected offsets, you will need at least four clear white pine or Alaskan cedar battens ten feet in length and ¼ x ¼" in section. These will be used for springing the curves of the body plan, so they should be of the finest grade of lumber possible, free of knots, straight grained and smoothly cut. If you are lofting to Design Offsets, you will also need at least four additional ¼ x ¼" battens scarfed to a length 20% longer than the proposed vessel. These will be used for springing the Buttocks, Waterlines and Diagonals. Springing "slow" curves, such as those at the garboard (turn of the bilge near the keel), may not be achieved with the wooden battens, so you will need to purchase a 4' acetate architect's spline from your local drafting supply. Don't try to make your own out of Plexiglass, as it will be too brittle, but you might be able to cut one from PVC sheet.

You will need a hammer and a box of 1" nails for holding batten and spline positions, a 50' tape measure marked in feet and inches, a 12' measuring tape refill and a chalk line. Purchase a box of No. 2 black lead pencils as well as two complete sets of colored pencils and a box of colored Magic

Markers. Correcting errors by erasure is very tedious and messy, so I recommend using a ½" brush and white poster paint. As it will seem that your pencil sharpener is never where you want it, you can save yourself a mile of walking by mounting your sharpener on a scrap of 2 x 6" lumber and taking it wherever you go. You will need a long (8') and a short (3') straightedge and these are best made out of 6" and 3" widths of Formica or Masonite. Cut and plane them very accurately, sighting the drawing edge for precision. You may choose to mount handles on these, which may be of wood, countersunk-screwed from the underside.

You will also need a good architect's scale of the appropriate size, a large pair of dividers and a large compass. You will be able to save a lot of wear and tear on your body by investing in a pair of kneepads—the kind used by masons and tile layers.

A large (8') and a small (3') triangle will be needed and may be constructed out of a sheet of ply or made up of ply or Masonite strips. To lay out an accurate 30°-60° triangle, use the "3, 4, 5" method; i.e., the base leg to equal three units, the side leg to equal four units, the diagonal leg to equal five units. It makes no difference what these units represent (meters, feet, inches, half feet, etc.) as long as you are consistent.

Once your triangles have been constructed or cut, check for accuracy by setting the base against your base-line batten and drawing a line along the side leg. Now, turn the triangle over and superimpose it onto the line just drawn. If the triangle and the line do not exactly match, carefully plane the base leg of the triangle to correct the discrepancy. Check and recorrect the triangle accuracy until it is perfect.

Tick sticks are an absolute necessity in lofting. They are strips of wood cut to a 1/8 x 3/4" section and of 6' to 10' length. Their purpose is twofold: They take the place of the architect's dividers and they record vital dimensions. Each tick stick yields eight measuring surfaces, i.e., four edges at each end; and the marks placed thereon should be of the same color as the particular line they represent. You cannot have too many tick sticks, so cut at least eight, then sand and paint them flat white. (Fig. 72)

The Floor

Because lofting will serve as a basis for patterns as well as a working surface, plywood or composition board sheets laid end to end and edge to edge should be used as the loft platform. This will allow the driving of nails for the purpose of bending splines or to hold bending blocks.

If you are lofting from Design Offsets, the size of the drawing surface should be about 20% longer than the intended vessel and wide enough to extend from the base line to a few feet below the keel (or vice versa if your base line is under the hull).

If you are lofting from Corrected Offsets, you need only reproduce the body plan and a few mold and hull structures, so your drawing surface need be only about 50% wider than the intended beam and high enough to extend from the base line to a few feet below the keel. This type of lofted drawing is called a screive board. (Fig. 71)

It is best to lay the platform onto a hard, flat surface such as a concrete floor or level parking lot, but these areas may not be readily obtainable. Lofting on soft ground or even a carpeted floor always causes problems because the platform may not lie flat, thus causing visual distortion while

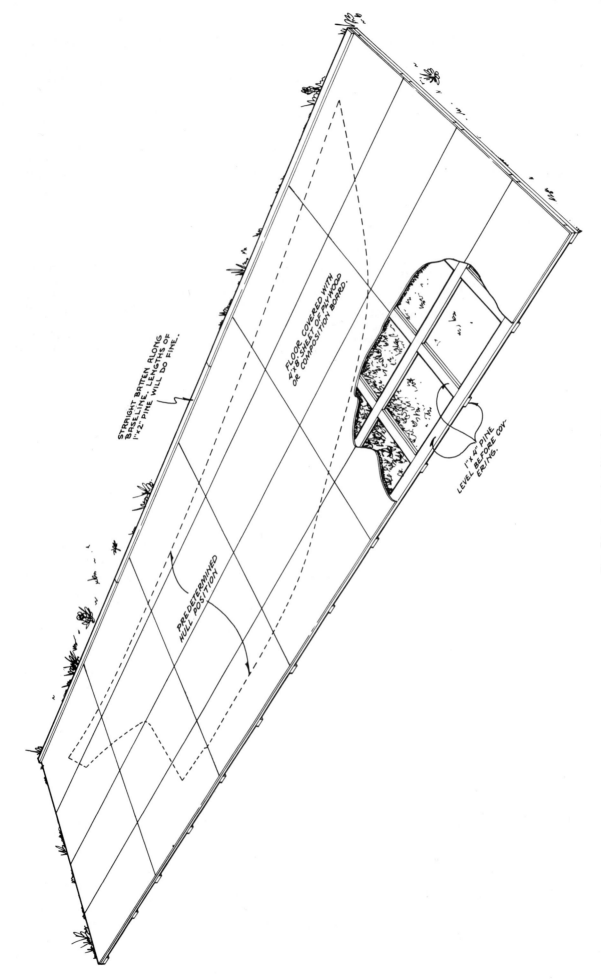

STRAIGHT BATTEN ALONG
BASELINE. LENGTHS OF
1"×2" PINE WILL DO FINE.

FLOOR COVERED WITH
4'×8' SHEET OF PLYWOOD
OR COMPOSITION BOARD.

PRE-DETERMINED
HULL POSITION

1"×4" PINE
LEVEL BEFORE COV-
ERING.

Fig. 73. A good lofting floor.

working the fastenings loose through excessive movement. If open ground must be used, be sure to lay 1 x 6″ (or so) boards liberally under the platform sheets and joints to prevent sagging. Once all sheets have been arranged and their edges firmly butted together, drive corrugated fasteners every few inches along the joints. Give the entire floor a couple of coats of flat white paint or primer. (Fig. 73)

Offsets

"Offsets" is simply the term referring to the measurements of the hull lines where they cross or intersect the hull grid ordinates. (Fig. 74) There are normally three divisions of offsets; i.e., half-breadths (distances from center line to the sheer, waterlines and center ghost), heights (distances from the base line to the sheer, profile and buttocks), and diagonals (oblique distances from the center line to the hull at each station). Other hull form measurements may be required to fully define the yacht's shape but may not appear on the Table of Offsets because of inconvenience. These dimensions usually relate to the longitudinal measurements of the profile taken at the waterline intersections. These dimensions will appear on the Lines Drawing itself. On the drawing may also be such graphic definitions as the sheer, keel, "flat," etc., which help clarify hull development. (Fig. 70)

The normal manner of listing offsets is to depict feet, inches and eighths of an inch. An offset written as "4.3.4" would actually be measured as 4′3½″. If the designer is measuring on a large scale drawing where extreme accuracy is more feasible, he will often interpolate to the nearest 1/16″ by placing a "+" after the offset figures. Thus, 5.10.7+ would be measured as 5′10 15/16″. In my work with amateur builders, who may not be accustomed to the standard offset numbering procedure, I have preferred to write the offsets as they would commonly be measured by any carpenter. So, 5′10 15/16″, then, would be written into the table exactly as is. I have found that this practice reduces errors that many times have been made out of the split-second confusion over the more accepted offset figure system.

The Base Line

The base line will be clearly indicated on all of your hull construction drawings. It represents the reference from which all vertical offsets have been taken as well as many of the internal structural measurements. If the base line is improperly placed onto the lofting, you will run the high risk of ultimately building irregularities into the hull and complicating future construction.

The base line will lie along the upper or lower edge of the floor (depending upon the location chosen by the designer). Don't simply trust the straightness of the platform edge to establish this line. Rather, drive nails at the extreme ends of the floor and snap a chalk line. Check for accuracy by holding your head near the floor and sighting down the line. If corrections are necessary, mark them clearly. Now, nail a 1 x 2″ pine batten along the chalk line. The batten will serve as a straightedge for the base of a large triangle as well as a "stop" for tape measures and "tick sticks," so be absolutely sure the batten is straight and firmly fixed. (Figs. 73-75)

The Profile and Plan Grid

This refers to the network of straight lines which represent "cuts" or "planes" passing through the hull at different axes. The planes of the grid used in yacht design are: "Waterlines" cutting the hull on a horizontal plane and shown as horizontal lines on the grid in profile and body plan; "Buttocks" cutting the hull vertically parallel to the center line, and shown as horizontal lines on the plan view (half breadth) and vertical lines of the body plan; "Diagonals" which are shown as varying angles passing through the body plan only; and the "Stations" which are portrayed as vertical lines passing through the profile, plan and diagonal views. There may be any number of waterlines, buttocks, diagonals and stations but, generally, the larger the vessel, the more of each there will be.

It will make no difference in what sequence the lines of the grid are laid down but it is important that their nomenclature be marked clearly and at close intervals. In my own lofting, I use different colored pencils when laying down the grid to eliminate some of the confusion as the work proceeds in its complexity.

When the scale lines drawing is created by the designer, the views are usually kept separate with the possible exception of the body plan, which is often superimposed one over the other. (Fig. 70) When your lofting nears completion, you will have your hands full just keeping these views straight in your own mind. I have often seen as many as thirteen lines passing through only one square foot of lofting, all at different angles and meaning different things.

When drawing the spaces for the grid lines, don't count on a triangle for accuracy. Lay the proper intervals off with a tick stick, connecting a row of marks with a straightedge or snap a chalk line. Once again, sight all of the lines for absolute straightness. Check and recheck your work. Don't let so much as a 1/16″ error get away. It will surely haunt you later on!

Body Plan Grid

These are the vertical, horizontal and diagonal lines representing the drawing planes as seen from the ends of the vessel. The body plan may be positioned anywhere on the main lofting floor or drawn as a completely separate unit. If it is to be positioned on the main floor, choose a convenient station line to represent the body plan center line.

The buttocks will be measured outboard from each side of this center line at their respective distances.

The waterlines for the body plan need not be drawn separately, as they will be represented by the waterlines of the profile view, which also will pass through the body plan.

The Diagonal Grid (Body Plan)

Out of convenience alone, most designers draw diagonals through the body plan at 30°, 45°, and 60° to the center line because of their available triangles. This, in turn, makes it much easier for you to recreate these angles. Furthermore, most designers space their waterlines and buttocks at the same intervals, which automatically solves the laying down of 45° diagonals by simply drawing the line so that it bisects the waterline/buttock intersections.

In many cases, and properly so, the designer will be less arbitrary in his choice of diagonals. He may choose to draw lines which pass through the hull most nearly perpendicular to its surface. These can be diagonals of 18°, 27°, 52°, etc., which precludes the use of triangles for their solution. When this occurs, the designer will have dimensioned the

DESIGN OFFSETS
PENELLA-REGINA BRENDA

bruce bingham

3930 Jones Road   Walnut Creek, California 94596

TO SCALE

HALF BREADTHS

	AP	11	10	9	8	7	6	5	4	3	2	1	0
SHEER	2'9 5/16"	4'8 9/16"	5'2 9/16"	5'7 1/8"	5'11 7/8"	6'1"	6'0 1/4"	5'9"	5'3 3/16"	4'8 3/4"	3'9 3/16"	2'7 7/8"	
GHOST	0	0	0	0'1"	0'4 1/16"	0'6 3/4"	0'6 3/16"	0'4"	0'1"		00 5/8"	00 5/8"	
W.L. +4'	2'10 3/8"	3'8 3/4"	4'6 3/8"	5'2 3/8"	5'8 5/8"	6'1 1/8"	6'1 1/8"	6'0"	5'8 3/4"	5'2 7/8"	4'5"	3'4 1/8"	1'1 1/8"
W.L. +3'	2'7 3/16"	3'7 3/16"	4'6 1/4"	5'2 1/8"	5'8 3/4"	6'1"	6'1"	6'0 1/8"	5'8 3/8"	5'1 1/2"	4'3"	3'1 7/8"	1'6 5/8"
W.L. +2'	3'0 3/8"	4'3"	5'1 1/8"	5'7 3/16"	6'0 13/16"	6'0"	5'11 5/8"	5'7 3/16"	5'4 7/8"	4'11 1/4"	4'0 3/8"	2'9 3/4"	1'1"
W.L. +1'	3'3 3/4"	4'8 3/4"	5'5 5/8"	5'7 5/8"	5'10"	5'11 1/2"	5'9 7/8"	5'4 7/8"	4'7 7/8"	3'7 3/16"	2'3 3/8"	0'7 3/16"	
L.W.L.	0	3'5 7/8"	4'10 1/2"	5'5 1/2"	5'7 1/2"	5'6"	5'0 1/2"	4'2 3/4"	3'1 5/8"	1'8 3/8"	0		
W.L. -1'		0'10 5/16"	3'6 3/4"	4'8"	4'11 1/4"	4'11 1/16"	4'3 1/16"	3'5 3/16"	2'3 5/16"	0'10 1/16"			
W.L. -2'			1'3 1/16"	3'0 1/16"	3'5 5/8"	3'7 3/16"	2'11 1/16"	2'1 7/8"	1'0 9/16"				
W.L. -3'			0'4 1/2"	1'2 3/4"	1'9 13/16"	1'9 5/8"	1'5 3/16"	0'9 9/16"					
W.L. -4'				0'2 3/4"	0'6 3/8"	0'10 1/8"	0'8"	0'4 7/8"					
W.L. -5'				0'1 3/4"	0'5"	0'7 7/8"	0'7 7/16"						

DISTANCE FROM BASE

	AP	11	10	9	8	7	6	5	4	3	2	1	0
SHEER	2'2 3/4"	2'5 3/4"	2'8"	2'9 3/8"	2'9 1/2"	2'9 1/8"	2'8 3/8"	2'6 3/8"	2'3"	2'0 1/8"	1'9 1/4"	1'5"	1'1 1/8"
PROFILE	4'10 1/8"	5'10 1/4"	7'0"	8'4 1/2"	13'2 3/4"	13'1 3/4"	13'1"	13'0"	12'3 1/4"	10'9 5/8"	9'9"	8'9 9/16"	7'0"
1' BUTT	4'8 3/8"	5'8"	6'9 1/2"	7'11 1/2"	9'1 9/16"	9'1"	10'9 5/16"	10'5 9/16"	9'7"	9'0 5/8"	9'0 5/8"	7'10 1/4"	5'3 3/16"
2' BUTT	4'4 3/8"	5'5"	6'6 1/2"	7'7 5/8"	8'8 5/8"	9'5 9/16"	9'10 3/8"	9'10 1/4"	9'1 1/4"	9'3 1/8"	8'3 1/8"	6'6 3/4"	2'10 7/8"
3' BUTT	5'0"	6'1 5/8"	7'2 5/8"	8'3 5/8"	9'0 5/16"	9'3 3/8"	9'3 3/8"	8'11 3/4"	8'4 3/8"	7'2 5/8"	4'4 7/8"		
4' BUTT		5'5 3/16"	6'8 5/16"	7'9"	8'5 5/16"	8'8 5/8"	8'9"	8'3 5/16"	7'4 9/16"	5'0 9/16"			

DISTANCE FROM ℄

	AP	11	10	9	8	7	6	5	4	3	2	1	0
DIAG "A"	2'7"	3'8 1/4"	4'8 3/4"	5'6 1/4"	6'2 3/4"	6'7 3/4"	6'8 5/16"	6'7 1/2"	6'2 3/8"	5'5 5/8"	4'6 5/8"	3'4 1/8"	1'0"
DIAG "B"	1'3 1/2"	2'9 5/8"	4'0 1/4"	4'11 5/16"	5'10"	5'8 5/8"	5'3 5/8"	4'7 7/8"	3'8 5/16"	2'6 1/4"	1'0 1/2"	
DIAG "C"	1'0 1/4"	2'3 1/2"	3'5 5/8"	4'6"	5'5 1/2"	5'4 1/2"	5'1 3/8"	4'6 3/8"	3'8 3/8"	2'6 5/16"	1'1 1/8"	
DIAG "D"	0	0'4 1/2"	2'6 5/16"	3'3 1/4"	3'6"	3'4"	3'1 1/8"	2'3 5/16"	1'3 5/8"			
DIAG "E"		0'9 3/4"	1'6 1/8"	1'9 5/8"	1'11 1/4"	1'8 5/8"	1'3 1/8"	0'7 7/8"				

Fig. 74. Offset tables.

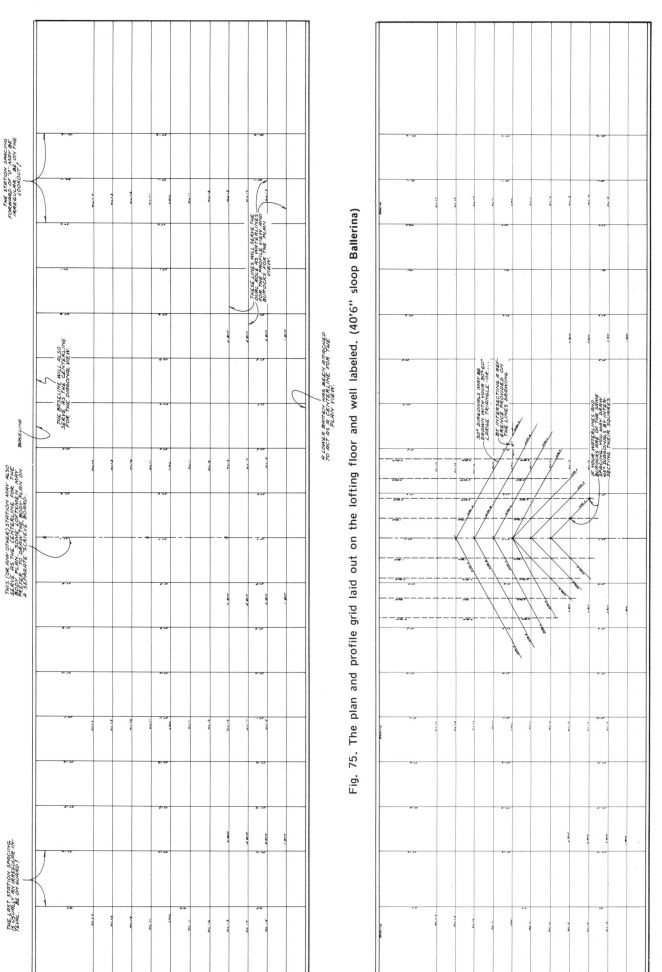

THE LAST STATION SPACING IS USUALLY ANY IRREGULAR IN TERVAL. MAY BE ON GUARD!

THIS (OR ANY OTHER) STATION MAY ALSO SERVE AS THE CENTERLINE FOR THE BODY PLAN. SOME LOFTSMEN MAY PREFER TO DRAW THE BODY PLAN ON A SEPARATE SCREENS BOARD.

BASELINE

THE BASELINE WILL ALSO SERVE AS THE CENTERLINE FOR THE DIAGONAL VIEW.

THE STATION SPACING FORWARD OF "O" MAY BE IRREGULAR, BE ON THE LOOKOUT!

THESE LINES WILL SERVE THE DUAL ROLE AS WATERLINES FOR THE PROFILE VIEW AND BUTTOCKS FOR THE PLAN VIEW.

A LOWER BATTEN HAS BEEN ATTACHED TO ACT AS PLAN VIEW.

Fig. 75. The plan and profile grid laid out on the lofting floor and well labeled. (40'6" sloop **Ballerina**)

30° DIAGONALS MAY BE DRAWN WITH YOUR 30-60° LARGE TRIANGLE -OR-...

BY INTERSECTING A REF ERENCE PROVIDED ON THE LINES DRAWING

IF YOUR WATERLINES AND BUTTOCKS ARE OF THE SAME SCALE YOU MAY INTER- 45° DIAGONAL MAY SECTING THEIR SQUARES.

Fig. 76. The body plan grid buttock and diagonal lines added to the lofting. (40'6" sloop **Ballerina**)

diagonal position as it relates to a grid buttock or waterline. It is only necessary, then, to reproduce this reference on your lofting, drawing a straight line through it to its termination at center. If, however, the designer has not terminated the diagonal on center at the intersection of a waterline, he will have indicated a second measurement. (Fig. 76)

Lofting from Designer's Offsets

Sheer and Profile (Side View)

Using the offset dimensions shown for the sheer under the heading "Distance from Base Line or Heights," mark the respective measurements at each station and the fore and aft perpendiculars. Lay a heavy batten accurately along these marks, driving nails at every other station and sight the line to detect possible unfairness. Do not be hasty in drawing the line, but view the batten from all directions, looking for hard or soft spots and any unnatural tendencies. Additional checking may be done by shifting the nail positions one station to the right or left or by removing selected nails at questionable positions. Don't trust the offsets on blind faith alone, and keep in mind the inherent errors in your designer's own work. His working drawings will be only as accurate as your lofting, so you must consider that you are both working as a team and that you are the extension of his efforts.

If your batten misses an offset mark by an eighth of an inch on a vessel of 30' L.O.A., you are within an acceptable tolerance, so don't be upset with your architect as an eighth is about as closely as he can read his scale. If, however, your batten misses a mark by a half inch, one inch, or even one foot, you can bet the designer misread the scale, which is quite common. When this occurs, completely ignore the respective offset and adjust the batten so that it looks correct.

When you are satisfied that your sheer is "sweet," draw the line heavily as it must not be readjusted under any subsequent circumstance. All other lines of your lofting must conform to the sheer, not vice versa. Once you have removed the batten, label the line clearly and mark the distance from base at each station onto a tick stick and set it aside for future use. (Fig. 77)

Now lay down the profile. From your Lines Drawing, indicate all longitudinal dimensions on their respective grid waterlines, then mark "distances from base line" as called for by the offsets. If any straight lines appear within the profile, draw these first, then spring a light batten connecting the remaining marks. Sight and view as before to detect errors. The profile will never be simply one continuous smooth curve, but rather a series of varying curves, reverses and straight lines. Try to get the "feel" of the profile from the Lines Drawing when judging for correctness and, once again, ignore what seem to be large errors. Now scribe the profile heavily, label it clearly and transfer the dimensions to a new tick stick.

It will be necessary to locate the forward edge of the rudder, which may correspond to the center line of the rudder stock or the aftermost of the keel deadwood. This line usually bears reference measurements on the Lines Drawing. (Fig. 77)

Sheer (Plan View)

Transfer offsets to the plan view, correct and draw in the same fashion as the other lines, being sure, once again, to label the line clearly and to mark a tick stick for future use. As with the sheer in the side view, you will not alter this line hereafter, so indicate it heavily. The forward end of this line will terminate at the ghost while the aft end crosses the aft perpendicular, not simply ending at the transom. (Fig. 77)

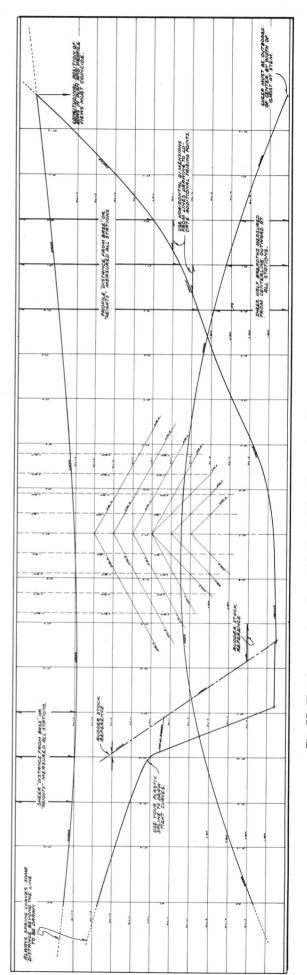

Fig. 77. The sheers and profile add to the lofting plan and profile views. (40'6" sloop Ballerina)

Sheer (Body Plan)

Now is the time to put the tick sticks into action again. Here we will use two of them at the same time, as the body plan sheer requires two dimensions simultaneously in order to establish one position. Place the tick stick for sheer/distance-from-base vertically against the base-line batten. Also hold the tick stick marked for sheer/half-breadths horizontally from the center line. Move the stick back and forth, up and down until respective station marks coincide. Place a small dot at the intersections. (Figs. 78-81)

Bow Ghost and Keel Flat (Plan View)

No boat of any distinction ever had a sharp bow, regardless of what many might think. A good boat should not "cut" the water but, rather, move aside in a manner yielding the least resistance. This is accomplished partly through the proper shape of hull as well as the development of a hyperbolic or elliptical stem section as seen in the plan view. Some designers may simplify this slightly by drawing the forward ends of waterlines as radii, and this is far better than a razor-sharp bow. Further, it is almost impossible to build an absolute point on a ferro-cement boat, so you can generally dismiss its use.

The most common method employed by designers for creating the ellipse of the stem and radii of the keel edges is by first indicating the stem and keel as being "squared" and having distinct corners. (Fig. 79) The corners, called "ghosts," are not actually constructed into the hull, but are used only as references for terminating stations, waterlines and diagonals. Once these lines have been squared in this fashion, the radii or ellipse will be drawn tangent to the lines and the "flat" to depict the true shape. (Fig. 80)

The keel flat is rarely a constant width, as it normally finds its narrowest point a short distance below the load waterline while being widest in the keel. Usually the flat will taper toward the stern, where it will coincide with the half-breadths of the hull along the trailing edge of the deadwood. If the vessel has a fin keel, however, the flat will converge with the center line, thus forming a sharp trailing edge. If the vessel is to have a barrel bow, the flat will flare outward above the waterline toward the deck. When drawing a clipper bow, the ghosts will remain parallel above the load waterline; I will deal with its development separately.

To draw the ghost lines in plan, transfer the half-breadths to the lofting, measuring away from the center line. Spring the batten, correct obvious errors, draw the line and label, then transfer the dimensions to a tick stick. All waterlines, drawn later, will terminate on the ghost, not at the center line. The ghost must also be represented in the body plan and may be transferred to it using the tick stick. (Fig. 81)

Flat or Ghost (Body Plan)

Using your tick stick for ghost half-breadths, lay off the proper widths from the center line for each station. As with the sheer, a second dimension will be necessary to establish the ghost heights so the tick stick for profile from base will crisscross the first while being held against the base-line batten. Forward ghost marks will lie to the right side of the vertical center line while aft ghosts fall to the left. In addition to the ghost half-breadths at stations, the designer may have indicated widths at various waterlines at the stem. If so, mark these dimensions from the center line at their respective positions. Now connect all ghost points with a fair line and indicate each position with the proper station number. (Fig. 81)

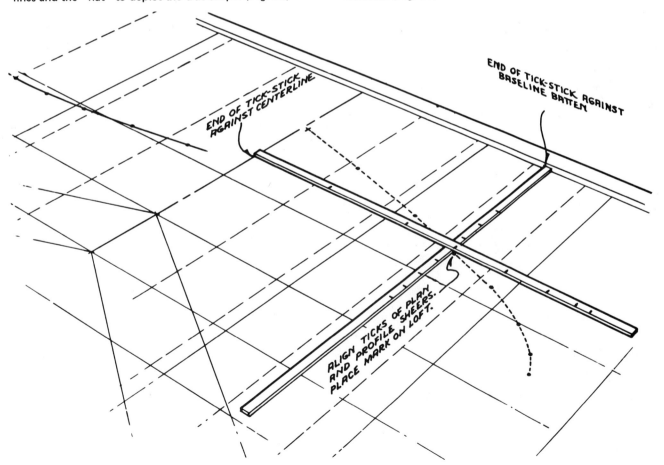

Fig. 78. Using two tick sticks simultaneously to establish a single point.

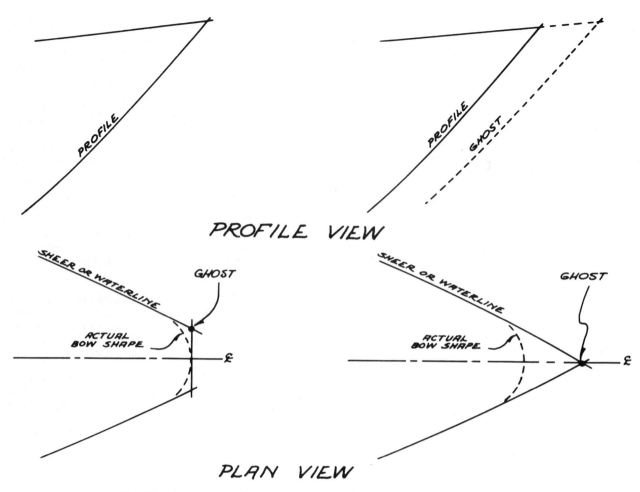

PROFILE VIEW

PLAN VIEW

Fig. 79. Method of determining and drawing the "ghost" near the bow.

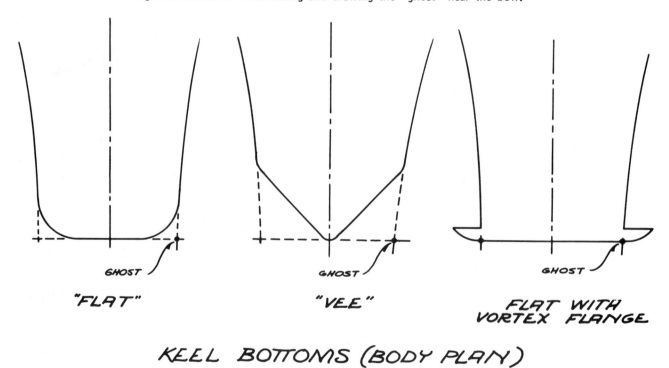

"FLAT"   "VEE"   FLAT WITH VORTEX FLANGE

KEEL BOTTOMS (BODY PLAN)

Fig. 80. Determining the ghost points for various keel shapes.

Stations (Body Plan)

All architects have a predefined shape of hull that they have pursued throughout its development. In order to dictate this development and that of all subsequent lines, they base their entire work upon the sheer and profile as well as performance characteristics common to amidship and end sections. These stations should be considered most important above all others.

To alter the shapes of these stations (sections), however minutely, would be to change all others and, hence, the very nature of the boat itself. Once the representative sections are drawn, they must not be modified in any way. If waterlines, diagonals or buttocks drawn later do not pass fairly through these sections, it must not be considered the fault of these sections but, rather, that of the other lines. These other lines, then, bear the responsibility of fitting the representative sections, not vice versa.

To draw the representative sections, mark off the distances of these stations for all waterlines, buttocks and diagonals. Using the prelocated sheer and ghost points, connect all marks with fair curves, using a small batten. As with the previous lines, "feel" these sections to your utmost, keeping your corrections as slight as possible by changing only those positions which conform the least. Those points requiring adjustment must be immediately corrected on your offset table. During this remeasurement, also change the figures required of the sheer and profile as drawn so they will read correctly for possible later use. Once the representative sections have been completed, very lightly draw all other stations into the body plan in the same manner, but do not correct their offsets. (Fig. 82)

The Load Waterline (Plan View)

As with all foregoing lines, we will proceed to the next most important, which also follows the sequence of the architect. This will be the load waterline as shown in the plan view. Its shape will be dictated by three primary points; i.e., the ghost at the stem, the midship section and its aft termination at the center line. As these points have been unalterably committed by previous lines drawn, they cannot change. Now, using a new tick stick, pick up the half-breadths of the waterline from the stations of the body plan, transferring their dimensions to the plan view. Lay a batten, making the necessary changes in the intermediate station positions (hopefully only slight), and draw the line. Using the same tick stick, remark the indicators in error and transfer the revised positions back to the body plan. Those stations which now do not conform to the new waterline marks must be redrawn lightly to bring them into conformation. In doing so, place a distinct dot at all station/waterline intersections to indicate that these points, too, are now committed and must not be changed.

As you can now see, every line in every view bears a dependent relationship with all others. If one line is changed, even slightly, it demands that all other lines reflect the change. If an alteration in one line to correct an apparent error forces a bump or hollow into another line, it means that your modification needs to be sought elsewhere. Sometimes the flaw will be quite obvious, but as the lofting becomes more complicated through the addition of new lines, errors will become much more subtle. As unscientific as it may sound, trial and error is the only way to proceed. Most often, the solution may be found by drawing a new complementary line; i.e., a cut which passes through a line in question as close as possible to the suspected error. For

example, if an unusual tendency or bump rears its ugly head in the vicinity of the two-foot buttock, then it would be prudent to draw the two-foot buttock line. Once done, the intersection (or failure to intersect) of the two lines will give you a stronger basis for making the proper correction or of believing the original position to be true.

The Middle Buttock (Profile)

Now you will begin filling in the blank spaces. I prefer to lay in the most representative buttock. This line will have as its basis seven committed points; i.e., its intersection at the forward sheer, forward representative station, forward L.W.L., midship section, aft L.W.L., aft representative station, and aft sheer. These points must accurately reflect the positions committed by the plan view and body plan and, once again, must not be modified. Secondary positions should also be transferred by tick stick from the lightly drawn body-plan stations to their respective vertical grid station lines. The buttock, when sprung and drawn, will appear in the side view. Once properly faired, correct any discrepancies in the stations to bring them into conformation. (Fig. 83)

Diagonals (Diagonal Views)

These lines are drawn in exactly the same manner at the waterlines except that their measurements are taken parallel to their respective angle from the center line at each station instead of horizontally. The position of the forward and aft ends must be determined by first transferring the height of the diagonal (where it intersects the center line) in the body plan to the profile; secondly by transferring the intersection of the diagonal height/profile intersection to the plan view. If the aft end of the diagonal passes through the transom, the line will be continued to the aft perpendicular.

The most common error of interpretation by amateurs when lofting occurs when drawing the forward end of the diagonal curve, as they usually bring the line fully into the center line. This error is most natural because the body plan usually shows the diagonal grid intersecting the center line. In essence, it does, but only after rounding the corner of the stem ghost, thence to the center line across the flat.

To locate the ghost point in the diagonal view, first transfer the height at which the diagonal passes through the ghost in the body plan to the profile. Now transfer the longitudinal position of the ghost/diagonal up to the center line of the diagonal view. Finally, transfer the distance from the center line to the ghost measured diagonally in the body plan outward from the ghost position of the diagonal view.

If this diagonal ghost measurement is not properly laid off from the center line, it could induce a significant error and distortion, especially if a barrel bow is being lofted.

The same situation exists at the stern if the thickness of the deadwood is being lofted instead of the full tapered rudder. Once again, do not bring the diagonal into the center line but, rather, set it off by the angular measurement of the body plan diagonal as it passes through the center line and deadwood.

The first diagonal normally laid down is the one that is most representative of the dead rise or straight lines of the bilge; i.e., the portion which lies between the chine and the garboard. Your primary reference points will be the ghost at the bow, the distance off at AP and the midship section. Of equal importance will be those points which lie close to the load waterline and middle buttock, as these lines have

(Text continues on page 95.)

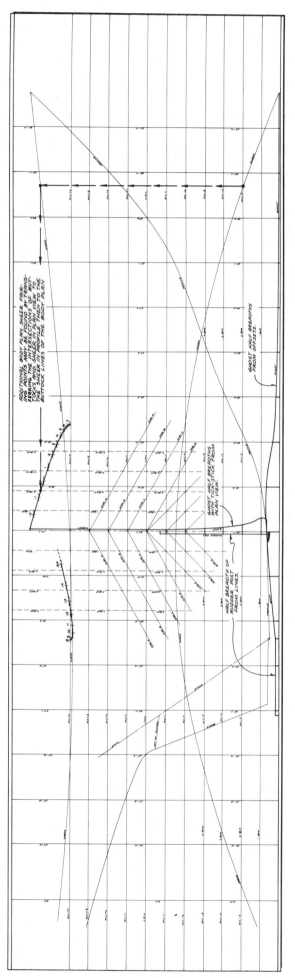

Fig. 81. The sheer and ghost added to the body plan and ghost drawn into the plan view. (Ballerina)

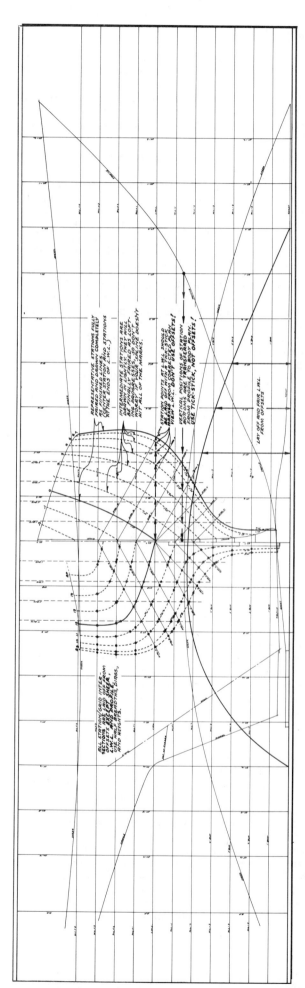

Fig. 82. The representative stations drawn darkly, the intermediate stations sketched lightly onto the body plan grid. The L.W.L. ticked off and faired into the plan view. (40'6" sloop **Ballerina**)

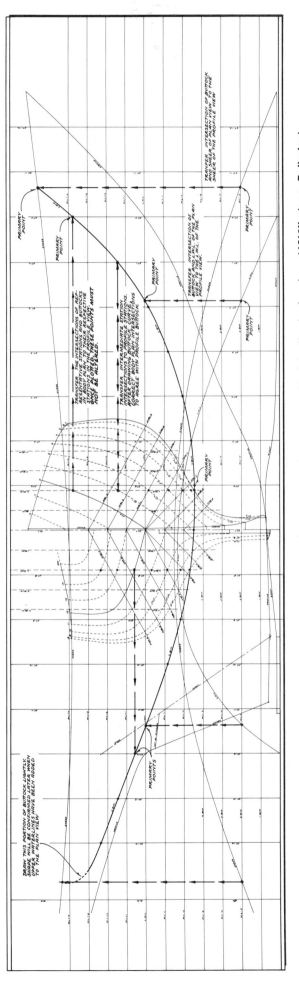

Fig. 83. The mid-buttock added to the profile view. Resulting points are committed in the body plan. (40'6" sloop Ballerina)

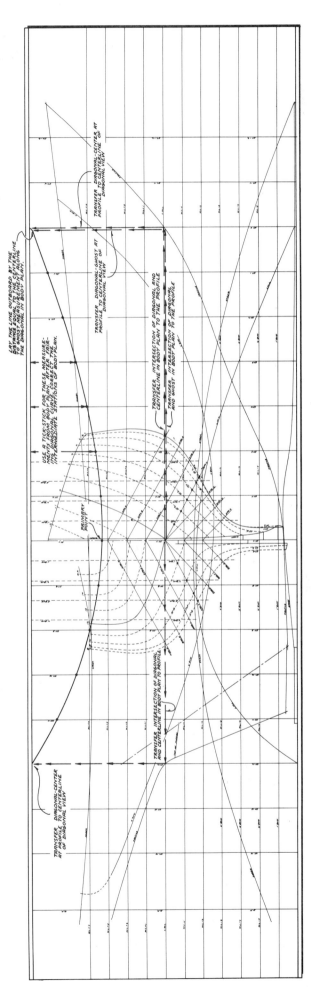

Fig. 84. The mid-diagonal added to the diagonal view. Resulting points are committed in the body plan. (40'6" sloop Ballerina)

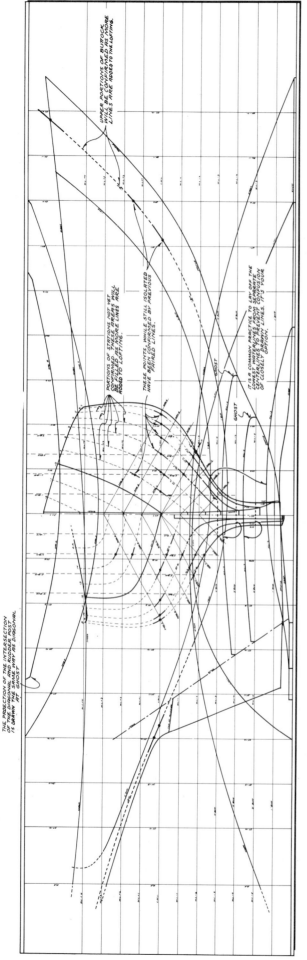

THIS PROJECTION OF THE INTERSECTION
OF THE DIAGONAL AND RUDDER POST
IS DRAWN THE SAME WAY AS DIAGONAL
AT GHOST

UPPER PORTIONS OF BUTTOCK
WILL BE CONFIRMED AS MORE
LINES ARE ADDED TO THE LOFTING.

PORTIONS OF STATIONS NOT YET
CONFIRMED. THESE AREAS WILL
BE FILLED AS MORE LINES ARE
ADDED TO LOFTLINES.

THESE POINTS, WHILE STILL ISOLATED
HAVE BEEN CONFIRMED BY PREVIOUS
FAIRED LINES.

GHOST

GHOST

IT IS A COMMON PRACTICE TO LAYOUT THE
LOWEST WATERLINES FROM SEPARATE
CENTERLINES TO PREVENT CONFUSION
OF CLOSELY DRAWN LINES. IT'S YOUR
OPTION.

Fig. 85. Adding lines to all views of the lofting. Note that I have chosen to work from the bottom of the vessel, upwards. The upper portion of the 1'
butt has not been committed as yet. The sequence of adding remaining line is a matter of personal preference or experience. (40'6" sloop **Ballerina**)

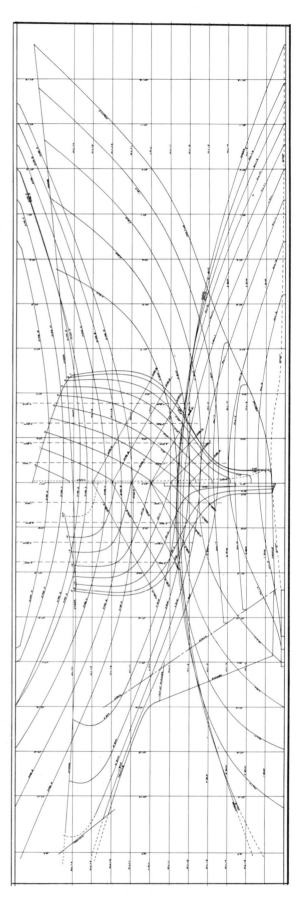

Fig. 86. All hull lines in all views have been fully faired and drawn. Every intersection of each line in the three different views corresponds to the
other. (40'6" sloop **Ballerina**)

(Continued from page 91.)

been previously committed and should be measured with a tick stick. All other points along the diagonal should represent measurements taken from the body-plan stations.

Once a smooth diagonal line has been drawn, transfer its corrected positions back to the lightly drawn remaining stations of the body plan and mark them clearly. (Fig. 84)

Shift of Emphasis

So far, you have concentrated only on the most important lines; i.e., those most representative of the hull shape. Those presently drawn are the sheer side view, sheer plan view, profile, ghost in body plan, ghost in plan view, L.W.L. in plan view, middle buttock in side view, and the primary diagonal. As these lines progressed, you have shifted from total dependence on the offsets to more dependence on the points dictated by the committed lines. As we proceed further, the offsets will become even less important until, finally, they will be of little value other than to check an occasional questionable area. The lofting, then, will solve more and more of its own self as it becomes more complex. I do not mean to say, however, that you are being given the liberty to redesign the boat. If the designer has been diligent in his accuracy when developing his Lines Drawing, then your lofting will agree within a very surprising tolerance. If the designer, in his haste, has been at all negligent, your lofting will detect his errors readily. Unfortunately, there is no way to judge the accuracy of the Lines Drawing in advance of lofting, but the trained eye can often feel inherent unfairness, particularly in rough stations. The most common shortcoming of some designers is that they lose their patience when nearing the completion of a Lines Drawing when there are only three or four lines left to do. If an intersection does not perfectly coincide with that of an opposing view, they will let the error pass if it is relatively small, thus counting on the fortitude of the loftsman. Be on your guard, then, and persevere because the responsibility for ultimate fairness weighs more and more heavily upon you.

Fairing the Remaining Lines

At this point, many loftsmen vary drastically in their procedural sequence and it is highly a matter of personal taste or experience as to which lines to proceed with next. Some might choose to "fill in" the areas of the lofting which are in the most need of commitment, while others prefer to work from the bottom of the hull upward. The latter approach is the one I prefer for my own work and it is much like squeezing the last bit of goo out of a tube of toothpaste. In essence, this solves the lines errors in much closer proximities rather than jumping around to widely different parts of the hull at the same time. The only problem with the "toothpaste" is that some of the lines (such as buttocks) may be committed only as successive segments at a time. This doesn't bother me particularly but some designers would rather work with full lines. I have found, however, that the "toothpaste method" of fairing is very fast and accurate whether on the board or lofting and that I am able to finish a drawing in about 60% of the normal time.

No matter which method you choose, it must be remembered that all lines of all views must interrelate and coincide, regardless of the amount of rework required to accomplish this. When drawing a waterline, for instance, it may dictate a small change in a buttock or two, which, in turn, demands an alteration within a station, thence causing further modifications of a diagonal. The first change must always result in fair curves of the other views. If not, you

must go at it again. Often, a minute alteration during the final phases of lofting will cause a change to be made in a line drawn days before. This is the plague of the loftsman and cannot be avoided.

Let us start with the lowermost waterline. This will commit the section through the keel and should be laid off with a tick stick from the body plan. Once drawn, proceed to the lowermost diagonal, then the next higher waterline. Continue to choose and draw successive lines upward into the body plan until the hull commitment fringes onto the area of the inboard buttock.

When drawing the inboard buttock, the entire line may be laid off using tick stick points from the body plan as well as the intersections of the waterlines drawn and its terminations at the forward and aft sheers. The lower portion of the buttock (that area committed by the existing lower waterlines and diagonals) should be drawn darkly, while the upper portion (less firmly committed due to fewer finished lines) should be kept rather light. The upper buttock will become fully committed later when successive diagonals and waterlines are added to the lofting . (Fig. 85)

As the successive fair lines are added and the resultant intersections coincide with each other, more and more of the body plan stations may be darkened in as their shapes are confirmed. As you proceed, you will see your drawing fill itself like a glass of wine and, hopefully, this will give you the encouragement to press on. (Fig. 86)

Judging Fairness

This is a most elusive term and one which defies all hard-and-fast definitions. To further complicate matters, not all lines of the hull need be "fair" in the normal sense. For instance, a vessel which is intended to have relatively hard turn of the bilge when viewed in section will produce a pronounced corner at the quarters of her waterlines. This proper abruptness will not follow the natural tendency of a fairing batten, nor will the curves of the stations. A vessel which has a very straight fair body (a station ghost extended beyond the turn of the garboard to the center line when seen in the side view) will naturally show an obvious soft spot in her lower diagonals as they pass through the garboard area or dead rise. Again, if the hull is to have hard bilges, this will be shown as a more sudden change in the shape of the buttocks as they cut through the vicinity of the chine. Vessels with very narrow stations forward will develop a strange inward kink in their upper diagonals.

Idiosyncrasies of projections such as these can lead even the best loftsman astray if he is not familiar with them. The tendency of most amateurs is to trust the batten's normal tendency regardless of what the offsets and Lines Drawing tell him and completely overlook the fact that fairness need not be simply a continuously decreasing or increasing curve.

When drawing my own lines, I am in the enviable position of being able to interpret my own intentions, but you have to read between your designer's lines (no pun intended). One consideration that can be of tremendous value is that of building an accurate hull before lofting. Unfortunately, few amateurs can accomplish this with hairline precision.

The physical properties of a fairing batten really limit its natural use and cause unjust influences which must be overcome by driving extra nails, compressing or .tensing the batten ends or by tapering the batten section. Often, moving one station point slightly will cause the curve to change its character completely for as many as five stations distant.

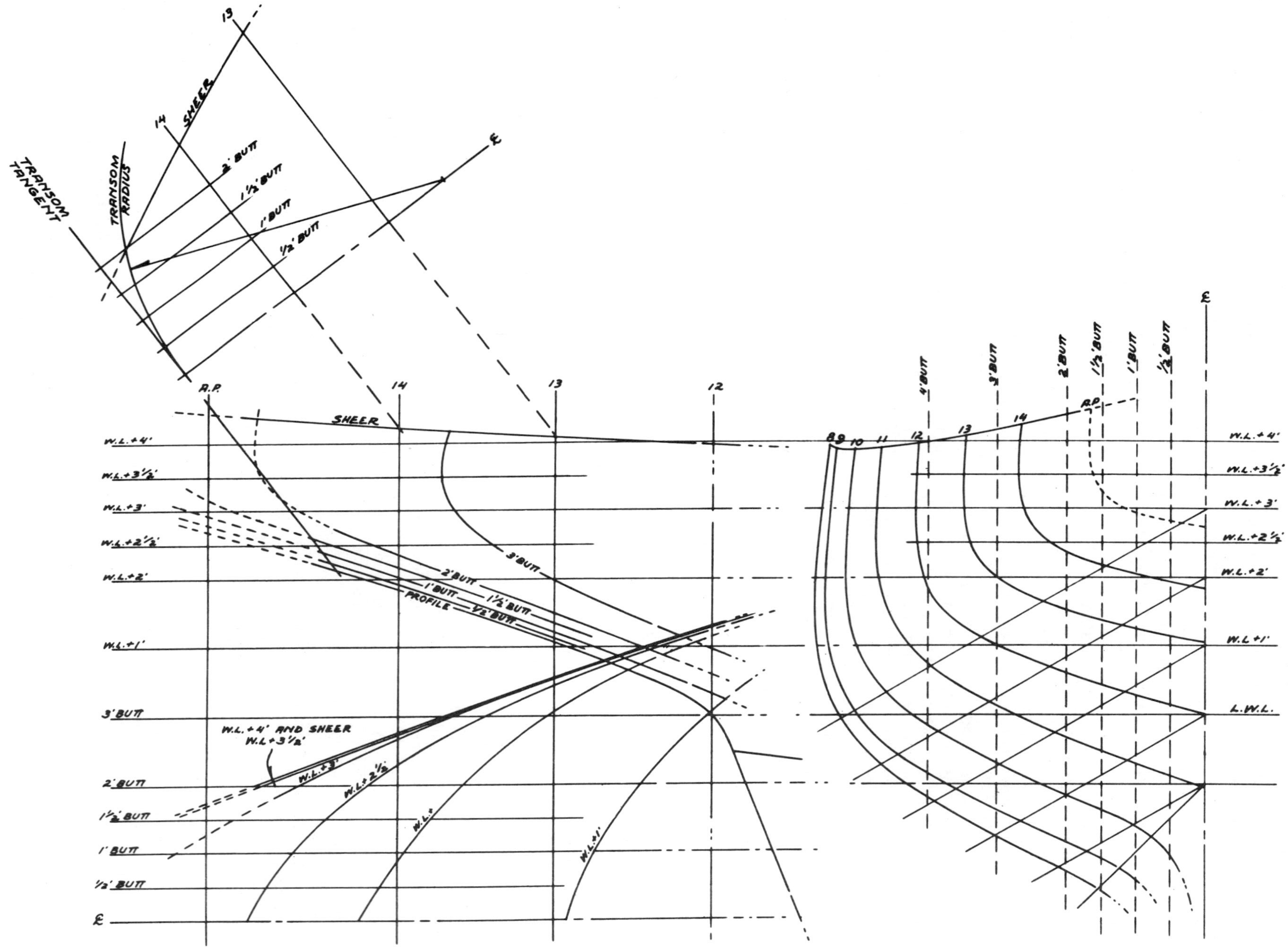

Fig. 87. Round counter transom lofting. Oblique radius projection. (40'6" sloop **Ballerina**)

I wish I could be of more help in guiding your sense of proper shape, but it can be gained only through experience and an acute awareness. To put it more bluntly, either you have it in you or you don't, all draftsmanship aside. I would strongly suggest that you read every book you can get your hands on that deals with hull development and lofting, in order to appreciate the differences in attitudes, approaches, and tricks of the trade.

Round Counter Transom Lofting Variations

At the outset, you would think that all rounded counter transoms are basically the same and, at a glance, they are. But there are two basic lofting methods available to you which will have an effect on the finished appearance.

Most yacht designers choose to curve the transom around a radius based on a horizontal plane (a constant radius being drawn for each waterline in the plan view). With this system, the shape of the deck at the transom in plan view also results in a radius.

The second method for lofting a rounded counter transom is that of developing a radius which revolves around an oblique axis which is parallel to the transom face. When projected onto the plan view, all waterlines and deck become elliptical rather than circular in shape. It is this method that I prefer for my own work, because the finished work on the completed yacht is more pleasant in appearance.

Round Counter Transom, Oblique Radius Projection

Because of the tight nature of curves involved in the transom lofting, you will find it a measurable benefit to add extra buttocks and waterlines to all lofting views. These lines need not extend the full length of the vessel, but only sufficiently to accurately represent the stern area. In the illustrated examples you will see that I have simply cut the buttock and waterline intervals in half. Using your tick sticks, pick up the new sections from the body plan, transfer them to the plan and profile views, and fair them in the same manner as the full-length lines.

Establish the transom angle in the profile view. Draw this line not only through the hull position but a considerable distance above or below. Now extend two or three of the aftermost stations above or below the hull. Take care that they are absolutely parallel to the transom face.

Draw a center line perpendicularly through the transom face extension, extending it to a distance which will accommodate the required transom radius. From this center line, carefully measure and draw buttock lines (as many as required to cover the finished transom).

With a tick stick, measure the sheer half-breadths in the plan view and transfer these measurements to the oblique station lines, laying them off from the oblique transom center line. Now fair the aftermost segment of the sheer line in the oblique view.

Measure the required transom radius from the transom face extension forward along the oblique center line. Draw this radius tangent to the transom face passing through the buttock lines and sheer. (Fig. 87)

Project the intersection of the transom radius and the sheer from the oblique view to the sheer of the profile view. This projection must be parallel to the angle of the transom and

will establish the corner of the transom at the sheer in the profile view. This point is now transferred vertically downward to the sheer of the plan view and horizontally to the sheer of the body plan. With your tick stick, make sure that the half-breadths of the transom sheer intersections are identical in the body plan and plan view.

Draw a series of parallel lines from the intersections of the buttocks and transom radius in the oblique view angularly toward their respective buttocks in the profile view. These buttock extensions must be parallel to the transom angle. The resulting intersections in the profile view will partially define the shape of the transom corner. Transfer the transom buttock corner intersections from the profile view vertically downward to their respective buttocks in the plan view and horizontally to their respective buttocks in the body plan. (Fig. 88)

Transfer the intersections of all oblique transom buttock lines and all waterlines in the profile view vertically downward to their respective buttock lines in the plan view. Now spring a batten through each set of buttock waterline intersections in the plan view. Place a distinct mark where these transom waterlines intersect the hull waterlines. These intersections will represent the corners of the transom at the waterlines in the plan view. Transfer these intersections vertically to the profile view. Also transfer the half-breadths of these intersections to their respective waterlines in the body plan. (Fig. 89)

You may now connect all transom corner points in all views with smooth curves, using your plastic spline.

Draw the transom crown from the sheer transom corner in the body plan. Now transfer the heights of the buttock crown intersections from the body plan to their respective oblique buttocks in the profile view. Connect these intersections and the transom sheer corner with a fair line.

Transfer the crown buttock intersections of the profile view vertically downward to their respective buttock lines in the plan view. Once again, connect these points and the transom sheer corner with a fair curve. (Fig. 90)

Round Counter Transom, Horizontal Radius Projection

You will find it extremely helpful to add extra buttocks and waterlines to the stern area of the vessel, as it will increase the number of fairing points required in developing the tight transom curves. In the illustrated examples you will see that I have simply cut the buttock and waterline spacing in half. These extra lines must also be drawn into the body plan.

Establish the transom angle in the profile view as indicated on your lines drawing. Now transfer the intersection of the transom face and sheer from the profile view vertically downward to the center line of the plan view. Then transfer the intersections of all waterlines and transom face to the center line of the plan view in the same manner. (Fig. 91)

Using the transom radius prescribed on your lines drawing, swing successive arcs through the transom waterline/center-line intersections in the plan view. Remember that each arc must be of the same radius and must extend fully outboard to its respective hull waterline. The transom-waterline/hull-waterline intersections will represent the transom corners in the plan view.

Now transfer these transom corners from the plan view
(Text continues on page 109.)

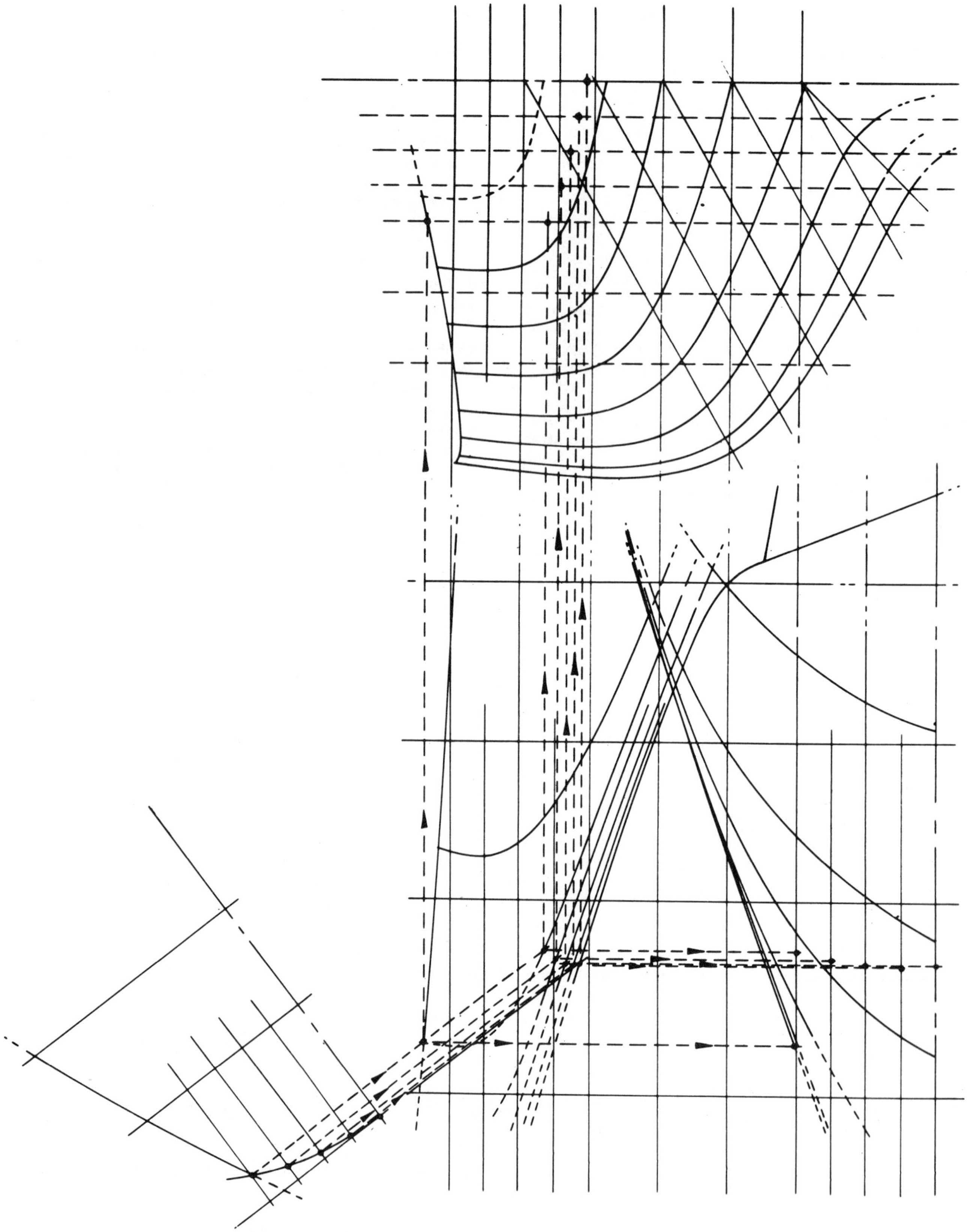

Fig. 88. Round counter transom lofting. Oblique radius projection. (40'6" sloop **Ballerina**)

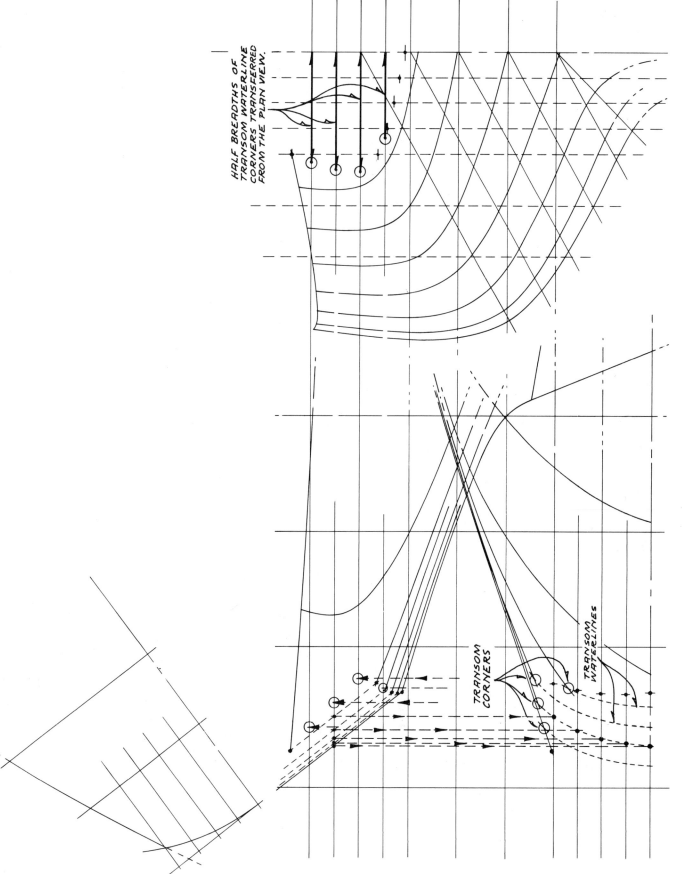

HALF BREADTHS OF
TRANSOM WATERLINE
CORNERS TRANSFERRED
FROM THE PLAN VIEW.

TRANSOM CORNERS

TRANSOM WATERLINES

Fig. 89. Round counter transom lofting. Oblique radius projection. (40'6" sloop **Ballerina**)

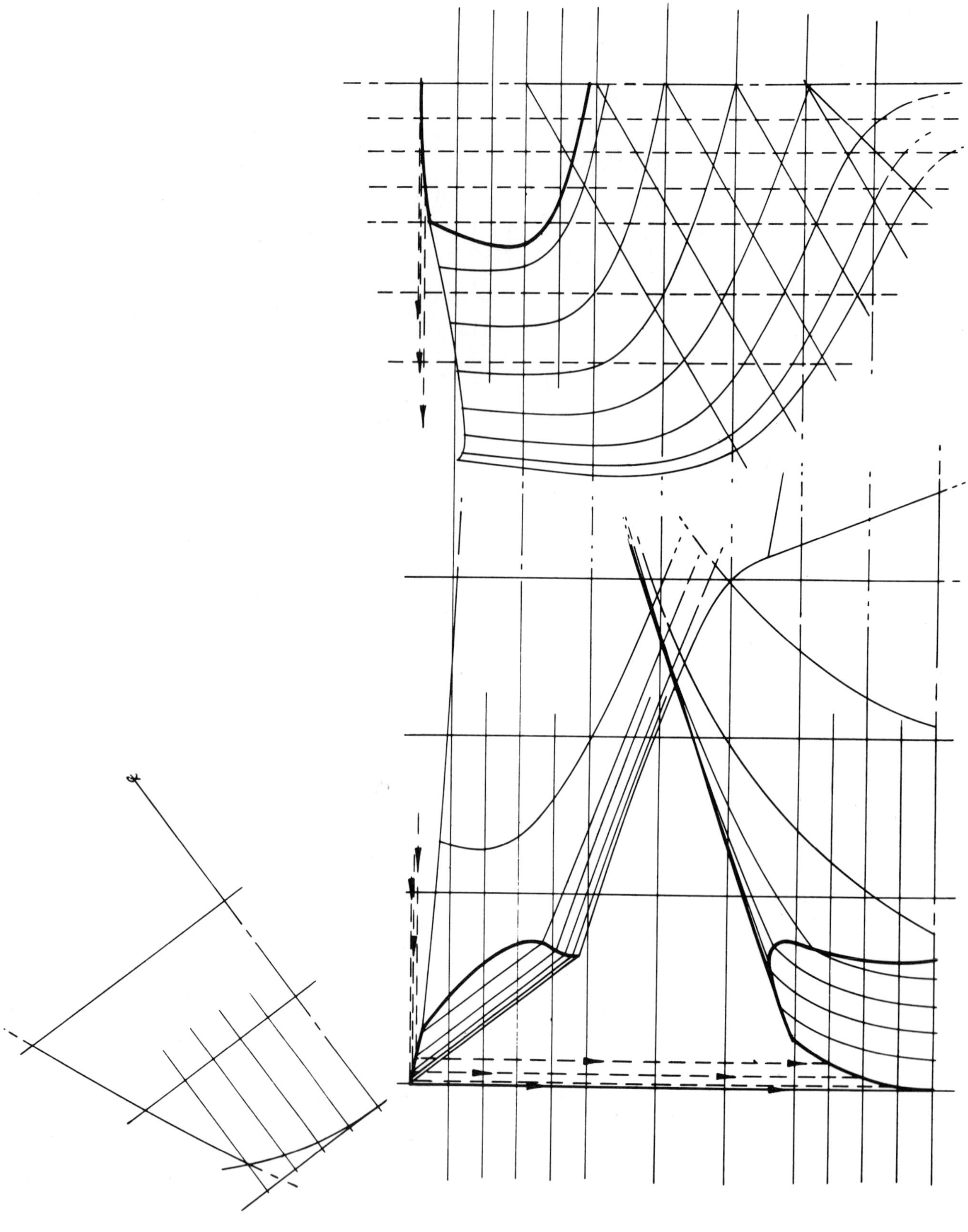

Fig. 90. Round counter transom lofting. Oblique radius projection. (40'6" sloop **Ballerina**)

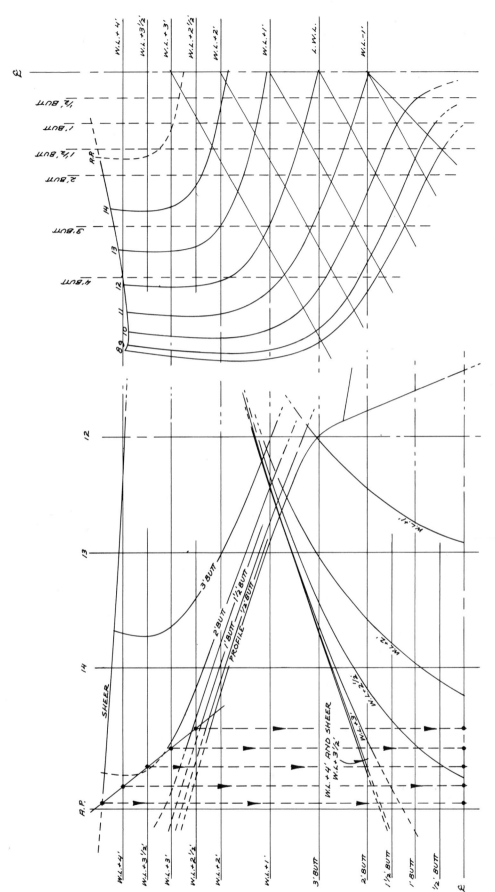

Fig. 91. Round counter transom lofting. Horizontal radius projection. (40'6" sloop **Ballerina**)

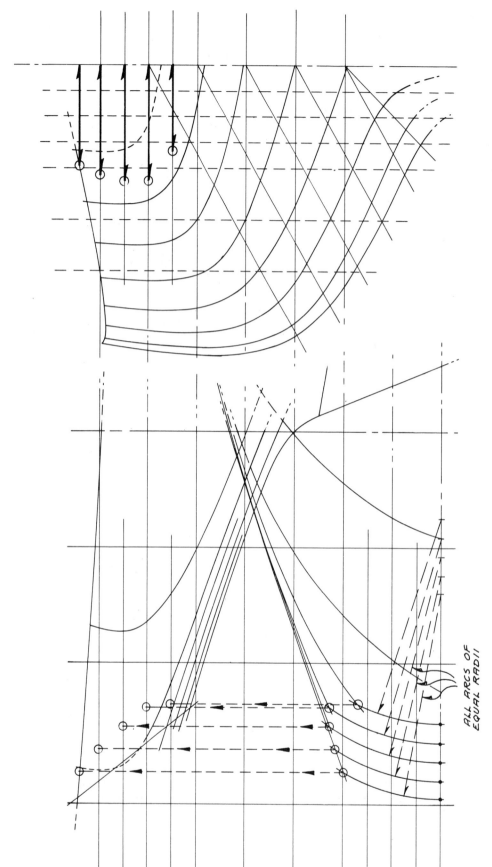

ALL ARCS OF
EQUAL RADII

Fig. 92. Round counter transom lofting. Horizontal radius projection. (40'6" sloop **Ballerina**)

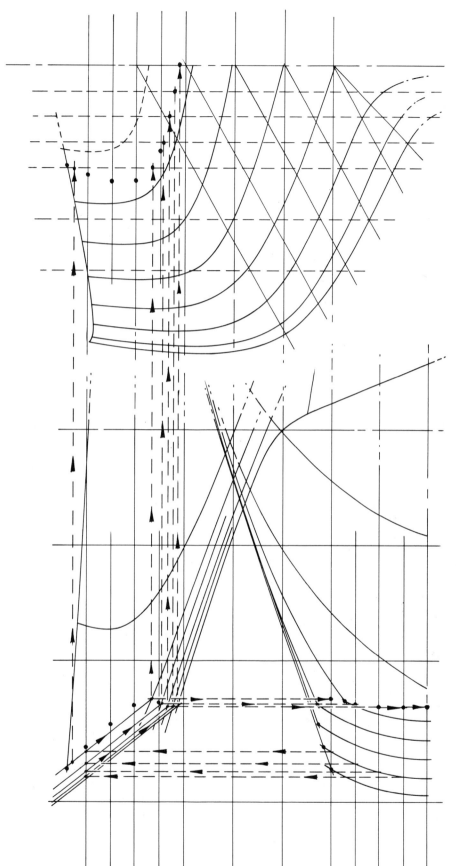

Fig. 93. Round counter transom lofting. Horizontal radius projection. (40'6" sloop **Ballerina**)

103

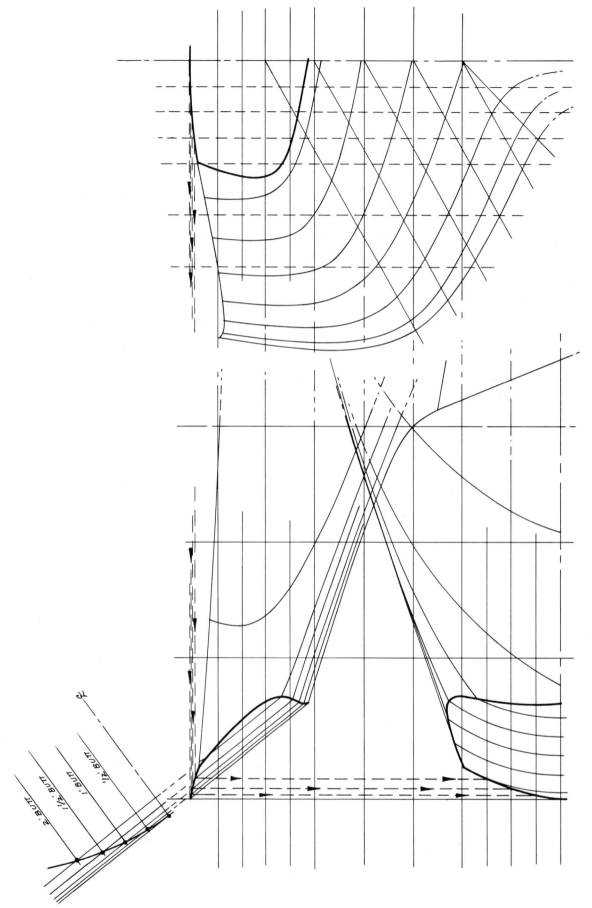

Fig. 94. Round counter transom lofting. Horizontal radius projection. (40'6" sloop **Ballerina**)

2" BUTT

1½" BUTT

1" BUTT

½" BUTT

℄

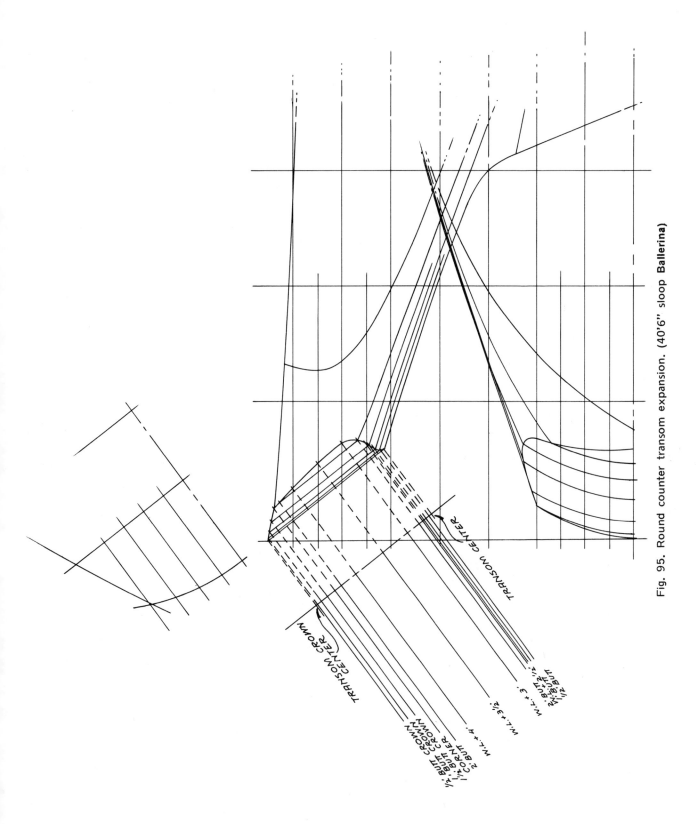

Fig. 95. Round counter transom expansion. (40'6" sloop **Ballerina**)

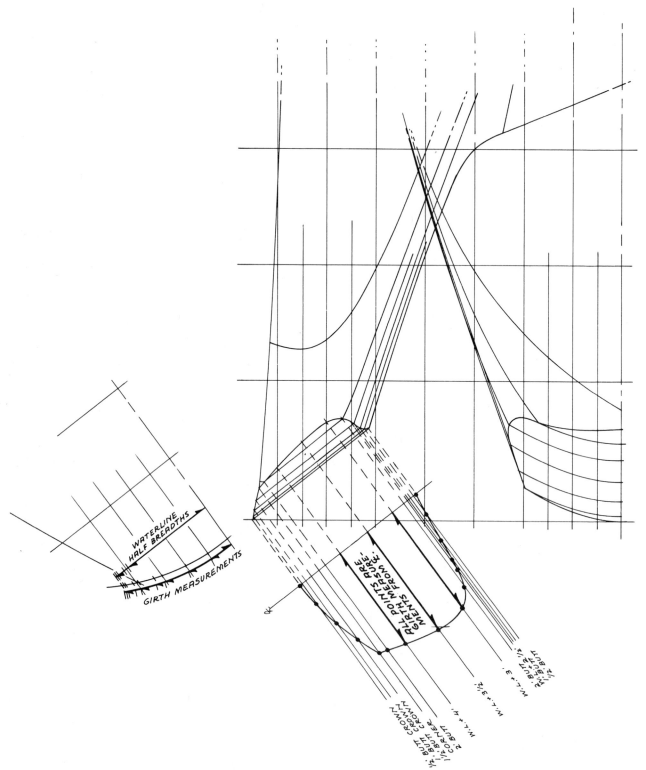

WATERLINE
HALF BREADTHS

GIRTH MEASUREMENTS

ALL POINTS ARE-
ALTH MEROM
GIRTHS MEASURE-
MENTS

BUTT CROWN
1/2 BUTT CROWN
1/2 BUTT CROWN
1/2 BUTT
1/2 CORNER
2 BUTT
W.L. + 4'
W.L. + 3'1/2
W.L. + 3'
1/2 BUTT 2'1/2
2 W. BUTT
1/2

Fig. 96. Round counter transom expansion. (40'6" sloop **Ballerina**)

106

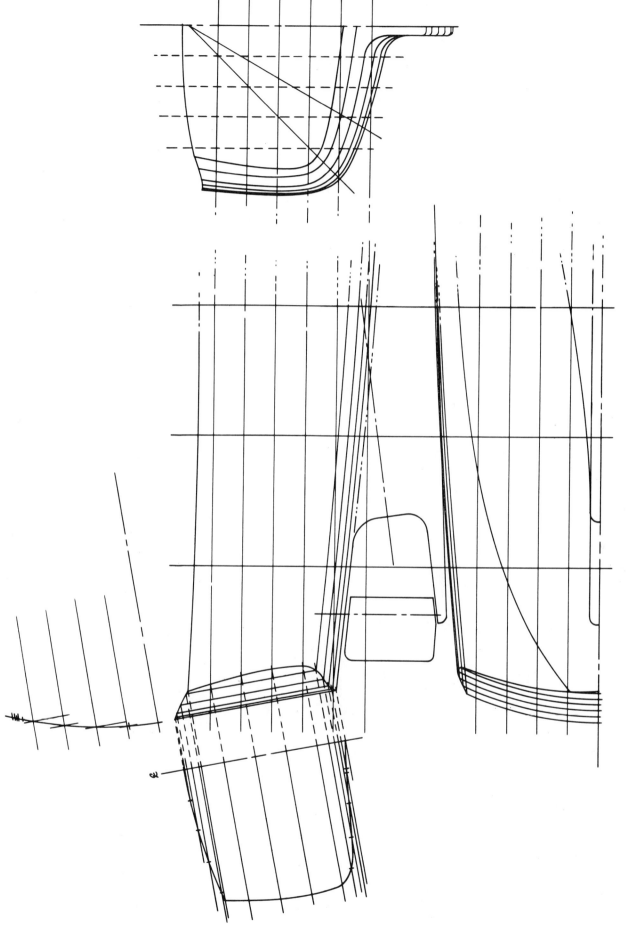

Fig. 97. Round canted transom expansion. (Powerboats)

107

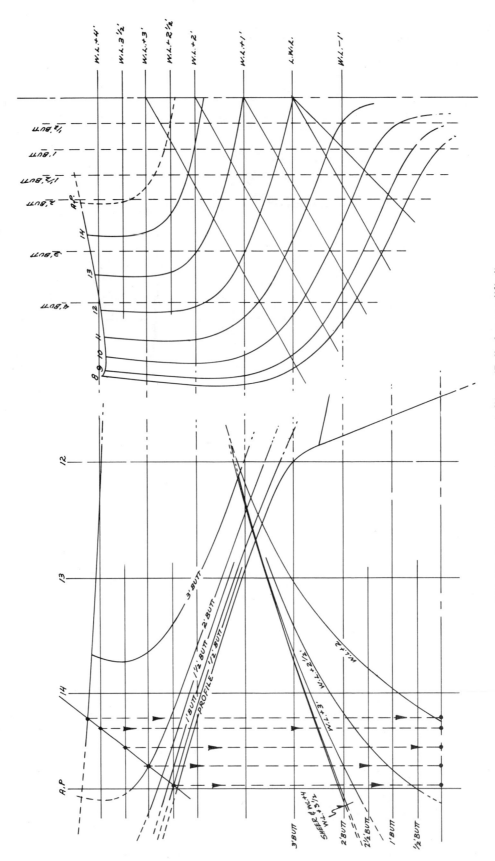

Fig. 98. Round reverse transom lofting. (**Ballerina** modified)

108

(Continued from page 97.)

vertically to their respective waterlines in the profile view. These points will partially define the transom corner in the profile view.

Using a tick stick, transfer the half-breadths of the transom-waterline corners from the plan view to their respective waterlines in the body plan. These points will partially define the transom in the body plan. (Fig. 92)

From any one of the transom radius waterlines in the plan view, transfer each buttock intersection vertically to its respective waterline in the profile view. Remember, it makes no difference which transom radius you use.

Draw a series of oblique buttock lines parallel to the transom face in the profile view, so that they pass through the waterline buttock intersections. Extend the oblique buttock lines downward until they intersect their respective hull buttock lines. These intersections will define the lower shape of the transom corner in the profile view. Project the buttock corner intersections vertically to their respective buttock lines in the plan view. These intersections will define the inboard shape of the transom in the plan view.

Transfer the heights of the buttock corners from the profile view to their respective buttocks in the plan view. These intersections will define the shape of the lower edge of the transom in the body plan. (Fig. 93)

Establish the transom camber in the body plan, extending it from the transom corner. Project the heights of the intersections of the crown and buttock lines in the body plan horizontally to their respective oblique transom buttock lines in the profile view. Connect these intersections with a fair curve to establish the transom crown in the profile view.

Now transfer the buttock/crown intersections of the profile view vertically to their respective buttock lines in the plan view. Connect these points with a fair line to establish the after edge of the transom.

Although the end product may look almost the same, you may have noticed that there were some reversals in the procedural sequence between the horizontal radius transom and the oblique radius. In the latter instance you began with an oblique transom section, but this time the oblique transom section will be a necessary part of the end product. This section is vital to the proper execution of the transom face expansion (to follow) and will be projected from the profile view.

First, extend the oblique transom buttock lines in the profile view some distance above or below the hull. These buttock extensions must be parallel to the transom angle. Draw a center line perpendicularly through the oblique buttock lines. From this center line, lay off and draw buttock lines at their proper intervals. Use as many buttocks as necessary to encompass the width of the transom. Mark the intersections where each transom face buttock passes through its perpendicular counterpart. Swing a curve through these intersections, drawing the line a short distance past the outboard buttock. This will complete the projection of the oblique transom section. You are now ready to begin the transom expansion. (Fig. 94)

Girth Measurements

There will be circumstances during your vessel's construction when it will be necessary to measure the actual finished length of a curved object which has been drawn only orthographically (profile, plan or end views). Such circumstances may occur when determining the cut length of a cabin side, rounded cabin end, or curved cockpit coamings.

Quite normally, you may think you could simply run your tape measure around the cabin or coaming carling. In the case of transom expansion, there is a tendency to straighten out the curved waterlines to ascertain half-breadths of the finished piece. Well, this can result in wasted time and lumber.

Girth is the distance along the surface of a curved object measured on a plane which results in the least distance between two given points. In the case of a cabin trunk face or coaming, the line of measurement will appear perpendicular to the angle of the surface when viewed from the ends. This measurement line, if drawn, will appear as a curve when the surface is flattened out. One type of notable girth measurement lines which may be familiar to you appears on maps and charts which show airplane routes as curves and are known to navigators as "great circles."

In the case of transom expansion, all girth measurements must be taken along the curve of the oblique section. This is accomplished, first, by driving nails into the lofting floor at the intersections of the curve at the center line and buttocks. Spring a batten around the nails, then mark it at the center line and the specific point, or points, being measured. When the batten is lifted and straightened, it will accurately reflect the true distance between all points along the transom surface.

All references to girth in the following pages should be measured in the foregoing manner.

Round Counter Transom Expansion

This is accomplished by first extending lines perpendicularly from the transom face from all points of the transom and crown. A center line will pass perpendicularly through these extension lines. (Fig. 95)

With your tick stick, transfer all waterline and sheer half-breadths to the oblique view and indicate these points on the transom section curve or radius.

Using a flexible tick stick (**see Girth Measurements**), indicate girth measurements of all buttocks and waterline marks. Transfer these girth measurements to their respective lines in the oblique expansion view. Connect all points with fair curves. The finished shape will accurately represent the actual configuration of the transom in a flat form and may be used confidently to cut a precision transom template. (Fig. 96)

Although my previous graphic examples have been of a typical rounded sailboat counter transom, the lofting and expansion of a rounded canted powerboat transom is executed in exactly the same manner. (Fig. 97)

Lofting a Rounded Reverse Transom

Add extra waterlines and buttocks to the plan view, body plan and profile view. These lines need not extend the full length of the vessel but should be concentrated in the stern area. They will add to the number of fairing points which are helpful in drawing the tight curves of the transom.

Establish the proper transom angle in the profile view. Transfer the intersections of the transom angle and waterlines (including the sheer) vertically to the plan view center line. (Fig. 98)

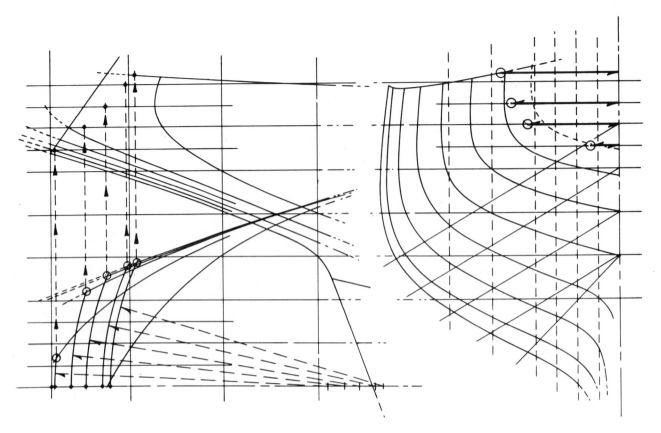

Fig. 99. Round reverse transom lofting. (**Ballerina** modified)

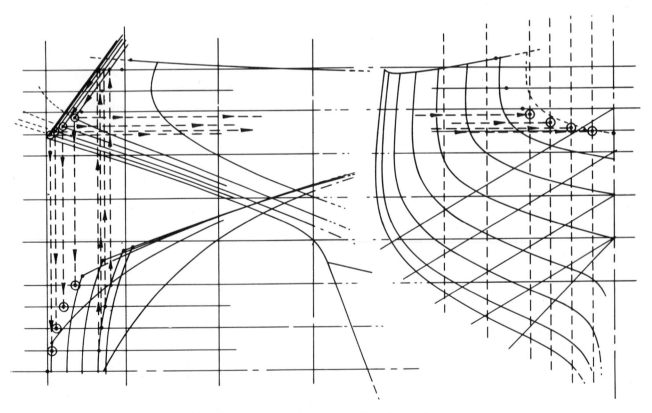

Fig. 100. Round reverse transom lofting. (**Ballerina** modified)

110

Strike arcs through the center-line/waterline intersections of the plan view using the radius prescribed in the lines drawing. Each arc must be of the same radius and must intersect its corresponding hull waterline. These waterline intersections will represent transom corner points in the plan view.

Project the waterline corner points from the plan view vertically to their respective waterlines in the profile view. These intersections will represent transom corner points in the profile view.

Transfer the half-breadths of the waterline corners from the plan view to their respective waterlines in the body plan. The resulting intersections will represent transom corner points in the body plan. (Fig. 99)

From any one of the transom radius waterlines of the plan view, transfer its buttock intersections vertically to its respective waterline in the profile view. Now draw a series of oblique buttock lines parallel to the transom angle so that they pass through these buttock/waterline intersections. The intersections created by the oblique buttocks and hull buttocks in the profile view represent the lower corner points of the transom.

Transfer the heights of the buttock corners from the profile view to their respective buttocks in the body plan. These points will represent the lower corners of the transom in the body plan view.

Transfer the buttock corner points of the profile view vertically to their respective buttock lines in the plan view. These points will represent the forward edge of the transom in the plan view. (Fig. 100)

Connect all transom corner points in each view with fair curves to complete the transom shapes.

Establish the transom crown from the sheer/transom corner of the body plan. Transfer the heights of the crown/buttock intersections of the body plan to their respective oblique buttock lines in the profile view. Connect these points with a fair curve to establish the transom crown in the profile view.

Transfer the crown/oblique buttock points from the profile view vertically to their respective buttocks in the plan view. This will establish the shape of the transom at the deck in the plan view. Your transom lofting is now complete.

Prior to expanding the transom to form a full-size working template, it will be necessary to draw an accurate transom section. This section must represent a plane passing perpendicularly through the transom face. To project this section, first extend all oblique buttock lines, transom face angle, and sheer corner some distance above or below the profile view. Now draw a center line perpendicularly through these oblique buttock extensions. Carefully measure and draw a series of buttock lines at their proper intervals away from the center line. Draw a fair curve through all corresponding intersections. (Fig. 101)

Round Reverse Transom Expansion

The expansion of the round reverse transom is identical to the expansion of a rounded counter transom. While the shapes look a great deal different at a glance, you can imagine that the variation is only a matter of rotating the axis of the transom angle. In all other respects the projec-

tions of these two types of transoms are handled in the same manner. (Fig. 102)

Lofting the Flat Raked Transom (Dinghy Transom)

Next to the plumb flat transom, the flat raked transom is the easiest to loft and build and is most often found on power boats or small sailboats of workboat derivation. This is not to say that the flat raked transom is inherently unsightly or bulky in appearance. Indeed, if the hull of the vessel is well rounded, especially the garboard area, a delightfully graceful stern can result. The classic Whitehall and Bahama dinghies have become loved by discriminating yachtsmen the world over because of their beautiful "wine glass" transoms. The flat raked transom was also the style of the noted gigs and longboats which have found such a prominent place in our nautical history. A well designed dinghy stern is truly worthy of the deepest pride and deserves careful development.

The hull lines of all views are carefully faired toward the aft perpendicular station in the normal manner. In the example, you will see the waterlines and stations have been drawn to the width of the keel and deadwood. (Fig. 103)

Establish the angle of the face of the transom in the profile view. Now transfer all intersections of the buttocks and waterlines with the transom face vertically to their respective lines in the plan view. Project the heights of the buttock/transom face intersections to their respective buttocks in the body plan, then transfer the half-breadths of the waterline/transom points from the plan view to their corresponding waterlines in the body plan. All resulting points in all views represent the finished shape of the transom at the hull. (Fig. 104)

Establish the transom crown from the sheer to the center line in the body plan. Transfer the intersections of the crown and buttocks in the body plan to the transom face line in the profile view. Project the buttock/transom crown intersections from the profile view vertically to their corresponding buttocks in the plan view. Spring a fair curve through all crown and sheer points of the plan view to complete the transom shape. (Fig. 105)

Flat Raked Transom Projection

Unlike a curved transom, it is not necessary to expand the flat transom. In order to evolve a working pattern, however, it is required that a view be drawn which is projected perpendicularly from the angled transom face. Remember that all views of the transom drawn thus far are seen from some angles other than "straight on" and do not depict the true orthographic shape required for molding.

Draw a center line which is parallel to the transom angle (or you may use the transom face in profile as your center line). Now project lines perpendicularly away from the transom angle from each and every intersection in the profile view. These intersections must include all waterlines, hull buttocks, sheer at transom, crown buttocks, etc.

Mark the half-breadth distances of all waterlines, buttocks, sheer at transom, and deadwood at their corresponding projected lines in oblique view.

Connect all resulting points with fair curves to evolve an extremely accurate finished transom template. (Fig. 106)

(Text continues on page 119.)

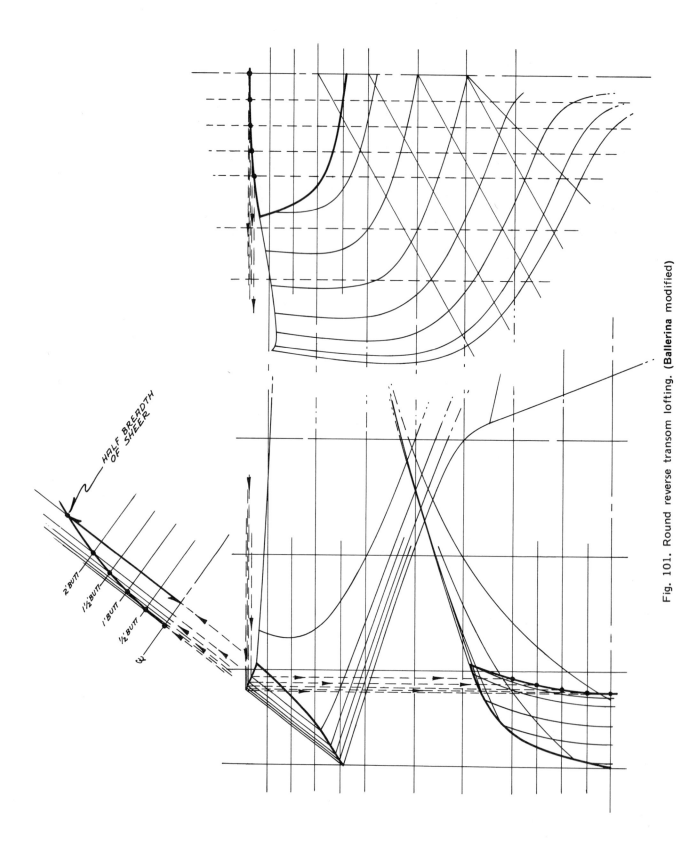

HALF BREADTH OF SHEER

2' BUTT
1½' BUTT
1' BUTT
½' BUTT
3

Fig. 101. Round reverse transom lofting. (**Ballerina** modified)

112

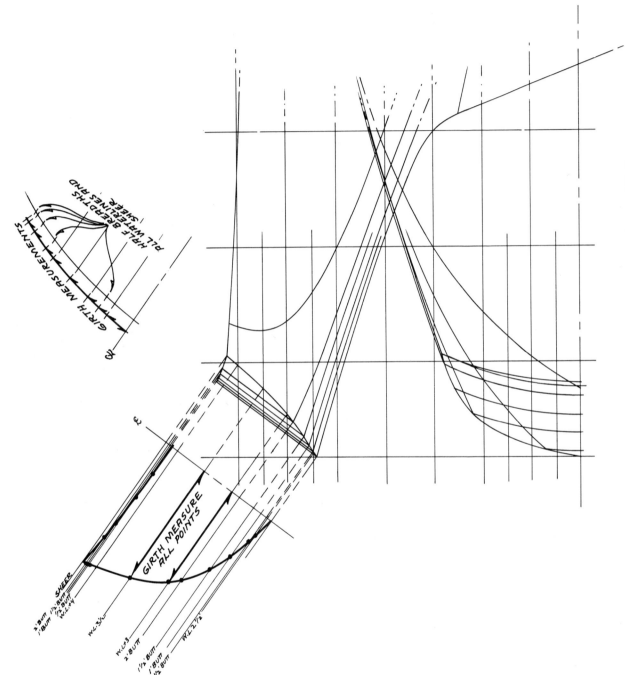

GIRTH MEASURE
ALL POINTS

SHEER

Fig. 102. Round reverse transom expansion. (Ballerina modified)

113

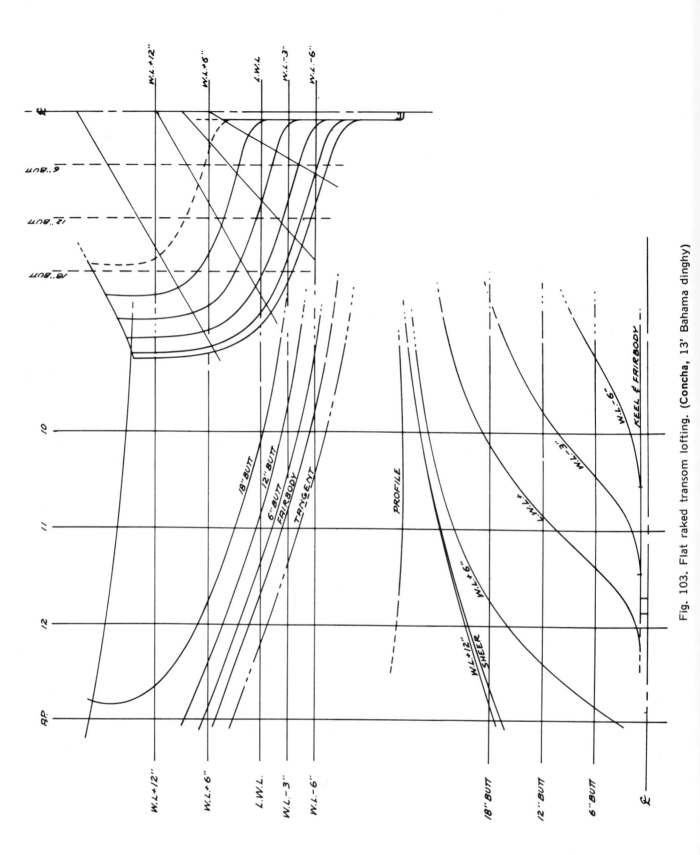

Fig. 103. Flat raked transom lofting. (Concha, 13' Bahama dinghy)

114

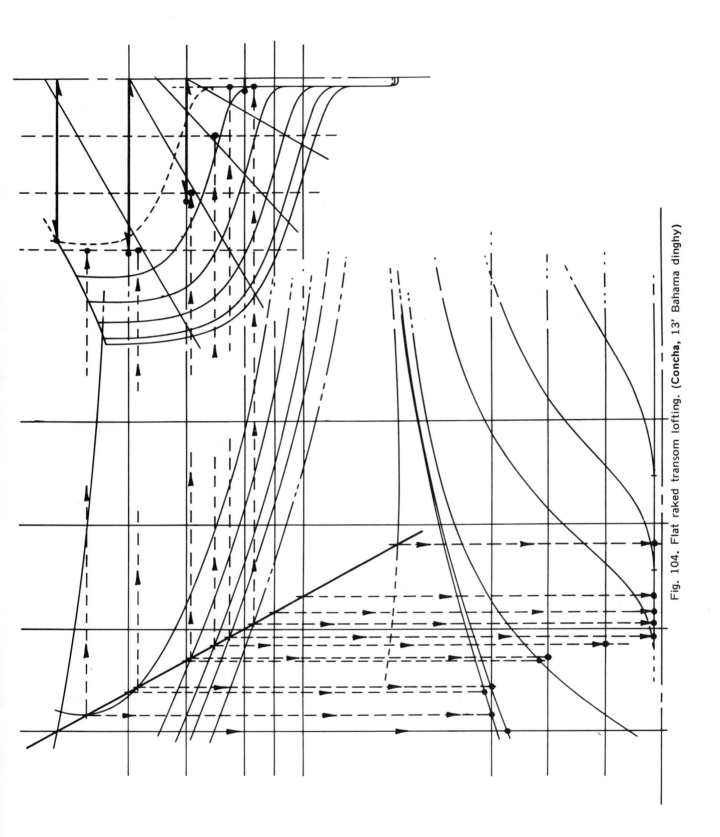

Fig. 104. Flat raked transom lofting. (Concha, 13' Bahama dinghy)

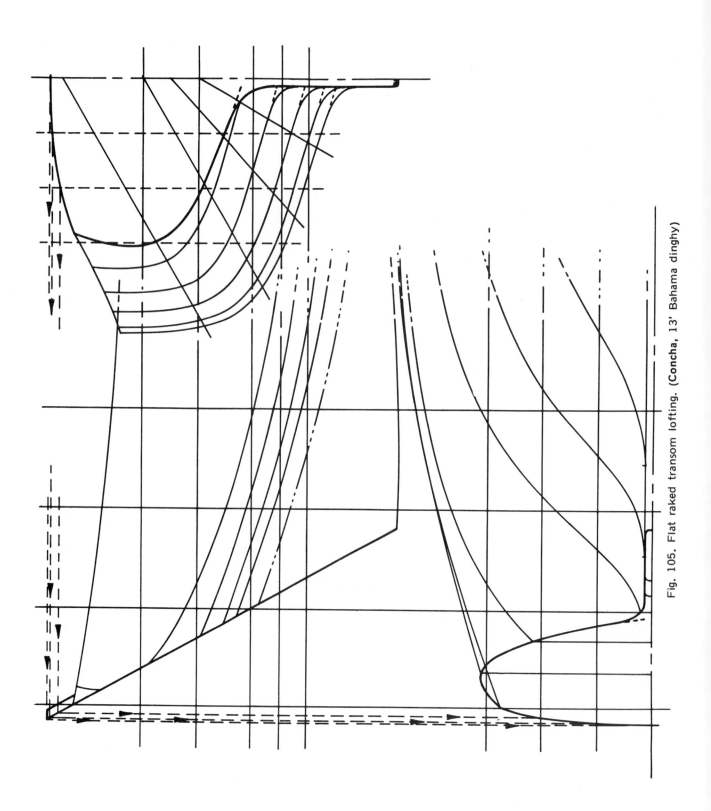

Fig. 105. Flat raked transom lofting. (Concha, 13' Bahama dinghy)

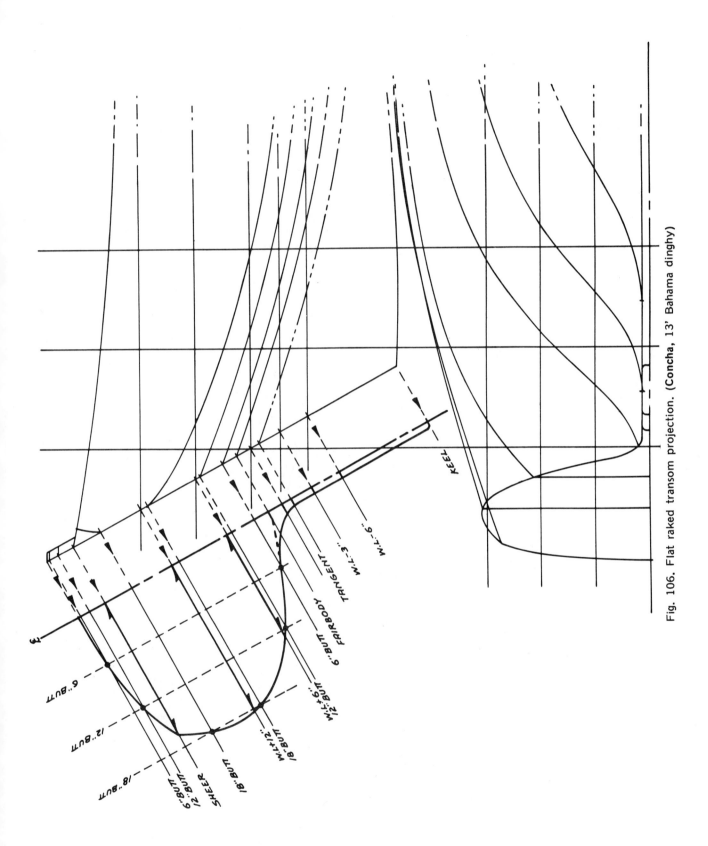

Fig. 106. Flat raked transom projection. (Concha, 13' Bahama dinghy)

117

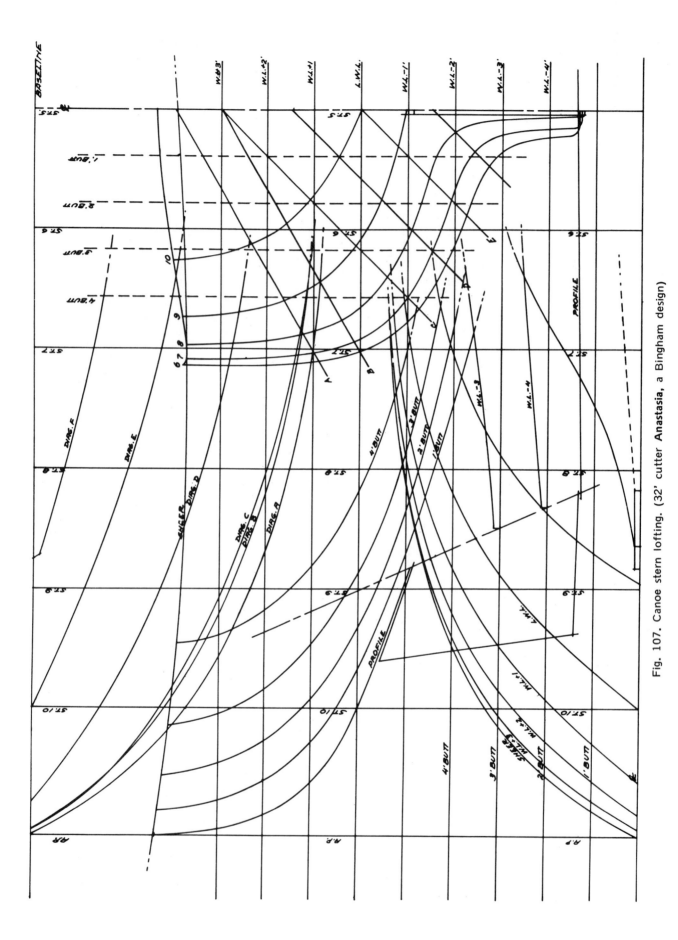

Fig. 107. Canoe stern lofting. (32' cutter **Anastasia**, a Bingham design)

118

(Continued from page 111.)

Lofting a Canoe or Cruiser Stern

These two types of sterns are almost identical to each other and are the most common found on double-enders. The canoe stern comes to a distinct point all the way up to the sheer, but the cruiser stern becomes elliptical near the deck where its lower pointed shape is transformed into a rounded shape at the center line in the plan view. Either the canoe or cruiser stern may be tumble-home in the profile (slanting slightly inward) but it is more common with the cruiser stern.

Lofting the canoe or cruiser stern is very simple, as all waterlines in the plan view are brought fully to the center line at the intersection of the profile point which is transferred from the profile view.

The buttocks of the profile view and stations of the body plan are faired in exactly the same manner as the forward portion of the vessel. Be careful that all line and grid intersections of each view correspond to each other (Fig. 107)

Lofting the Plumb Flat Transom

This type of transom is usually used on power workboats. This is the simplest transom to loft and build because there are no angles or curves to expand or project. The hull lines are faired to station "A.P." in all views in the normal manner. Because "A.P." usually represents the actual stern of the vessel, the finished transom shape may be taken directly from "A.P." as it appears in the body plan.

The Barrel Bow

The barrel bow is one which increases its stern radius as it rises toward the sheer. This radius increase is apparent in the flaring of the ghost lines in all lofted views.

The barrel bow is very modern and powerful looking but is not simply designed for aesthetic purposes. (Plate 6) It has the effect of creating the sailing characteristics of a vessel with longer overhangs but within a shorter L.O.A. It is often designed to provide flare, as in the example shown, while yielding a wider foredeck near the headstay where working space is always lacking.

Lofting the barrel bow is almost identical to the sharp bow except that the ghost lines are continuous curves instead of straight along the upper portion of the bow. Developing the ghost curve requires laying off the proper offsets in all views and fairing the points fully up to the deck or sheer. (Fig. 108)

The Clipper Bow

The clipper bow generally uses two separate lines in the profile view to define its shape; i.e., the hull ghost and the actual shape of the billet head. If the billet head could be removed from the vessel, it would result in a form much the same as a barrel bow because of the flaring characteristics of the ghost lines as they near the sheer.

With this in mind, first draw the vessel in the profile view without the billet head. Fair all lines completely in the normal manner developing the ghostline as described for the barrel bow.

Now lay the shape of the billet head into the profile view. While the upper flat surface of the billet is usually an extension of the deck line, be aware that it is not a hard and fast rule.

Transfer the waterline/billet head (finished profile) intersections vertically to the center line of the plan view then extend the straight portion of the ghost line forward until it reaches the end of the billet head in the plan view.

Fair the existing waterlines into their respective billet head/waterline intersections using liberally hollow lines. Each hollow line will become progressively shorter as you move downward where the billet head merges with the original profile.

To ensure that the hollow waterlines of the billet head are fair, it is a safe measure to add a series of closely spaced station lines near the bow in the profile view. These stations are then fully faired into the body plan while making the necessary correction in the waterlines of the plan view to ensure the proper shape. (Fig. 109)

Adding Stations for Bulkheads, Bulkhead Webs, Tank Baffles, Etc.

It is rare, indeed, when a bulkhead position will exactly correspond to the position of a regular lines station. This is also true of the mast step faces, transverse engine bed webs, tank beds and ends, or whatever other transverse structure is to be built into the vessel. If you are constructing your boat to the open truss method, or your plans require the prefabrication of these units for other construction systems, it will be necessary to establish their shapes accurately.

Locate the longitudinal position of the required member in the profile and plan views. Draw a vertical line through the profile and plan views just as if it were an extra station (which, in fact, it is).

Transfer the heights of the bulkhead/buttock intersections from the profile view to their respective buttocks in the plan view. Transfer the heights of the profile/bulkhead and sheer/bulkhead intersections to the body plan in the same manner.

Transfer the half-breadths of the bulkhead/waterlines of the profile view to their respective waterlines in the plan view.

Transfer the widths of the diagonals at the bulkhead to their respective diagonal lines in the body plan.

Connect all of the resultant points in the body plan, thus completing the accurate development of the desired bulkhead. (Fig. 110)

It is not always necessary to draw the bulkhead shape fully to the sheer line of the vessel in the body plan if the unit being drawn terminates a considerable distance below. In the case of tank baffles and engine bed webs, it may only be necessary to reproduce several vertical feet in the body plan. Do not forget, however, that you will reproduce a much truer shape if you spring your batten some distance above and below the actual drawn line required.

Extra Buttocks for Engine Beds, Tank Baffles, Etc.

If your vessel is to be constructed using the open truss system or pipe frame method, it will require that longitudinal structural units be accurately preshaped and fabricated. These units, then, will be built into the armature prior to the application of hull rods and mesh. The most prominent longitudinal structure to be considered are the engine beds. *(Text continues on page 126.)*

119

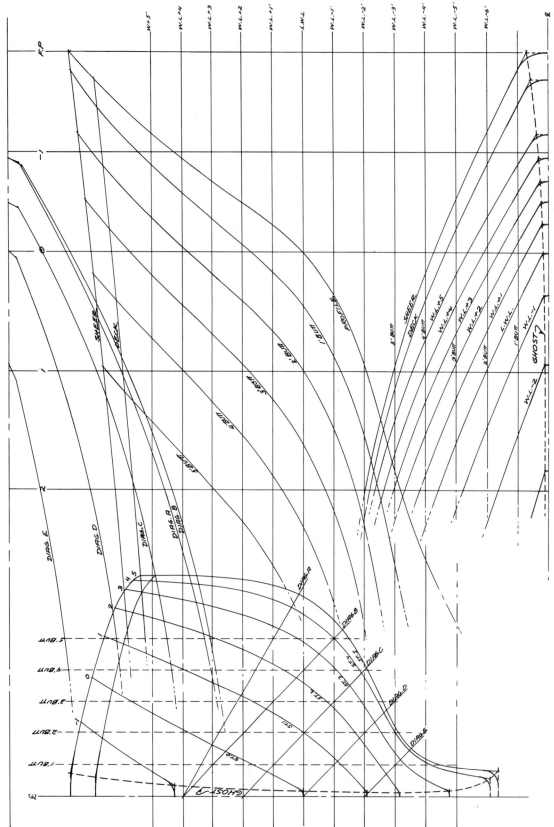

Fig. 108. A typical barrel bow development. Note the flaring of the ghost lines in all views. The radius of the stem is drawn after all lines have been faired. (49' ketch **Christina**, a Bingham design)

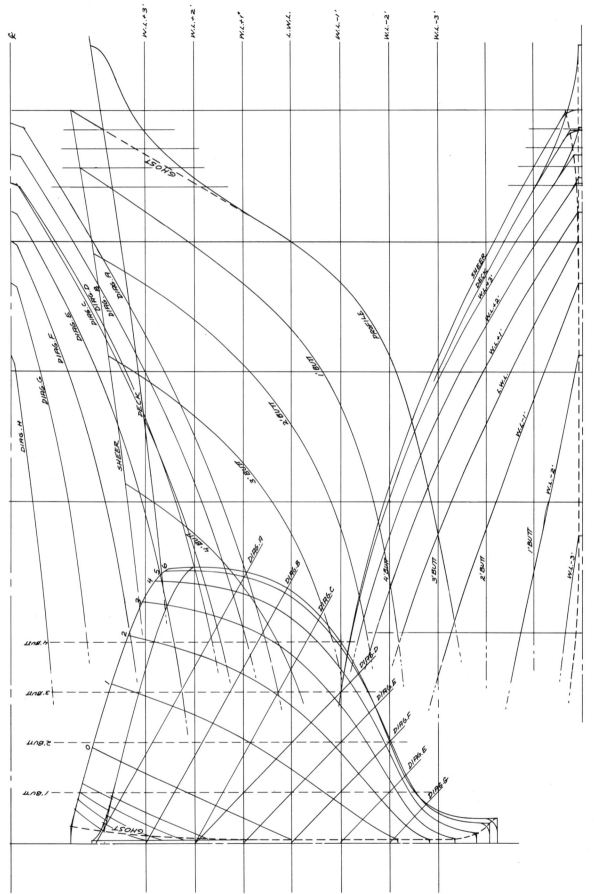

Fig. 109. A typical lofting for a clipper bow. Note that all hull lines have been first faired to the flared ghost line, then additional concave lines drawn outward to the billet head. Additional bow sections have been drawn in the profile and body plan to aid fairing. (32' hermaphrodite brig, Alicia, a Bingham design)

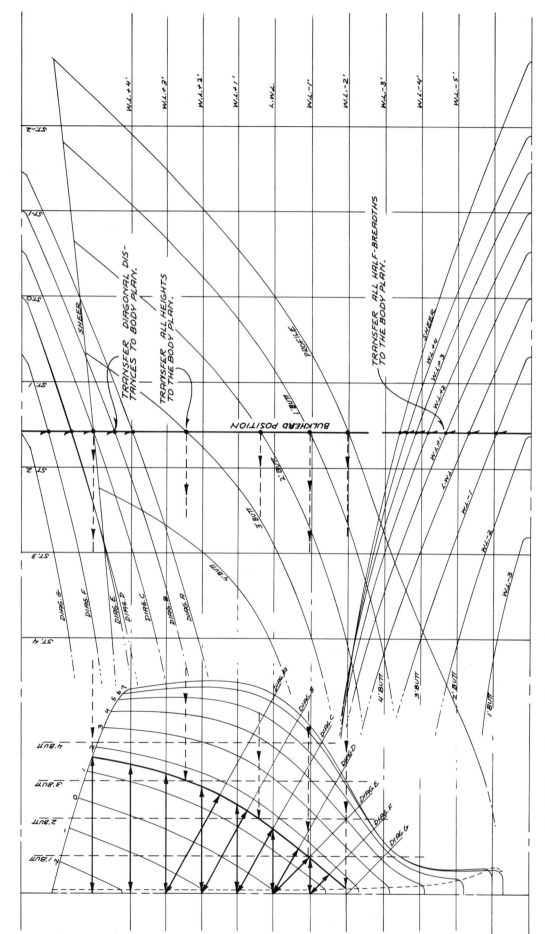

Fig. 110. Locating and drawing additional sections for bulkheads, tank baffles, etc. (40'6" sloop **Ballerina**, a Bingham design)

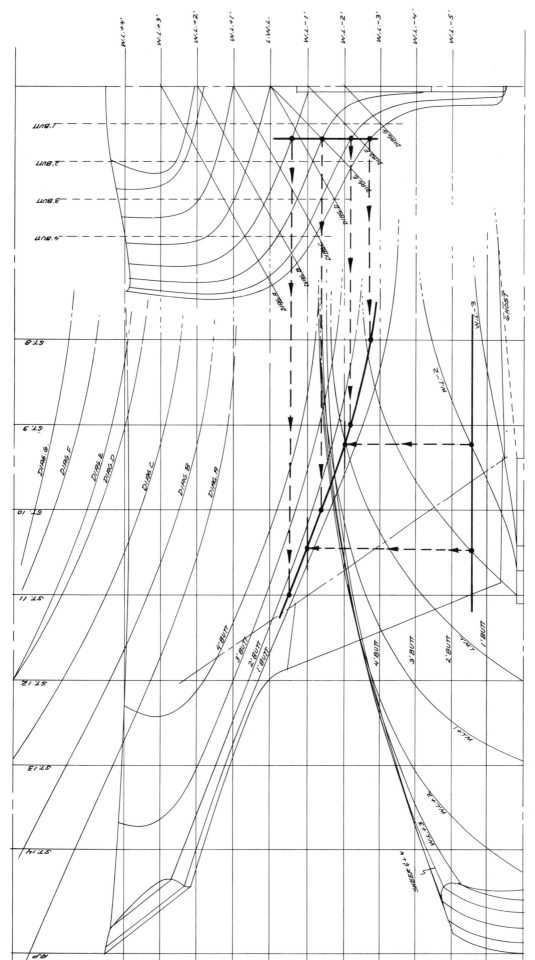

Fig. 111. Locating and drawing additional buttocks for engine beds or tank baffles. (40'6" sloop **Ballerina**, a Bingham design)

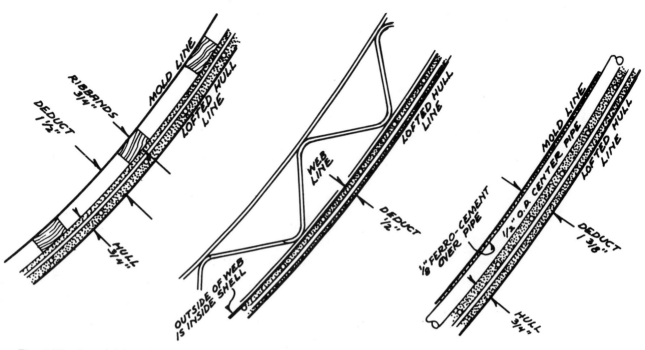

Fig. 112. Ascertaining the required lofted to mold line deduction using various construction methods. Proper thickness will depend upon the specific layup recommended by your designer.

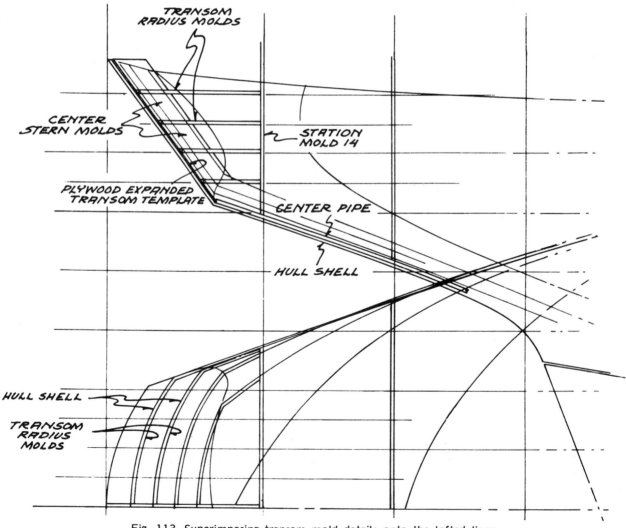

Fig. 113. Superimposing transom mold details onto the lofted lines.

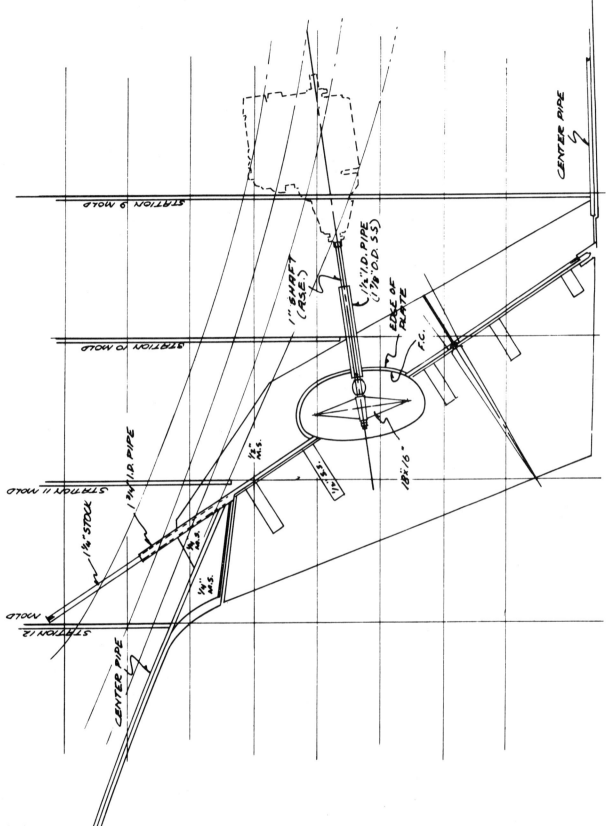

Fig. 114. Superimposing the deadwood, propeller shaft and log, rudder detailed and rudder shaft tube detail. Notice the section drawn perpendicularly through the rudder stock in order to aid the loftsman's understanding of the required structure. Also note that he has reproduced the propeller and necessary zinc anode in order to determine the proper shape and clearance in the propeller aperature.

125

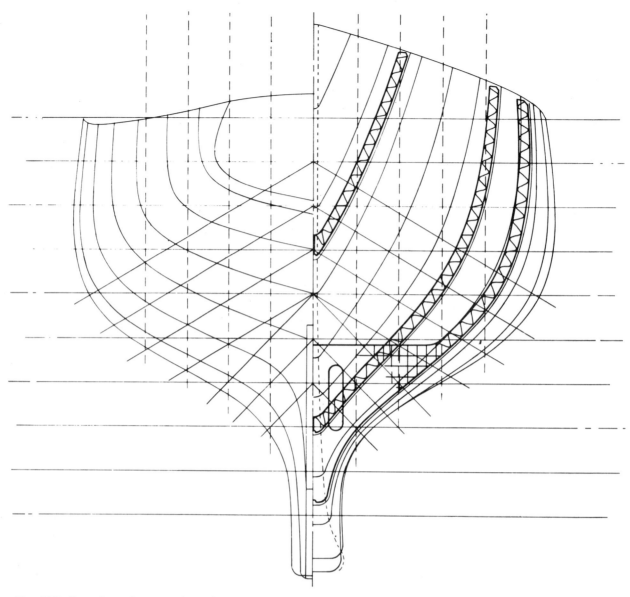

Fig. 115. Superimposing open-truss frames and tank baffle onto the lofted body plan. It is not necessary to indicate the wiggle rod when doing so, but I have shown it for clarity only.

(Continued from page 119.)

Because their inner faces are parallel to the buttock lines in the plan view and body plan views, their shapes at the hull may also be represented as buttock lines. It is this shape at the hull which is required prior to construction. No less important is the configuration of the longitudinal tank baffles.

Lay off lines in the body plan and plan view parallel to the center line at the prescribed distance outboard.

From the new buttock in the plan view, transfer its intersections with waterlines vertically to the corresponding waterlines in the profile view.

From the new buttock line in the body plan, transfer the heights of its station intersections to the corresponding stations of the profile view.

Spring a fair line through all resulting points in the profile view. This will complete the development of the new buttock line. (Fig. 111)

Deducting Thickness (The Spacing Wheel)

The sole purpose of the entire lofting procedure is to evolve hull structure or mold patterns which will result in a smooth hull shape and the proper fitting of interior units. Now that all hull lines have been faired in all views, it is time to begin using the lofting for which it was intended.

In all probability your lines have been drawn to the outside of the hull surface. In this instance, it is quite obvious that some modification must occur to the lines so they will be representative of the mold or web shapes which lie some distance inside.

You must first, of course, determine the distance from the outside of the hull at which your actual working line will occur. In the case of truss or pipe frame construction, the thickness deduction from the lines may equal the thickness of the hull. If you are to be constructing your vessel over a wood mold, however, you will have to deduct both the hull thickness and mold ribband (batten) thickness to determine the station mold shape.

126

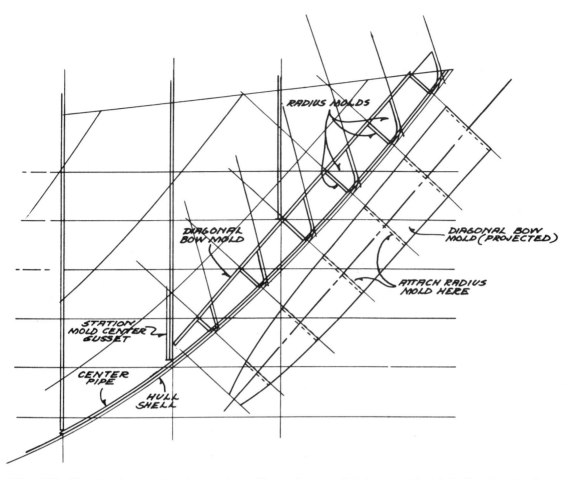

RADIUS MOLDS

DIAGONAL BOW MOLD

DIAGONAL BOW MOLD (PROJECTED)

ATTACH RADIUS MOLD HERE

STATION MOLD CENTER GUSSET

CENTER PIPE

HULL SHELL

Fig. 116. The development and superimposition of a parallel bow mold detail. Try to visualize how all of the parts will fit together to form the proper shape.

The quickest and most convenient way of making the thickness deduction is by using a spacing wheel. This is simply a circle cut from a piece of plastic or Masonite. The diameter of the circle must be exactly twice the required thickness deduction. A hole is drilled at the center of the circle.

To use the spacing wheel, spring a batten around the required shape using plenty of nails. **Important:** The batten must lay completely to the outside of the line. Now place a pencil through the hole in the wheel and run the wheel along the inside of the batten. (Plate 33)

The spacing wheel will not account for the bevel of the stations at the hull, but on cruising boats the error will be minor and insignificant, so don't worry about it. In the case of high performance racing boats, the bevel should be accounted for.

Keep in mind that the thickness deduction may not always be the same everywhere. For instance, the mold thickness at the transom face may be equal to a ¼" sheet of plywood, whereas the rest of the hull may be ribbanded with ¾" stuff. (Fig. 112)

Superimposing Mold and Structural Detail

As I previously mentioned, the purpose of the lofting is to develop workable forms for molds or hull structures that will allow you to construct the fair shape of the hull or interior units. In this area it will be necessary for you to completely understand all facets of the Mold Construction drawing or Webs Assembly drawing provided by your architect. It is wise to study these drawings for several weeks in advance of your superimposing mold or structural detail onto your lofting. You must be absolutely sure of every last detail and be completely familiar with how the various pieces will fit together and attach. To stumble ahead without this degree of understanding is to risk major complications and errors in the drawing of these full-size patterns.

It is obvious that the work to follow will require a great deal of simple draftsmanship, as many of the mold or structural parts will require reproduction in all views in order to determine that they will work or fit together properly. I can give you no hard and fast rules as to how these units should be drawn, as each designer has his own preferences as to the development of mold structures and all web detail. If you have not been given a mold drawing by your designer, you should completely review the next chapter in order to visualize what will be required for forming the hull form. I have included several examples of typical mold details as they would be seen on the lofting. These should serve as indicators as to how to proceed. (Figs. 113-116)

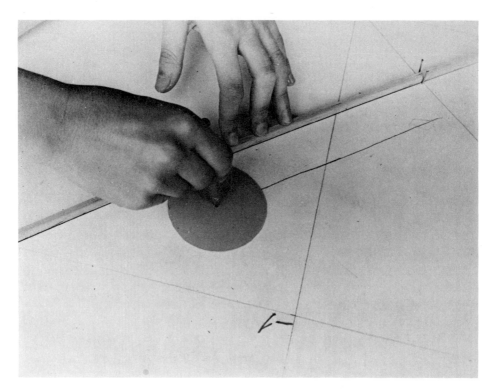

Plate 33. Using a spacing wheel to deduct hull shell and mold thickness from the lofted station lines.

Plate 34. The author corrects offsets from **Andromeda's** completed scrieve board. Note that each station has been faired with individual battens, which can be moved from time to time to eliminate unnecessary drawn line changes.

section three

setting up

scaffolding or skids

Choosing the Site

In choosing a building site, it is imperative that the ground be as solid and as nearly level as possible, for it will ultimately be required to bear a tremendous concentrated weight which must not be allowed to shift or settle during any stage of construction. The worst possible condition would be an empty field in which water is able to stand, forming mud pockets, as even concrete bearers would not be able to take a firm footing. The ideal site, of course, would be a paved parking lot, driveway or warehouse with permanent cement flooring. Care must be taken, even in these latter instances, to establish a perfectly level datum line so that the scaffold or skid base may be wedged or blocked to correct any inconsistencies in the supporting structure. Remember . . . the ground may become water-saturated during curing.

Base Timbers or Sleepers for Scaffolding

If the boat is to be built on open ground, the position of the scaffolding or skids should be preestablished in order to locate the positions of the base timbers. By sighting a level down the length of each base position, scrap lumber blocks or cement supports should be placed at distances approximately 2' apart so that their tops coincide perfectly with this level line. As this procedure must occur on both sides of the scaffold, leveling must not only occur lengthwise but transversely as well. It is upon these supports that the entire weight of the scaffolding structure and the finished boat will rest, so the larger the footing of each support, the better.

Scaffold (Upright Construction)

The actual erection of the scaffolding may occur in two different ways, depending partly upon the builder's available space and the number of helping hands. The first method is that of hoisting the individual station uprights, one after the other, until all stations have been erected, aligned and braced with diagonals. The second method is that of prefabricating the entire sides of the scaffold and raising into position, as two units, after which they will be transversely connected and diagonally braced.

Method 1

Once the datum supports have been positioned, place the scaffold base timbers on top of the supports, putting each plank end to end to achieve the required scaffold length on each side. It is best to preestablish (at least roughly) the station positions so that they will not fall at the butt joints as each joint must receive a butt block which may interfere with placing uprights. Once the scaffold base timbers have been laid they should be adjusted inward or out until they are the proper distance apart and parallel for their full lengths. Now mark the station positions along one of the scaffold base timbers.

In order to ensure that the station positions on the opposing base timbers are absolutely perpendicular, diagonal measurements must be made as indicated in the illustration to establish at least one station on the opposite base from which all other stations may be measured.

Each upright unit is now prefabricated by laying down the upright timber at the prescribed distance apart, attaching each end with a spanner cleat and cross span respectively. The upright unit is now carefully squared and diagonally braced.

After all of the upright units have been constructed, position them on their respective station marks, raise to the vertical position by using a plumb bob or level, and diagonally brace permanently fore and aft. (Fig. 117) After all station units have been erected, sight along the under sides of all the cross spans to make sure that they are in line and level relative to each other. If a visible error exists, it must be corrected by whatever measures necessary. Upon satisfactory alignment, header timbers are placed along the tops of the station uprights to lock the entire structure together. (Fig. 118)

Method 2

The base timbers for one scaffold side are butted end to end to achieve the proper scaffold length, then attached together with butt blocks. The position of these joints and butt blocks must not interfere with station positions. The base timber is now marked with the station positions and turned on edge, allowing the upright timbers to be attached to the base timber at their respective station positions. The header timbers are now attached to the tops of the station uprights (being sure to maintain the exact station spacing as was used on the base timber) and the entire scaffold side is squared and diagonally braced. (Fig. 119)

Once completed, the scaffold side must be positioned so that the base timber rests on the edge of the support blocks where it will be raised to the vertical and transversely braced with diagonal timbers. This entire procedure must now be repeated for the opposing scaffold side. (Fig. 120)

After both scaffold sides have been erected and aligned transversely so that all stations are perpendicular to each other, the upper cross spans will be attached as well as the individual station diagonal bracing. (Fig. 121)

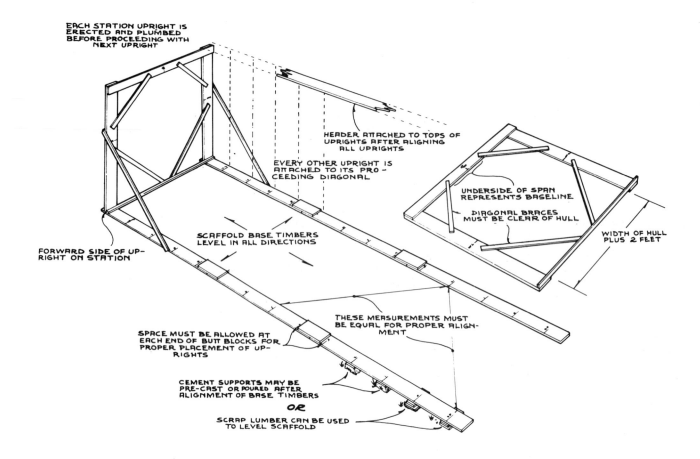

EACH STATION UPRIGHT IS
ERECTED AND PLUMBED
BEFORE PROCEEDING WITH
NEXT UPRIGHT

HEADER ATTACHED TO TOPS OF
UPRIGHTS AFTER ALIGNING
ALL UPRIGHTS

EVERY OTHER UPRIGHT IS
ATTACHED TO ITS PRO-
CEEDING DIAGONAL

UNDERSIDE OF SPAN
REPRESENTS BASELINE

DIAGONAL BRACES
MUST BE CLEAR OF HULL

WIDTH OF HULL
PLUS 2 FEET

SCAFFOLD BASE TIMBERS
LEVEL IN ALL DIRECTIONS

FORWARD SIDE OF UP-
RIGHT ON STATION

THESE MEASUREMENTS MUST
BE EQUAL FOR PROPER ALIGN-
MENT

SPACE MUST BE ALLOWED AT
EACH END OF BUTT BLOCKS FOR
PROPER PLACEMENT OF UP-
RIGHTS

CEMENT SUPPORTS MAY BE
PRE-CAST OR POURED AFTER
ALIGNMENT OF BASE TIMBERS
OR

SCRAP LUMBER CAN BE USED
TO LEVEL SCAFFOLD

Fig. 117. Constructing and erecting scaffold uprights one at a time. This system seems best suited when space is at a premium.

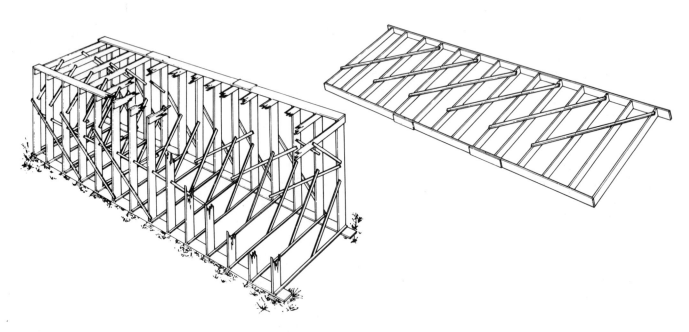

Fig. 118 (left). A completed scaffolding cut away to clearly show all parts.
Fig. 119 (right). A complete scaffold siding being constructed on the ground. It has been carefully aligned prior to affixing diagonals.

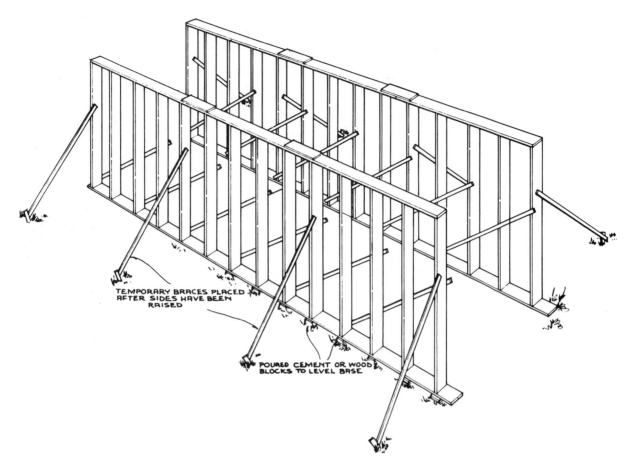

TEMPORARY BRACES PLACED AFTER SIDES HAVE BEEN RAISED

POURED CEMENT OR WOOD BLOCKS TO LEVEL BASE

Fig. 120. Both scaffold sides have been erected and aligned for plumb and parallel. Note the blocking under the base timber to achieve perfect leveling.

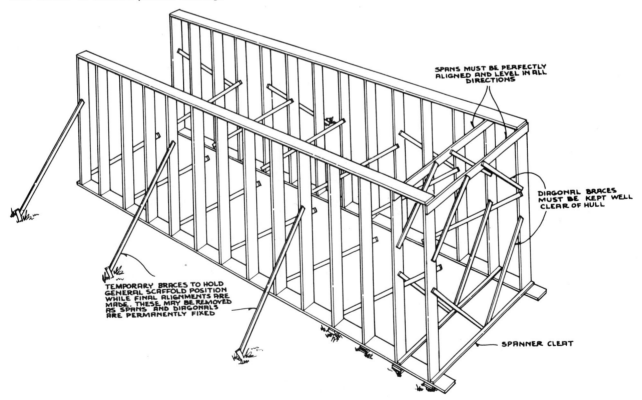

SPANS MUST BE PERFECTLY ALIGNED AND LEVEL IN ALL DIRECTIONS

DIAGONAL BRACES MUST BE KEPT WELL CLEAR OF HULL

TEMPORARY BRACES TO HOLD GENERAL SCAFFOLD POSITION WHILE FINAL ALIGNMENTS ARE MADE. THESE MAY BE REMOVED AS SPANS AND DIAGONALS ARE PERMANENTLY FIXED

SPANNER CLEAT

Fig. 121. Cross spans and diagonal braces being attached to the scaffold siding.

Scaffold Alignment

It cannot be overstressed that the ultimate product of building the scaffold is perfect alignment. Regardless of what the remainder of the scaffold does in terms of warping or settling, it is the ultimate position of the cross span with which we must be most concerned. It is from these cross spans that the vessel's station molds will be hung, and critical measurements taken. The under edges of the cross spans usually represent the baseline (as seen in most of the construction drawings) and it is from this baseline that all vertical relationships are derived. It is wise to measure and indicate the L.W.L. from the baseline on each station upright as well as to mark the center line on all cross spans, as they will supply constant check points throughout the building project. (Plate 35)

Scaffold Bracing

Caution should be employed when attaching transverse diagonals so that they do not interfere with the hull at each respective station. It is impossible to overbrace, even at the risk of having to reposition as the hull takes form, and it makes little difference as to whether braces lie in- or outboard of the scaffold.

Working Platform or Stage

Whether building a large hull or small, every inch of the vessel should be conveniently within an arm's reach without tiresome extensions, particularly during the stapling and wire-tieing stages. On boats of 45', two tiers will most generally be required but on smaller yachts the working platform may be lifted at the ends to compensate for the sheer curve.

The stage base may be either of 2" x 4" cantilevers attached to each upright or "H" frame extending fully to the ground level and should be wide enough to support at least two 2" x 12" planks. It is important not to squeeze the stage at the beam positions as it will hinder free movement of the worker as well as cause difficulties in passing material. (Fig. 122) In the forward and stern sections where there may be excessive flare, the stage should be somewhat wider than amidship to allow working along the sheer without having to bend backwards while still being fairly close to the hull at its own plane. At least 6" of space should be left between the stage planking and the hull so as not to interfere with wire-tieing and plastering, and somewhat more would be ideal for the clearance of the mesh rolls. (Plate 36) There should be a ladder or stairway to the stage positioned at each end of the vessel or on opposite sides of the hull amidships. In addition to the stage the builder should also construct ladder rungs between a pair of scaffold uprights on each side of the boat to allow easy entry into the mold.

Vertical Measurements

In all of my designs as well as those of most other architects, an overhead baseline is incorporated to allow an accessible reference point for taking measurements inside of the hull. If the baseline is under the vessel, interior measurements are complicated by virtue of the hull creating its own impenetrable obstacle. As I have already mentioned, the scaffold cross spans may be used as a constant reference plan whereby the builder may drop a tape or plumb bob to any given point.

The most convenient method, however, is that of affixing a

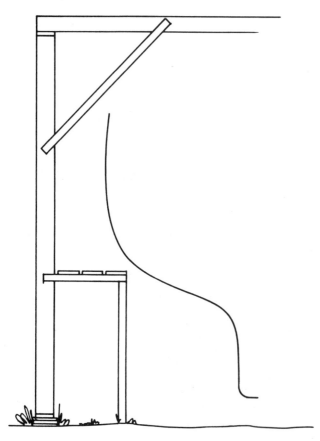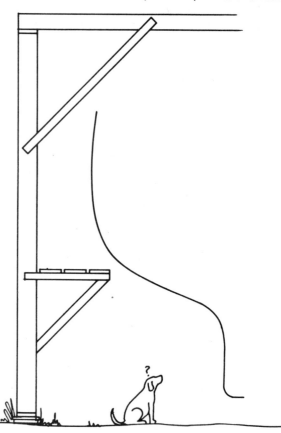

Fig. 122. Scaffold staging.

Plate 35. The scaffolding for a Bingham-designed 46' **Capella** nears completion as longitudinal and transverse diagonal braces are installed. All scaffold parts can be clearly seen.

Plate 36. Plenty of unobstructed working space is vital to good workmanship and safety. This **Doreana** stage is a prime example.

135

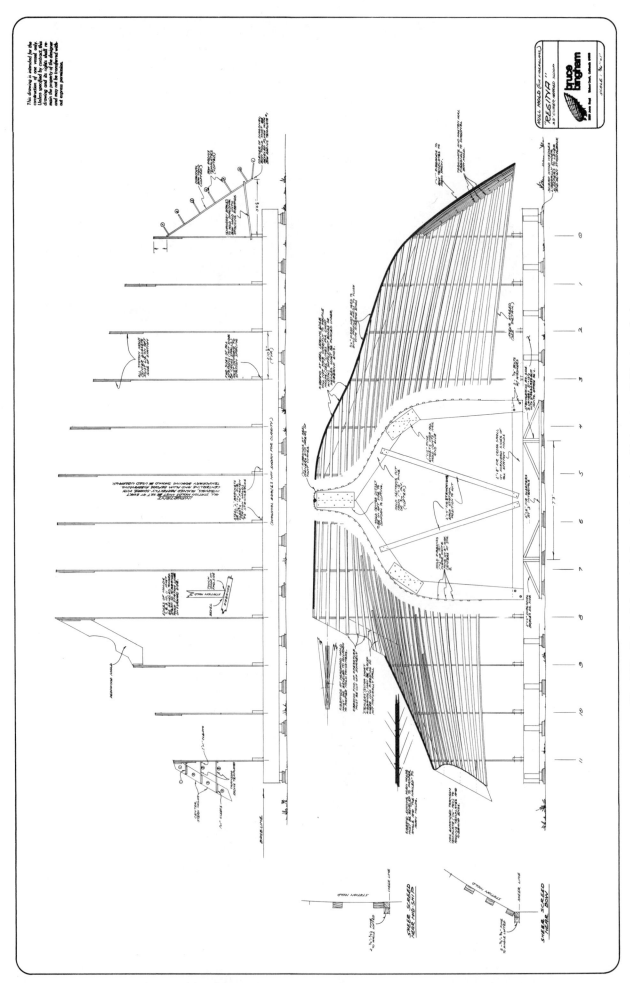

Fig. 123. Regina's inverted strongback and mold drawing. A Bingham design.

series of yardsticks onto one of the scaffold uprights so that their marks accurately represent the distance to the baseline. A long, clear, flexible plastic hose is now partially filled with water and food coloring and one end temporarily attached to the scaffold upright so that the water level in the hose corresponds to the desired measurement to be transferred to the hull. The loose end of the hose may now be carried to any position inside or outside of the boat for the purpose of transferring the given dimension. No matter how high or low, to what distance or position the hosing is carried, the water level will remain constant to its pre-set position. This simple measuring device may be used in such ways as positioning thru-hulls, marking joinery, drawing the waterline, establishing sole beam positions or however else the builder desires and can save hundreds of hours of time. It should also be noted that a set of yardsticks may be installed at some point inside the boat to eliminate needless travel during interior phases of construction.

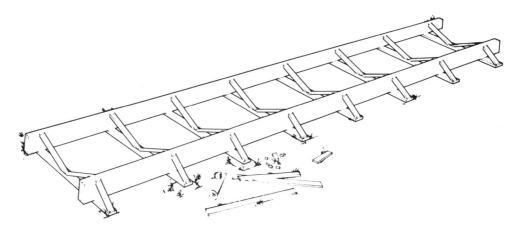

Fig. 124. A completed strongback, sometimes termed as "skids."

Plate 37. A strongback or skid substantially constructed to receive the inverted station molds.

137

Skid (Inverted Construction)

This term refers to the supportive structure comprised of transverse sleepers and longitudinal strongbacks upon which will rest the station frame cross-spalls. For skid construction we will first lay support blocking at close intervals directly under the centers of each strongback position. These supports must be perpendicular to each other to allow for the resting of the transverse sleeper timbers. Once these have been positioned and aligned for level in all directions, the strongbacks are attached, edges up, and held in the perpendicular position through the use of short diagonal braces. Because it is the strongbacks which will ultimately bear the entire weight of the hull, they are generally constructed of two layers of 2″ timbers. (Figs. 123 & 124) The station positions are now carefully marked along the upper strongback edges taking measures to ensure that they are perpendicular to the center line. (Plate 37)

For very small boats (less than 10,000 pounds), a strongback system may not be necessary, most particularly if the vessel is being built on paved ground. In this instance, the station mold timbers will simply be extended beyond the sheer to a point corresponding to a baseline. As each station mold is erected, keel up, support blocking and wedges are inserted under the mold timbers. As successive stations are raised, they are blocked in a like manner, leveling and plumbing by sighting along the edges of the station cross-spalls. Assurance of squaring to the center line is accomplished by duplicating diagonal measurements from the sheers of station pairs. (Plate 38)

Scaffold Variations

If two or more vessels of the same type are to be constructed on the same side, it is quite possible to combine these scaffolds so that they share common sidings, therefore eliminating a considerable duplicate effort. (Plate 39)

When building small vessels (under 10,000 pounds), it is quite possible to substitute the wood scaffold system with that of pipe and specially designed connectors. This is particularly advantageous when building open-truss and pipe-frame constructed hulls. (Plate 40) Pipe scaffolding, however, is not recommended for the use of wood station systems as their attachment to the cross spans is complicated.

In Chapter 8, I indicated that the scaffolding can be totally eliminated for the construction of vessels under 20,000 pounds by supporting the bottom of each station on a well-supported strongback; the molds hold a proper alignment through the liberal use of longitudinal and transverse diagonal bracing. (Plate 28) The greatest care, however, should be exercised in choosing the building site as ground settling would be absolutely disastrous.

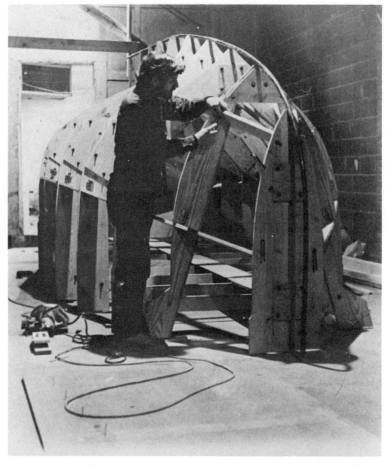

Plate 38. A Bingham-designed **Flicka** constructed in the inverted position, using a level floor as the construction base. Note station extensions.

Plate 39. Two scaffolds joined by a common side may reduce cost of construction slightly.

Plate 40. Three open truss armatures being constructed simultaneously using a specially designed welded pipe scaffolding. (Courtesy, The Continental Corporation)

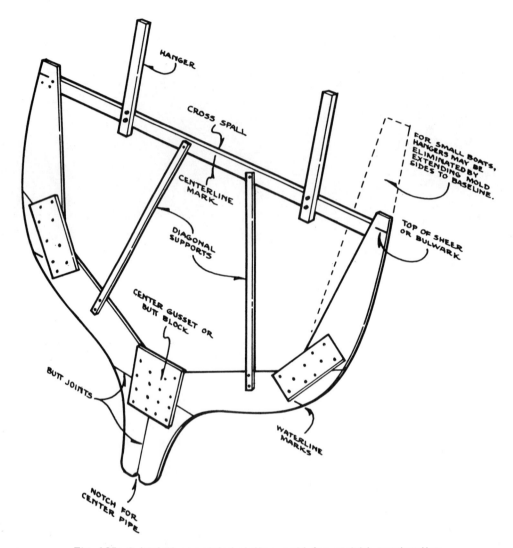

HANGER

CROSS SPALL

CENTERLINE MARK

DIAGONAL SUPPORTS

CENTER GUSSET OR BUTT BLOCK

BUTT JOINTS

NOTCH FOR CENTER PIPE

FOR SMALL BOATS, HANGERS MAY BE ELIMINATED BY EXTENDING MOLD SIDES TO BASELINE.

TOP OF SHEER OR BULWARK

WATERLINE MARKS

Fig. 125. A typical completed station mold for upright construction.

12

wood molds

Stations

The station molds are the prime dictating factor for the finished shape of the hull and must be constructed accurately and should be of such strength to allow for the support of the builder as well as the progressively increasing weight of steel and concrete. They may be built of 1″ x 12″ pine or scrap plywood and most generally incorporate used lumber. It should be noted that if plywood is employed, you may run into some difficulty as staples may not adhere well to the end grain. The construction procedure may take one of two basic forms; i.e., joining to the station segments after shaping, or joining before shaping. The following paragraphs will describe each method.

Cutting before Joining

The relatively straight curves of the forwardmost stations will usually only require one plank upon which you will transfer the station shape including the sheer, waterline and center-line marks. Once so marked, cut away the excess wood, being sure that you have allowed no less than four or five inches of width at the narrowest portion of the curve. If you inadvertently cut to the inside of the mold line, it will require that you build up this deficiency with plastic wood or whatever else will serve this purpose. If you have cut to the outside of the line, plane or sand to the correct shape. By running your hand lightly along the finished cut edge, you may discover slight flaws not visible to the eye. Continue to fashion the station until it feels absolutely smooth. Remember that any errors in the station mold, however slight, will be mysteriously magnified upon the application of your rod and mesh. Once done, carefully cut the joint at the bottom center then replace the mold atop the lofting again to check for accuracy. If you find an existing error in the position or angle of the center line, it must be immediately corrected. When you are satisfied with the finished shape, use it as a template for scribing its opposing side onto another plank. When cutting the duplicate section, remember to allow for the pencil thickness.

Carefully measure from the lofting the width of the sheer to the edge of the mold at the station being constructed. By doubling this dimension, the total distance between the sheer marks may be determined prior to attaching a cross-spall. The center of the mold at the keel is now butted together and aligned, then joined with a plywood or 1″ x 12″ gusset or butt block. If the vessel being built requires a center pipe or other backbone structure, it may be necessary to cut a notch at the base of the mold to receive this unit. Before you set the finished mold aside, be sure that it bears indicators at the waterline and sheer as well as the center line on the cross-spall.

When the shape of the station precludes the use of single plank to achieve the entire curve, it will be necessary to cut and attach them in progressive segments. I feel it is best to start at the keel laying the first plank so that it covers as long a station portion as possible. Transfer the shape and fashion as previously described, allowing the inner edge of the plank to exit the station shape. Once done, replace the station segment atop the lofting in order to scribe the plank edge which will be transferred to the next segment as a cut line for joining. Successive segments are transferred and fashioned in the same manner, taking care to cut the lower end to correspond with the preceding segment edge. The segments are once again aligned on the lofting and the two pieces attached with a gusset or butt block. To complete the shape of the entire station, continue this procedure using as many planks as necessary.

After you have faired the station mold and have checked it for accuracy, duplicate and join the opposing station segments in the manner previously described. (Fig. 125)

Cutting after Joining

This method simply requires the butting of progressive planks so as to completely cover the required station line allowing, of course, for a minimum width at the narrowest point of at least four inches. Transfer the mold line, with all indicators, to the rough wood station shape and cut as required for absolute accuracy. The opposing station sides will be joined with a gusset at the center as well as a cross-spall from sheer to sheer. In all respects the finished product will be the same as if the segments had been cut before joining and the system has the advantage of not having to be as meticulously fit, but the drawback seems to be having to cut an immense length of timber in one shot with the chance of running into nails along the way. Flopping the stations over for duplicating is usually a two-man job. (Fig. 125)

Beveling

Because only one side of the station mold can represent the actual station plane, it is necessary to bevel the edges of the station molds when they are positioned on the side of a diminishing curve as it is obvious that the mold lumber would then lie slightly outboard of the desired shape. The angle of the bevel need not be accurate so long as it's equal to or exceeds that caused by the tapering of the hull. This may best be accomplished after the station molds have been erected as it will allow you to spring a temporary batten as a check for this inherent discrepancy. For this purpose, I strongly suggest the use of a belt sander or electric plane. (Fig. 126)

Beveling may be avoided by making a slight alteration in the station mold positions when they are erected. Normally, the forward sides of all station molds are usually placed exactly on the station line. If, however, the after station molds are moved forward by a distance of their own thickness, then the aft sides of the molds will represent the station and will, thus, be located on the station line. Personally, I have no preference one way or the other, but I feel that it injects the possibility of getting one's numbers mixed.

to use a variable radius from one station to the other so as to intersect a given height and sheer. While it may require the additional work of plotting the individual deck shapes, the result will always be a more pleasant deck appearance.

Variable Radius Camber

This may be established accurately and quickly using two well-known methods. The first requires that a pencil be attached to a length of long wire (not string) which is used as a "compass," allowing a circle to be passed through the

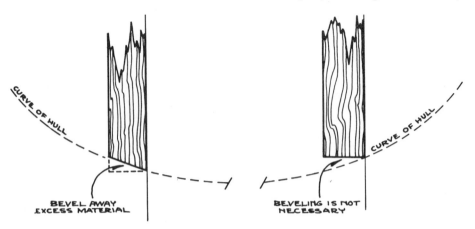

Fig. 126. Necessity for beveling station molds.

Cement Deck Mold (Upright Systems)

If the vessel is to have a cement deck, a camber mold will replace the normal cross-spall and is to be aligned so that it will intersect the sheer or deck line at the sides of the mold. (Fig. 127) In some plan packages, the development of the crown curve is described as a constant shape throughout the length of the vessel, but any professional builder will tell you that this crown system will ultimately produce a flat spot at the foredeck unless the height of the curve is very slight. For this reason, I have chosen in most of my designs

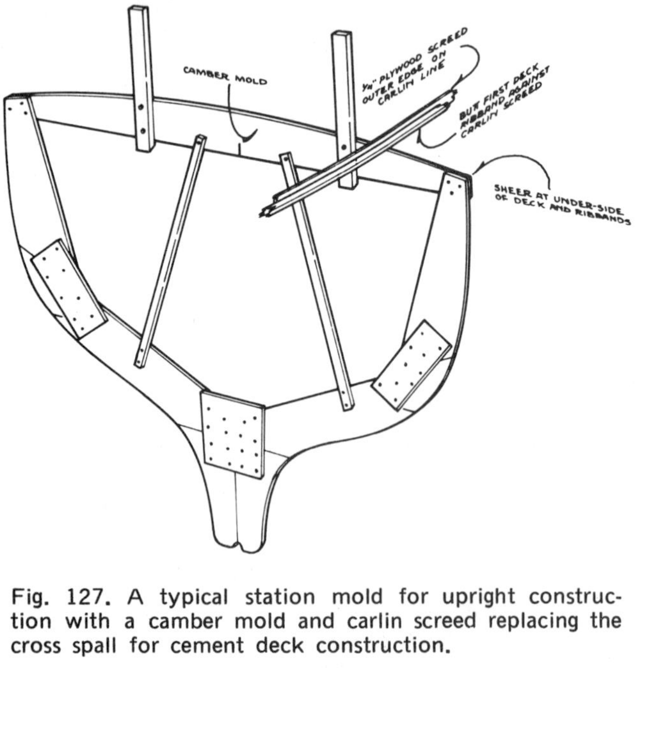

Fig. 127. A typical station mold for upright construction with a camber mold and carlin screed replacing the cross spall for cement deck construction.

center crown height mark and the sheer. In large boats, it may require a wire as long as 15 or 20 feet and the adjustment of its proper length must be done on a trial and error basis. Another system is that of driving nails corresponding to the proper width of the sheers. Two boards are placed against the two nails so that they cross at the center crown height where they will be temporarily nailed to the established angle. By placing the pencil point at the intersection of the two boards, the boards may be slid along the nails from side to side, thus scribing a perfect circular segment. This must be done for each individual crown position. (Fig. 128)

Constant Deck Camber

This may be drawn in the following manner: Scribe a horizontal line representing the beam at the widest point of the boat; then divide the line into four equal segments with vertical lines. From the intersection of the horizontal line and center vertical, swing a radius which represents the height of the crown as recommended by your architect. Divide the base of one side of the radius into four equal segments; then divide the quarter perimeter into four equal segments. Connect the base marks with the perimeter marks, thus producing a series of lines being vertical at the center while increasing angularly outboard. Measure the lengths of these connecting lines, mark to mark, and transfer the dimension to the respective segment verticals. As you now have a height at center as well as decreasing height marks on the vertical segments, they may be interconnected with a curve which intersects the sheer baseline at the maximum beam mark. This curve should be duplicated for its opposing side. (Fig. 129)

The finished pattern may be used to establish the crown at any other position along the deck. It may represent the open edges of bulkheads, molds, either upper or lower edge of tapered deck beams (if wood decks are to be used), or in

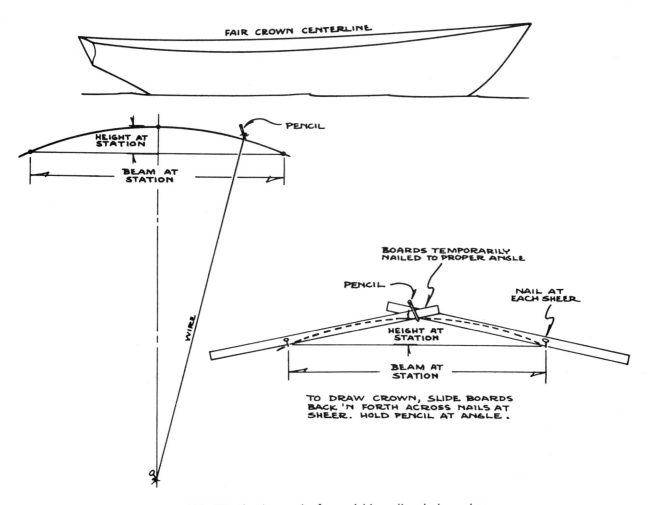

FAIR CROWN CENTERLINE

HEIGHT AT STATION

BEAM AT STATION

PENCIL

WIRE

BOARDS TEMPORARILY NAILED TO PROPER ANGLE

PENCIL

NAIL AT EACH SHEER

HEIGHT AT STATION

BEAM AT STATION

TO DRAW CROWN, SLIDE BOARDS BACK 'N FORTH ACROSS NAILS AT SHEER. HOLD PENCIL AT ANGLE.

Fig. 128. The development of a variable radius deck camber.

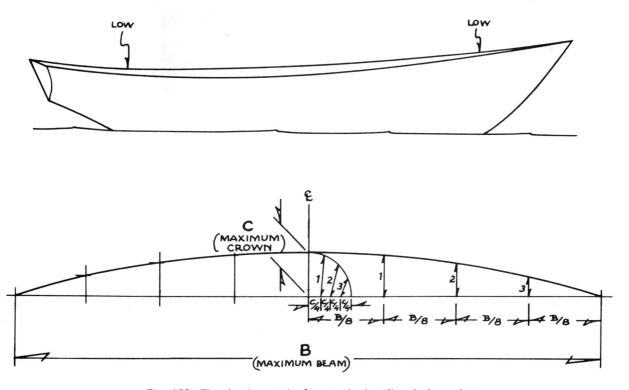

LOW   LOW

℄

C (MAXIMUM) CROWN

1 2 3   1   2   3

$\frac{C}{4}$ $\frac{C}{4}$ $\frac{C}{4}$ $\frac{C}{4}$

B/8   B/8   B/8   B/8

B (MAXIMUM BEAM)

Fig. 129. The development of a constant radius deck camber.

143

any other manner which requires the use of a crown shape. To establish the curve at a position other than the widest beam, simply measure the half-breadth from each side of the pattern center to the intersection of the curve. (Fig. 130)

Cement Deck Mold (Inverted Systems)

The construction of the cement deck mold for upside-down construction is considerably more complicated than that of the upright method. This will consist of building the station

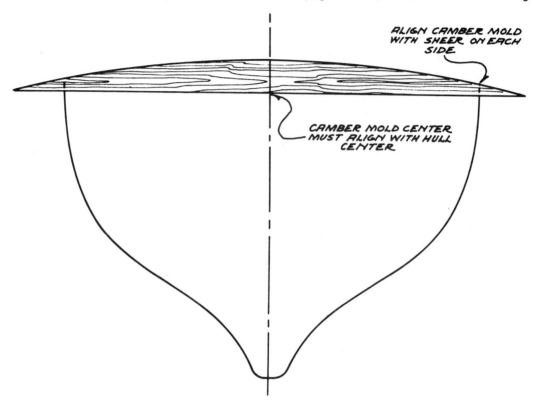

Fig. 130. Using the constant camber mold to find the deck crown at any position of the hull.

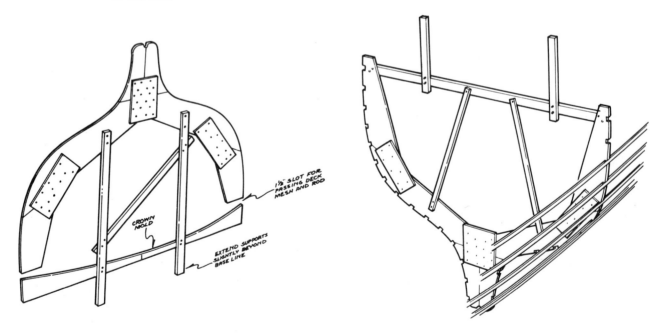

Fig. 131. A typical completed station mold for inverted hull and cement deck construction.

Fig. 132. A wood station mold notched to receive ribbands.

in the normal way, terminating the upper edge of each station at the underside of the deck line. A concave deck mold is now shaped (as previously described) for the given station and attached to the cross-spall with spacers or butt blocks so that a gap of approximately 2" exists between the camber mold and the hull mold at the sheer. This gap will allow for the eventual insertion of mesh and rod, which will be blended into the hull armature. Once the stations have been erected onto the skid or strongback, a 1" x 2" carlin screed is attached per plan and the concave deck shape planked or covered with plywood to achieve a smooth outer surface. This will be explained more fully within this chapter. (Fig. 131) (Plate 41)

Notched Molds

These are sometimes preferred by builders for reasons I have never understood. It does facilitate the flush attachment of mold ribbands which results in a beautiful structure but adds little to strength or function. By prelocating batten positions in this manner, it precludes the ribband from taking its own natural shape which introduces some strange visual eruptions, thus impairing your perspective when fairing your mold. This notching or rabbeting takes a great deal of time and should be reserved only for the construction of wooden boats with bent frames. Even here, this exercise is of highly questionable value. (Fig. 132)

Small Vessels

On very large hulls, the station molds are most generally suspended from the scaffolding by means of timbers attached to the station cross-spall, as the distance from the base to the sheer may become quite excessive amidships. (Plate 48) In smaller vessels, however, this distance may be only a foot or two, thus creating unnecessary work. My

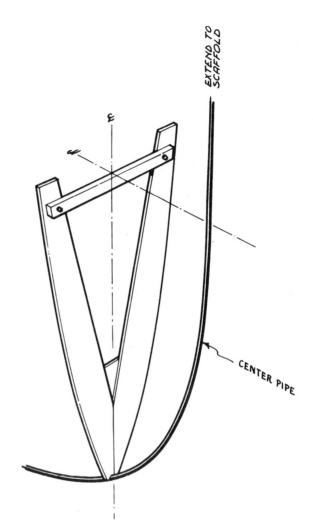

Fig. 134. A pipe may be fairly bent for the construction of a spoon or plumb bow on small vessels without the necessity of using a stem mold.

suggestion, therefore, is that the station mold timbers be continued above the sheer to allow their direct attachment to the scaffold cross spans. (Plate 42) When the small vessel is built upside down, the station mold extensions may be cut flush with the baseline allowing them to be set directly on top of base support timber or support blocks, thus precluding the necessity for a strongback or elaborate skid system. (Plate 38)

Spoon Bow

If the bow of a vessel comes to a point, or nearly so, the bow mold may be simply a piece of wood cut to the shape of the profile and attached perpendicularly to station "O" with the upper end held at a given distance forward by diagonal braces from the cross-spall of station "O". (Fig. 133) If the vessel uses a center-pipe or bar, it will be prebent to the hull profile, and attached to the center bow mold with wire only **after** the application of the two center mesh strips (described later).

If the vessel is quite small (say, under 25'), it is possible to eliminate the attachment of the wood bow mold. This may be accomplished by accurately prebending the center-pipe to the shape of the stem. The upper end of this pipe, when installed, should extend a considerable distance above the

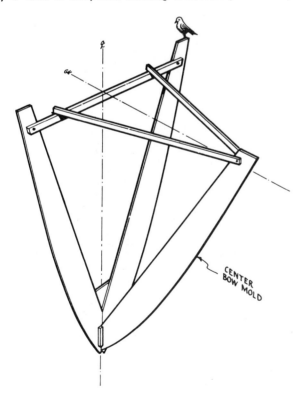

Fig. 133. A typical mold for the construction of a pointed spoon bow.

Plate 41. An inverted deck mold, planked, covered with poly and ready for rod and mesh. Station molds will be attached after completion of deck so as not to hamper construction.

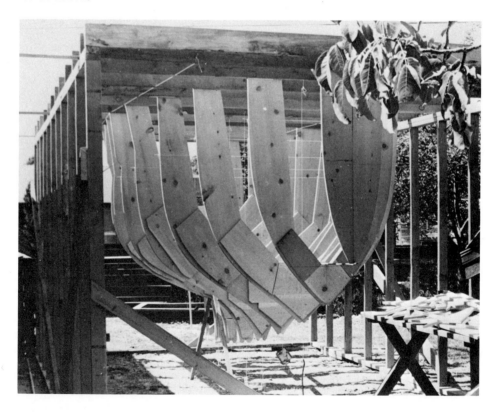

Plate 42. The 20' **Flicka** station molds are extended beyond the sheer to act as mold hangers, eliminating more complicated suspension. A Bingham design.

sheer with provisions for bracing or wiring in order to prevent its movement. (Fig. 134)

Barrel Bow

In the case of the construction of a barrel bow, we face a much more complicated procedure. A diagonal center mold must first be cut, which will look like a very narrow station mold. At predetermined intervals, radius templates will be attached perpendicularly to the center mold which shall establish the changing sections of the stem. The entire bow mold assembly must be "toenailed" to the bottom of station "O" and held to the proper angle by diagonal braces running from the bow mold assembly to the cross-spall on station "O". Vertical ribbands will be nailed to the radius

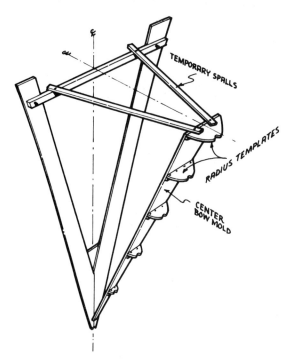

Fig. 135. A typical barrel bow mold.

templates for the full length of the stem assembly, while the hull mold is terminated at the diagonal bow mold. (Fig. 135)

Clipper Bow

The construction of a clipper bow is the combination of the spoon and barrel. A center profile mold is cut and attached to a transverse diagonal form. Radius templates dictating the round of the bulwark above the billet head are then fastened to the diagonal mold. (Fig. 136) Short vertical staves are attached to the radius templates while the major hull ribbanding is attached to the longitudinal center profile or stem mold. (Fig. 137)

Dinghy and Flat Transoms

These configurations are, by far, the easiest to build as they are simply extensions of the hull taking the form of a normal station which may be plumb or raked. In order to provide the smoothest possible flat surface, the transom mold may be constructed of a solid sheet of plywood, covered with wax paper to provide an easy separation of the concrete, assuming that the mold is left in the hull on

plastering day. This will raise an obvious objection such as the difficulty of assuring penetration against a barrier. It

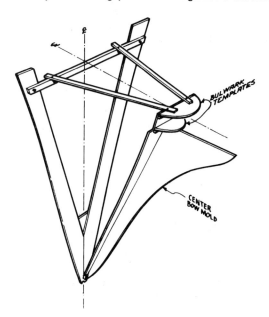

Fig. 136. A typical clipper bow mold.

will be necessary then to vibrate the mortar thoroughly in this area which may still leave inadvertent voids to be back-plastered on a later occasion. (Plate 43)

To avoid the penetration problem, the dinghy transom mold may be constructed in the same manner as all previous stations. Once the armature has been well tied, this mold can be removed entirely from the hull and steel angles wired temporarily to the inner face to prevent distortion during plastering.

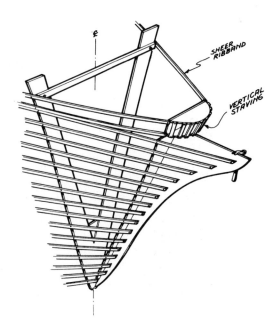

Fig. 137. Ribbanding a typical clipper bow.

147

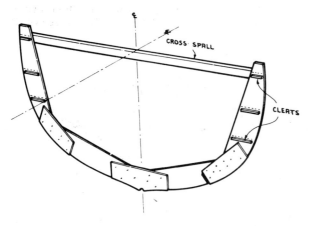

Fig. 138. Cleats being attached to the aftermost station mold to effect the attachment of the transom radius templates.

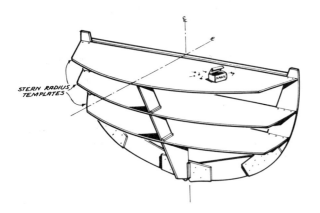

Fig. 139. Transom radius templates attached to the aftermost station mold. Note the center transom mold providing for the proper horizontal alignment of the radius templates.

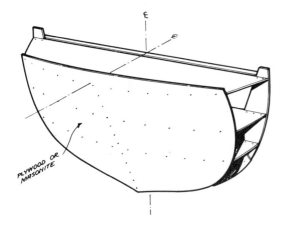

Fig. 140. The expanded plywood transom template attached to the finished transom mold.

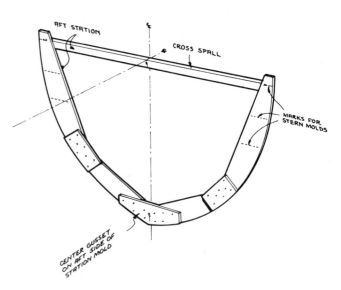

Fig. 141. The locations marked for the attachment of the horizontal stern template cleats onto the aftermost station mold.

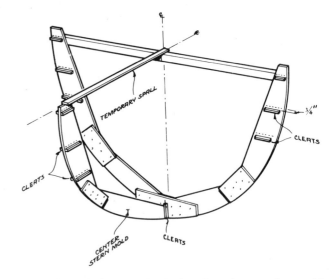

Fig. 142. Attaching and aligning the stern center mold.

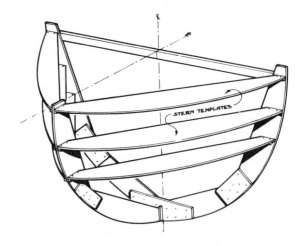

Fig. 143. Attaching the stern templates to the center mold and the aftermost station mold.

148

Counter Transom Sterns
(Long, Short, Plumb, and Reverse)

This mold construction constitutes several horizontal templates bearing the transom radius and corners which are attached to the aft side of the aftermost station. With the use of corner cleats these templates most generally occupy the positions of the hull waterlines which intersect the stern and are held in the horizontal positions by the use of vertical center spacers. Once the transom templates have been attached and aligned, a ¼" plywood form is sprung to the established radius and nailed into position. This form may extend approximately ¾ of an inch beyond the corners of the radius templates for the purpose of receiving an attachment of butt ends of the hull ribbands. Some architects call for vertical staving to be attached to the radius templates instead of a plywood sheet but it will make little difference in the overall effect. (Figs. 138-140) (Plate 44)

If the design calls for a bulwark which lies on the same plane as the transom angle, no additional construction will be required. If, however, the design calls for a reverse bulwark angle, it will be necessary to cut and attach several small vertical templates reflecting this angle. These molds will be spaced longitudinally across the top of the uppermost transom radius template. A narrow strip of plywood is then sprung around the shape of the bulwark templates, thus producing the finished transom sheet. (Plates 45 & 46)

Many architects and builders prefer to prefabricate a pipe frame shaped to the transom corner with additional horizontal spanners bearing the proper transom radius. This structure may be aligned in position upon the aftermost station mold with the use of wood bracing, or may be welded directly to the center pipe and held in its proper position with temporary pipe diagonals.

Cruiser and Canoe Stern

This construction consists of several horizontal templates bearing the stern waterline shapes attached to the aftermost station. They are held level through the use of a number of center vertical spacers. (Figs. 141-143) (Plate 47) The ribbanding and fairing of these configurations will generally require that the hull ribbanding be terminated at the last mold at which point diagonal or vertical staving will be applied. In order to create a smooth blend between the two ribband networks it may require that they be saw-scored on their inner face to facilitate their natural adjustment to fairness.

The Knuckle Counter Stern

This type of aft section was most commonly seen on the clipper ships of the late 1800's and is particularly handsome on large yachts of a character nature. Its mold construction consists of several horizontal templates attached at the hull waterline positions held level with vertical center spacers. A series of small vertical templates are attached to the uppermost waterline mold so that their axes are roughly perpendicular to the shape of the sheer as seen in the plan view. The portion of the hull below the knuckle will be ribbanded in the normal manner, although liberal saw-scoring may be required. The portion of the stern above the knuckle, however, will require that horizontal ribbands be bent around the stern radius. Due to the degree of this curve, the ribbands in this area may require lamination. (Fig. 144) A variation to this procedure is to cut a waterline mold representing the sheer. Vertical staves then connect the waterline mold and the stern knuckle mold. (Fig. 145)

Hanging the Molds

Once all of the stations, bow and stern molds have been constructed and properly beveled, they must be suspended from their respective scaffold cross spans. Attach tight-strings across the waterlines and center lines of each mold for the purpose of sighting transverse and vertical alignment. Lift each station mold with ropes, measure from the baseline to the waterline and adjust the mold to its proper height. Temporarily nail the mold hangers to the cross-spall.

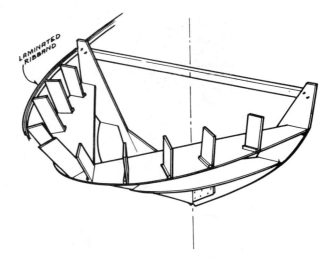

Fig. 144. Construction of a knuckle counter stern (method 1).

It may take a considerable amount of leveling and adjustment to position the stations properly; do not drill the hanger bolt holes until you are completely satisfied with your positioning. To prevent the station mold from swaying fore-'n-aft, attach scrap diagonal bracing from one station to the next, being sure that each is perfectly plumb when doing so. (Plate 48)

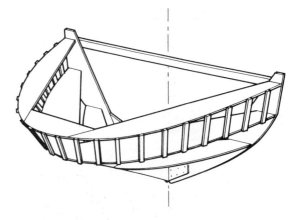

Fig. 145. Construction of a knuckle counter stern mold (method 2).

Raising the Stations (Inverted Systems)

If a small hull is being constructed, the station mold will have been extended and cut at the baseline. They may be raised one at a time to the vertical, plumbed over center line, adjusted perpendicularly and temporarily placed in position. Succeeding stations are erected in the same manner, using waterline and center-line tight-strings for

Plate 43. The "dinghy" type transom on the Bingham-designed **Flicka** is the easiest of all to build, as it requires no radius templates, only a flat surface. This builder could have used a solid sheet of plywood.

Plate 44. The popular Benford 17' catboat employs the most common counter stern with a slight radius. Note transverse radius templates at 90° to transom, which will be cut flush with the sides of the hull. What a beautiful little ship she'll be!

150

Plate 45. **Doreana's** reverse transom bulwark templates and plywood mold (viewed from port side forward).

sighting alignment through the mold. Place wedges under the mold if necessary to correct heights or transverse level. (Plate 38)

If large vessels are being constructed, the molds are erected in much the same manner, except that the station cross-spall will be positioned at the baseline. Each mold is inverted, the cross-spall set atop the skid or strongbacks and the station temporarily braced to the vertical. (Plate 37) When all stations have been erected in this manner, adjust for proper alignment and plumb, sighting along the water-line and center-line tight-string. When you are completely satisfied with your positioning, toenail the station cross-spalls to the strongbacks to prevent fore-'n-aft sway, fasten scrap diagonally, brace from mold to mold.

The Deadwood

This is the area at the aft and narrowest point of the keel. In wooden hulls it is always constructed of solid timber as the reduction of the vessel's section precludes the possibility of planking.

The substantial construction of the deadwood serves several important functions: one is that of aligning the shaft log and of supporting stern and rudder hardware, to withstand the impacts of grounding and of supporting a major weight of the vessel when dry-docked. The importance of the dead-wood area has not been minimized through the advancements in ferro-cement construction.

Because the deadwood weldments are generally of an integral nature, they must be installed within, and as a part of, the mold system and hence its discussion in this chapter.

Plate 46. The **Doreana** elliptical counter transom with reversed bulwark. This hull is ready for mesh and rod. A Bingham design.

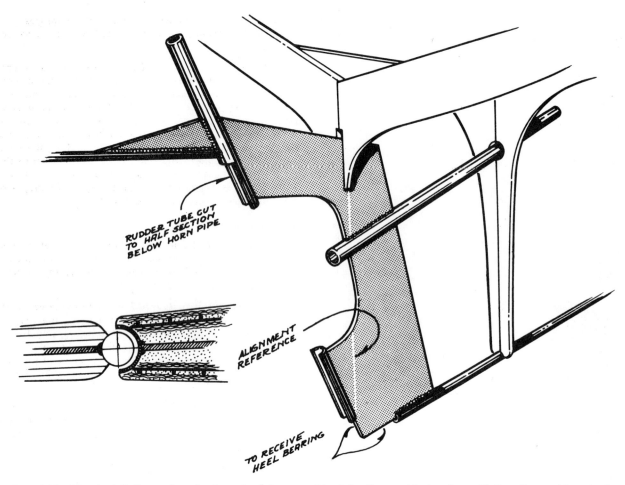

Fig. 146. The installation of a deadwood plate assembly into the mold structure. Notice the rudder stock pipe out into a half section below the hull. The finished deadwood section reveals the least possible amount of deformation at the rudder stock for increased steering efficiency.

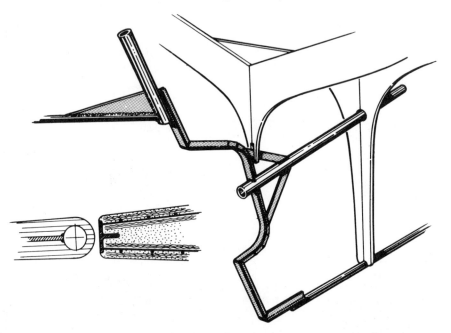

Fig. 147. "T" bar deadwood weldment, most commonly found on ferro-cement vessels. A laminated plywood rudder blade has been faired to a rounded shape, to decrease the water turbulence between the deadwood and the rudder's leading edge.

In my own designs I prefer to use a steel plate weldment, cut to the profile and the shape of the propeller aperture. This plate is usually fabricated from three separate sections; i.e., the portion below the stern tube, above the stern tube and a small piece aft of the rudder post. When the entire weldment is assembled, it will house the stern tube and rudder tube in one prealigned unit.

Because the deadwood plate usually intersects a station, that mold will have to be notched to allow the passage of the plate. Alignment marks transferred from the lofting are used to facilitate proper positioning with the mold system. On large vessels, the deadwood unit may weigh several hundred pounds thus requiring both suspension from the scaffolding and blocking beneath so as to avoid the placing of excessive weight on the mold system. Once the deadwood has been installed, the center and horn pipes (or other center structures called for by your designer) will be welded to it, creating one continuous member from stem to stern.

You will notice in the illustrations that the rudder tube is cut into a half-section below its protrusion from the hull. By doing this, the tube will serve as a fairing for the leading edge of the rudder post. This will reduce resistances measurably while increasing the steering forces and sensitivity. (Fig. 146) (Plate 49)

Some architects have elected to fabricate the deadwood section by using angle iron, re-bar or reinforcing rods. In the latter case, its construction will not proceed until such time as the armature rods are placed on the mold. The angle iron and re-bar, however, should be considered as a part of the mold (as with the deadwood plate system) and will require installation at this time. (Fig. 147) (Plate 50)

Center-Pipe

Prior to applying the first mesh to the hull mold, it is generally required that a pipe backbone be installed along the vessel's profile. Its purpose is not considered as structural once the hull has been plastered but it does provide a stiff center member over which the mesh and reinforcing rod may be easily and fairly bent. Without the center-pipe, there is a high risk of creating corrugations between the station molds along the vessel's profile. The center-pipe also provides a method for attaching certain integral units to the hull armature such as the stemhead, deadwood and rudder tube.

It is quite obvious that if the center-pipe is simply covered with concrete on plastering day, there is a possibility that cracks will develop along the center of the hull. Upon severe impact, the pipe may even break through from the inside of the shell. It is necessary, therefore, that the concrete layer covering the pipe be reinforced to prevent these happenings.

Prior to installing the center-pipe, two narrow strips of mesh should be bent into a long "U" section and installed along the center line, pushing the mesh firmly into the center-pipe notch at each mold. The center-pipe, carefully bent to proper shape, is now placed into the mold notches and over the center mesh strips. As it will be impossible to purchase the pipe in a continuous length, short sections must be shaped, installed and butt-welded as separate operations. Judge the pipe carefully and correct as necessary to produce long, flowing curves, free of bumps, hollows and flat spots. At the point where the center-pipe overlaps the deadwood plate, weld for a permanent installation.

Upon placing the center-pipe, lace the inner mesh strips tightly to the pipe member with soft wire. Now fold the protruding mesh edges back against the station mold and staple. (Fig. 148) (Plate 51)

Some designers and builders elect to use a reinforcing bar as a center member, rather than pipe because it is generally easier to bend. The use of a bar will not add or detract from the strength of the center member and can be considered as interchangeable with the center-pipe specifications. In any event, it must not be installed without first providing for the inner mesh covering as previously described.

Plate 47. The cruiser stern mold construction of the 46' Bingham-designed **Andromeda** ketch.

"Bundle of Rods" and Channel Center Members

Many designers have elected to use other systems for providing strength along the profile of the hull. Jay Benford has coined the term "bundle of rods" to describe his system. In essence, it is a network of longitudinal re-bars vertically and transversely laced with "wiggle rods" or trusses. The arrangement of these rods is always carefully drawn out by the architect and results in an extremely strong unit.

The disadvantage of this system is that it takes a considerable amount of time to construct and may incur distortion unless you are a proficient welder. I will not attempt to fully describe the procedure of fabrication here, as it is always spelled out clearly by the architect, but you can refer to the illustration to form a general idea of its makeup. (Fig. 149) (Plate 52)

It should be pointed out that the bundle-of-rods system is usually associated with the open-truss construction method, but it can work equally well with wood mold systems.

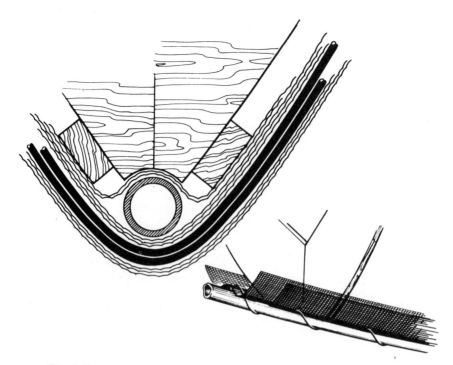

Fig. 148. Installing the center inner mesh strips and center pipe.

Plate 48. The Bingham 46' **Doreana** ketch requires a more substantial mold suspension system than would a much smaller vessel. Here, 2" x 6" hangers are connected to mold cross-spalls, thence to the scaffold cross-spans. Note mold bilge bracing.

Steel channel has been used quite successfully on many ferro-cement designs and has the advantage over the bundle-of-rods system in that it does not require the amount of welding and, consequently, is somewhat faster to construct and considerably stronger. It precludes the possibility of tapering the base of the keel toward the deadwood, however, and should be considered primarily for use on very small hulls and powerboats. Most often, the channel only extends along the base of the keel where a center-pipe then continues forward from the beginning of the rise of the profile. Once again, this center structure system can be successfully employed on wood mold construction systems. It should not, however, replace the system specified by your architect. (Figs. 150-152)

Ribbands (Partial Mold Systems)

Discounting the cost factor, only "A" grade lumber should be purchased. Knots, sap pockets, or severe checks will cause ribbands to over-bend and may incur breaks or cracks. The best lumber is white cedar or white pine. Extremely soft woods (such as redwood) should be avoided and at no time should you reduce the thickness called for by the architect.

As it is impossible to purchase wood stock in lengths that allow a continuous run along the entire mold, several joints will be required. Each joint must be carefully positioned so as not to fall on a station mold or within very close proximity. The joints, from one ribband to the next, must also be well staggered so that joints will not be concentrated in any one area of the hull. The best joint for this purpose is a long diagonal scarf, and it is well worth your time.

As it is extremely difficult to cut a long angle accurately in a thin piece of wood, it is best to clamp about a dozen ribbands together face to face, drawing the desired angle across the edges of all the ribbands at once and cutting them in one swipe. As each bundle of ribbands is cut for scarfing, be careful to maintain the same angle at all times. The angular joints may now be clamped and glued. Little purpose will be served by driving nails or screws into these joints if the angular cuts have been smooth and free of "hair." (Fig. 153)

Another method for forming ribband joints is to glue and nail butt blocks to the backs of butt joints. As the butt block will tend to stiffen the ribband for a short distance causing undesirable flattening between stations, the butt blocks may have to be tapered in order to allow the ribband to take its natural, fair line. When considering the time savings in the method described to cut scarf joints, I feel there is little to be gained by using the butt joints. (Fig. 154)

The spacing of the ribbands should be as close as possible at center station without causing them to overlap at the ends of the vessel. All ribbands along the outward curvature of the hull will inherently converge fore-'n-aft and you are cautioned against fighting this tendency.

It is best to lay the first ribbands along the sheer with their upper edges (upright construction) in line with their respective sheer marks. As no two builders will exactly duplicate their efforts, slight variations may become apparent requiring minor adjustments in height. These discrepancies can be more quickly detected if you sight closely along the ribband, as it will have the effect of foreshortening perspective. The next ribband to be attached should be approximately along the turn of the bilge, and it should be laid in such a way as to reflect the feel or flow of the vessel. All successive ribbands may be applied in whatever sequence you desire, taking care to fasten them alternately, one side then the other, to eliminate distortion caused by concentrated bending stresses. (Plates 26, 30 & 53)

Ribbanding Tight Curves

On some hulls, the curvature of their sections may cause great difficulty in bending ribbands. This may be solved by sawing the ribband vertically along its center so as to produce, essentially, two ribbands in one, which will be sprung and fastened simultaneously. (Fig. 155) If you prefer, you may use two separate strips of wood, each being half the thickness of the ribbands normally required. These pieces need not be glued together.

Plate 49. **Doreana's** deadwood plate installation clearly showing ribband positioning, half-pipe rudder post, horn weldment and prop shaft log.

Saw-Scoring

As I explained in Chapter 10, all curves continue invisibly past the point of their actual termination. This continuation must always be assumed in order to create the proper shape near the ends of all lines. When applying ribbands, however, it is quite impossible to extend them beyond the hull itself, to help shape the hull properly. The result in most amateur construction, therefore, is that their hulls tend to become straighter at their ends than desired. This flatness can usually be detected by sighting closely along each batten. The most notable areas of this problem occur between the forwardmost station mold and the stem, and the aftermost station mold and the stern.

The flattening of the ribbands in these areas can be corrected by "saw-scoring." This is easily accomplished by first

(Text continues on page 160.)

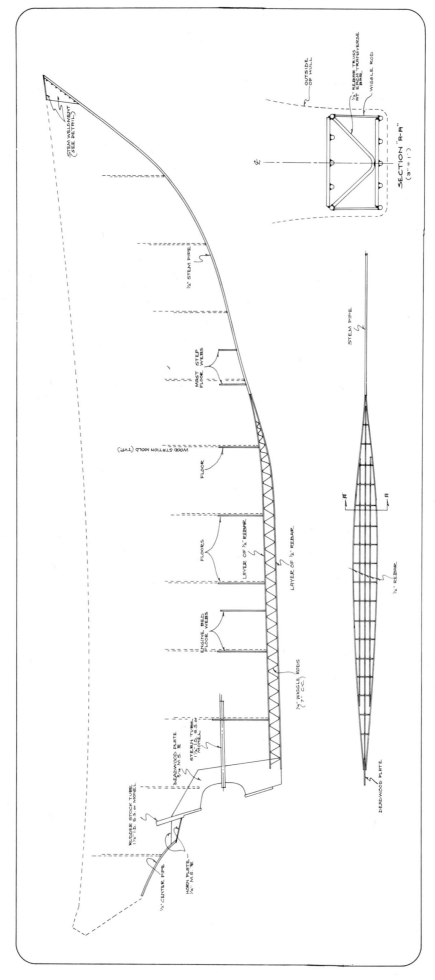

Fig. 149. A typical "bundle of rods" center member on a fairly large hull.

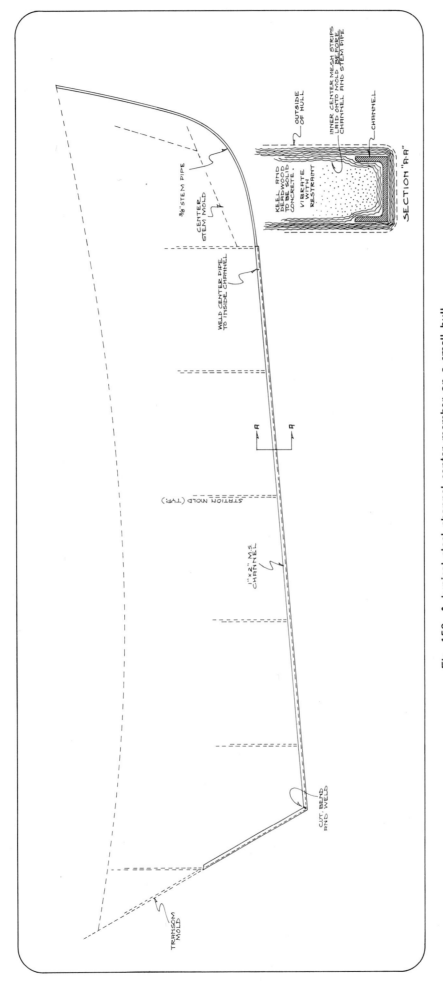

Fig. 150. A typical steel channel center member on a small hull.

157

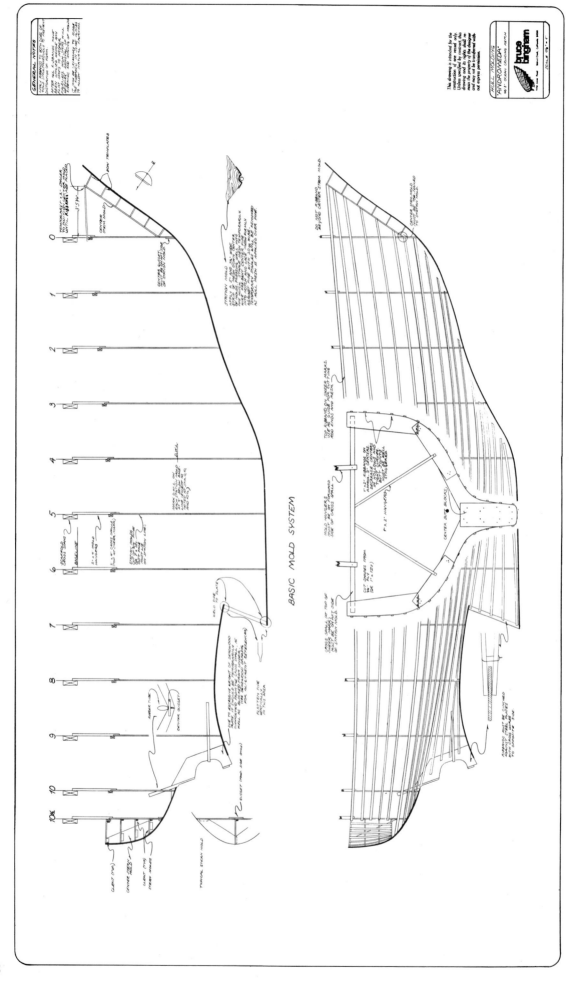

Fig. 151. Andromeda's upright mold drawing.

Plate 50. A "T" bar used on this Sampson double-ender forms the rudder post at the deadwood, the upper shape of the propeller aperture and the stern center member.

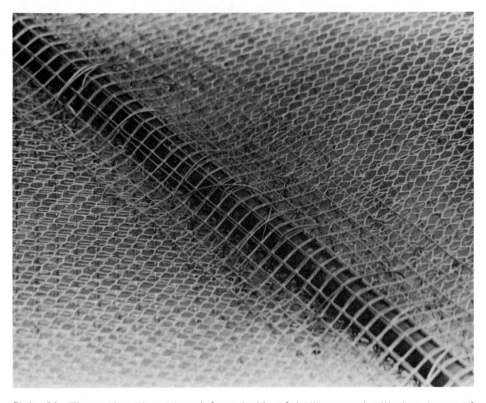

Plate 51. The center pipe (viewed from inside of hull) covered with two layers of mesh and wire laced securely.

(Continued from page 155.)

setting your circular saw to a depth of about ¼". Saw a series of vertical cuts into the batten in question from the inside of the hull. As you do this, the batten will begin to visibly spring outboard. The position and frequency of the saw scores must be determined by the individual circumstances. Generally speaking, the deeper or more closely spaced, the more rapid the curve of the ribband will become. Never saw-score the ribband on the outside of the curve as it will inevitably cause splitting or breaking of the ribband. (Fig. 156)

Plate 52. A typical "bundle of rods." Keel weldment on an open truss constructed vessel. Note that the outer mesh layers have been laid prior to setting up the stations and keel member.

Ribband Corrections

If for any reason the ribbands create an obvious flat spot in the hull at a station, they should be pried from the molds in that area and wood chips of the required thickness inserted to give the ribband a firm footing against the station mold. It would be wise to lightly glue these wood chips in order to prevent the possibility of their falling out. If on the other hand the ribbands form a high spot in the hull, the ribbands must be pried away from their respective molds and the molds shaven or notched to the required depth, allowing the ribbands to fair properly.

As it is extremely difficult to judge absolute hull fairness at the ribband stage, it would be prudent for you to cut a long, thin batten which can be laid against the hull vertically and diagonally to detect any subtle bumps and hollows. While these foregoing measures may seem tedious and time-consuming, I cannot overemphasize the fact that it is at this stage of construction that we commit the finished hull to its final shape. As I tour cement construction sites, the most obvious flaws are shadows and divots in the concrete surface, and they will become even more apparent when these vessels are painted. These can have no

effect other than to impair one's impression of what could have been a fine yacht, as well as lowering its potential market value.

You should always remember that the quality of your efforts will depend greatly upon your patience and sense of self-criticism. Many home-built boats have become slipshod and depressing failures simply because the owner has fallen victim of his own anxiety. While it may require some humility, it will never hurt to have an outside experienced eye pass judgment upon what may not be obvious to you alone.

Cross Tying

There may be occasions when some of your mold ribbands cannot be nail-attached to any part of the mold, thus causing them to spring outboard of their desired shape. This distortion may be corrected by passing a loop of string or wire around the battens on each side of the hull and drawing them up snugly into their proper positions.

In the area of the deadwood and near the upper portion of the deadwood where the ribbands will require severe reverse curves, cross tying may also be employed to help form the proper shapes. (Fig. 157)

Stealers

Because it is most desirable to allow the ribbands to take their most natural shape, some areas of the hull will result in abnormally wide ribband spacing. Short extra ribbands, called stealers, may be inserted to fill these discrepancies. The most common areas for the use of stealers are in the deadwood and at the turn of the bilge amidships. Feel free to use them wherever you think the mold requires more definition. (Fig. 158)

Square Ribbands

If you reach an area where a ribband must be bent severely in two directions simultaneously, or where the vertical curvature of the ribband causes one of its edges to bulge outboard, square lumber must be used. While it might require the removal of a ribband already fastened, it is not worth avoiding the little bit of extra time involved.

Cedar Molding

As described earlier, the cedar or closed mold system required that the station framing be constructed in the normal way, then completely planked over as if you were building a wooden hull. (Plate 54) Some builders prefer to use ¾" x 3" cedar or pine strakes while others rather strip-plank with ¾" x ¾" battens fastened edge to edge. Regardless of the system, only finishing nails should be employed and they should be countersunk deeply into the lumber to allow planing to fairness. It will not be necessary to glue any of the strakes to the stations, but it is wise to scarf-joint the lumber in order to produce continuous lengths. Because of the tapering of the hull shape toward its ends, it will be necessary to taper some of the planking while adding "stealers" amidships (additional planking used to increase girth discrepancies). Do not bother over inaccurate joints and seams, but concentrate on allowing the strakes to bend naturally. "Saw-scoring" from the inside of the mold (described earlier in this chapter) may be necessary to produce the desired fullness in the bow and stern areas. Do not plank over the center line if a longitudinal pipe is called for by the architect.

Plate 53. Ribbands applied to each side of a Samson **Sea Mist** simultaneously. Note that each ribband has been sprung fairly so as not to distort the builder's judgment.

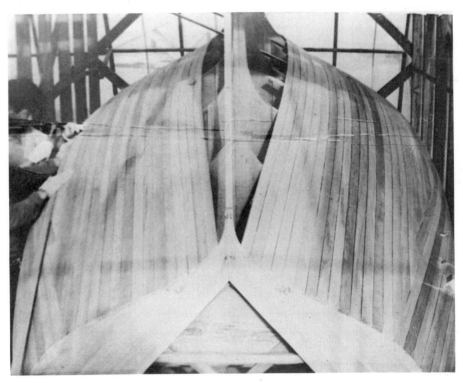

Plate 54. A **Flicka** mold being "cedar" planked at the Marine Crafts International yard in Pusan, Korea. (A Marine Crafts Photo)

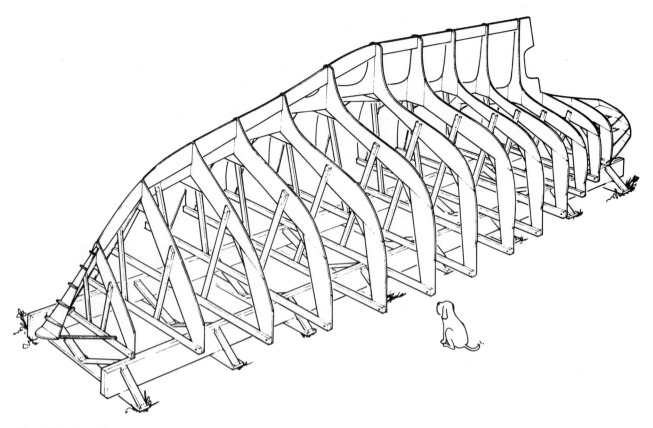

Fig. 152. A partially completed inverted hull mold. Note additional profile templates between stations. These are not always necessary and may not be called for by your architect.

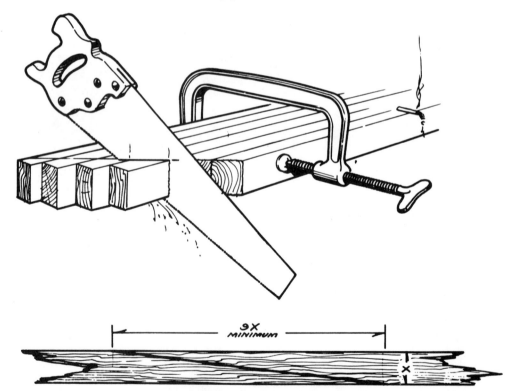

9X
MINIMUM

Fig. 153. Cutting long scarfs in thin lumber.

Upon completion of the mold, it will be necessary to sheath the entire structure with a vinyl film to prevent the mortar from seeping into joints and seams. Take great care to smooth out all visible wrinkles in the vinyl and scotch-tape every overlap. The mold is now ready for meshing.

inevitable rust bleeding, the difficulty of shaping and consequent distortion, and the possible separation from the concrete upon impact or stress. Because many designers have specified the use of steel sheer screeds, I will discuss their installation in Integral Structures.

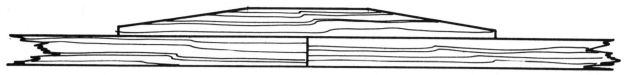

Fig. 154. Ribband butt joint and block.

Before proceeding headlong into this construction method, be particularly aware that cedar molding creates tremendous difficulties on both plastering day and during curing, as it is almost impossible to produce 100% mortar penetration and vinyl barrier traps water and air pockets during setting. With only few exceptions, the scores of hulls built

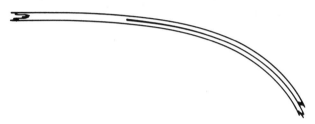

Fig. 155. Rip bending ribbands.

to this system betray massive voids on the inside of the shell when the mold is removed and successful repair is somewhat questionable. If the mold planking is left in the hull as a liner, suggested by some designers, these voids will not only go without correction but will become eventual moisture traps for the influence of ultimate rot.

Cedar or closed molding should therefore be restricted to the builders of mass-produced hulls equipped with high-pressure Gunite or mortar pumping systems. It cannot be guaranteed that penetration will be assured through vibrating, as repeated motion within the mortar breaks down its delicate crystalline form so necessary for strength.

Wood Screeds

In essence, a screed is a strip of wood (or metal) used in masonry to establish the finished thickness of a plastered edge. It also serves to create the proper smoothness along the edge surface. Some builders prefer to use metal screeds. I, personally, dislike their use along the sheer because of

If you are (and I hope you are) providing for a wood sheer screed on your wood mold (unless you are planning open-upright plastering), you have several alternative configurations from which you may choose, but several points must be kept in mind. If your vessel is to have a bulwark and cap, the edge of the hull should be horizontal along its entire surface. (Fig. 159) If a wooden deck is to be laid overlapping the edge of the vessel's hull, the screed must be angled to the deck crown at the sheer. This angle will vary according to longitudinal position. (Fig. 160) Obviously, a vessel being constructed with a cement deck will not use a sheer screed unless it is to have a bulwark.

The construction of the sheer screed at the bow and stern is most practically accomplished with plywood sheet, as it may be difficult to produce the proper shapes in solid lumber. (Fig. 161) The screeds along the sheer, however, are usually built up of two layers of battens of proper thickness. Prior to attaching the screeds, they should be coated with melted paraffin to enhance a clean separation from the concrete.

Technically speaking, porthole, thru-hull, bolthole, scuppers and hawsehole blanks are screeds. They may be installed onto the mold at this time (unless you plan to remove the mold for open-upright plastering), or sometime during the laying of mesh and rod. For their construction and attachment, see Integral Structures.

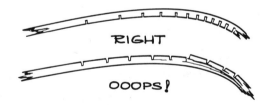

Fig. 156. Saw scoring to provide for tight bends in ribband lumber.

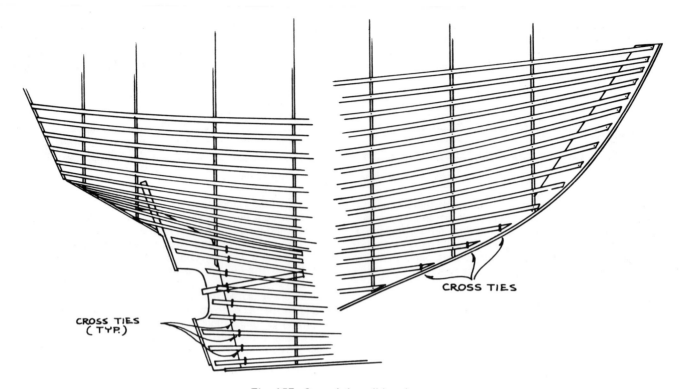

CROSS TIES

CROSS TIES
(TYP.)

Fig. 157. Cross tying ribbands.

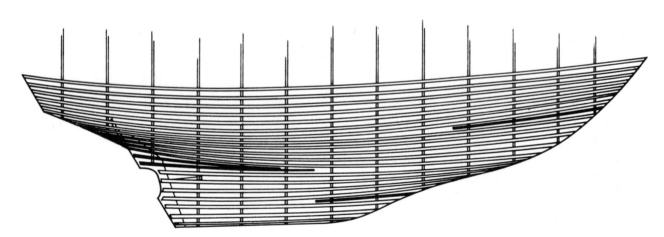

Fig. 158. Stealer ribbands.

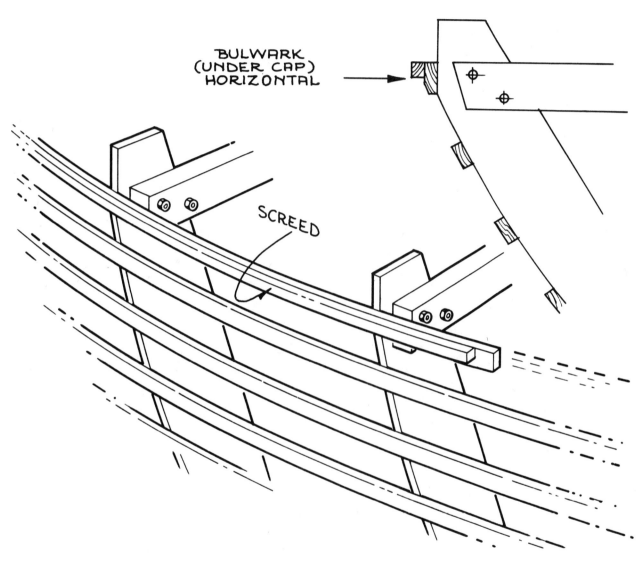

BULWARK
(UNDER CAP)
HORIZONTAL

SCREED

Fig. 159. Wood bulwark screed.

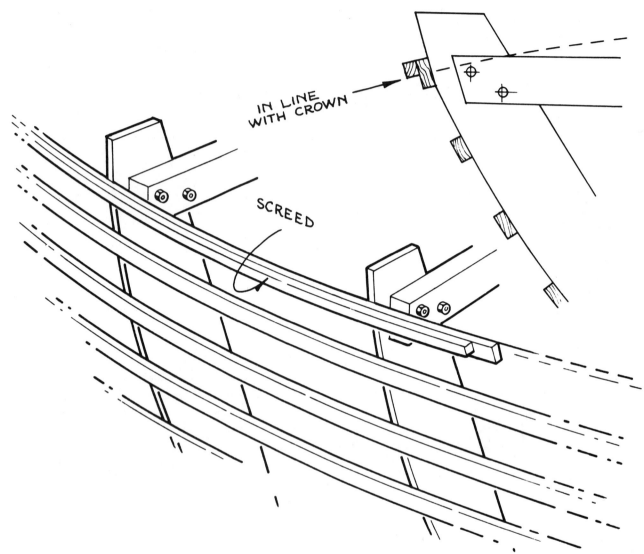

IN LINE WITH CROWN

SCREED

Fig. 160. Wood screed for use with wood decks.

166

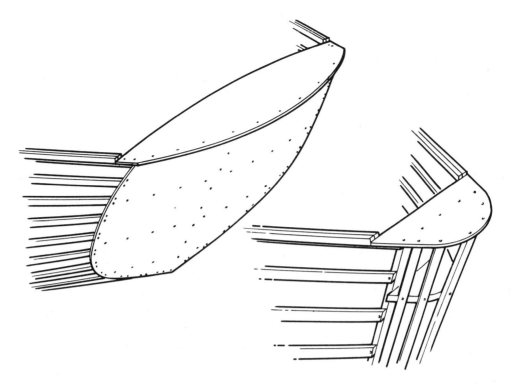

Fig. 161. Plywood sheer screeds at bow and transom.

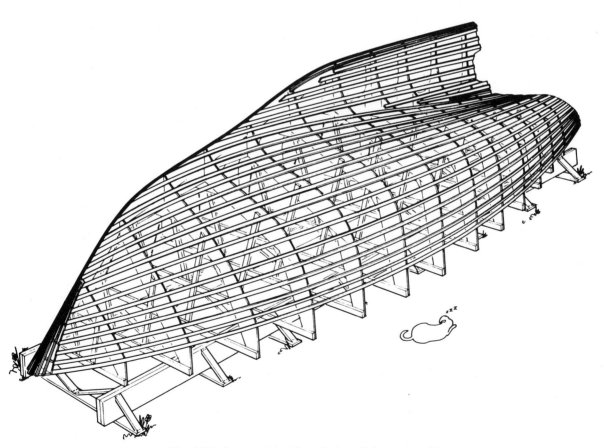

Fig. 162. A completed inverted partial wood mold.

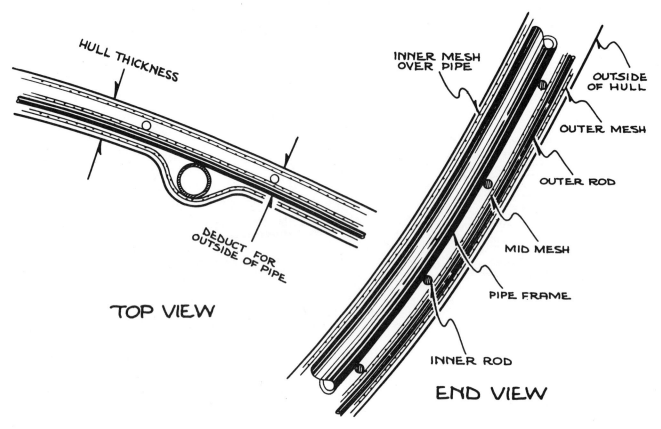

HULL THICKNESS

DEDUCT FOR OUTSIDE OF PIPE

TOP VIEW

INNER MESH OVER PIPE

OUTSIDE OF HULL

OUTER MESH

OUTER ROD

MID MESH

PIPE FRAME

INNER ROD

END VIEW

Fig. 163. Determining proper line for the outside of the pipe frame must be preceded by an accurate sketch of the hull shell section. This vessel will be using mid mesh.

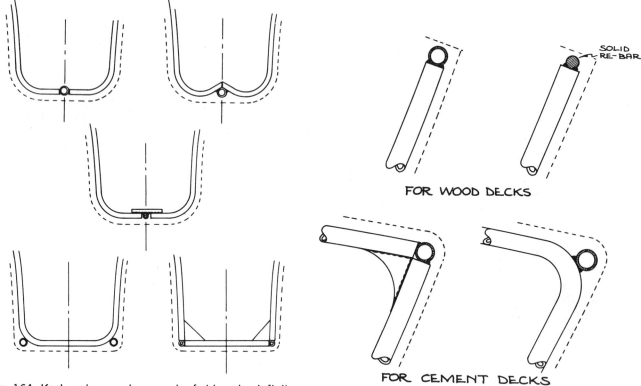

Fig. 164. Keel center member may be fashioned as infinite variety of ways, each producing comparable strength. They are the most popular and easiest to accomplish by the amateur.

SOLID RE-BAR

FOR WOOD DECKS

FOR CEMENT DECKS

Fig. 165. Sheer pipe and sheer bar arrangements.

pipe frames

When ferro-cement was introduced to North America by John Samson in 1967, pipe frame construction became an early standard. It was a fairly easy system, even more basic than wood mold construction, so there is no reason to question its popularity.

The system does require a considerable amount of welding and the bending of large pipe stations accurately is no mean feat. The pipe stations usually remain in the vessel (although it is not necessary) while serving little or no structural purpose. The resulting visible pipe frame ridges are somewhat unsightly in a fine yacht and are difficult to fit joinery too. With these considerations in mind, pipe frame construction still remains a valid and viable system which can lend itself to virtually any design.

Determining the Bend Line

If you have purchased plans specifically for the purpose of pipe frame construction, the definition of the lofted lines will be clearly spelled out for you, i.e., outside of pipe or inside of pipe. If, on the other hand, you have purchased plans designed for wood mold construction, the loftings of which are usually drawn to the edge of the station molds, you will have to determine the pipe position so that it will not change the proportions of the boat. Before making this determination, you must first decide whether the pipe frames are to remain in the completed hull. (Fig. 163)

Keep in mind that an enlargement of only ¼" to the lines of the finished hull can result in several thousand pounds of added displacement.

Choice of Pipe

In terms of pipe diameter, the size of the station frames and degree of trussing are the determining factors. In general, you may safely use the following chart if you prudently use temporary internal stiffening.

Vessel Displacement	Pipe Size (Nominal In.)
Under 10,000	3/8"
10,000-25,000	1/2"
25,000-40,000	5/8"
40,000-65,000	3/4"
65,000-80,000	1"

The type of pipe most common and suitable for yacht construction is black iron pipe. It has not been softened by a galvanizing process and, therefore, once bent, maintains its shape quite readily. Black iron is relatively easy to cut and weld without undue distortion, and is available at any reputable piping supply house.

Bending Pipe

I feel that bending pipe to station shapes is best done a little at a time along the entire length of the pipe, repeatedly reworking the shape until you have closed in to the lofted line accurately. It will make no difference whether you use a hydraulic bender or a bar type of bender but it is imperative that the finished station frame lies perfectly flat when at rest. Any twisting of the frame will inevitably cause distortion. (Plate 55)

Center Members

The center member of most pipe frame hulls consists of a center stem pipe at the bow and fore foot and a "bundle-of-rods" arrangement along the base of the keel. (Fig. 149) (Plate 52) This type of construction usually produces a vertically flat keel with sharp corners. While these may be good from the standpoint of windward performance and hauling, they tend to produce severe stress risers along their edges.

A single center pipe arrangement seems most practical and easy to install along the vessel's profile. There are several systems for its installation and may be interchanged according to the designer's or builder's preference. (Fig. 164) It must be pointed out that, if the pipe frames are to be left in the finished hull, it is imperative that the center pipe be completely enclosed with several layers of mesh. To not do so would surely result in cracking along the unreinforced side of the cement.

Sheer Member

It will be necessary to attach a steel sheer member interconnecting all pipe frames, whether the vessel is to have a cement deck or not. This member may be of a continuous pipe equal to the diameter of the station frames. If this system is chosen, you must allow for its attachment when determining the cut position of the upper end of the pipe frame. (Fig. 165)

If you desire, you may use "L" (angle) steel along the sheer. It is fairly easy to spring bend without twisting distortion. (Fig. 166) Avoid using a pipe bender to shape "L" stock, however, as it is almost impossible to produce a smooth curve this way. Flat steel bar is absolutely out because it distorts beyond any reasonable control when bending on the plane of its long axis.

Deck Beams

As you already know, I cringe at the thought of using cement decks on sailboats unless **FER-A-LITE*** or other light

*FER-A-LITE is a registered trademark of Aladdin Products Inc., Wiscasset, Maine.

aggregate compounds are used. Such cement decks may be laid up over pipe deck beams carefully bent and welded to the upper ends of the pipe station frames. The beams should be marked at the carlin positions but not cut until they have been attached to the frames and the frames erected and aligned. A steel plate gusset at the beam/frame corners will stiffen these beams measurably. After erecting the frames, the sheer and center pipes attached, the beams may be cut and the carlin screeds attached. To aid in aligning the deck beams it will help if you lay a section of pipe along the entire length of deck, clamping each beam to it while making the necessary beam adjustments. (Fig. 167)

For the installation of wood deck beams, a triangular bracket is simply attached to the upper end of each pipe station frame. This bracket should be predrilled to receive the beam bolts. (Fig. 168) This bracket will not be covered with cement and, therefore, it will be required to slit the inside mesh to allow its passage of each bracket.

Sole Beams

I cannot recommend using steel angle as sole beams because it complicates the attachment of the plywood flooring. I believe the most logical approach is the predrilled triangular

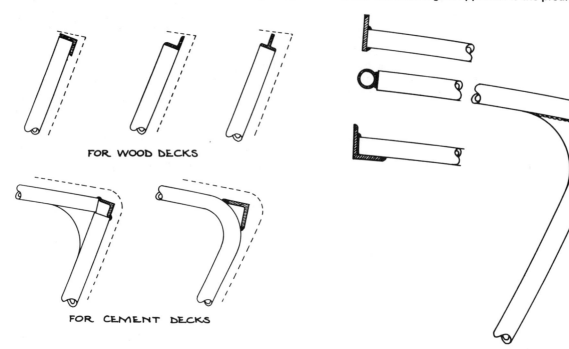

FOR WOOD DECKS

FOR CEMENT DECKS

Fig. 166. Sheer iron arrangements.

Fig. 167. Cabin and hatch carlins for cement decks.

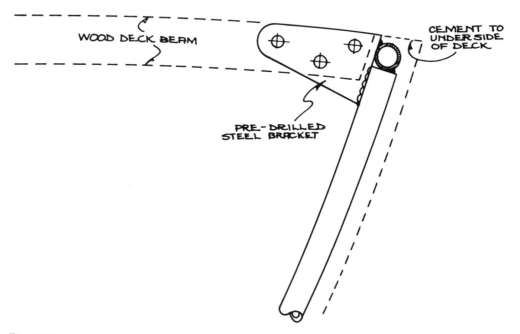

WOOD DECK BEAM

CEMENT TO UNDER SIDE OF DECK

PRE-DRILLED STEEL BRACKET

Fig. 168. A pre-drilled steel bracket is welded to the pipe frame to receive a wood deck beam. Notice the upper edge position relative to the finished cement.

beam brackets welded to the pipe frames. Wood sole beams may then be bolted to these brackets. (Fig. 169) For method of locating beam brackets see Integral Structures.

Deadwood and Shaft Installation

The deadwood may be constructed of plate steel (Fig. 146), steel "T" bar (Fig. 147), or of pipe. In the latter instance, two layers should be used to provide for the proper thickness of the keel trailing edge as well as fairing around the propeller shaft exit. (Fig. 170)

Transom Frame

The flat raked transom, plumb flat transom or reverse flat transom pose little problem in construction. A pipe frame may be simply bent to the projected transom shape including the transom crown. The finished frame is positioned at its proper angle within the hull's framework, welded to the center member and sheer pipe. (Fig. 171)

If you are constructing a curved transom, the procedure is a little more complicated. You must first construct a wood mold representing the radius of the transom. This may be built using two plywood radii templates interconnected with wood slats. The width of the wood mold must be slightly more than the depth of the transom on center.

Now cut a paper or plywood pattern exactly to the shape of the expanded transom (deducting the proper hull thickness, etc., if necessary). Bend and attach this template to the radius mold, being sure that the center of the expansion is exactly perpendicular to the axis of the mold radius.

Bend the transom frame piping to fit the expanded transom pattern on the radius mold. The pipe radius and mold radius must match. (Fig. 172) Once all parts are properly bent and welded together, the transom frame may be installed into the hull framework.

Floors and Tank Bulkheads

These installations are very simple with pipe frame construction but are not unlike the open truss method. The favor here is that the required welding will not tend to distort the pipe frame shapes as with truss frames. It is primarily a matter of welding cross rods from one side of the pipe frame to the other, then adding vertical rods as necessary. Provisions for baffle swashes and limbers must be made by using shaped steel flat bar screeds or short pipe sections. (Fig. 173)

Engine Beds

Many configurations are available for these units but surely

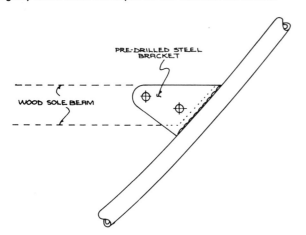

Fig. 169. A pre-drilled steel bracket is welded to the pipe frame to receive a wood sole beam.

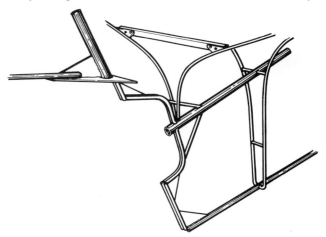

Fig. 170. A typical pipe deadwood arrangement.

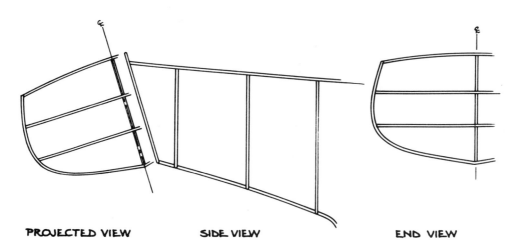

PROJECTED VIEW   SIDE VIEW   END VIEW

Fig. 171. A flat raked transom pipe frame. Note that it is the projected view which must be used for the proper fabrication.

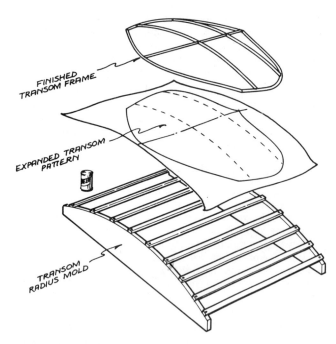

Fig. 172. Building a rounded counter transom pipe frame requires the use of a temporary transom radius jig in order to produce the proper curve. The same method is used when building to the open truss or web system.

you should follow your designer's recommendation. Basically, the engine bed shapes are prebent pipe, which are welded to their respective station pipe frames, then laced with reinforcing rod and covered with mesh. The beds may be transversely braced with gussets or floors. Provisions for mounting may be accomplished by installing predrilled steel angle or wood timbers. (Fig. 174)

Mast Steps

These units do not differ a great deal from the engine beds in concept. Once again, there is an infinite variety of configurations which may be successfully employed and it is best to follow your designer's recommendations. They may be a framework of pipe, webbed with reinforcing rod then poured solid with concrete; constructed as a steel plate weldment fitted to the pipe frames; or steel brackets welded to the pipe frames drilled to receive a wood timber structure. (Fig. 175) Other possibilities may be seen in Integral Structures.

Chain Plates

While several variations are also shown in Integral Structures, one of the simplest approaches for pipe frame con-

struction is that of welding the mild steel chain plates directly to the pipe frames. This may not always be possible if the frames do not occur in the chain plate position. Provisions must be made to prevent the pipe frame from breaking loose from the hull armature, as its transverse reinforcement will normally be negligible. This may be accomplished by providing for short doubling rods which are passed over the pipe frames with their ends worked into the hull armature. It may be necessary to notch the outboard edges of the chain plates to allow for the passage of the double rods or they may be welded directly to the pipe frames. (Fig. 176) (Plate 56)

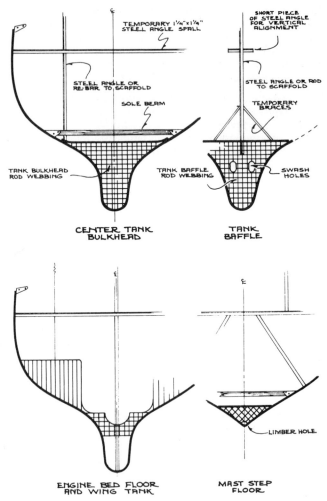

Fig. 173. Typical floors and baffles. Notice the combinations of pipe and rod being used as well as the variations in the frame hanging system.

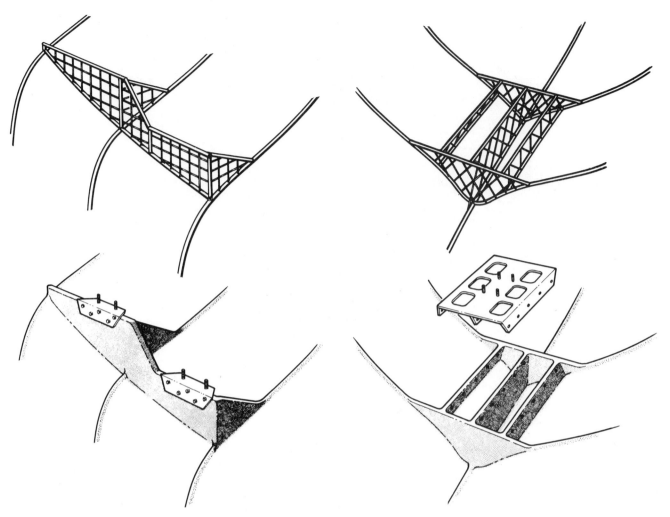

Fig. 174. A well engineered armature for a pipe frame engine bed provides for thrust and vibration in all directions. Top view shows metalwork prior to covering with mesh.

Fig. 175. A hefty mast step built as a box girder comprised of pipe and rod. Mast compression can be evenly distributed over a large area of the hull in this manner. Many other variations are possible.

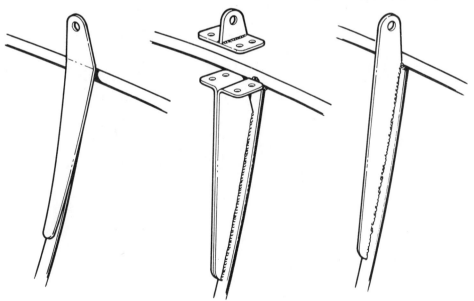

Fig. 176. If the architect thinks well ahead into the construction when drawing the sail plan, he may evolve a rig whose shrouds fall conveniently onto pipe frame positions. Chain plates which are welded directly to pipe frames are undoubtedly the strongest that can be constructed for ferro-cement.

173

Plate 55. Bending pipe frame with a custom-made hydraulic bender. (Great Lakes Ferro-Cement Corp. Photo)

Plate 56. A **Flicka** being built by the pipe frame method. Hull has been aligned and is waiting the attachment of the sheer bars. (Great Lakes Ferro-Cement Corp. Photo)

174

web or truss frames

The open truss frame construction system varies widely from the pipe method, although many want to argue this point. Unlike the pipe frame system, web framing becomes a necessary structural part of the finished hull. This occurs because the frames act as integral stiffeners, which allow the shell of the vessel to be of somewhat lighter construction. During building, the frames also serve as forms for shaping the vessel's shell.

The web or truss frame system is particularly suited to vessels with relatively large expanses of flat surface, such as found on most powerboats. This is especially true in barge construction where a web frame system is absolutely essential. (Plate 57) One of the principal drawbacks, however, is the fact that web frames do infringe on the useable space of the vessel. I rather doubt that the slight weight saving derived through this form of construction ever really pays for itself on a cruising vessel. But if a racing design is being built, the sacrifice of space may be well served. If the vessel's hull is well rounded, the strength

of the vessel will be inherent in both her compound shape and the laminate of the armature. Many designers still prefer web construction in light of this and their plan packages go to great lengths in describing the construction procedure. I will, however, go into some detail here in order to point out the processes of fabrication, as well as some of the pitfalls and complexities.

Frame Layout

Each web frame or bulkhead will be comprised of reinforcing rod defining the perimeter of the inside of the hull, the underside of the deck, and the inner edge of the frame. In some cases the inner edge of the frame may be constructed of flat steel bar in order to produce a sharp square edge. This will be clearly called out on your plans. Within the perimeter's shapes, notches will be provided for receiving such longitudinal structures as the sheer pipe or bar, cabin or hatch carlin bars, and keel center pipe. These notches may be

Plate 57. An inverted open truss armature for a 54' houseboat hull being constructed at the Betts Yard in Alameda, California. The truss system lends itself well to this relatively flat-bottomed configuration.

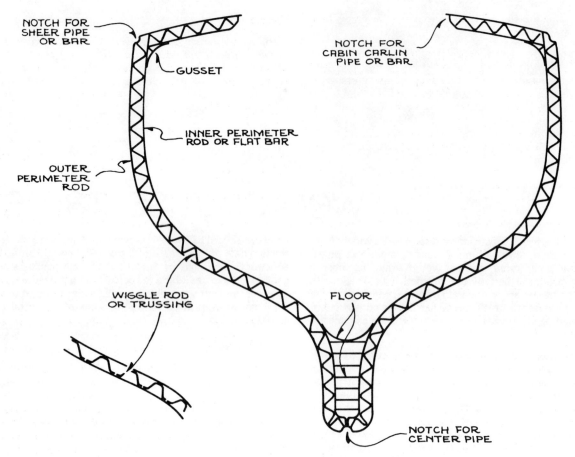

NOTCH FOR
SHEER PIPE
OR BAR

GUSSET

NOTCH FOR
CABIN CARLIN
PIPE OR BAR

INNER PERIMETER
ROD OR FLAT BAR

OUTER
PERIMETER
ROD

WIGGLE ROD
OR TRUSSING

FLOOR

NOTCH FOR
CENTER PIPE

Fig. 177. Typical web frame layout without tank, bulkhead or machinery mount provisions.

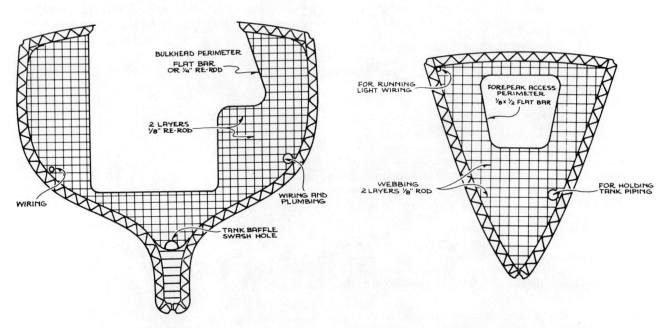

BULKHEAD PERIMETER
FLAT BAR
OR ¼" RE-ROD

2 LAYERS
⅛" RE-ROD

WIRING

WIRING AND
PLUMBING

TANK BAFFLE
SWASH HOLE

FOR RUNNING
LIGHT WIRING

FOREPEAK ACCESS
PERIMETER
⅛ x ½ FLAT BAR

WEBBING
2 LAYERS ⅛" ROD

FOR HOLDING
TANK PIPING

Fig. 178. Typical web frame layout with bulkhead and tank baffle on station.

Fig. 179. A full bulkhead built into a station frame. Note the installation of passages and opening.

176

formed in many different ways but usually are simply bent into the perimeter rod.

Those frames whose lower sections traverse an open bilge area must be fitted with limbers to provide for the free flowing of bilge water (unless the frame also serves as a tank end). These limbers may be short sections of pipe inserted through, and welded to, the frame webbing, or they may be made by inserting waxed wooden blanks into the webbing prior to plastering. Holes should also be provided in the frames to allow for the passage of the ship's piping and electrical wiring. (Figs. 177 & 178)

Tanks, Baffles, Floors and Bulkheads

Within the web frame system we must consider more elements than simply the framing of the vessel. Other units may be constructed in the same manner as, or part of, the hull framing. Full bulkheads, for instance, will traverse the entire width of the hull and will be provided with openings for doors, piping, limbers, etc. (Fig. 179) Partial bulkheads will simply be extensions of the inner frame perimeter to include the bulkhead shape, again providing for openings. The transverse members of the engine beds and mast steps may also be constructed as portions of web frames or as separate units, but in the same manner. (Fig. 180) In all cases, the construction procedure will be essentially identical, and once you have mastered the construction of a web frame, these other units will pose few problems. (Plate 58)

Webbing

You now will encounter many different methods recommended by different designers to accomplish essentially the same end. Jay Benford, for instance, calls for the liberal use of "Wiggle Rod," which is basically the bending of reinforcing rod into a zigzag pattern so it will lie between, and make contact with, the inner and outer frame perimeters (Plate 59) This system is fast and very strong. Richard Hartley, on the other hand, recommends individually prebent truss rods, which I believe takes more time and involves many more small pieces to accomplish the same purpose as wiggle rod. (Plate 60) I believe that the two systems are interchangeable and should be more a matter of the builder's personal preference. It should be noted, however, that whether you choose wiggle rod or individual truss sections, these systems are only applicable where parallel frame perimeters are encountered. (Fig. 177)

Where large web surfaces are called for, however, the filling in of the web may be accomplished by vertical and horizontal rods, double diagonal rods, single layer vertical rods, or single layer horizontal rods. (Plate 61) The system will be clearly called out by your designer and is determined on the basis for which the given member is to serve. Unfortunately, there is no hard and fast rule for this. (Fig. 177)

In cases where a floor, bulkhead, etc., lies on the same plane as a frame, it is most usual to first fabricate the frame in its entirety, using wiggle rod or trusses, then add the additional section to the completed frame. (Plate 62)

Before removing the finished web or frame from the scrieve board or welding floor, it will be necessary to attach spalls (spreaders) and other bracing in order to prevent any possibility of distortions while the unit is being moved or as weight is being added to the vessel's armature. Usually this bracing is fabricated from 1" x 1¼" or 1¼" x 1¼" steel angle.

Wiggle Rod

It should be pointed out once again that wiggle rod serves the same purpose as individual trusses, and you have the option of either method. In my opinion, wiggle rod is far easier, because it does not take the degree of accurate bending as do the individual trusses, nor does wiggle rod require the amount of cutting and welding.

Wiggle rod may be prefabricated in lengths as long as desired on a jig consisting of a steel base plate and a series of steel pegs or tapped 3/8" bolts. (Fig. 181) To bend the wiggle rod is simply a matter of placing a section of pipe over the rod for use as a handle. As the rod is bent back and forth, the pipe handle is slipped along the rod. When bending the wiggle rod, take great care to ensure that the radii of the bends are as sharp as possible. Once a section of wiggle rod has been completed to the end of the jig, remove it from the jig and place the remaining straight portion of rod at the beginning point. Continue bending and repositioning the rod until you have completely bent the remaining portion of steel.

Trusses

As I mentioned previously, the individual trusses and wiggle rod may be interchanged without fear of sacrificing the strength of the vessel or the configuration of the web frames. Making up trusses is a tedious procedure requiring muscle and accuracy.

Making up trusses will first require the cutting of several hundred lengths of reinforcing rod of the prescribed diameter into 9½" or 10" sections. Each one of these rod sections must be premarked to indicate the proper bend points. As each design varies considerably, the bend points will generally be specified by your designer. Generally speaking, however, they are marked off as follows: 1", 3½", 1", 3½", 1". (Fig. 182)

In order to speed up the bending process, it is helpful to draw a full-size pattern upon which to check your work as you go. Each bend will approximate a 45° angle. The bending of the

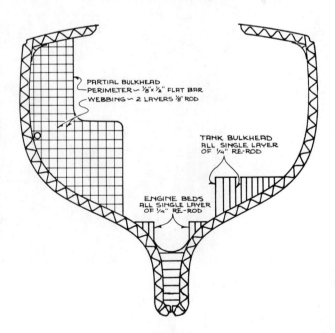

Fig. 180. A partial bulkhead, tank end and engine bed gussets built into a station frame.

Plate 58. The web frames and forepeak bulkhead clearly show wiggle rod and webbing. The inner frame and opening perimeters have been finished with steel flat-bar.

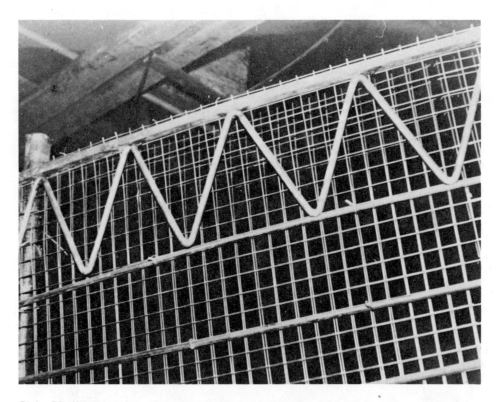

Plate 59. Wiggle rod as used along the bulwark of a Benford 14' tug. The wiggle rod as used in frame construction is virtually identical. (A Jay R. Benford Photo)

trusses should be done in a vise using a short section of ½" pipe as a bending handle. If you are unable to execute the desired bend at exactly the premarked point, do not worry about it, as it is the width of the truss that is the most important. Final adjustments can be made in each truss in order for it to fit the pattern or prescribed width before going on to the next truss.

Plate 60. Typical framing with individual trusses as found on most Dick Hartley designs. Notice that the hull rods are being wire-tied to the frame (not welded) in order to minimize distortion.

Welding Up Frames

To eliminate the confusion of crossing lines, each frame structure should be drawn out on your scrieve board one at a time as you go. Indicate the perimeter of the frame at the inside of the hull clearly, then carefully lay out the inner perimeters, taking care to accurately lay off all openings, swash holes, limbers, etc. (Fig. 183)

Drive a series of closely spaced 2" nails along the line representing the hull perimeter of the frame. You may now begin bending the frame perimeter rod against the row of nails. As you do, place a second nail on the inner side of the rod to hold it firmly into position. You may not need a bending tool for this operation as most hull curves are generally quite soft. When you reach the turn at the lower portion of the keel, however it will be useful to apply a bending tool. The notch required on the center line to receive the keel centerpipe will have been preformed. For this reason, it is wise to consider beginning the formation of your perimeter rod at the lower portion of the station at the center line so that the keel pipe notch will be accurately located. It should also be pointed out that the closer you

space your nails along the station line, the more accurate will be the frame shape and the less chance there will be for distortion during the welding process. (Fig. 184)

Now you may begin attaching the wiggle rod or truss. If you are using wiggle rod, it is laid out flat and pressed against the outer perimeter frame rod, lightly tack-welding each point of the wiggle rod as it touches the perimeter rod. You will have no trouble forming the wiggle rod to the shape of the frame. The wiggle rod will continue around the entire frame perimeter. If you are using the individual trusses, they are positioned against the perimeter rod one by one with their "feet" facing outboard. Lightly tack-weld each of the feet to the perimeter rod, then continue placing successive trusses in the same direction. Do not overweld.

As you reach the upper or lower ends of the frame, you may be required to cut short individual pieces of rod section in order to completely and uniformly fill in any areas which are not conveniently reached with the trusses or

Plate 61. A web bulkhead with transverse and vertical rods in position. This builder is using 3/8" solid steel reinforcing bar for frame and opening perimeters. This is simply one of an infinite number of variations which can occur with web or truss construction.

wiggle rod. This may be done somewhat at random. The important thing here is to provide triangulation of the steelwork. You should be cautioned against removing any of the nails holding the frame in position during the welding, because heating and cooling during welding may ultimately distort the finished frame shape. It is wise to consider placing weights upon the frame as you go, such as concrete blocks. (Fig. 185)

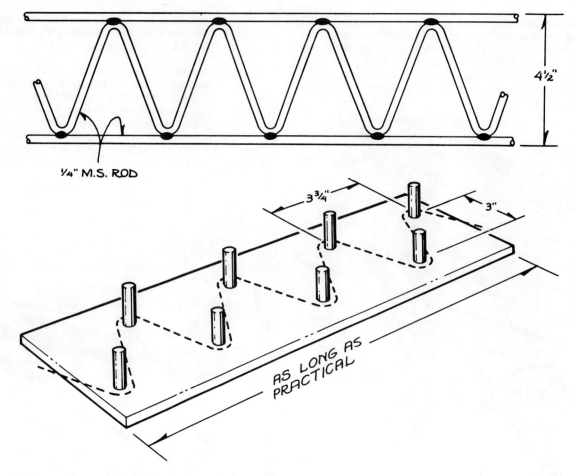

¼" M.S. ROD

AS LONG AS PRACTICAL

Fig. 181. A simple jig of steel plate and short 3/8" rod pegs used to fashion "wiggle rod." The completed frame width will vary from boat to boat.

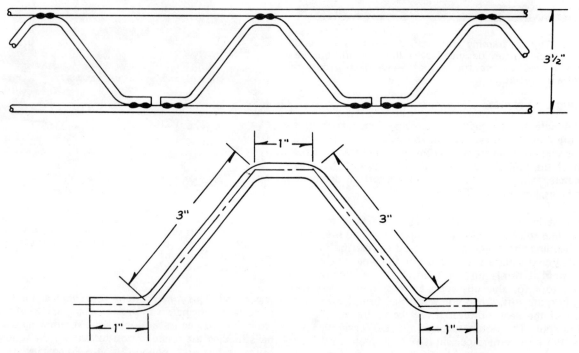

Fig. 182. Individual ¼" rod trusses bent to form a 3½" frame section. A 3¾" web would be created by increasing the diagonal legs to 3½".

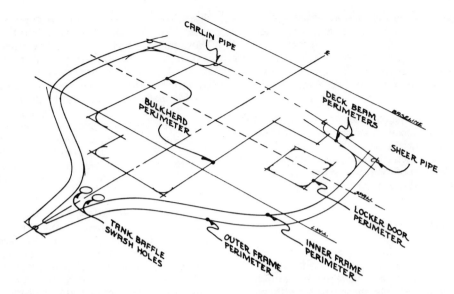

Fig. 183. Accurately laying out all components of frame or bulkhead is imperative before bending or welding.

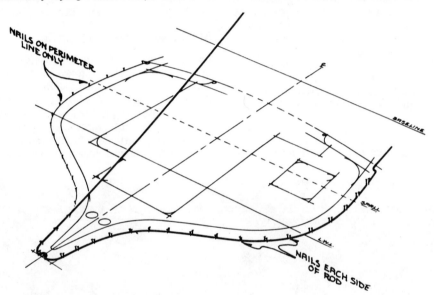

Fig. 184. A single row of nails is driven along outer frame perimeter; the rod is bent against the nails; then a second row of nails is driven to hold the rod in position.

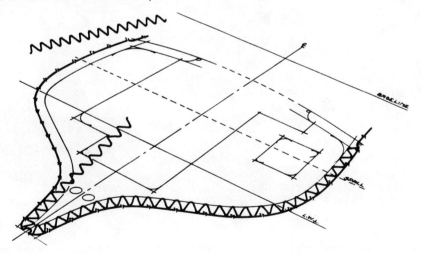

Fig. 185. Wiggle rod (for trussing) is pressed against the outer perimeter rod and tack-welded at all touch points.

You may now begin installing the inner frame perimeter. This may be either reinforcing rod or flat bar, as recommended by your designer. This is a very simple matter of forcing the rod or flat bar against the peaks of the wiggle rod, lightly tack-welding these points as you go. The inner perimeter will surround the entire frame from sheer to center to sheer. (Fig. 186)

Unless the frame is to receive beams, floors, bulkhead extensions or other integral units, it may be braced or stiffened in preparation to receive mesh. Carefully indicate all markings which will be necessary during construction, such as the waterline or positions of longitudinal members. Without these marks it will be difficult for you to align the frame properly from your scaffolding.

Joining Half Frame Sections

If you have constructed your frame in half sections, they may be lightly tack-welded at the center, then adjusted on the scrieve board for proper alignment. Move the sheers inward or outward so as to duplicate the sheer half-breadth on the opposite side of the scrieve board. Do this very accurately. In order to maintain this width, place a 1¼" x 1¼" steel angle across the frame from side to side and weld it to the frame perimeters. In order to aid in the proper vertical alignment when erecting your frames, the steel angle should be located at the same vertical height on all frames. This will facilitate sighting down the steel angles from one end of the boat to the other. This alignment will become more readily apparent during the actual construction of the hull.

Deck Beams

Those frames which are required to have only partial deck beams, i.e., extending to the cabin carlin only, will have had these beam sections fabricated into the total frame shape. Those frames which require full deck beam widths, however, will require their attachment at this time, if not previously done. The deck crown should be drawn out in some convenient area on your welding floor. The beam section itself is made up in exactly the same manner as the frames. Once the beam section has been assembled it is placed upon the completed web frame, marked accurately, and cut to fit from sheer to sheer, then welded to the two frame halves. (Plates 63 & 64) Be sure to mark the center line accurately. It must be noted here that provisions should be made at the sheer for receiving the sheer pipe or other required sheer member.

The inner corner, where the deck beam and inner frame perimeter meet, should be liberally radiused by adding a prebent section of flat bar or rod. Several trusses should then be located within the radiused area to give it additional strength. This radius will eliminate the critical stress riser, which may occur in this area. It should be pointed out at the same time that whenever sudden changes of direction occur within the frame shape, the inner perimeter should always be well radiused in the same manner. (Fig. 187; Plates 63 & 64)

Bulkheads, Floors and Baffles

As previously mentioned, the construction of these units will first require the fabrication of the full or partial frame, using wiggle rod or trusses. This will define the shape of the unit at the sides of the hull. The inner configuration is now carefully bent to shape using flat bar or reinforcing rod, and welded to the frame. Now the unit is filled with webbing of straight rod sections laid vertically, horizontally, or diagonally in the prescribed number of layers. In the interest of preserving the

Plate 62. A truss frame floor constructed at the same time as the frame itself. Here the center-pipe is being installed. Note the notches at the bottom center of each frame.

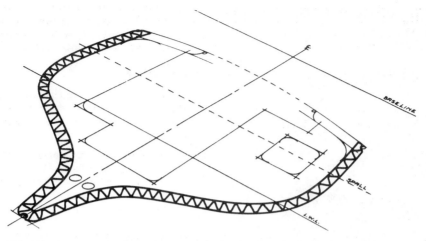

Fig. 186. The inner perimeter rod is sprung against the wiggle rod (or trusses) and welded at all touch points. This stage will complete the frame if wood deck and beams are to be used (see Fig. 168).

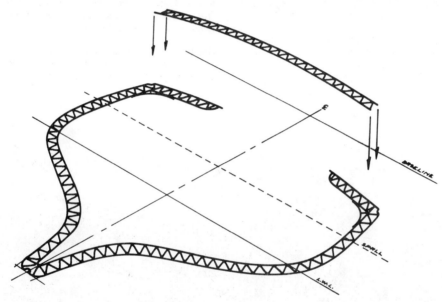

Fig. 187. If cement or **FER-A-LITE** deck is to be used, the deck beams (fabricated in the same manner as frames) are attached to the frame.

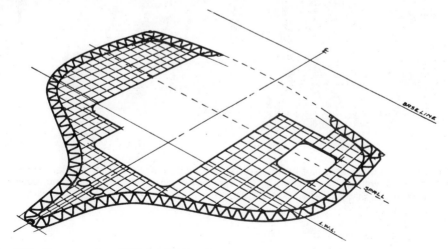

Fig. 188. If a cement or **FER-A-LITE** bulkhead is to be constructed, the rod webbing and bulkhead perimeter may be added to the frame simultaneously. For large expanses, it is wise to use two layers of small diameter rod.

Plate 63. A typical web frame installed in a Benford 50'
ketch. Notice the liberal radius at the corner of the deck
beam and frame. The overhead steel I-beam has been employed as a means for the laying of a wooden deck. (A Jay
R. Benford Photo)

maximum amount of strength and eliminating undesirable
distortion, try to keep your welding to a minimum if at all
possible. When webbing with two layers of rods, it is far
better to wire-tie the rod intersections. (Figs. 188 & 189)

Applying Mesh

You will find that it is much easier to apply the required
number of mesh layers to the framing before hanging your
frames from the scaffold or while applying the mesh to the
hull shell.

Mesh panels are laid economically atop the finished frame
and carefully cut to a proper fit. Be sure to leave about three
inches of excess mesh beyond the outer frame perimeter. This
excess will be folded perpendicularly to the plane of the
frame and will eventually be covered over with the inner hull
mesh layers. In order to accomplish the folding back of the
mesh, slits must be cut at fairly close intervals fully to the
outer frame perimeter.

If two or more layers are required on each side of the web, it is
desirable to shorten the excess to create a tapering effect.
This will eliminate a sudden bump when the inner hull mesh
layers encounter the frame areas.

The most economical use of your mesh may require many
small pieces. This will have little effect on the finished
strength of the frame, but you should take great care to

(Text continues on page 189.)

Plate 64. A typical web half-beam. Notice the modification of the normal wiggle rod
treatment as well as the use of a flat bar frame perimeter. The radiused corner is
employed to eliminate a serious stress-riser.

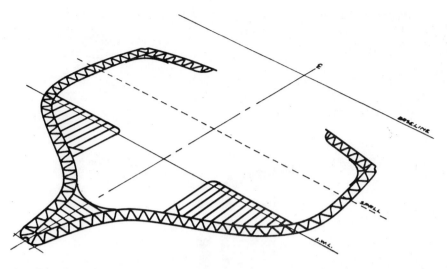

Fig. 189. Floors, baffles, and tank bulkheads are fabricated in much the same manner as large bulkheads. Only a single layer of rods is usually applied.

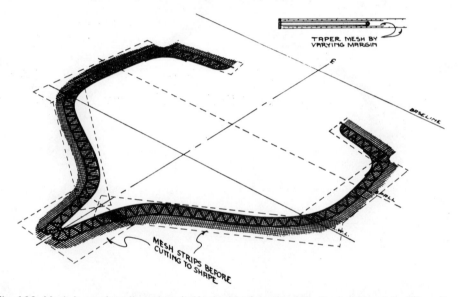

TAPER MESH BY VARYING MARGIN

MESH STRIPS BEFORE CUTTING TO SHAPE

Fig. 190. Mesh is most easily cut and attached to frames while the weldment is still on the scrieve board. This will stiffen the unit markedly thus eliminating undue distortion when moving. The mesh is allowed to overrun the frame perimeter to serve, in part, as an attachment device to hull mesh.

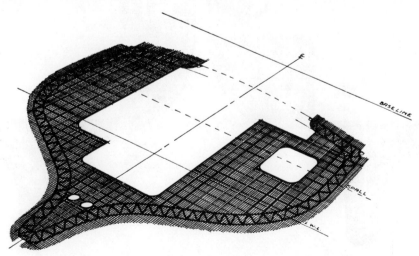

Fig. 191. Mesh is applied to full bulkheads while on the floor.

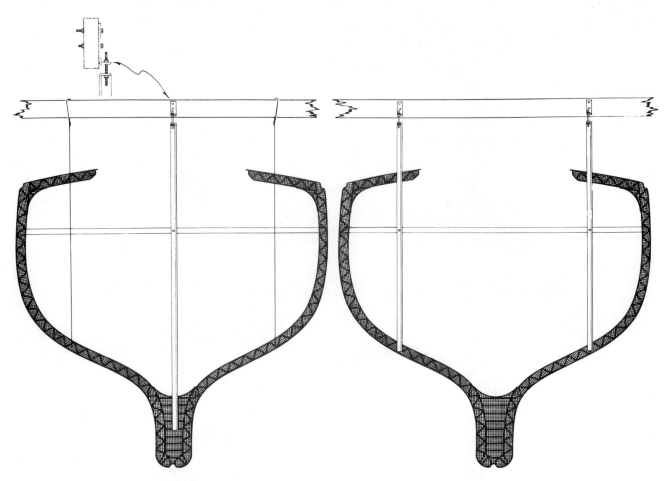

Fig. 192. Hanging a web frame from a single center iron and outboard leveling rods.

Fig. 194. Hanging a web frame using two irons.

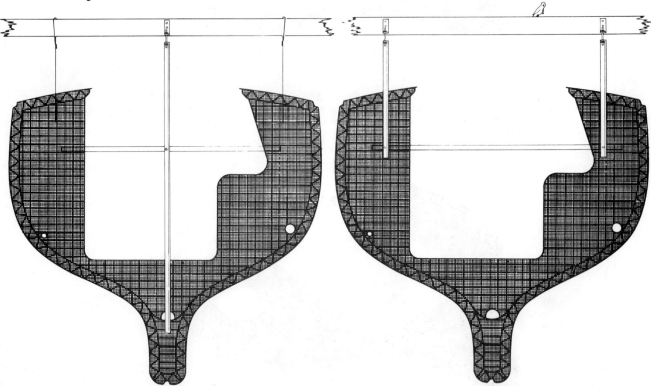

Fig. 193. Hanging a web bulkhead from a single center iron and outboard leveling rods.

Fig. 195. Hanging a web bulkhead using two irons.

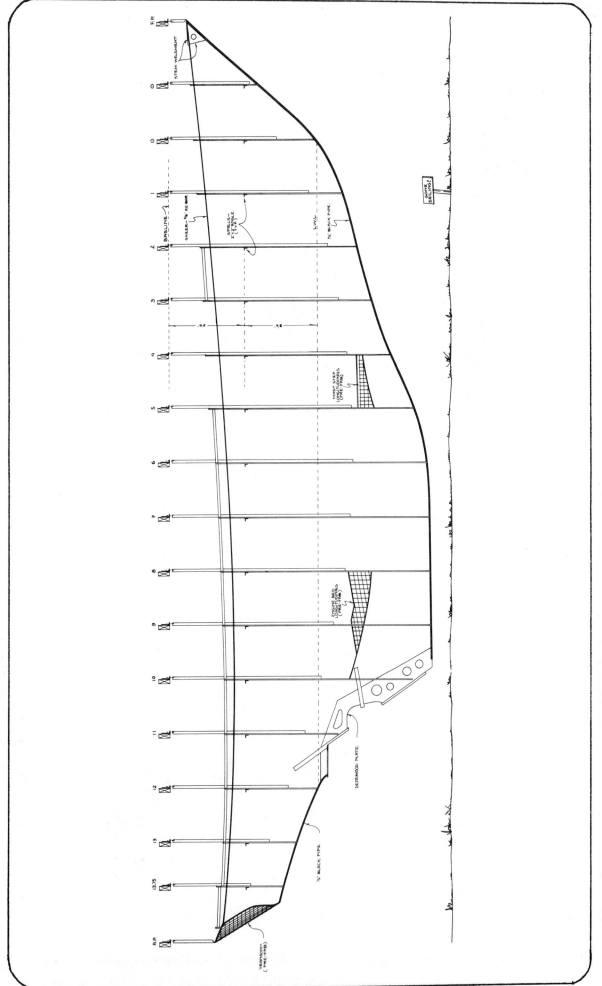

Fig. 196. A typical truss frame hanging plan of the major longitudinal structures; i.e., sheer bar, carlins, center-pipes, engine beds and mast step, deadwood weldment and horn plate.

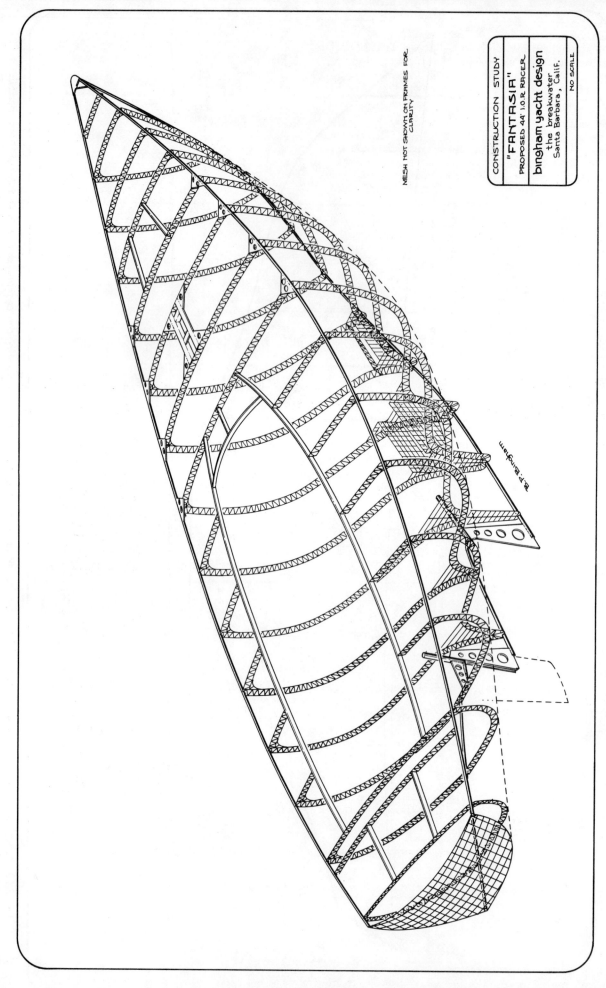

MESH NOT SHOWN ON FRAMES FOR CLARITY

Fig. 197. A typical truss frame hull armature before the application of hull rods and mesh. All parts of the hull construction can be seen in this illustration.

(Continued from page 184.)

ensure that the mesh butt joints are well staggered and do not occur within the same vicinity from layer to layer. (Figs. 190 & 191)

Hanging Frames

As I previously indicated, it is desirable to locate a transverse 1¼" x 1¼" steel angle spall from one side of each frame to the other. This spall will not only be used for alignment, but will be the prime means for supporting the weight of the frame during construction. Additional diagonal bracing may also be added to help distribute this support more uniformly around the frame.

There are many ways available for hanging frames and bulkheads. Dick Hartley recommends that a steel angle be attached to the frame and to the spall, then to the scaffold where a vertical adjusting bolt is provided. The transverse leveling of the frames is accomplished by dropping reinforcing rods from the scaffold cross-spans downward to the spall where it will be tack-welded upon proper alignment. (Figs. 192 & 193)

It is also possible to use two steel angles attached to the scaffold cross-spans and downward to the frame spall. Each of these vertical angles, then, will be fitted with individual adjusting bolts. (Figs. 194 & 195)

Once all stations and bulkheads have been properly aligned

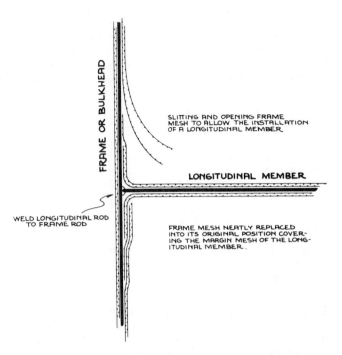

Fig. 198. The method for attaching longitudinal members (engine beds, tank baffles, etc.) to transverse frames or bulkheads.

Plate 65. Longitudinal members for an engine bed which have been prefabricated and installed into an open truss armature.

189

from the scaffold cross-spans, it will be necessary to block up their lower portions to relieve the frame of unnecessary weight. This blocking can be temporarily removed from time to time to allow for working under the keel, such as installing the center pipe. Be cautioned, however, against removing all of the blocking at the same time. (Fig. 196)

Installing Longitudinal Members

The center pipe or other keel member may be prefabricated atop the lofting, prior to its installation onto the frame-work. The prefabrication may have to occur in short sections in order to facilitate easy handling. It will be raised to the hull from lines passed over the scaffold cross-span and welded to its proper position at each frame bulkhead, etc. A temporary longitudinal spall may be required from the fore-most station cross-spall to the stern in order to facilitate proper bow alignment. The skeg assembly, deadwood plate, or deadwood weldment may now be raised to its position and welded to the center members or as required by your designer.

Before installing the sheer bar, it is imperative that you square each frame perfectly to the center line. This transverse alignment may be held by welding sections of rod from each side of the cross-spall of one frame to the center of the cross-spall of its adjacent frame. These alignment rods should occur between each frame for the continual length of the vessel. The sheer bars may now be raised into position and fitted into their respective notches provided at each bulkhead and frame. I do not recommend attempting to prebend the sheer bar. It is far better to spring it naturally during installation. When installing the sheer bars, you should be particularly cautioned against installing only one side at a time; rather, the bars on each side of the vessel should be installed simultaneously. This will prevent any possibility of stressing or distorting one side of the vessel.

I believe that it is prudent to install the cabin carlin bars and hatch carlin bars at this time. It will help stiffen the deck beams remarkably. These units are installed in exactly the same way as the sheer bars. (Fig. 197)

Longitudinal Bulkheads

The units to which I refer primarily are those of the center tank baffles, engine bed members, etc. These units are prefabricated on your welding floor in exactly the same way as the bulkheads. They will be covered with mesh in a like manner while also allowing for the folding back of a mesh margin to be wired into the hull shell. The units will be raised to their proper positions between their adjoining frames or bulkheads. (Plate 65) The mesh on the frames or bulkhead should be slit and pulled back slightly to provide for the welding of the transverse bulkhead perimeter to the rods of the frame or bulkhead. The mesh margin of the longitudinal member will be placed under the mesh of the bulkhead or frame. The mesh point, then, will be securely wire-tied. Additional details for other integral structures will be found in later chapters. (Fig. 198)

Modifications

As you may have noticed in the accompanying photographs, the examples do not necessarily agree with the procedures that have been explained here. There are many viable alterations to the basic web system which may be detailed by your architect. The opportunities for variation are virtually infinite.

meshing

Since the introduction of ferro-cement to naval architecture, chicken wire or aviary netting has been the mesh most used throughout the field. Usually it has been applied as four layers on each side of the rod laminate. When cemented, this results in a shell thickness of about ¾" to 7/8" weighing from 10½ to 13 pounds per square foot (based on two layers of ¼" rod).

Square Welded Mesh

While the weight of the foregoing systems is well within reasonable tolerances and the mesh is very easy to apply, it is the ultimate strength with which you must be most concerned. Based on hundreds of tests by scores of architects as well as numerous universities, it has become apparent that the chicken wire layup leaves a great deal to be desired in terms of bending, compression and tensile strength. The result of this testing was the introduction of a ½" square 19-gauge welded mesh with a light copper flashing. It proved to be far superior in all respects. Compare its attributes against chicken wire:

1. The mesh is manufactured of a higher tensile steel wire.
2. The wires are of a slightly larger diameter (19-gauge versus 22-gauge).
3. The mesh is stiffer both laterally and diagonally.
4. All horizontal wires of the mesh are welded on the same side of their perpendicular counterparts, not woven (Fig. 199), allowing consecutive mesh layers to be interlocked or nested for reduced thickness.
5. The increased steel content of the square-welded mesh layup allows a reduction in the number of layers necessary and consequently the cost of the mesh layup becomes comparable to that of chicken wire.
6. There is a remarkable increase in strength of the armature laminate thus allowing a notable reduction in thickness of the hull shell as well as a lighter resultant weight.
7. Because of the square mesh stiffness, it has the tendency to lie more fairly, whereas chicken wire is easily distorted by the slightest errors within the hull mold.

The copper flashing found on much of the square welded mesh is applied as a temporary protective coating while the mesh is in storage or shipment. As the mesh becomes exposed to weather, the copper will usually disintegrate, leaving the exposed steel wire. If the copper coating does not disintegrate, as in dry climates, an acid wash may be applied to the hull armature to ensure that all copper has been removed since the adherence of cement to copper is very poor.

Another version of the square welded mesh has become popular with ferro-cement boatbuilders. It is identical to the preceding copper washed mesh in all respects except that it is galvanized instead of copper coated, and should be considered by you as the first choice for your hull construction. The properties of galvanizing will protect the metal from oxidation not only during the curing process but also in the event of hairline cracks developing in the finished hull allowing salt water to seep in and attack the mesh. The latter condition can cause a dangerous impairment of strength as the rusting of the mesh along the length of a crack often results in a complete breakdown of strength. The galvanized coating, of course, will prevent or retard this effect.

Mesh Quantity

Many designers have listed the quantity of mesh required for complete construction within their plan packages. More normally, however, only the hull area is listed. The hull area, of course, does not take into account the surfaces of decks, tanks, etc., so these units will have to be estimated by yourself prior to purchasing your material. If your architect has listed neither the mesh quantity requirement nor the hull surface area, you may refer to the chart for a close approximation according to varying hull types. (Fig. 200)

Considering Other Mesh Sizes

At this writing, very little has been said by the promoters of ferro-cement concerning the use of other square mesh types, but it is worth the discussion. I cannot recommend that you attempt to use any heavier mesh than the 19-gauge when applying it to a normally rounded hull of less than 60

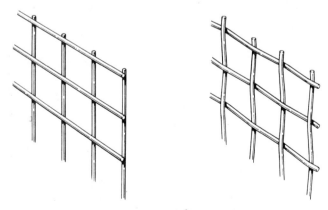

Fig. 199. The proper square mesh to use is the welded version (left). Woven square mesh (right) should be avoided as it increases shell thickness.

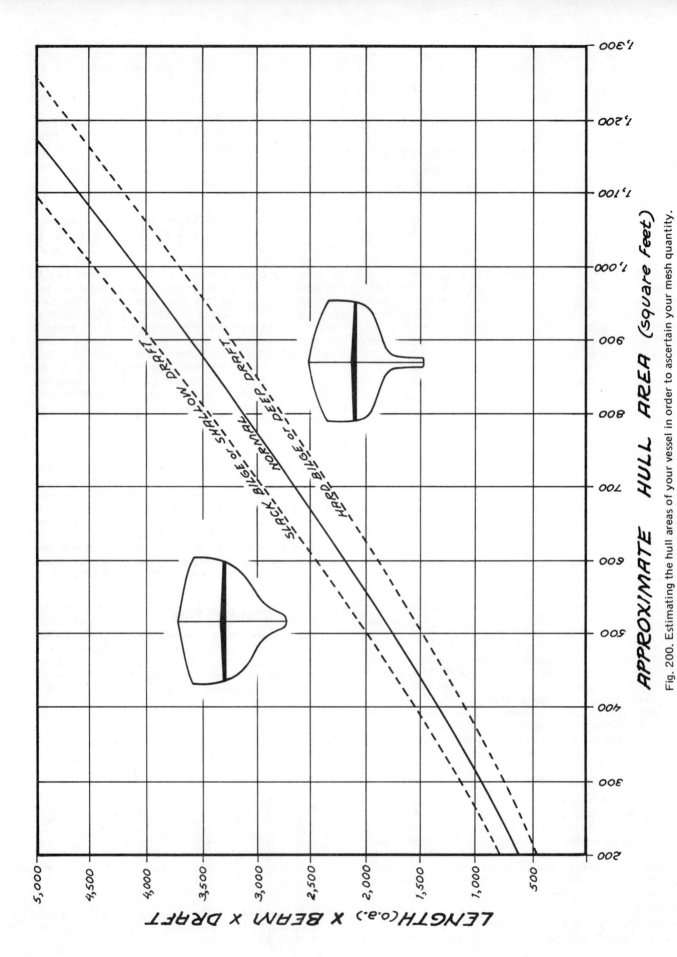

APPROXIMATE HULL AREA (square feet)

LENGTH (o.a.) x BEAM x DRAFT

SLACK BILGE or SHALLOW DRAFT

NORMAL DRAFT

HARD BILGE or DEEP DRAFT

Fig. 200. Estimating the hull areas of your vessel in order to ascertain your mesh quantity.

192

feet L.O.A. When considering relatively flat surfaces, however (such as barge or houseboat hulls, some powerboat hulls, decking, tank components and bulkheads), it would be perfectly acceptable to reduce the number of mesh laminates by substituting the 19-gauge wiring with 16- or 14-gauge square mesh. By doing so, a tremendous amount of work may be eliminated while producing a more naturally fair structure. The heavier mesh, being stiffer, tends to lie more flatly, thus reducing distortion.

In many cases, reinforcing rod may be eliminated completely through the substitution of equal thicknesses of heavy mesh. This will rarely cause mortar penetration problems as would be expected by using many more layers of a finer weave.

In areas of the hull, deck or cabin where non-structural concrete work is required, such as cockpit coamings, etc., a reduction in shell thickness and weight may be achieved through the use of a more closely fabricated mesh than ½". One-fourth inch, 19-gauge, square welded mesh is available universally and can be employed to reduce thicknesses significantly while maintaining an extremely high metal/mortar ratio. Because of its close weave, it is also very conducive to the shaping of small or finely sculptured parts.

Galvanic Action During Plastering

It is known that a galvanic current (electrolysis) does occur as a result of the zinc coating being immersed in wet mortar. The electrolysis causes the formation of minute bubbles of hydrogen which are trapped and expand within the concrete, causing cracks in the new hull unless certain precautions are taken. The addition of chromium trioxide to the mixing water eliminates this problem and will be dealt with further.

The galvanized square welded mesh, never to be confused with the most commonly available woven hardware cloth, is purchased in rolls of 200 feet, 2 or 3 feet in width. These rolls should be left wrapped until the day they are to be to be used.

Most builders prefer to apply layers of mesh in the same direction, horizontally, while others have preferred vertical or diagonal application. Because my recommendation is to apply reinforcing rods diagonally, I encourage only the vertical and horizontal mesh layup as the combination obviously introduces a multidirectional metal structure. If you prefer, you may combine both the vertical and horizontal systems.

Center-Pipe Meshing

It is important to note here that if your vessel has been designed with a center-pipe arrangement along the profile, it must be installed prior to laying any hull mesh panels. (Plate 66) The installation of the center-pipe has been fully described in Chapter 12 (see Fig. 148).

Horizontal Paneling

Because of the weight of lengths of mesh panels when applying mesh, it is absolutely imperative that a slight tension be applied horizontally to the mesh as it is being stapled or wired to the hull form. This can be effected by the use of "stretch blocks" as suggested by Jay Benford. This is basically a wooden clamp affair attached to each end of the mesh of the precut mesh panel. The "stretch block" of one end is secured to the scaffold or other firm structure

while the other end is fitted to a block and tackle and, again, to the scaffold or structure. The mesh panel is carefully lifted to the proper height against the hull, using hoisting lines as necessary to prevent severe sagging, while you take great care not to bend the mesh. The mesh is now stretched tightly with the block and tackle so that it fits snugly against the hull for the full length of the vessel. (Fig. 201)

You may prefer to simply unroll a precut length of mesh along the side of the hull, stapling as you go wherever it touches the mold in order to hold in its initial position. This system actually works just as well as the stretching method and does not require extra gear. I would suggest that the mesh be rolled in such a way as to bend away from the hull whenever possible as it will cause a more uniform fairness as successive staples are driven. (Fig. 202)

Mid-Mesh

The optimum situation in ferro-cement construction is that of evenly distributing the steel content throughout the concrete, eliminating concentrations. This is best achieved through the use of many successive mesh layers without the application of reinforcing rod. Unfortunately, this is rarely possible except under rigidly controlled circumstances and should not be considered by the amateur except for cabin structures, tanks, small detail and very small hulls. Substituting a rod with four or five additional layers of mesh increases cement penetration problems unless high-pressure pumping or Gunite equipment is used on plastering day. Reinforcing rod does serve the vital function of helping to fair the hull, particularly with open truss or pipe frame construction where no actual mold is used, and it creates the desired laminate effect so necessary for bending resistance.

Of course, reinforcing rod is absolutely necessary at any time you use chicken wire mesh. The rods will then serve the prime function of bearing the stress and compression loads.

An efficient compromise between "rod" and "no-rod" systems is available to the builder. One or two of the required mesh layers may be sandwiched between the reinforcing rod thus filling the otherwise unreinforced space in the rod area. For example, if the mesh/rod combination called for is three mesh layers on each side of two opposing rods, it may be satisfactorily altered to read: two mesh, one rod, two mesh, one rod, two mesh. This modification has been avoided by the amateur builder solely on the basis that it is more time-consuming even though the resulting structure is more consistently strong.

It is also important to note here that moving rods away from the neutral axis causes them to bear a greater percentage of the loads on the hull. This effect becomes far more noticeable when using chicken wire mesh.

It is also possible to reduce the rod thickness by incorporating additional mesh layers accordingly. The variety of combinations is endless (see Fig. 209).

Some builders suggest that three layers of mesh may be stretched and applied simultaneously. I do not believe that this saves any labor but instead complicates the procedure of cutting darts (fairing wedges) in the mesh as it is almost impossible to get to the undermost layers for this purpose. It also makes it extremely difficult to perfectly fair the underlying layers. *(Text continues on page 198.)*

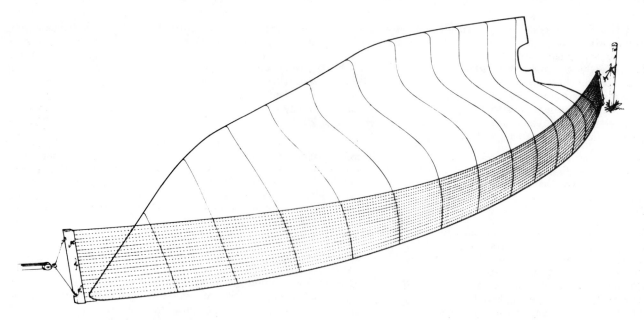

Fig. 201. Lifting a premeasured mesh panel against the hull using wooden stretcher and tackle.

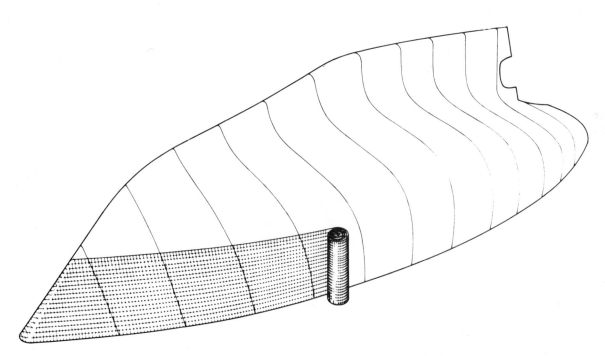

Fig. 202. Rolling a premeasured mesh panel onto the hull.

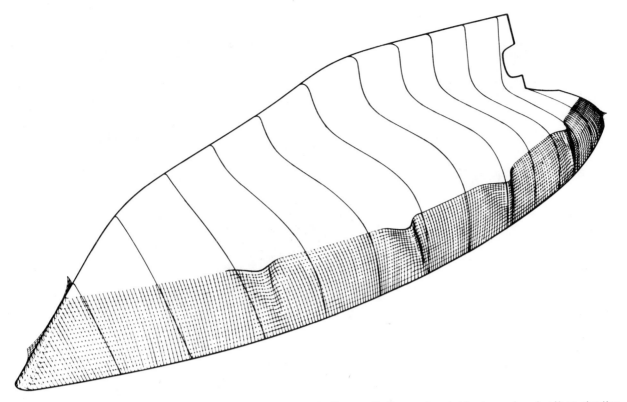

Fig. 203. After stretching or rolling the mesh panel onto the hull, smooth the mesh wrinkles toward each other, stapling as you go.

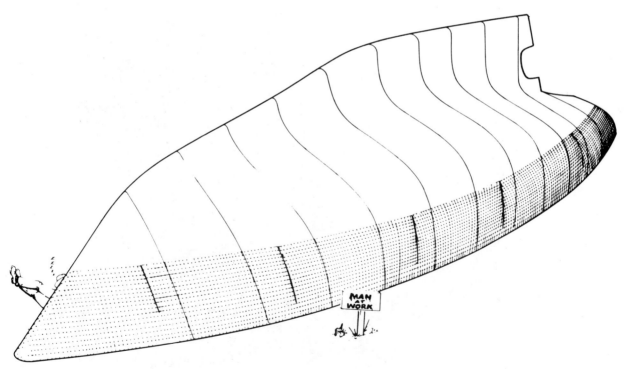

Fig. 204. The mesh wrinkles have been darted, thus allowing the remaining mesh areas to be stapled fairly to the mold.

195

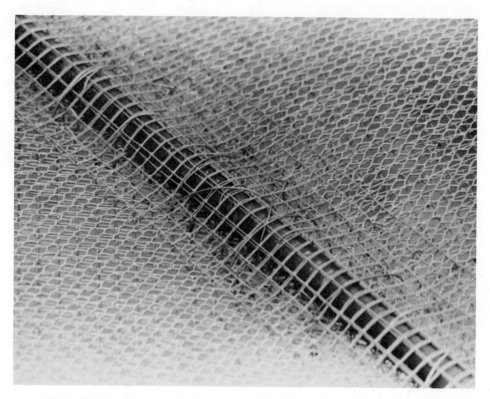

Plate 66. A typical center-pipe installation as seen from the inside of the hull. Careful inspection will reveal the two inner mesh layers. Note the wire lacing which crisscrosses the pipe angularly to the center line.

Plate 67. Wire-tying, a seemingly endless and thankless task, can take several months of spare-time labor on a large hull. Virtually anyone with boundless patience can do it and it can be confidently entrusted to the entire family.

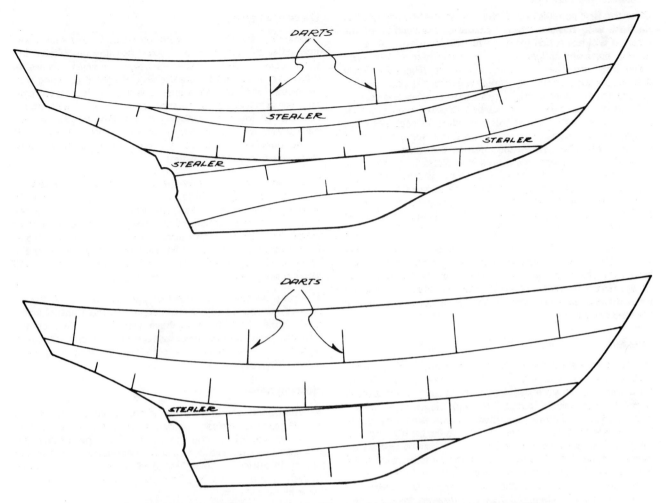

Fig. 205. The use of stealers in order to fill unavoidable gaps between mesh paneling.

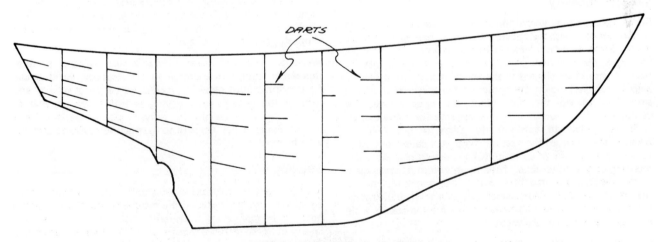

Fig. 206. Vertical mesh paneling.

(Continued from page 193.)

Once the first panel is raised, the edge should be adjusted to lie flush with the sheer line. Because of the rigidity of the mesh, it will not touch the mold or web framing in all areas, so the apparent bulges should be smoothed toward each other until the sheer line tends to distort. (Fig. 203) When this occurs cut a dart in the bulge, relieving the compression on the mesh. Once done, a triangular overlapping section will result; this overlap should be cut away leaving a butt joint in the mesh. Repeat this procedure at each apparent bulge. At the same time the bulges are being smoothed toward each other, staple the mesh to the mold or wire to the web, but only enough to hold it in place. (Fig. 204) The next lower panel of mesh to be laid is handled in an identical manner, taking care that the edge of the second panel butts evenly with the first; this may require some tailoring. Succeeding panels are applied in the same manner until the entire hull has been covered.

When you apply the second layer of mesh, the upper edge of the first panel should be positioned a foot and a half below the sheer so that it completely covers the butt joints of the first mesh layer. The void along the sheer will require an additional narrow panel. Now the vessel is completely covered once again.

Stealers

It will not always be possible or convenient to cover every inch of the hull with regularly shaped mesh panels. It will be entirely up to your own ingenuity to judge what is most economically feasible, but it is my experience to allow the mesh panels to follow their most natural tendency rather than in straight lines so as to cut down on darting and jogging. This procedure often results in leaving uncovered wedge- or crescent-shaped areas at the ends or middles of the panels. These uncovered areas must be filled in with scrap mesh cut to the proper shape. These are commonly called "stealers." Their use does not weaken the hull shell and may be used throughout the armature to cut down on time and waste. (Fig. 205)

Vertical Paneling

Some builders may prefer to use the vertical system instead of horizontal meshing because it avoids having to handle the extreme lengths of panels and is popular for installing the inner mesh during open truss or pipe frame construction. It should be started at the sheer by applying a row of staples (or ties) down the center of the mesh strip as it is smoothed against the hull. When the mesh crosses the center line, continue up the opposite side for a few inches. It is impossible to continue this strip for any distance because the angle of the keel is such that it causes the mesh to veer off in strange directions. In order to complete the meshing to the other side of the boat, you must start again at the sheer, down the hull, butting the new panel to the first at the keel. The first panel may take some adjustment but succeeding panels will be easier. Again, you staple or tie only up the center of the mesh,

As with the horizontal system, any bulges in the mesh must be "herded" toward each other by smoothing with the hand. Darts must be cut in the bulges and the triangular overlapping sections removed.

Additional panels are applied in a like manner taking care to stagger the butt joints and darts. When the second or third layers of mesh are attached, begin on the opposite side of the vessel from the preceding layer so that the butt joints at the keel do not occur in the same area. (Fig. 206)

Diagonal Paneling

During my travels from yard to yard, I have seen more innovative procedures than I could possibly describe within these covers. One notable variation, however, comes to mind that I feel worth mentioning. Diagonal paneling may be chosen with no detrimental effect on a vessel's strength and its application is claimed to have eliminated many hours of fitting while reducing shaping dart lengths. While I have previously stated that mesh is available in 3' widths, it is also possible to purchase a 2'-wide roll which seems to be more practical for this system.

The panels are applied in the same manner as the vertical method except that the lay of wire is at approximately 45 degrees, alternating its direction from layer to layer. As the mesh rounds the bottom of the keel, it may be allowed to follow its most natural direction with minimum alterations. An occasional stealer may be required if the course of the panel tends to separate from its adjoining neighbor. All other procedures are virtually identical.

The only objection to diagonal paneling is that of the increased number of joints required within the finished armature. While these seem to have very little effect on the ultimate shell strength when butts have been carefully overlapped, it does create the problem of smoothing mesh layers. (Fig. 207)

Nesting Mesh

As mentioned previously, one of the greatest attributes of the 19-gauge, ½"-square welded mesh is that all horizontal wires are welded on the same side of the perpendiculars. You can use this feature to a great benefit because it allows you to interlock or "nest" the mesh layers in such a way as to reduce the thickness of the shell while it maintains the same metal content.

From the illustrated examples, you can see clearly that when two layers of mesh are applied to each side of the re-rod, the reduction in total hull thickness can be more than 1/16" when mesh is properly nested. (Fig. 208) When using three layers on each side of reinforcing rod, the reduction is about 5/32". If at first your mesh does not nest properly, turn the panel over.

Although this may seem insignificant on the surface, remember that any increase in construction weights decreases the ballast capacity, raises the center of gravity and reduces the vessel's sail carrying ability. If we are able to reduce the shell weight by only ¾ pound per foot, the savings could easily amount to as much as 1,000 pounds on a 45-footer.

Stapling

If an upright or inverted wood mold system is being used, attachment of the mesh to the mold is accomplished with a heavy-duty manual or pneumatic staple gun. The first few layers of mesh should require only ½" staples, but as the armature thickness increases during construction, the staple length may reach ¾" or better, so be sure to purchase or rent a stapling gun which can accommodate these sizes.

In previous chapters I have spoken of methods for fairing the mold but even if these instructions are followed closely, distinct irregularities can occur during the meshing phase of construction. One of the most notable but least considered errors is that of driving the staples across the wrong mesh wire. On observing a sample mesh strip, you will see that it is divided into an upper plane (wires running in one direction) and a lower plane (wires running in the perpendicular direction). If the staples are driven across the upper wire, it will add a slight additional thickness to the finished hull as well as cause the staple ends to interfere with the proper nesting of successive mesh layers. It is important that you be consistent in this regard, as it only takes one improperly placed staple to produce a visible wave in the armature. (Fig. 210)

If a partial or full mold is employed (the mold remains in the hull during plastering), the staples are left in the armature to hold it firmly in position. When an open mold system is used, however (the mold totally removed before plastering), all of the staples must be cut or withdrawn in order to facilitate the dismantling of the stations and ribbands. While this may sound very tedious, the cutting or pulling of staples actually proceeds very quickly **but only if**

they are visible. For this reason I suggest that the staple strips be sprayed with shocking pink or green paint before their insertion into the stapling gun.

Wire Ties

All open construction systems (truss pipe frame or wood mold) require that the entire armature be firmly wire-tied to achieve suitable rigidity of the finished form. (Plate 67) In the case of the open wood mold method, wire-tying occurs prior to the removal of staples but with truss and pipe systems it is done from the laying of the first reinforcing rod. Ideally, all rod intersections receive a tie passing through all mesh layers but some builders have found it sufficient to tie as few as every third. I cannot recommend this, however, as wire-tying also serves the important purpose of compressing the mesh layers firmly together to prevent bumps and hollows in the finished hull.

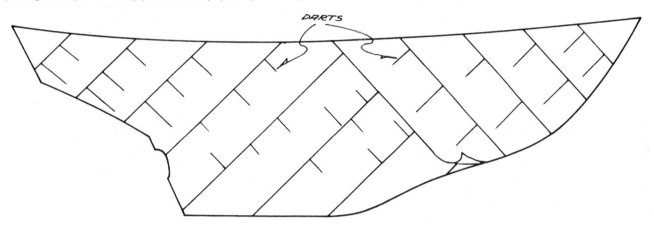

Fig. 207. Double diagonal mesh paneling.

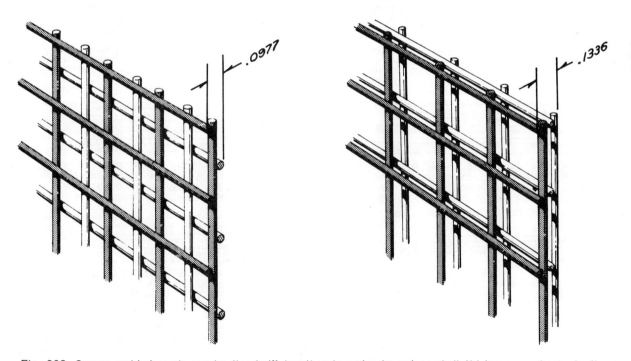

Fig. 208. Square welded mesh can be "nested" together in order to reduce shell thickness, as shown in the example on the left. Note here that the two mesh layers have been misaligned to create a uniform coverage. Two mesh panels which have not been properly nested (right) increase shell thickness markably. Note here that the two mesh panels have not been misaligned, thus causing greater expanses of unsupported concrete.

199

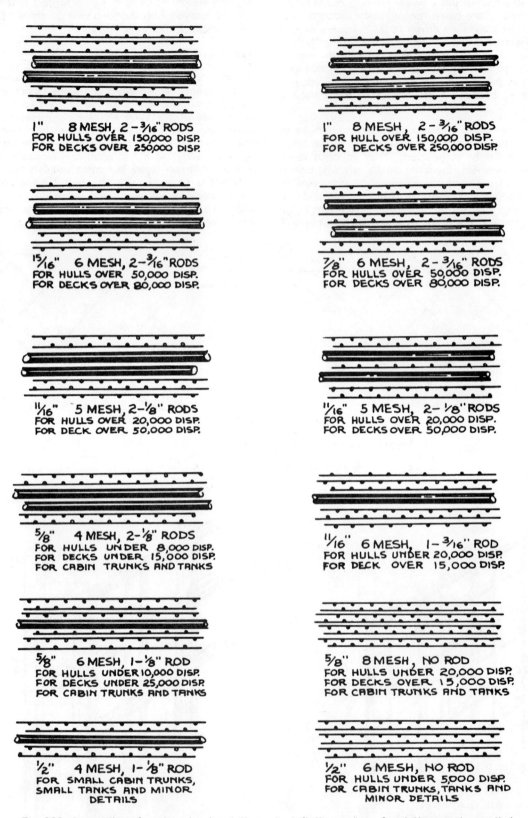

Fig. 209. A sampling of mesh and rod variations. An infinite number of varieties may be applied according to the specific circumstances. Be sure to check with your designer before employing an unusual combination.

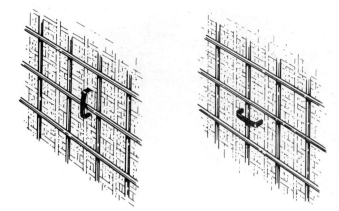

Fig. 210. Improper stapling (left) may interfere with successive mesh layers. Properly stapled mesh (right) may alleviate inadvertent bumps.

As with staples, there is a right and wrong wire tie. As I have described previously, the loop passes over the lower mesh wire in order to reduce thickness and to aid in the proper nesting of successive mesh layers. This is particularly important when affixing the last mesh layer, as an improperly placed tie may protrude above the hull surface creating the probability of exposing minute points of steel as the finishing trowel skims the hull. It is also important that the twisted ends of every wire tie be pushed back into the armature to avoid the same problems. (Fig. 211)

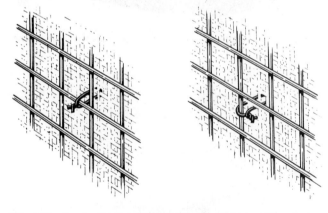

Fig. 211. An improper wire tie may interfere with successive mesh panels, thus increasing thickness (left). The proper wire tie (right) is wound to the lower mesh wire and is turned back to avoid inadvertent bumps.

The wire ties most generally seen are of a 19- or 20-gauge steel, purchased in 10- or 25-pound wire rolls. To mass-produce prebent wire ties, wind successive loops around a piece of wood or steel measuring 3" x ¼". Remove the loops as a single bundle and cut into two sections across the flat portion. (Fig. 212) Commercially manufactured wire ties, prebent and cut, are also available in keg lots. One of the most impressive suggestions to reach me has been to use women's corrugated hairpins (not bobby pins). They are manufactured of an annealed steel, available in several weights and lengths, and are lacquer-coated to retard premature rusting. The corrugations of the haripins allow you to pre-position hundreds at a time from the inside with little concern for their falling out. This eliminates the necessity for a second person to be positioned on the outside of the hull for the twisting operation.

Several mechanical systems are available to you to aid in wire-tying. Aircraft wire twisters can be purchased at most automotive supply stores but seem to be of little advantage. Jay Benford suggests welding Vise Grips to a drill bit coupled with a variable torque release bit. The power source used for this gimmick is a variable speed drill. The unit compacts the mesh while twisting the wire and breaks off the excess. Instructions for making and using the wire-tie tool are thoroughly described in *Practical Ferro-Cement Boat Building*, by Jay Benford and Herman Husen (copyright 1970, Jay R. Benford & Assoc., Inc., Seattle, Wash. 98103).

The only drawback to the wire-tie tool is that it is a two-handed unit and may take up as much time as it saves.

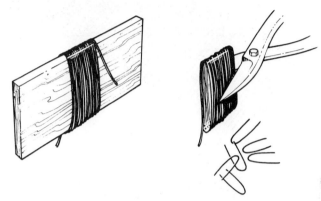

Fig. 212. Making your own wire ties.

I think it is purely a matter of personal preference as to which method each builder uses. If I had a choice, I would simply use a pair of common dikes. With these, the wire is simply grasped ¼" above the mesh, twisted approximately ¾ of a turn (or until proper compacting is achieved), then a quick squeeze nips the tie to a convenient length. Always push the tie back into the mesh so as not to interfere with the next mesh layer or plastering.

Jogging

Because of the inherent resistance of square welded mesh to form itself to a compound shape, darting does not solve all of the fairing problems. After the darts have been cut, low bumps will be readily apparent along the hull. The method

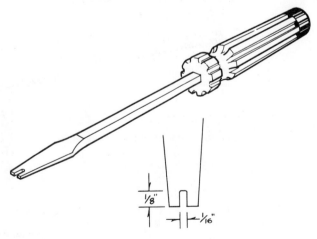

Fig. 213. A mesh "jogging" tool fabricated from a common screwdriver.

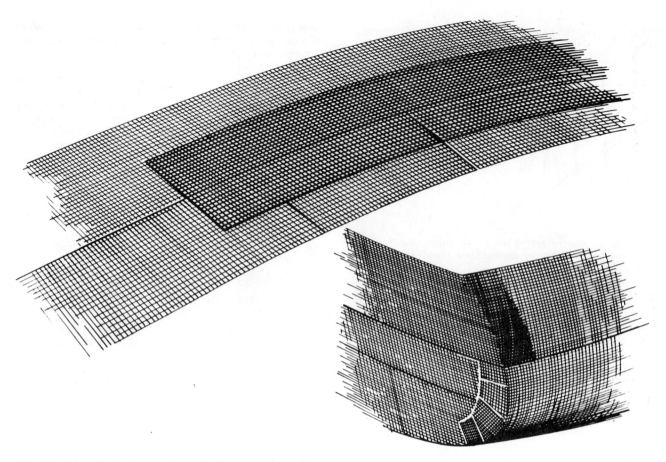

Fig. 214. Overlapping mesh layer. Note that the butt joints are always well staggered. The corners, such as the edge of the transom, denote an alternating butt direction.

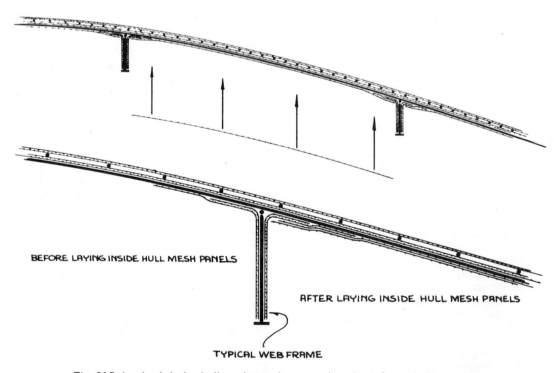

BEFORE LAYING INSIDE HULL MESH PANELS

AFTER LAYING INSIDE HULL MESH PANELS

TYPICAL WEB FRAME

Fig. 215. Laying interior hull mesh panels on a web or truss frame hull armature.

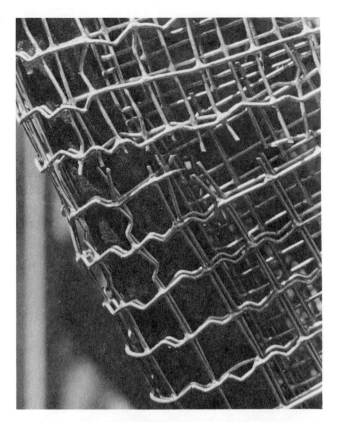

Plate 68. "Jogging" helps to form the mesh into the compound shape at **Flicka's** stem. Tuning the successive mesh layers over the entire surface can ensure the ultimate fairness.

for removing these bumps is called "jogging," which simply means that the mesh wires are crimped or twisted angularly, thus shortening the mesh while increasing its tension in a given area. (Plate 68) The jogging may be done with a standard screwdriver with a 1/8" notch cut into its blade. (Fig. 213) Simply insert the screwdriver, give it a slight twist, and go on to the next square. As each jog is made, you will actually see the bump change its shape. Jogging goes very quickly and you will be amazed at the degree of fairing control achieved by this means. In order to prevent bumps from becoming magnified through successive mesh layers, each layer should be jogged to "perfect" smoothness before proceeding to the next layer. When web or pipe frame construction is being employed, jogging may be done from inside or outside of the hull. Take care not to overstress the mesh as it may cause flat spots. Check your work continually by running a wood or plastic spline along the hull at various angles.

Joints, Laps, Corners

Obviously, the more consistent the mesh distribution, the stronger and fairer will be the finished hull. It is imperative, therefore, that mesh butt joints be well staggered from layer to layer within the armature. (Fig. 214) The farther apart the better. It was once thought that folding a mesh panel in half to create two layers would speed up its application, but it actually has the effect of concentrating the layer joints as well as making it almost impossible to fair out the underlying layer. (Plate 69)

Any overlapping mesh, however small the area, unless called for by the design, will surely create a ridge or bump. If so

much as one loose wire is allowed to stray over an adjacent panel, it will make itself known in the form of a bump on the finished hull.

When meshing over outside corners such as the transom edge or bow, always extend the mesh layer fully around and beyond by at least 6". This may require the cutting of small darts to cause the mesh edge to lie flush. When applying successive mesh layers, stagger the direction of the edge to avoid concentrations of joints.

As you encounter "inner" corners especially during truss and pipe construction or when constructing bulkhead webs, the same changing of directions of laps must occur as previously described. The problem of working mesh doubly into the angles and crevices is not easy but may be solved by first impressing the mesh part of the way with the rounded end of a ball peen hammer, then finishing it off with a common pizza cutting wheel. Extra wire ties may be necessary to hold the mesh firmly in position.

Divots

Upon completion of any layer of mesh, look for severe indentations which exceed 1/16" regardless of the size of the area involved (not including dimples caused by wire ties or staples). These must not be corrected by attempting to pull the mesh outward as you will lose the advantages of wire compaction. In many instances, however, an additional layer of mesh may be secured to the divot by carefully cutting a patch somewhat smaller than the indentation. Before installing the patch, bend all of the wire ends slightly inboard so that they will not protrude above the surface or interfere with additional mesh layers or plastering. After wiring the patch in place, check this work with the fairing spline to insure against overbuilding. Additional wire ties in the patch area may be required to prevent excessive thickness.

Plate 69. Four mesh layers butt-joined at exactly the same point—from two directions! Extreme errors such as this can render a vessel totally unsafe and almost worthless.

Applying Mesh to Inside of Truss Armature

If you have chosen the open truss construction method for your vessel on the basis of ease of construction, you are going to be in for a great shock. While meshing the outside of an open truss hull is essentially the same as with any other system, applying mesh to the inside of the hull is extremely difficult, painstaking and frustrating. The reason for this is simply that the mesh panels between truss frames will rarely allow their being precut in a straight line (except in those areas most nearly amidship). You will also find that applying mesh to the inner side of the keel is almost impossible unless you have bamboo arms.

In most cases the mesh panels between the truss frames will require either cutting the panel to a curve or "S" shape or darting the panel considerably as the mesh is applied.

As previously mentioned, your truss frames should have been mesh-covered prior to their raising to the scaffolding. The overlapping mesh margin must have been folded back flush with the hull surface prior to installing interior panels. The required number of interior mesh panels must overlap onto the frame mesh margin and should butt up against the web frame as closely as possible.

Most often the interior mesh paneling of an open truss vessel is applied after all of the outer hull mesh layers have been attached and faired. The inner mesh panels must be wired fully through the armature upon their installation. Although absolute fairness is not mandatory on the inside of the hull, you should strive for the same degree of armature compactness as with the outside mesh layers. Needless to say, it will not hurt your pride nor the appearance of the vessel if you are able to achieve a smooth interior contour. (Fig. 215)

Fishhooks

This is a term applied to loose wire ends along cut lines and stray ties which may impede plastering, cause exposed steel or create bumps as a result of additional thickness. If you run a gloved hand over the hull, the fishhooks can be quickly detected. These must be cured immediately by tapping them into the hull or running them down gently with the round end of a ball peen hammer. Keep in mind that the armature dictates the shape of the finished hull. Unfairness is rarely corrected successfully through the addition of excess mortar. Grinding and filling is a tedious procedure which should be avoided on the finished hull and every effort should be made toward perfection at this stage of construction. It is still possible to correct shallow spots by inserting wedges between the mold and armature. Exercise all of the patience you can muster at this point and do not let your anxiety or eagerness get the best of you.

Plate 70. Fred Bingham justifiably admires a remarkably fair meshing on Don Terry's Samson **Sea Smoke**. Note the careful attention paid to butting panels properly. This vessel is the standard of quality upon which the future of ferro-cement must be built. Later, plastering was equally fine.

reinforcing rod

For the past three or four years here in North America it seems to have been the general rule to use mild steel rod (under 80,000 psi) for lineal reinforcing between the mesh laminates. It has also been accepted, with little or no question, that a ¼" diameter rod is the only rod to use. The time is long overdue to dispel these hardbound fetishes!

Have you ever tried to straighten out a bent coat hanger? Then you know that it can only be done after hours of work. Imagine inadvertently bending or kinking a long reinforcing rod just as you are about to fasten it to your hull mold. It's going to take some doing to get it smoothed out; otherwise you run the risk of building a hump into your new boat. This is one of the basic problems with mild or soft steel. While it is certainly easy to bend, especially in areas of tight curves, it can cause headaches when trying to create a long smooth surface. Soft rod also has a way of scalloping outward between ties or staples, thus causing unfairness as well as weak spots due to the lack of compaction.

Because of the low tensile strength and lack of stiffness of mild steel, the rod diameter used for ferro-cement must be significantly larger than that of higher yield materials thus creating a greater shell thickness and total weight of metal and cement in order to achieve the same strength characteristics.

Many architects now design hulls specifically suited for the application of high-carbon, hard-drawn or tempered spring steel rod. These compositions range from 110,000 psi to 250,000 psi, clearly almost twice to four times stronger than mild steel. The ideal strength, however, seems to be 130,000—160,000 psi. Above this the rod becomes difficult to work and even brittle when bent. The general term commonly applied to these categories is "high tensile" rod and their characteristics are stiffness and resistance to bending. When cut with bolt shears, they resound with a distinctive "ping" and, during application, lie more smoothly over irregularities which may exist in the mold system.

Only a short time ago, ferro-cement opponents (and proponents) claimed that small boats could not be built of concrete. Through the use of square welded mesh and small diameter spring steel rod (or no rods at all) we have now successfully produced scores of beautiful vessels under 20 feet in length. The example of my father's 12' Bahama sailing dinghy, Concha (Fig. 216), will also attest to the advancements made since the publishing of the first construction techniques. While it has also been said that ferro-cement multihulls were quite impossible to successfully construct, this has long been proven to be a promotional fallacy.

High tensile rod is an absolute necessity when building to the open mold, truss and pipe frame systems as the tendency to fairness provides considerably fewer supportive members than mild steel. To attempt these systems with a softer material is to run the risk of corrugating the finished hull with no possibility of remedy.

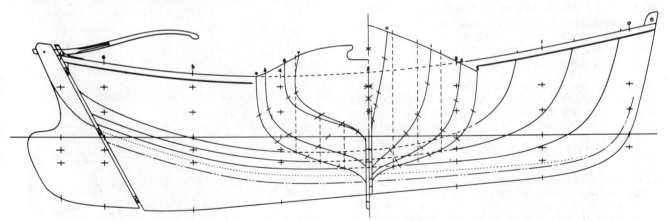

Fig. 216. Fred Bingham's 12' Bahama Dinghy, Concha, is an excellent example of an extremely small ferro-cement design, long thought to be impossible. While her displacement is extremely heavy (750 pounds) compared to most other 12-footers, it is the nature of the Bahama Dinghy to be so. In service, these dinghies were either ballasted with rocks or loaded down with shellfish. In their original form, they were excellent sailers in heavy seas and make ideal workboats.

Fig. 217. A simple experiment using a piece of cloth to help visualize diagonal stresses when torque is applied.

Rod Directions

Most ferro-cement boats constructed up to 1972 have been built purely on the basis of ease of construction. It was simpler to run rods fore 'n aft instead of athwartships, so that's the way it was done with little or no regard for the stress or compression lines of the vessel. For instance, the major strains subjected to a deck structure are transverse compression and bend, yet most ferro-cement builders have run the reinforcing rods lengthwise. In the areas of the chain plates, the hull is under tremendous vertical shock loads, but few vessels have compensated for them.

The diagonally oriented construction approach is by no means a new theory or of my own innovation and can be clearly borne out through a study of the past century of marine design. The problems of reducing twist with a wooden hull girder have long been solved through the installation of double-diagonal steel or bronze strapping systems. The rules for this diagonal strapping have been meticulously spelled out by Nathaniel Herreshoff, Nevins and even by Lloyd's. Besides the strapping systems for hulls, there is the proven superiority of double-diagonal planking, diagonally molded plywood and diagonally oriented fiberglass shell layup. Personally, I find it extremely difficult to ignore these graphic records of structural improvement regarding torsional rigidity.

The greatest single fatigue factor in any ship is its continual twisting about its longitudinal axis as it rolls through a seaway. It is this very twisting that causes the seams of wooden hulls to open, of joinery to break apart, of porthole frames and decks to leak and to bring about the ultimate failure of the hull shell strength. The stress and compression on a vessel's structure are rarely static or consistent. They are constantly changing direction and degree. A single experiment may be tried to help visualize the stress lines of a flexing plane. Stretching a piece of cloth flatly from hand to hand, move each end up and down in opposite directions. You will see by the wrinkles in the cloth that the stress actually occurs diagonally, never perpendicularly or along the thread of the cloth. (Fig. 217)

In order to best compensate for the multi-directional stresses about the vessel, it is my feeling that a 45 degree diagonal lay of reinforcing rod is the most logical and strongest approach. (Plate 71) At the instant the structure is under tension or compression, both rod layers (assuming two layers are used) bear the load. If the structure is being bent, regardless of direction, both rod layers combat the action simultaneously. No single rod layer is left to do the whole job alone, no matter how short the period of time may be.

The double diagonal rod system also has the important benefit of more uniformly fanning out the rigging stresses into the hull shell while softening many of the tight vertical curves encountered during construction when applying the common vertical and horizontal rod system (one of the former objections to the use of high-tensile steels).

Quite obviously, there is a point where your convenience and patience may override the structural benefits of diagonal reinforcing. Typical is the construction of ferro-cement decks bearing wide carlins or in the development of concrete cabin trunks, as they would require bending and attaching additional scores of short rod lengths. For my own boat, I would consider my labor as much less significant than the added torsional strength but many will not feel as I do in this regard.

Compression Side vs. Tension Side

Very little has been said concerning which direction of rod should lie on which side of the matrix even though it can make as much difference in strength as 30% (bending resistance). It has long been known that a hull shell has a very definite compression side (inboard) and tension side (outboard) and that these stress lines vary in intensity and direc-

Plate 71. Diagonal reinforcing rods may be clearly seen as the first layer is applied to a 20' **Flicka** sloop. A Bingham design.

206

tion as the ship flexes. Because the compressive strength of reinforced concrete is approximately ten times that of its tensile strength, it is quite logical to concentrate the greatest possible tension girder toward the outside of the hull. Because a normal hull will flex far more along a vertical axis rather than horizontal, the obvious conclusion is that the horizontal rods (when used) should lie on the inner side of the armature.

Unfortunately, most wood mold systems using horizontal ribbands discourage the first application of horizontal rods, as it then creates a fairing problem. This problem is magnified when only one or two mesh layers lie between the rods and the mold. One answer to this, of course, is to lay the inner rods on a diagonal orientation. It will represent a loss of strength over a first horizontal layer but a very definite gain over both layers of horizontal and vertical.

This same problem may be related to deck construction. Here we find the tension; thus the transverse rods should also be to the upper side of the armature.

Rod Diameter vs. Rod Spacing

A few designers, realizing that a small hull would have been grossly overbuilt with two opposing layers of ¼" rod, have reduced the structure to only one parallel rod layer or they have opened the distance between rods (thus introducing larger areas of nonreinforced concrete between inner and outer mesh). The stronger and more logical solution for reducing thickness and weight, therefore, is to recommend

Plate 72. Drawing steel rod from the outside of a coil into the homemade rod straightener.

Plate 73. The rod straightener, simply constructed from three V-belt pulleys. Note the screw adjustment for one of the pulleys on the side of the assembly.

207

smaller rod diameters while closing up the rod spacing considerably to maintain an optimum cement/steel ratio.

Straightening Rod

When your reinforcing steel is purchased, it will arrive in bound coils (usually 200 pounds) of several hundred feet each. Rod lengths should never be pulled from the center or top of the coil as it will cause the rod to emerge as a spiral that will cause severe twists and kinks. Rod must be removed only by unrolling from the outside of the coil and this is best accomplished by setting the coil on a revolving horizontal spool.

Because of the circular bend of the rod caused by the manufacturer's coiling, it requires straightening before applying to the hull. This is done with a "rod straightener," logically enough, which is an affair constructed of three pulley sheaves bolted to a steel plate. (Plates 72 & 73) The sheaves are arranged in a triangular pattern with one being adjustable (inward or out) to control the degree of pressure required. As the bent rod passes between the sheaves, it will emerge as a workable shape.

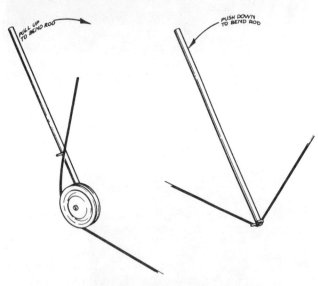

Fig. 218. Homemade rod and bar bending irons.

Bending Rod

As previously mentioned, there will be areas of tight curves to which a non-prebent rod will not conform due to its inherent stiffness. It is only rarely that hard-drawn, spring steel can be bent tightly by hand except in the smallest diameters and this will hold true for large mild steel and re-bar as well. My suggestion is that you construct portable bending irons of the following type: for large curves, bolt a 4–6" pulley to the end of a 3' length of ¾" or 1" pipe. Weld ¼" or 5/16" dowel or bolt to the same side of the pipe, approximately 4" above the pulley. To bend rod, place it against one side of the dowel and into the opposite side of the pulley. Working the iron back and forth along the length of the curve; exert leverage against the rod until the desired shape is achieved. For tight bends, weld two bolts or dowels to each side of a 3' section of pipe. To use this iron, place the rod against the outside of one dowel (away from the direction to be bent) and to the inside of the other, then draw the pipe handle back until the desired bend is achieved. (Fig. 218)

Diagonal Rod System

With the diagonal rod system almost all rods on the vessel can be laid as continuous lengths around the entire hull without "marrying." After completing the inner mesh, begin installing rods amidships to establish an angle of approximately 45 degrees. Place the upper end of the rod 1/16" below the sheer with its loose end passing under the keel, stapling as you go. Press the rod firmly against the hull, bending outward as necessary to eliminate pulling away at the turn of the bilge. Before fastening the rod around the corners or edges of the keel, prebend the approximate radii required. When the rod is properly fitted, continue stapling under the boat and up the other side of the hull, once again terminating the rod 1/16" below the sheer marks.

After your first layer of rod has been fastened to the mold or web system, use a wooden batten or plastic spline to check for any unfairness. Run the batten in all directions and angles over the entire hull and if you find an occasional rod protruding beyond the others, "jog" it for additional tension or until fair.

Attaching the second layer of rod proceeds identically to the first but at a perpendicular angle. Under no circumstances should any of the rod intersections be welded to avoid wire-tying-or stapling. Welding has the effect of corrugating and shortening or lengthening the rods undesirably, thus distorting the hull. (Plate 74) The tremendous heat created by the arc also burns the carbon of high tensile steels, weakening the rod severely. Those situations which require light tacking will be discussed under separate headings in the next chapter. Upon completion of the second layer, check your work with the fairing batten once again and make whatever corrections are necessary to achieve a flawless surface. (Plates 75 & 76) Remember at all times that any bump or hollow in your metalwork will be visibly apparent after plastering.

Horizontal and Vertical Rod (Wood Molds)

If you choose this system it makes no difference where you begin attaching. It is imperative, however, that you work on both sides of the hull simultaneously to avoid distortion of the vessel. To attach rods only on one side at a time will cause even the strongest mold to twist to some degree.

After the inner mesh layers have been carefully faired, the vertical rods are attached to the mold. Align the upper ends about 1/16" below the sheer ribband, running the rod down the hull, under the keel and up the other side to just below the opposite sheer. In the areas of tight curves, the rod should be prebent to prevent its tendency to spring away from the mold, especially at the garboard (turn of the bilge) and edges of the keel. If the rod lies slightly away from the hull at any point, drive additional staples to draw it snugly to the mold. (Plate 77)

When all vertical rods have been placed, run a wooden or plastic spline in different directions and angles over the entire hull to make sure that your work is perfectly fair. If you find an occasional rod which seems sprung outward or loose, it should be "jogged" for additional tension or until fair.

The placing of the horizontal rods immediately follows the attachment of the vertical rods or mid-mesh layers (if applicable). The first rods generally placed are those at the sheer that are positioned approximately 1/16" below the top of the upper ribband. As with the vertical rods, always work from one side of the vessel to the other, spacing successive rods as evenly and fairly as possible in accordance with your plan specifications. If rods are allowed to rise and fall in a zigzag fashion along the hull, it may severely impair your visual judgment of fairness, and your project makes a good impression if rods are beautifully fair when curious visitors make their calls. (Fig. 220; Plates 78 & 79)

It is imperative that you staple (wood mold systems) or wire-tie (pipe and truss frame systems) as often as possible to prevent rods from scalloping between fastenings. It takes little pressure to push the armature outward as much as a quarter of an inch so be very critical on this point when looking over your hull. When you press at various points along the hull there should be no visible inward movement. If there is, drive more staples (see Plate 77).

Under no circumstances should any rod intersections be welded for the purpose of hastening the construction by the avoidance of wire-tying. Weld beads have the effect of drawing adjacent steel around themselves, thus scalloping and corrugating the rods undesirably (see Plate 74). This scalloping shortens the effective lengths of the rods, both new and previously placed. Welding also weakens the higher tensile steel by burning up its carbon content. There are, of course, situations that will require light tacking, such as in the deadwood area, but this should have no effect on the finished strength when properly designed. These conditions will be discussed under separate headings in the next chapter.

Spacing Rods

Your design package should state the maximum parallel distance between rods although slight variations are generally acceptable especially in very tight or complicated areas

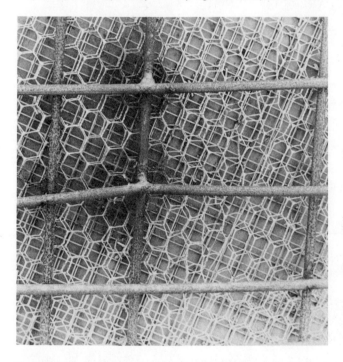

Plate 74. Note the kink in the rod caused by the heat of welding. Successive welds can distort the entire hull, add to shell thickness, or weaken the armature when high-tensile rod is used.

Plate 75. The obvious bumps along this vessel's keel are purely the result of the impatient builder wanting to "get on with it." Each phase of construction must be correct before proceeding to the next stage.

where you will have to apply your ingenuity and judgment. Most rod spacing will be consistent throughout the armature, however, and may be achieved neatly with the use of spacing hooks or irons. (Fig. 221)

The spacing hook is a short piece of scrap rod (preferably mild steel) bent into the shape of a "c" and is visually restricted to the attachment of horizontal rods. In use, six or more are placed at an even distance along the previously installed rod. The new rod is lifted into, and suspended in position by the hooks, then stapled to the mold (or wired to the truss or pipe frames). Once the new rod is completely attached, the last rod pair is squeezed together in order to remove the hooks. The obvious disadvantage of spacing hooks is that the application of rod may proceed only from the top down which may not always be desirable; vertical spacing with this method may be almost impossible unless the one hook alone is slid along the rods as attachment proceeds.

The spacing iron is usually a steel bar with two pairs of short metal dowels welded to it. The distance between the dowel pairs represents the desired rod spacing. To use the spacing iron, the upper dowels are placed over the previously attached rod while the lower pair receive the new one. As the new rod is attached to the mold or armature, the spacing iron is slid along at a short distance ahead of the work. The spacing iron may be used for any rod direction or angle.

Rod Stealers

Because of the tapering of the hull at its ends and the expansion amidships, it is impossible to maintain an even spacing throughout the entire length of the hull and there will be areas where a change of rod angle will be desirable such as the upper portion of the aft end of the keel. Stealers or short intermediate rod lengths may be attached

Plate 76. Grace in steel. **Flicka's** diagonal rod system at the turn of the bilge.

Plate 77. The outward scalloping of these reinforcing rods caused by too few staples will only add to the finished hull weight and the extreme difficulty in achieving a smooth appearance.

Plate 78. The armature of the 32' hermaphrodite brig, **Alicia,** being constructed by Norman Bonnenburger, is a prime example of an impeccably executed vertical/horizontal reinforcing rod system. A Bingham design.

to the mold or armature to fill in open spaces created by the changing hull shape. Although the stealers do not run continuously through the length of the vessel, they do not jeopardize strength. One end of each stealer may be married to the regular hull rods if the builder desires (see Fig. 220 and Plate 79).

Cement Decks (Upright Systems)

As I indicated previously, my rod direction preference is transverse for direction. In this way the vertical hull rods, or the outer diagonal rod layer, may be bent inward at the sheer ribband and continued across the deck mold ribbands. Where the rods meet a hatch or cabin carlin screed, they should be cut approximately 1/16" short. If, however, the rods are to reach the entire distance from sheer to sheer, it is best to marry the rods approaching from the opposite side. Be sure that these joints are well staggered in their positions. If you intend to construct ferro-cement hatch coamings, mast collars, etc., simply bend the rod to the vertical position and nip off 1/16" below the finished height. (The construction of carlins is fully discussed in later chapters.)

When you build to the pipe or open truss systems, the rods most usually run longitudinally. The first rod is most generally placed along the carlin. Successive rods are spaced outboard until reaching the sheer. As the rod ends reach the edge of the deck they should be bent over the edge, cut approximately 8 to 10" below the sheer and wired into the hull armature, flush with the second rod layer. (Plate 80)

Cement Decks (Inverted Systems)

If your boat is not to have a bulwark, bend the second layer of diagonal hull rods (or verticals) inboard and insert through the slot provided between the hull mold and deck crown mold. Staple the rods transversely across the deck mold, cutting slightly short of the carlin, or marry to their opposing counterparts as described for the upright method.

A variation is that of first laying rods completely across the deck mold, then bending them upward at the sheer and stapling to the hull armature for a distance of about 1'. This method seems to be somewhat easier as it provides a more accurate sheer line.

If your boat is to have a bulwark, the second layer of diagonal hull rods (or vertical rods) is bent at the sheer, inserting through the slot between the sheer and deck mold (if hull has been preformed) and cut short approximately 1' inboard. (Plate 81) The deck rods are now stapled fully across the deck mold, allowing their ends to protrude through the slot at the sheer. The deck rods are now bent downward against the bulwark mold and cut 1/16" short of the bulwark line. Longitudinal rods may be called for along the bulwark armature, and these will be applied in the same manner. (Plate 82)

Marrying Rods

The reinforcing rod need not be of continuous single lengths so don't worry if they fall short of the required distance. Additional rod may pick up where the other ends by marrying the two together for about one foot and joining them with three or four firm ties. If marrying is necessary for successive parallel rods, stagger their positions well to prevent weak areas.

Stem Rods

As your vertical or diagonal rod application reaches the angular rise of the bow, it is quite possible to bend the rods under (or over) the stem continuing up the other side. Some builders, however, will prefer to prebend "V" or "U" shaped rod sections, fixing these in place along the stem prior to fastening the hull rods. The hull rods, then, are married to the stem rods and carried up to the sheer. The advantage of the latter system is that the marrying of the rods adds significantly to the amount of steel in this vulnerable area. It also gives you more control over the fairness of the bow profile. Separate stem rods may be used regardless of the rod direction employed.

Plate 79. Bonny's **Alicia** already shows a remarkable fairness which will pay off on plastering day. A Bingham design.

Corner Rods

In areas such as the sharpest turns at the bottom of the keel and at the transom corners, there are always a few obstinate rods which refuse to conform with the others. They may not be readily apparent at a glance, so hold a wood batten against the hull profile to check for these critters. The corner rods which lie outside of the fair line may usually be corrected by bending the "V" section up or down until they align with the others. Because there are few lines on the hull that are as noticeable, take your time to work these areas properly. (Figs. 222 & 223)

Cross Wiring

In the thinnest portion of keels and skegs vertical rods will

(Text continues on page 215.)

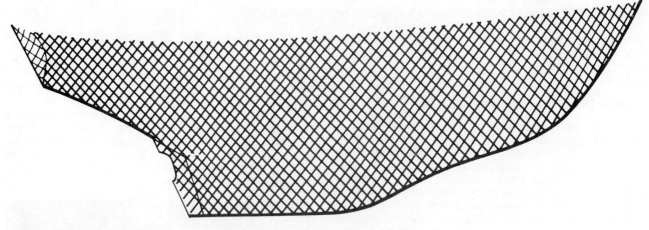

Fig. 219. A typical diagonal rod system.

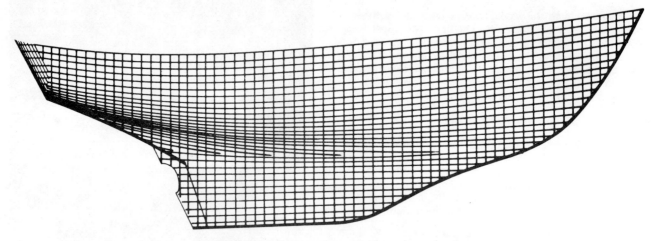

Fig. 220. A typical horizontal/vertical rod system.

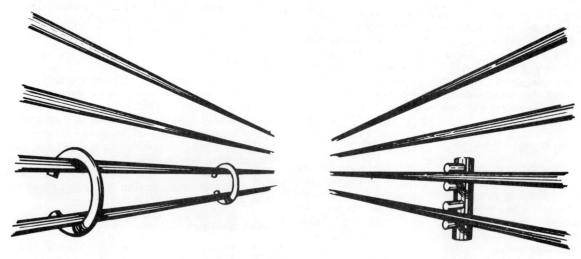

Fig. 221. Rod spacing irons.

Plate 80. A beautifully executed double deck-rod system on an open-mold hull armature. Fairness is exemplary.

Plate 81. Longitudinal deck rods bent downward to reinforce the bulwark. Transverse rods (when laid) will bend upward to become integrated with hull armature.

213

Plate 82. Deck rods protruding through slot between hull and bulwark. Alternating rods will be turned up and down to form a homogeneous structure of hull, deck and bulwark.

Plate 84. Tedious patching in order to correct armature irregularities.

Plate 83. The price of impatience.

Plate 85. The objective and critical builder's eye could have prevented this unfairness.

(Continued from page 211.)

tend to bow outward as a result of the tight bend required under the center line. Often no amount of stapling or wire-tying will solve this problem and the result will be an abnormal, undesirable thickening of the hull shape. To correct this, pass a long loop of wire over the rods on one side of the hull, through the armature and mold and wire-tie to the rod on the opposite side of the hull. If continued turning of the cross tie results in breaking of the wire, it may be necessary to substitute picture hanging cable as it is considerably stronger. If it is possible to reach the cross tie from the inside of the mold, a stick or scrap rod may be inserted into the loop and used as a tourniquet to help in drawing the bowed hull to its desired position. After the hull has been plastered the cross ties may be cut away or left in place as the circumstances dictate. (Fig. 224)

Jogs and Sisters

Occasionally some rods will protrude and bow outward beyond their adjacent neighbors, creating a wave in the hull surface. If additional stapling or wire-tying does not correct the problem, it can occasionally be forced to one side or the other to relieve compression, but this may become a tedious operation if many such areas require correction. The answer to this, then, may be the use of the jogging iron, consisting of a 3' pipe handle to which two steel studs are attached. The studs are placed on each side of the rod in

question while the handle is pulled slowly toward you. This causes the reinforcing rod to take on an "S" shaped curve which places tension on it—in effect, shortening the length of the rod slightly. Jogging proceeds very quickly and you can accurately tune and fair the hull in this way. (Fig. 225)

If a particular rod is creating a flat spot, however, it must be corrected by cutting to relieve tension, then interconnecting the loose ends with a short "sister" rod.

Judging Your Work

For years ferro-cement proponents have ranted and raved over the superior quality of this marvelous construction technique. Concrete boatyards have emerged like mushrooms while tens of thousands of amateurs have undertaken the production of their own dream yachts. Visions have been painted of achieving quality matching the most exemplary vessels today while claiming easy construction and vast cost savings.

Upon observing objectively existing ferro-cement examples, however, the very opposite seems to be most apparent. I am not being overcritical when I state that the majority of concrete yachts in existence today present a junky, bumpy and slipshod appearance when docked next to the average production vessel. While your budget may be a major contributing factor in the outcome of your efforts, the greatest

Plate 86. Just a little more effort could have improved this vessel's appearance and performance.

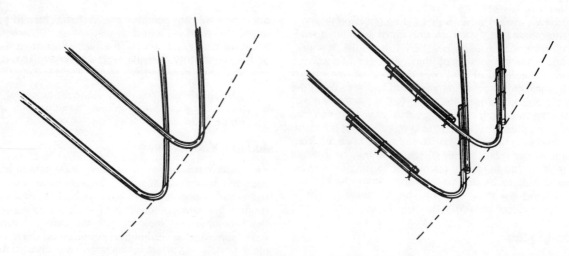

Fig. 222. Treatments of stem or transom corner rods.

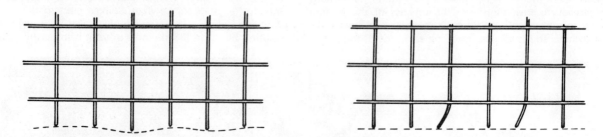

Fig. 223. If stem or transom corner rods do not lie in a fair line, those protruding above the others may be bent upward or downward for correction.

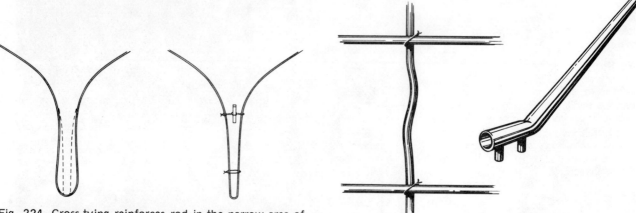

Fig. 224. Cross-tying reinforces rod in the narrow area of the keel to correct undesirable bulging beyond the predetermined shape.

Fig. 225. A rod "jogging" iron.

216

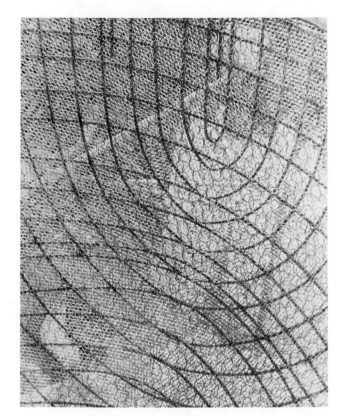

Plate 87. The true beauty to be hidden within. Untold hours have been invested in this armature to ensure a fine finished product.

failures have occurred as a result of the lack of patience and observation.

If the ferro-cement builders today would only make occasional field trips to local marinas to see how a good boat should look, they would surely return with a healthier and considerably more critical view of their own standard of quality. There is no reason for any ferro-cement hull to be apologetic for its lack of perfection nor should the material itself be used as an excuse for the builders lacking craftsmanship.

No fiberglass, wood, steel or aluminum boat ever became smooth and glossy overnight. It took manufacturers countless hours and untold sweat and exertion to evolve their flawless products. They had to overcome many of the same problems which you are now encountering. The fairing and smoothing techniques described in this and other chapters must be religiously and patiently adhered to if you expect your own boat to emerge with pride, value and the dignity she deserves.

Ferro-cement, in truth, does not require the years of practice demanded of the experienced ship's carpenter, but if you think the material allows complacency or forgives sloppy work, you may be in for a rude awakening. If you are depending upon plastering over your divots and bumps at this stage, the result will require torturous grinding and filling to cover up all that could have been solved at the outset.

Some builders seem to be totally blind. They can run their hands over the armature, completely insensitive to waves and obvious eruptions of steel and mesh. For this reason I have included a Rogue's Gallery of miscues, faux pas, and technological abominations. (Plates 83—86) While I have run the risk of offending the sensitive natures of some well-intentioned builders, their examples will serve as my sacrificial lambs for the benefit of you, the builder. Do not be discouraged by what you see, but rather be inspired to do a better job and continually strive toward unbiased and truly reflective observation. (Plates 87 & 88)

Plate 88. A proud **Andromeda** glistens in the morning sun in anticipation of the day's plastering. The builder's hours will be well rewarded.

The following chapter, **WIRE PLANK,** is condensed from *A Revolution in Ferro Construction,* by Platt Monfort, copyright 1973, Aladdin Products Inc., Wiscasset, Maine. Edited by Bruce Bingham.

wire plank

What Is It?

WIRE PLANK refers to a system of eight parallel continuous wires held in place with widely spaced light-gauge cross wires. The material is supplied in 500-ft. rolls and is the invention and product of Aladdin Products, Westport, Maine. It provides an alternative for shaping hulls. This system enables application of steel wire reinforcement in continuous strips transversely, diagonally, and longitudinally on a hull armature.

When the hull is completed, there are no joints, butts or overlaps in the wire. Every strand is full length, and as these strips are only 3½" wide, they are easily shaped in compound curvature.

The large diameter wire used in **WIRE PLANK** provides more than double the steel content provided by ordinary chicken-wire mesh; up to 14 per cent by volume and 70 pounds per cubic foot. This content increase is somewhat lower when compared with 19-gauge square welded ½" mesh. This high steel content provides greater strength; therefore hull thickness can be reduced. Mortar penetration into **WIRE PLANK** is considered better, or at least equal.

Mechanics Of Open Grid Frame Design

The design philosophy of this system is based on applying a rigid mesh of heavy-gauge continuous wires over the outside of a rod framework.

The rod framework is made sufficiently stiff to carry the stress of impact and bending loads. The mesh provides a tough, high-strength diaphragm over the gridwork of the rod frame. (Fig. 226)

The resulting mesh produced by the multilayer diagonal stripping with **WIRE PLANK** brand wire provides a continuous three dimensional structure. There are, of course, no butts, darts or overlaps inherent with roll material to reduce strength.

The rod framework is located on the inner side of the panel. In this location, it carries efficiently the tension in bending instead of being buried in the neutral axis of a sandwich of lightweight mesh.

The cement bears very little stress. Its main function is simply to keep the water out. The part of the cement that ordinarily encased the rodwork (essentially a heavy core material between the mesh layers) is eliminated, resulting in a great deal of weight reduction.

WIRE PLANK is a registered trademark of Aladdin Products Inc., Wiscasset, Maine.

When using only a light rod framework, high tensile pre-stressed rod, ¼" diameter, is desirable to use longitudinally because of its very high strength. The fact that it is supplied in straightened condition makes it lie into fair hull curves like battens. Spacing width of the longitudinal rods is variable to suit the weight and duty of the boat.

When this high tensile, high carbon steel is to be used, it should be understood that no welding techniques are to be used for assembly. All joints should be made with ties, lacing, or similar methods.

The transverse frame rods are made from mild steel such as concrete re-bar or pipe, or truss frames (unless the vessel is built over a wood mold). The size to choose will vary with the displacement of the vessel. It should be as large as practical. This will be dictated by your designer.

Importance of Design

If you have already purchased a ferro-cement plan package for a vessel engineered for chicken wire or square welded mesh, do not attempt to convert the design for the use of **WIRE PLANK** without professional assistance. Your architect will be thoroughly familiar with the stress data necessary for this conversion and will work directly with the **WIRE PLANK** manufacturers to evolve the proper arma-

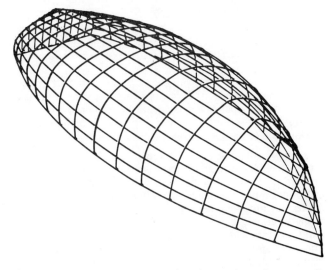

Fig. 226. A typical open grid rod armature for a small **WIRE PLANK** hull.

ture. One of the prime considerations when attempting such conversions is the weight difference. This is not a matter within the normal capabilities of the amateur builder as it will affect displacement, trim and stability.

The 13' Peapod (Plate 89) was constructed of **WIRE PLANK** brand mesh and plastered with Portland cement to a shell thickness of only 1/8". This boat weighs 250 pounds and has withstood severe abuse without any sign of weakness or failure. A 10' dinghy was then built using both materials in a composite construction. This boat weighs only 148 pounds and is extremely rugged. (Plate 90)

Basic Types of WIRE PLANK Layup

Three layup arrangements are used. The sandwich type is closest to typical ferro-cement construction.

The open grid frame is the lightest and least expensive type of construction. The reason for this is the fact that a lot of

the opposite sides of the **WIRE PLANK** skin in the same direction. This will produce a balanced layup—similar to the outer skins of plywood.

WIRE PLANK Specifications

Type Designation	W16x.437	W14x.437
Wire size	16GA.	14GA.
Wire Spacing	7/16"	7/16"
Layup width	3½"	3½"
Wt. per sq. ft.	.29 lbs.	.47 lbs.
Layer thickness	.062"	.080"

WIRE PLANK brand mesh is constructed of eight parallel hard-drawn mild steel wires having a tensile strength of 100,000 to 120,000 psi.

Scantlings

It would be far too complicated to spell out the specific

Plate 89. A beautiful little Peapod dinghy constructed with **WIRE PLANK** armature and troweled with ordinary Portland cement. Upon completion, she was only 1/8" thick and weighed a mere 250 pounds. Built in 1969, she has withstood severe abuse with no sign of weakness. (An Aladdin Photo)

the rigidity is derived from the rodwork and that is the cheapest item to buy. It is obvious that less mortar is used because of the open space between the rods, and this contributes to the weight reduction. It does add some effort in working the cement smoothly over the rods.

The simple shell has been limited to the smaller boats because of cost. Both simple shell and sandwich have an advantage of a smooth inner skin, which in turn relates to utilizing simple construction methods such as open type molds. All but the open grid frame systems may be applied to large vessels being constructed on the common wood mold as normally found with other ferro-cement vessels. The open grid system may be applied in much the same manner with smaller hulls, providing the mold is removed prior to plastering. (Fig. 227)

Wherever possible, when using more than two layers of **WIRE PLANK,** it is advisable to orient the outer layers of

layup for every vessel type and size. Many elements must be taken into consideration which are felt to be beyond the capabilities of most amateur builders. It is recommended, therefore, that you either consult your designer or Aladdin Products for their recommendations as to the specific **WIRE PLANK** combinations for your vessel. (Plate 91) I have, however, included rough figures for the quantities of **WIRE PLANK** necessary for the construction of normal hulls. (Fig. 228)

Application of WIRE PLANK

WIRE PLANK may be used equally well whether you are laying a vertical/horizontal rod system or a double-diagonal one. In either event, your first layer of **WIRE PLANK** will be laid at a 45° angle to the rod system.

Each **WIRE PLANK** strake should be premeasured from sheer to sheer or bow to stern, then cut to the required

220

angles. Lay on the first strake of **WIRE PLANK** diagonally at 45° to the keel, near the middle of the boat, between bow and stern. Measure the required length of each strake using a tape measure stretched from sheer to sheer in the desired location. Cut the length needed at an angle of about 45° on each end to match the angle at which the strake meets the sheer. **WIRE PLANK** is straight and quite stiff. It is advisable to pre-bend it to the approximate curvature of the hull before applying it. It then conforms easily to the shape required without undue stress or distortion of the rod framework. The small cross wires which extend beyond the edges of the **WIRE PLANK** should be placed toward the outside of the hull on the first layer of **WIRE PLANK**. For the second and subsequent layers these crossties are placed toward the inside of the boat. This leaves a smooth, closely spaced, uninterrupted layer of parallel wires on the outside surface of the hull. The crosstie wires are always buried on the inside, between the layers of mesh.

the direction of the wire. The second layer, which is applied at right angles to the first, will smooth these ripples and result in a very fair hull form. In applying **WIRE PLANK**, it will be very helpful to utilize small clamps (clothespins will do) to hold the strip in position while it is being tied in place.

As work progressess it will generally be necessary, due to the compound curvature of the hull, to vary the spacing slightly between the strips to compensate for this curvature. Usually the ends will have to be brought together and more space left toward the middle of the strips. Further compensation can be accomplished by working small joggles or kinks in the cross wires. This shaping is important to keep the strakes running fair.

On certain portions of the hull, where there is sharp curvature, the strakes have to be twisted. This may result in a tendency for one side of a group to lift away from the rod

Plate 90. A classic 10' Whitehall dinghy constructed with a **WIRE PLANK** armature and troweled with **FER-A-LITE** brand mortar. At 148 pounds, she would make an ideal tender or weekend car topper. (An Aladdin Photo)

A handy method of cutting the **WIRE PLANK** to the length required is to make up a measuring board (marked off in feet) to the maximum length required. Place this on the floor and unroll the wire on top of the board and cut it with heavy tinsnips. A turn of soft wire around the coil will keep it under control during this operation. Tick marks along the board as you go make it easy to judge the length of each strake from the one previously applied. As the first strake goes on, it will take what looks like a lazy "S" curve twist as it goes around the bilge. Don't try to straighten this out. The following strakes will also fall into the same curvature quite naturally and lie parallel to the adjacent wires. Wire ties can be used to secure the **WIRE PLANK** to the rodwork. The light 20-gauge crosstie wire of the **WIRE PLANK** can be used to tie the groups together side by side and to the rodwork. The ends of the wires are left to extend beyond the sheers until the entire layer is installed.

You will notice that when the first layer of **WIRE PLANK** is installed there will be slight undulations perpendicular to

frame. A pull on the wire ends of the strake to "skew" the wire will make it lie smoothly.

In an area where the wires meet a terminating member (such as the stem) at an acute angle, a special treatment of the ends of the **WIRE PLANK** is called for. At the point where the wire crosses, each strand is bent at right angles to the terminating member and then wrapped around the member to anchor it securely.

Lacing

Annealed 22-gauge wire can be used to sew the **WIRE PLANK** to the rod frame instead of using individual wire ties when constructing very small hulls, such as dinghies. Lengths of about 1' long are best to work with. A hook is bent on one end of the wire, pushed through the mesh, passed around a rod, and pulled back out. Grab the end of the hook with pliers and pull the wire tight. The wire is soft and can be laced around the rod and each strand of **WIRE PLANK**. After a number of pulls, the wire will become work-hardened and brittle. At this point simply cut off the wire, make it fast, and start a new lacing.

A two-man team can speed up the lacing process a great deal by having one man inside and the other outside the hull (open mold system).

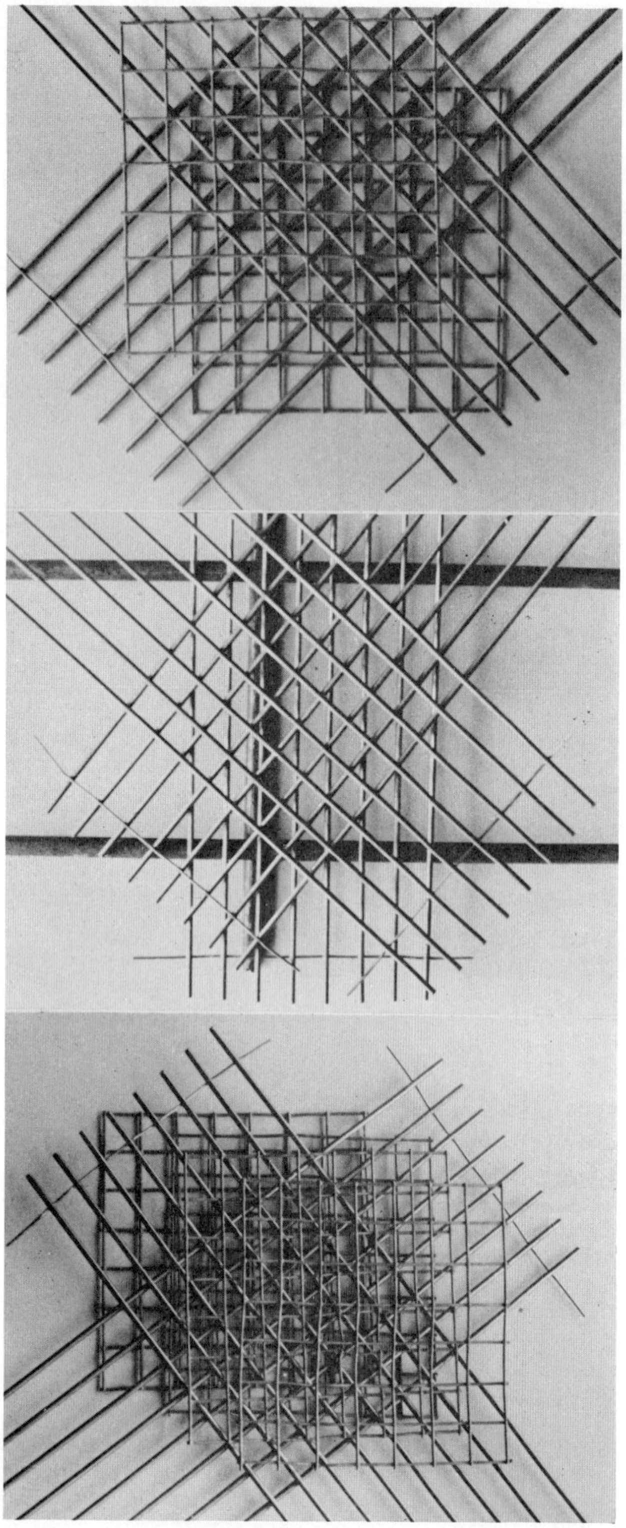

The Second Layer of WIRE PLANK

The second layer of mesh is applied at 90° to the first (on a two-layer layup for dinghies). Again, the first strip is started near the middle of the boat. With this layer, a considerable amount of tying is accomplished by utilizing the cross wires

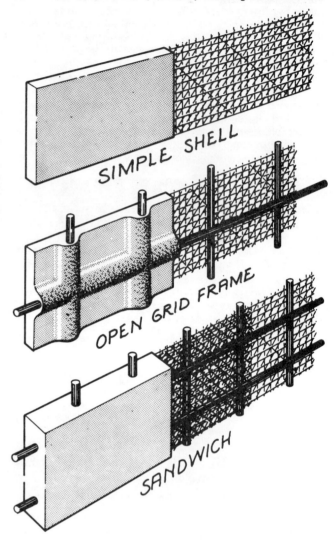

Fig. 227. Common **WIRE PLANK** layup methods.

Plate 91. A sampling of possible **WIRE PLANK** armature variations. Notice the use of rod and 19-gauge square welded mesh in conjunction with **WIRE PLANK**.

BOAT L.O.A	APPROX. HULL AREA (Ft.²)	LAYERS WIRE PLANK T.M.	ROLLS WIRE PLANK T.M.
10	55	2	1
15	100	2	2
20	200	3	4
30	450	4	12
40	800	4	22
45	1000	5	28
50	1250	5	43

Fig. 228. Typical **WIRE PLANK** quantities required for average cruising hulls.

of the **WIRE PLANK**. The strakes are pulled tight to the first layer and sufficient side-to-side tying is performed to properly align the strips. Remember to place the **WIRE PLANK** crosstie wires toward the inside of the boat.

WIRE PLANK End Treatment

Where **WIRE PLANK** ends at a terminal member like the stem, an alternate treatment to welding or wrapping turns is to carry the wire around to the member in an overlap. When doing this, be sure to fold the wires so that the ends of the overlap lie like fingers between the wires of the strips on the other side of the boat. This will eliminate any build-up of thickness of bumps.

Spot-welding

Use of the spot-welder will greatly simplify and speed up the process of fastening the **WIRE PLANK** during construction of the armature.

These machines are rather expensive but if a secondhand unit is obtained, it can be sold later to another builder after the project is completed. These tools vary in form and weight; some are self-contained units and others work with flexible cord electrodes. The air-cooled models are limited as to how many welds that can be made per minute by the overheating of the electrodes. Welding is easily accomplished by pressing the two electrodes against junction points of the rods or wires to be fastened. The switch is pressed and the two welds are made in one quick pulse. This can be repeated rapidly along a frame rod from spot to spot.

In its essence spot-welding sounds very simple. Well, it is, if you begin with the right equipment and know a few basic rules and tricks. I cannot go into all the details here, but if you intend to pursue the **WIRE PLANK** system, you may obtain comprehensive guidance from Aladdin Products Inc. (manufacturers of **WIRE PLANK**) in a profusely illustrated working manual. It would be well worth the small investment to obtain a copy for its many clarifications and diagrams.

Rusty Rods

Spot-welding will not work on rusty steel. It is necessary to use bright, shiny steel for the rodwork if spot-welding is to be utilized to fasten the **WIRE PLANK** to the frame. If an occasional spot of rust is evident on the rods, a smooth file or emery cloth can be used to burnish the steel. If the rod framework becomes thoroughly rusty, a light scuffing with a disc-sander will clean up the rust speedily. Should you wish to spot-weld to pipe frames, disc-sanding is also advised. Do not overdo the sanding and develop a flat surface on the rodwork as the welding works best between two round surfaces.

Fairing the Hull

When all the required layers of **WIRE PLANK** have been applied, there will probably be spots which are not fair and smoothly contoured. In some places where the **WIRE PLANK** has been forced into place, there will be a slight excess in length of the wire which can be "shortened" by putting numerous little jogs or kinks in the wire. In other

Plate 92. A Bingham-designed 20' **Flicka WIRE PLANK** armature ready for plastering, constructed by Ferro-Boat Builders of Edgewater, Maryland. Notice that the **WIRE PLANK** has been applied to what would have been a normal ferro-cement inverted partial wood mold. This seems to be the most sensible application for hull construction other than very small dinghies. (A Ferro-Boat Builders Photo)

places there will be little valleys and depressions which can be pulled out to some extent and faired smoothly. (Plate 92)

Integral Structures

These units may include mast steps, engine beds, beam brackets, chain plates, stem-head unit, portholes, etc. They are installed in much the same manner as with the normal rod and mesh systems. You may refer to the Integral Structures chapter for suggested makeup and construction of these units.

Rejuvenation of Old Wooden Hulls

Many older wooden boats have attributes not found in modern crafts built to provide maximum accommodations at minimum cost. Hull form, beautiful interior joiner work, fine equipment and spacious, well-designed accommodations are often found in older boats whose hulls have started to show their age. Many of these can and should be saved.

WIRE PLANK and **FER-A-LITE** can do it for you. First,

remove all moldings or rub rails that protrude from the hull contour. Now remove all paint from the hull using a heavy-duty sander with coarse sandpaper. This provides a good tooth for bonding. Next apply a thick coat of polyester or epoxy resin to provide a barrier between the salt-soaked wood and the wire mesh.

Then simply staple **WIRE PLANK** brand mesh onto the hull using heavy galvanized or stainless steel staples. The mesh should be applied diagonally as with the new construction and at least two layers should be used. It is desirable to use thin staples on the last layer to minimize any bumps in plastering. Larger boats will require more layers; three or four should be sufficient up to 50'. More will be needed for larger or heavier boats.

Staples can be driven with a hand-held power stapling machine or with a manually operated stapler. When the wire is well stapled to the hull and all faired with no humps and bumps, apply **FER-A-LITE** brand mortar. Trowel it well into the mesh and firmly down to the wood. Apply newspapers to the mortar before it gels, as previously described. This will provide a very smooth surface. (Plate 93)

Plate 93. One of the most remarkable advancements and future possibilities in the ferro-cement field is the refurbishment of existing old wooden or steel vessels. This is accomplished by stapling or stud welding light **WIRE PLANK** or mesh armature to the vessel, then troweling with **FER-A-LITE** or common Portland mortar. The vessel shown had lost its commercial certification because of severe rusting and overpatching. After a troweling with Portland mortar, the vessel gained approximately 6,000 pounds which lowered her flotation only 2''. She is now back in profitable service after a minimal investment.

integral structures

When the boat is working its way to weather during high winds and heavy seas, its motion will become extremely violent. Thrashing, pitching and sudden rolls may continue for days on end and each spasm and shudder places tremendous strains upon the rig, the hull and interior components. On large boats the stress upon the chain plates may reach 10,000 pounds each, while the compression loads under the mast can be several times that figure. The pitching vessel will transmit tons of shock force to the stem and the hull and the battering of the rudder can create hardware and hull fatigue in short order. With each change in motion the normal static loads may be multiplied many times and the hull structures must be built to absorb this continual workout for decades to come.

At this point you have successfully shaped the form of the hull but have had little opportunity to express your creative urges. But now, through the fabrication of the integral structures, you will be required to indulge your sculptural prowess, ingenuity and foresight. The initial procedures for the installation of mast steps, engine beds, chain plates, stem heads, sole beam attachments, mounting brackets for machinery, etc., usually proceed after the completion of the second hull rod layer but before applying outer mesh layers. The majority of the integral structures require the passage of reinforcing rod from the hull armature to the structure and should be attached in such a way so as to become a part of the hull proper. The purpose is to transmit stress, compression or vibrations over the largest possible hull area.

If you are building to the open or partial wood mold system, the construction of the units may require cutting away sections of the mold ribbands to allow accessibility from within. If you are building to the cedar or closed mold system, completion of the integral units will not be possible until after plastering, so "L"-shaped "starter" rods must be inserted by the drilling of holes fully through the mold at the positions required. With the open truss construction method, portions of the integral structures (such as tank bulkheads) may already have been included in the frame construction.

It is impossible for me to verbally describe the step-by-step procedure surrounding every circumstance as each builder may vary slightly in his hull armature construction. I shall, therefore, depict through illustrations the general nature of each unit and you will clearly see that many fabrication variations are possible. I show the diagonal hull rod system in all cases but the procedure is not changed if you choose to use vertical and horizontal hull rods. For the sake of clarity, I have not drawn the mesh layers in many of the examples.

Rod Types

Because of the amount of bending required to create sharp definitive detail, it is often most practical to use a mild steel rod for the construction of integral structures. Its use is particularly important in those areas where welding is necessary. If welding is not called for, however, and if only a few soft bends are encountered, use a high tensile rod. Because the ends of the rods of integral units are generally sandwiched between the hull mesh layers and lie on the same plane as those of the hull armature, their diameters should be the same as those of the hull armature.

Complete Unit Construction (Open or Partial Mold)

Because these building systems have been designed for internal accessibility, you can completely fabricate all of the integral structures in their entirety prior to plastering or the removal of the mold. Upon carefully locating the area of the given structure, cut loose and remove from the hull those mold ribbands which are in the way of the unit, thus allowing total workability. All of the rods for the integral structures are usually bent and installed in one piece.

It must be noted that in the illustrations some of the integral structural rods lie in the same plane and direction of the first hull rod layer while others are placed parallel to and on the plane of the second rod layer. This has the effect of more satisfactorily distributing the stresses throughout the thickness of the hull shell, while contributing to multidirectional levers.

When installing the structural rods which are to lie within the inner hull rod layer, you may find it necessary to pull the outer rods slightly away to allow for easy insertion. In cases where the structural rod enters the hull and terminates within, it will generally be of an "L" or "V" shape installed from the outside of the armature. If the rod is to be attached to the hull at both ends, however, it may be more practical to insert from the inner side of the hull, pushing both ends through then bend back to a distance of 6" to 9". Once the basic shape of the unit has been formed, additional reinforcing may be added as required. (Fig. 229)

Starter Rods (Inverted Partial or Full Cedar Molds)

Because of the difficulty in maintaining proportions and judging perspective when working on the inside of an inverted mold, the risk of mismeasurement plus the physical demands on the builder, I do not recommend that complete unit construction be attempted with these systems. When you build on a full cedar mold, it is impossible to remove the hull shell if each complete unit of integral construction is attempted without first cutting away some of the mold

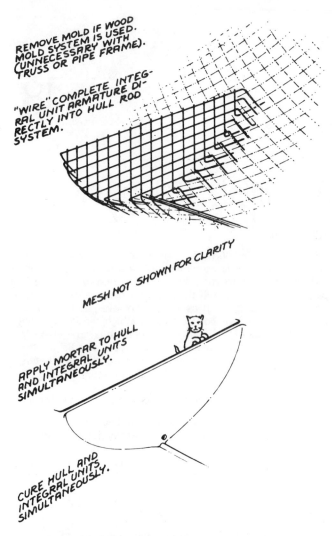

REMOVE MOLD IF WOOD MOLD SYSTEM IS USED. (UNNECESSARY WITH TRUSS OR PIPE FRAME).

"WIRE" COMPLETE INTEGRAL UNIT ARMATURE DIRECTLY INTO HULL ROD SYSTEM.

MESH NOT SHOWN FOR CLARITY

APPLY MORTAR TO HULL AND INTEGRAL UNITS SIMULTANEOUSLY.

CURE HULL AND INTEGRAL UNITS SIMULTANEOUSLY.

Fig. 229. Typical complete integral structure unit construction used for open mold system.

planking. The most accepted procedure, therefore, is simply the insertion of "L"-shaped starter rods to which the remainder of the integral unit is "married." Insertion of these "L"-shaped rods requires drilling through the mold, pushing through the vinyl vapor barrier, or cutting away portions of the mold planking. The rods should extend into the hull at least 6" (the more, the better). Once again, carefully study the illustrations, noticing the plane and direction of the ends of the starter rods within the hull armature. When you remove the finished hull from the cedar mold bend the starter rods downward or cut away portions of the planking to assist the separation. (Fig. 230)

Chain Plates

Chain plates which are simply bolted to the outside of the hull have become recognizable trademarks of most amateur-built yachts. They are clumsy in appearance, create a constant maintenance problem and are rarely considered within the acceptable aesthetic norm by discriminating yachtsmen. There are some exceptions to this rule; for example, as when used aboard workboats or those dictated by nautical tradition such as heavy schooners and other character boats.

Providing for these exceptions, the most professional appearance is achieved by installing interior chain plates

rather than simply bolting them to the outside of the hull. Unfortunately, the passage of the plate through the deck often promotes small leaks, if not bedded properly, and construction may be somewhat time-consuming. Each of the chain plate variations shown may be successfully applied to all hull construction methods and are interchangeable. Regardless of the system chosen, all metal parts should be prefabricated prior to the construction of the chain plate attachment as this will provide for accurate templates for the location of bolt holes and the accurate development of required angles. The angles of the chain plates and/or connecting webs should always be in a direct line with the shrouds or stays when seen in profile. Details concerning the fabrication and installation of the metal hardware are dealt with in Chapter 25.

The simplest of all chain plate designs is that of attaching a steel gusset of flat plate directly to the inside or outside of the hull. This will require the installation of steel bushings into the hull armature at each through-bolt location. The purpose of the bushings is to create a larger bearing against the concrete to prevent cracking and enlargement of the

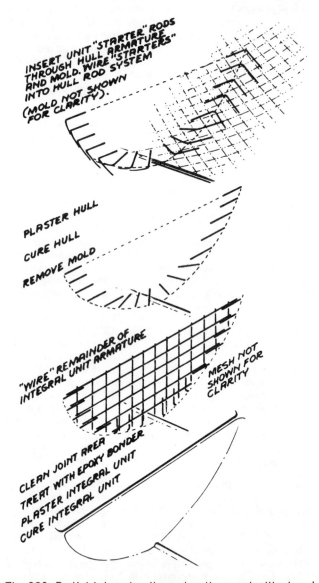

INSERT UNIT "STARTER" RODS THROUGH HULL ARMATURE AND MOLD. WIRE "STARTERS" INTO HULL ROD SYSTEM (MOLD NOT SHOWN FOR CLARITY).

PLASTER HULL

CURE HULL

REMOVE MOLD

"WIRE" REMAINDER OF INTEGRAL UNIT ARMATURE

MESH NOT SHOWN FOR CLARITY

CLEAN JOINT AREA

TREAT WITH EPOXY BONDER

PLASTER INTEGRAL UNIT

CURE INTEGRAL UNIT

Fig. 230. Partial integral unit construction used with closed or "cedar" mold construction.

bolt hole during its length of service. The diameter of the bushings should allow for a snug passage of the through-bolt while the length should be slightly less than the finished hull thickness. This will prevent interfering with plastering while causing a thin layer of mortar to cover the bushing end, thus sealing against rusting and bleeding. Notice the use of the "V"-shaped secondary rods which serve to distribute the shock of the rigging load well into the armature.

Always remember that any exterior chain plate system will cause some form of staining or bleeding down the side of the hull regardless of the type of metal used. An exposed bolt head, even when galvanized, will eventually rust and will always project a clumsy appearance.

Another ingenious system is that of forming a yoke of stainless steel wiggle rods, each leg of which is installed several feet into the hull armature. It should be noted that this system may be fabricated of two rod sections laid side by side for an increased strength. The attachment of the turnbuckle is by means of the stainless steel strap toggle and its finished appearance is clean and functional. Its use, however, should be restricted to vessels under 40'. Replacement or repair of this type of chain plate is almost impossible and should be a consideration of choice. (Fig. 231)

Bobstay Plate

This unit is for the purpose of attaching the lower wire from the bowsprit or boomkin. Once again, there are a dozen methods for its satisfactory installation. The most popular approach is that of fastening two separate plates on each side of the hull using through-bolts. The two plates are bent together at their intersection and continue upward as one. Although this fixture may be functional, it is awkward and very amateurish and will eventually become a rust trap if constructed of galvanized steel.

The equally strong and best looking choice calls for a triangular plate (preferably stainless steel) to be welded to the hull's center pipe. This requires the splitting of the mesh layers for the passage of the plate through the hull shell. It may also be necessary to notch this plate in order to clear any reinforcing rod which passes through its location. Once the plate is installed, the mesh layers are pressed firmly against it. Plastering will completely seal the weld and there will be no possibility of its being dislodged. Other bobstay installations are illustrated. (Fig. 232)

The Stemhead

One of the most concentrated stresses on the entire hull will be the foremost part of the bow as it must not only withstand the strains placed upon it by the forestay and Genoa tack but must also be constructed to absorb impacts of collision or docking incidents. Every time the boat lifts itself to a head sea, the aftward thrust of the mainmast will cause immense shock loads on the bow hardware while straining the hull itself. You can see, then, that it is imperative that these intermittent tensions be transferred over the largest possible area.

The stemhead installations illustrated are designed specifically to create back pressures against the armature and result in an interwoven rod structure at the center of the boat. You will notice in each case that connecting rods are used to firmly attach the hull center-pipe to the shell system. Without these connecting rods, fatigue would even-

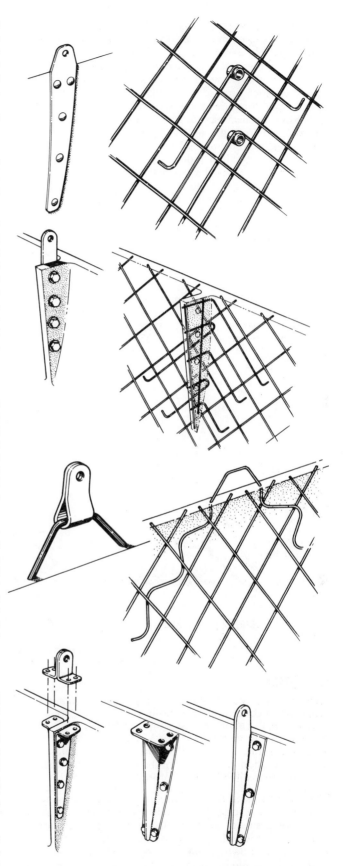

Fig. 231. Typical integral chain plate construction. Notice how the shroud strains are distributed over the widest possible area of the hull armature.

227

tually take its toll by tearing the stemhead and center-pipe completely out of the vessel. The connecting rods may be passed over the center-pipe or welded to it as "L"-shaped sections. In the first instance, the connecting rods are inserted from the inside of the hull as "U"-shaped sections. With their ends protruding outboard, they will bend back against the armature and be wired into place. If the "L"-shaped connecting rods are employed, they are prebent and installed from the outside of the hull. In either case, the connecting rods must lie on the same plane and direction as the outer hull rod layer. Any number of combinations may be derived from the illustrated systems. The use of a rod center-web, however, has been designed specifically for cedar mold construction and may not be plastered until the hull mold has been removed. Those stem systems which use an open horizontal underdeck plate should not be drilled for hardware until after the decking has been laid. The fabrication of external hardware is discussed in Chapter 26.

Because of the severe nature of the bends required, only mild steel connecting rods should be used. Upon fabricating the stem installation, additional inner mesh layers should be applied to fully cover the center-pipe, welds and connecting rods along the center line. (Fig. 233)

into the mast step system are now wired to the transverse rods. This network should also be as close together as workability permits. The next mast step rods to be installed are inserted longitudinally, being laid atop and spanning across the transverse network. As you can see, this armature system fully accounts for shock loads in all directions.

Because the finished mast step is to receive a steel socket weldment, provisions must be made for its attachment. This will be done by inserting several coarse-threaded galvanized or stainless steel rods which are carefully positioned, using the mast socket weldment as a template. To prevent the threaded rod positions from shifting during the pouring of the step, cross-wiring may be required. For additional strength the lower ends may be bent upward to form a hook. This will also prevent them from turning out of the concrete. The lengths of these threaded rods will depend entirely upon the depth of the hull at the step unit. The longer the better.

It is imperative that the mast step not be poured until after the full curing of the hull as the excessive weight may cause severe sagging. Prior to pouring, however, temporary plywood molds should be placed at each end of the mast step in order to prevent concrete run-off and the cured hull in

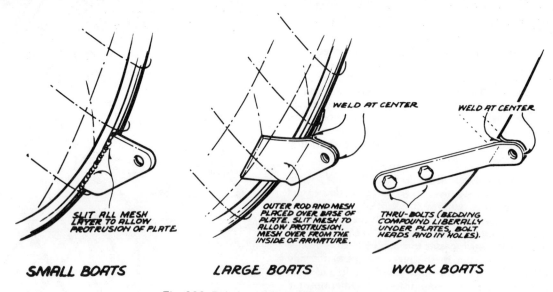

SMALL BOATS — SLIT ALL MESH LAYER TO ALLOW PROTRUSION OF PLATE.

LARGE BOATS — OUTER ROD AND MESH PLACED OVER BASE OF PLATE. SLIT MESH TO ALLOW PROTRUSION. MESH OVER FROM THE INSIDE OF ARMATURE.

WORK BOATS — WELD AT CENTER. THRU-BOLTS (BEDDING COMPOUND LIBERALLY UNDER PLATES, BOLT HEADS AND IN HOLES).

Fig. 232. Bobstay plate installation variations.

Mast Steps

Again I remind you of the enormous compression and side forces being concentrated at the base of the spar. It is the function of the mast step not only to hold fast the base of the spar, but to distribute its load over the widest possible area of the hull. As with all of the other structures, this load distribution will be by means of wiring the internal structural rods directly into both hull armature rod layers.

The transverse step rods are the first ones in. Bend an angle at one end only; pass the rod fully through the armature to the opposite side of the hull, then bend the protruding end back against the hull. In each case the direction of the bend may be up or down or fore-'n-aft and the length of the rod portion within the hull armature should be at least 6". The transverse rods should lie on several planes and are to be placed as closely as possible while still allowing for workability. Vertical rods extending from the bottom of the hull

the area of the step should be treated with an epoxy bonder to strengthen the joint between the new and old mortar. (Fig. 234)

Engine Beds (Solid)

The engine beds must be constructed in such a way so as to absorb the thrust of the propeller in both the forward and reverse directions, the continual high-frequency vibration during running, the severe rocking of the engine during idling and low-speed operation, as well as having to support the engine while the vessel is at an extreme heeled angle. It should be remembered at all times that the construction of the vessel must be substantial enough to withstand decades of service, and to bear the cumulative factors causing metal and concrete fatigue. The strongest type of engine bed is of two solid reinforced concrete pedestals installed approximately at the turn of the bilge on each side of the center

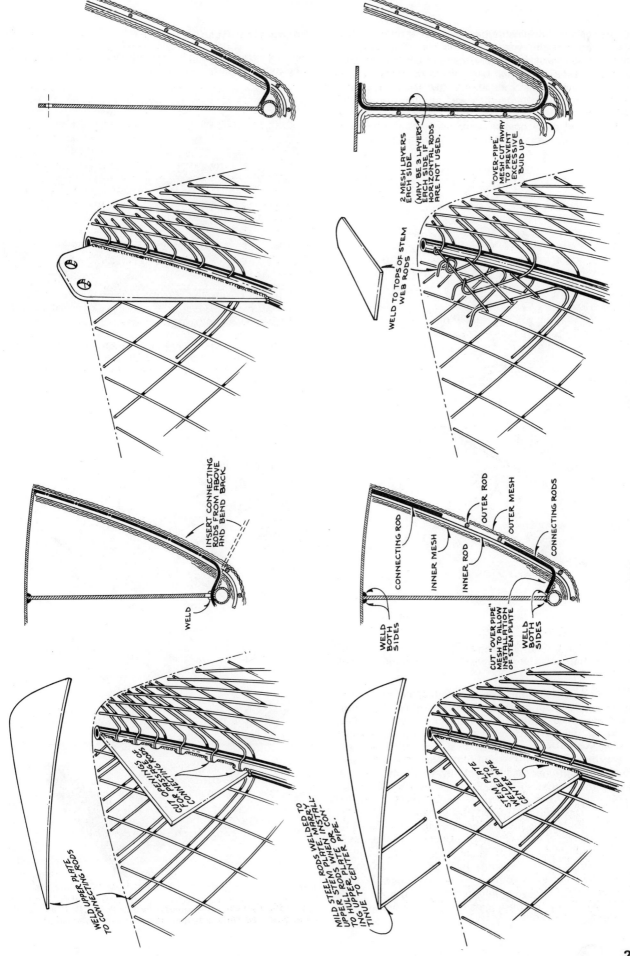

2 MESH LAYERS EACH SIDE
(MAY BE 3 LAYERS EACH SIDE IF HORIZONTAL RODS ARE NOT USED.)

"OVER-PIPE" MESH CUT AWAY TO PREVENT EXCESSIVE BUILD UP

WELD TO TOPS OF STEM WEB RODS

INSERT CONNECTING RODS FROM ABOVE AND BEND BACK

WELD

CONNECTING ROD
INNER MESH
INNER ROD
OUTER ROD
OUTER MESH
CONNECTING RODS

WELD BOTH SIDES

CUT "OVER PIPE" MESH TO ALLOW INSTALLATION OF STEM PLATE

WELD BOTH SIDES

WELD UPPER PLATE TO CONNECTING RODS

CUT OPENINGS OF FOR PASSING RODS

MILD STEEL RODS WELDED TO UPPER PLATE. MARRY UPPER RODS TO HULL RODS, INSTALLING TO UPPER PLATE CONTINUE TO CENTER PIPE.

STEM PLATE WELDED TO CENTER PIPE

Fig. 233. Stemhead installation variations.

229

line. As seen in the illustration, reinforcing rods are first installed transversely with their ends "married" into the hull armature. Vertical rods are then prebent and inserted from the outside of the hull being wired to the transverse. Horizontal longitudinal rods are inserted from the aft end of the engine, again prebending and pushing through the hull armature. You will notice that this procedure causes the engine bed to become an actual part of the hull armature system and that its reinforcing is multidirectional.

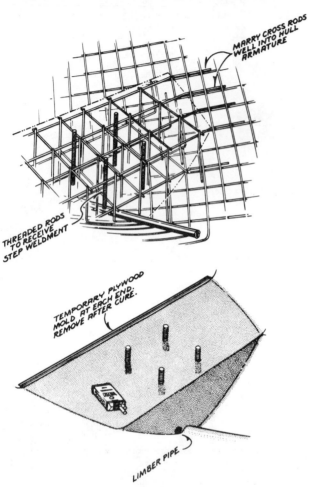

Fig. 234. Typical solid ferro-cement mast step construction. Note that the mast step rods are married well into the hull armature.

The reinforcing rods should be as close together as possible, leaving only enough space to allow for workability during construction. After the engine mounts have been accurately shaped, it will be necessary to install pairs of coarse-threaded galvanized or stainless steel rods for the purpose of attaching the engine mountings. These must be very accurately positioned and temporarily wired into place to prevent them from shifting during plastering. Allow approximately 3" of exposed rod above the plane of the engine bed. The solid reinforced engine bed should not be poured until after the hull has been plastered and cured, as this will prevent the possibility of sagging under the excessive weight. Prior to plastering the engine beds, it will be necessary to coat the concrete joint area with an epoxy bonder. Cover the exposed threaded bars with tape. Pouring of the concrete will also require temporary plywood molds to be installed around the engine bed shape to prevent run-off. (Fig. 235; Plate 94)

Engine Beds (Web)

This construction entails the fabrication of two vertical, longitudinal, reinforced concrete members which will have

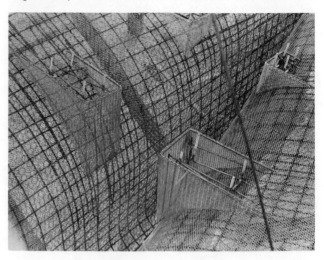

Plate 94. A variation on **Andromeda's** solid ferro-cement engine bed construction. Notice the extensive rod work and positioning of the engine mount bolts within the mesh screed work.

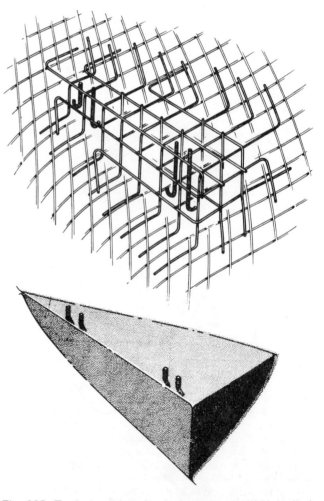

Fig. 235. Typical solid engine bed construction. Note that all engine bed rods are married well into the hull armature.

a finished thickness of 3/4" to 1".. These longitudinal members are, in turn, connected to web frames, concrete bulkheads (if applicable) or partial ferro-cement floors.

As previously described, the ends of all of the engine-bed web rods must be thoroughly worked into the hull armature so as to form a truly integral unit. In all cases where a corner is encountered during the bed construction, extremely generous radii must be applied; no sharp angles may be tolerated as they will become stress-risers and a continual source of trouble.

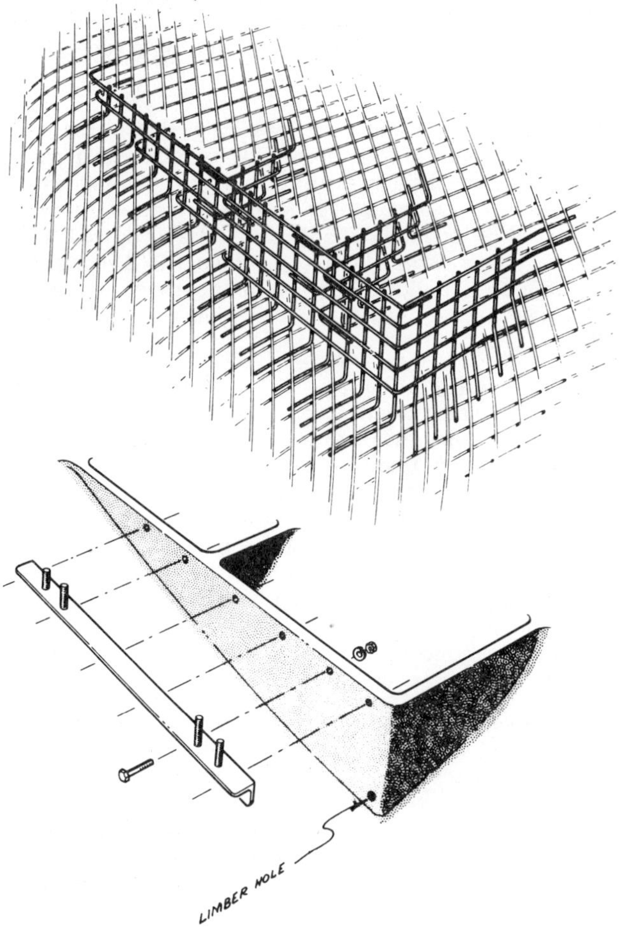

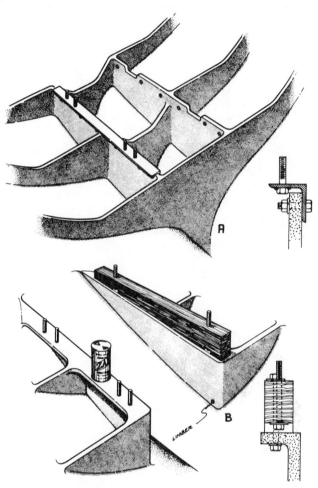

Fig. 236. Basic web engine bed construction.

In order to prevent the collection of water between the engine beds and the hull, it is important to install wooden dowels at the forward ends of the engine beds so as to form limber holes in the ferro-cement when the plugs are removed. (Fig. 236)

The actual configuration of the web engine bed may take many different forms which will depend upon the hull shape and engine type. The connection of the engine may also take several forms. Having prelocated bolt-holes in the engine-bed web with waxed wooden dowels, you may bolt-attach substantial angle irons along the entire length of the inner sides of the engine beds. The angle itself will be pre-drilled for the engine mount bolts which are tapped into the angle with their lower ends spot-welded for permanence. The upper exposed length of the mount bolts should be at least 3". (Fig. 237A)

A slight variation to the mounting previously described is that of creating a horizontal flange along the upper edges of the engine-bed webs in lieu of the steel angle. Prelocated bolt-holes along this flange will allow for the installation of oak timbers. These timbers, in turn, will be predrilled for the insertion of the engine mount beds. (Fig. 237B)

When building to the open-truss or pipe-frame method, several designers suggest that the longitudinal engine-bed webs be prefabricated and installed into the frame system prior to attaching hull mesh and reinforcing rod. This will

Fig. 237. Typical web type engine installations. Many other variations are possible.

accomplish the same effect as the construction I have described. (Plate 65)

There are many other engine mounting possibilities which may be derived as combinations of the systems illustrated.

Bulkhead Webs

I have described several alternative systems for applying bulkheads in other chapters. One of the cleanest, easiest and strongest of these methods is that of shaping a generous fillet on each side of the bulkhead at the hull by means of an epoxy grout. Scores of civil engineers have even built bridges using epoxy fastenings, but amateur ferro-cement builders lack confidence in its immense strength.

I cannot understand why so many amateur builders have such an incessant fear of their bulkheads loosening from the hull for they are primarily compression structures being subjected to negligible tension or fore-'n-aft impacts. The bulkheads themselves are subjected to bending, but they are most often stiffened by the joinery which is attached to them.

The most notable stress which the bulkheads must withstand is that of vertical sheer caused by the continual twisting of the hull. While I have no argument that many additional bolts through a bulkhead and web will certainly hold up under this sheer, any vertical movement (however small) will eventually cause these bolts to enlarge their through-holes, thus creating the possibility of loosening. I

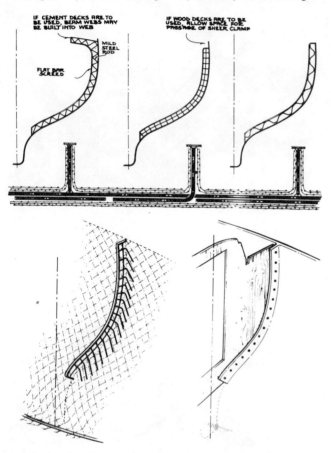

Fig. 238. Typical bulkhead attachment web construction.

personally feel that web and bolt bulkhead attachments are unsightly, cumbersome, and very amateurish though they require a great deal of valuable time to construct. They will, of course, allow you the liberty of less accurate fitting of the woodwork, but this will add little to the quality of your yacht.

If you are still apprehensive or believe that an occasional elephant may crash through the interior of your boat, you may, of course, proceed with the installation of ferro-cement webs. These are accomplished by accurately marking the bulkhead positions on the armature with chalk. Remember that the web will sit on the fore or aft side of the actual bulkhead position. Bend the required number of "L"-shaped rods being 3—4" on one leg and 6—9" on the other. Insert the short leg through the armature (or mold if applicable) and wire or staple the long leg to (and flush with) the inner or outer hull rod layer. Continue to insert successive rods along the web location at distances of about

4" apart. Remember that if a timber sheer clamp for a wood deck is to be installed in the finished hull, allow for its clearance at the sheer when constructing the bulkhead web.

If you are constructing your hull to the cedar or partial mold systems, the completion of your bulkhead web may not proceed until after curing the hull and removal of the mold unless you cut away a part of the mold in the way of the web position. You may complete the bulkhead web by tack-welding a 1/8" steel screed along the inner ends of the web rods. If you prefer, you may substitute the bar with a 3/8" reinforcing bar. An additional reinforcing rod will be welded to the centers of each web rod, interconnecting them in a series.

Two layers of mesh will now cover each side of the web weldment, wired through from side to side while overlapping the mesh layers several inches onto the hull armature. I would suggest prelocating your bolt-holes by inserting waxed wooden dowels through the bulkhead web armature. These dowels should protrude 1/8" beyond the mesh on each side. With the open mold systems, plastering of the web may be done at the same time as the hull. (Fig. 238)

If you are building to the cedar or partial mold systems and have not completed the web construction, this will proceed only after removing the mold from the hull. After adjusting the web starter rods to proper shape, weld a flat steel screed to the ends of these protruding rods, then attach the center connecting rod to each. Cover the bulkhead web with two layers of mesh on each side, overlapping on the hull for about 3—4". Prelocate the bolt-holes by inserting waxed dowels through the web, allowing 1/16" protrusions. Prior to plastering the web, coat the hull at the web attachment with an epoxy concrete bonder. The bulkhead web is now ready for plastering. (Fig. 230)

Several other web attachment methods have been illustrated, but it must be remembered that those systems which require the welding of the hull rods to the web structure may cause severe distortion and resultant unfairness. A significant loss of hull strength will also occur if high tensile hull rods are being used. Consider the heavy price you may ultimately pay for the sake of questionable building ease.

Cement Bulkheads

As I have just indicated under **Bulkhead Webs,** I feel that there has been a severe overemphasis on both the strength required of bulkheads and their attachments. For this reason, I feel that the use of concrete bulkheads for the purpose of increased strength over wood is totally absurd. After personally viewing hundreds of concrete boats, I have yet to see a handful of good-looking smooth fair cement bulkhead installations. While most builders have employed this construction in the name of ease and cost savings, they seem to have completely discounted the effects of weight on the boat's stability, the hundreds of tedious hours involved in wire-tying, as well as the complications of attaching joinery. The weight problem may be solved through the use of high tensile steel, **FER-A-LITE,** or lightweight aggregates, but few builders are willing to spend the extra money for these benefits. I suggest, therefore, that you review Chapter 4 on stability before proceeding with this construction. But if you are still convinced that you will produce a better boat, you may proceed in the following manner.

Carefully position a temporary plywood or plank form within the armature at the forward or aft side of the finished bulkhead location. This form will aid in assuring you that your finished work will be a flat surface. Staple mesh layers to the form. Now lay horizontal or diagonal rods against the bulkhead form, inserting their ends through the armature and bending back against the hull (or marrying to the starter rods). When your first rod layer is complete, lay a second rod layer vertically or diagonally opposed to the first and in the same manner. When the rod application is complete, cover the bulkhead armature with the outer mesh layers, overlapping 4—6" onto the hull armature.

In order to hold the bulkhead weight to a minimum while not sacrificing strength, it is far better to use two layers of a small rod diameter rather than one layer of larger size rod. Use high tensile rod if possible and limit welding to the attachment of screeds only. Cement bulkheads rarely need to be the same thickness as the hull unless they serve to partition cargo spaces or large fishholds. On vessels under 30,000 pounds, the average bulkhead thickness need not exceed ½". On vessels up to 60,000, 5/8" bulkhead thickness will suffice if built properly. (Fig. 239)

As I have previously mentioned, when constructing relatively flat surfaces, reinforcing rod may be completely elim-

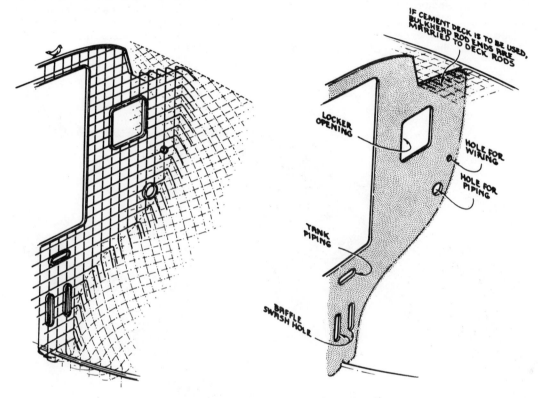

Fig. 239. Ferro-cement bulkhead construction.

Weld a 1/8" steel screed bar to the ends of the reinforcing rods around the perimeters of doors or other openings. I would suggest that the corners of these openings be constructed as radii to prevent stress-risers. If the bulkhead is to extend fully to the bottom of the hull, always make provisions for limbers or bilge drain holes. It will also be necessary to allow for the passage of plumbing, electrical wiring, tank piping (if applicable) or control cables. Once again, if a timber sheer clamp is to be installed for wood decking, an opening must be constructed through the bulkhead at the sheer for its passage. Prelocate as many bolt positions as possible by inserting waxed dowels through the bulkhead armature. Most joinery cleats, however, may be epoxy-attached without mechanical fastenings. When plastering the bulkhead, the wood form may remain in position but great care must be exercised to provide total mortar penetration, as your visibility will be minimal. You may elect to remove the wood form and this will require that the entire bulkhead armature be fully wire-tied together at each rod intersection to prevent any possibility of distortion. In this event it is also wise to attach a temporary timber stiffener to the armature to eliminate bending and excessive movement during plastering.

inated by substituting the conventional armature systems with additional layers of mesh to the required thickness. If this thickness gives concern for adequate penetration of the mortar, fewer layers of 16-gauge or 14-gauge mesh may be used with complete success, while producing a consistently smooth finish with far less work.

Beam Brackets

I have noticed that many concrete vessels have utilized longitudinal deck beams lying on transverse risers. I object to this practice for the same reason I have voiced on the use of only longitudinal deck rods or solely vertical bulkhead rods. The application of transverse rather than longitudinal beams under the sole not only supports the sole but stiffens the hull against inward pressures.

Several methods are available for the attachment of these sole beams, but I will only concentrate on two systems; i.e., an epoxy grout fillet or small ferro-cement fastening brackets.

If you prefer the fillet method, no allowances are required during the construction of the hull armature. If you prefer the bracket system, however, the construction must take place at this time. For open mold systems, prebend the required number of "V"-shaped rods, each leg being approximately 10" long. Insert each leg of the "V" rod through the armature from the inside the hull so that 5—6" protrudes through the armature. Bend the exposed rod ends fully back against the hull so that they lie flush with the hull rod layers. Now tie the short rods firmly in position. Cover each bracket rod with two layers of mesh on each side, overlapping the mesh approximately 3" onto the hull armature. Install a waxed wood dowel bolt-hole blank into each bracket. The finished brackets are ready for plastering and may be done at the same time as the hull. (Fig. 240A)

If you are building to a cedar or partial mold system, the beam bracket will require the insertion of two starter rods

the accommodation plan indicates the sole to be angular, the location procedure will differ considerably.

The forward and aftermost vertical beam positions are usually called out on the construction drawing and their respective brackets must be installed on both sides of the hull in the normal manner. Temporary beams are then clamped to these brackets in their proper locations. Two longitudinal timbers are placed on top of these beams to represent the straight sloping sole line (it may be necessary to prop up the longitudinals to prevent sagging). To determine the vertical positions of any given interim bracket, a straightedge is aligned to the underside of the longitudinal timbers and extended to the hull where the remaining bracket positions may be marked. Remember that the upper side of the sole bracket should coincide with the upper edge of the sole beam. (Fig. 241)

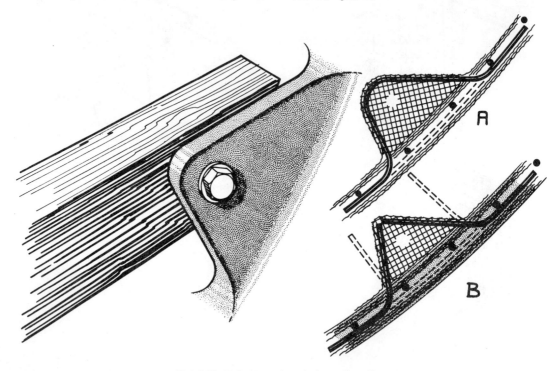

Fig. 240. Sole beam bracket construction.

at each location unless you cut away sections of your mold. The inner legs of these rods should be 3½—4" long and arranged in such a way so as to allow their ends to meet when they are bent toward each other. Do not bend them together at this time, however, as it will impede the eventual removal of the mold after plastering the hull. The outboard legs of these rods should point in opposite directions, being firmly wired to the hull armature flush with the outer hull rod layer. After the hull has been plastered and the mold removed, bend the inner rod ends together and lightly tack-weld. Cover each bracket with two mesh layers on each side, locating waxed wood dowels for bolt-holes in each bracket. Coat the bracket area with an epoxy bonder, then plaster the bracket. (Fig. 240B)

Beam Bracket Location

We have assumed to this point that the sole of the boat is perfectly level and that all base line measurements to the brackets will be the same. These measurements may be transferred to the hull dropping a tape measure from the scaffold cross-spans or by using the water level. If, however,

Accessory Machinery Mounts

The installation of such equipment as bilge pumps, water pumps, small generators, battery shelving, etc., may take the form of ferro-cement structures or their mountings may be timbers epoxied to the hull. Of course, the latter will require no attention during the construction of the armature. Ferro-cement mounting, however, may take many different forms, and I will describe a few in order to stimulate the imagination and ingenuity, providing that you think in terms of minimizing unnecessary weights. Some mounting may be constructed in much the same way as the web engine beds and must be shaped to suit the equipment, such as a generator installation. A rudder stock thwart may be supported at each side of the hull with brackets constructed in the same manner as for the sole beams. Vertical mountings may resemble small bulkhead webs in shorter sections. Provisions for supporting heavy navigation electronics may be made by literally constructing ferro-cement shelves or shelf brackets. The same premise may also be used for supporting batteries, life rafts, or other extremely heavy gear.

If the locations for these items have not been pinned down by your designer, predetermine their locations and critically judge the merit or drawbacks of building such permanent structures for their support. (Fig. 242)

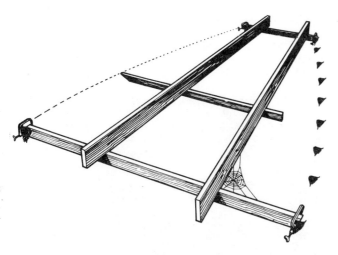

Fig. 241. Method of locating vertical locations of sole beams when the sole of the vessel is not horizontal.

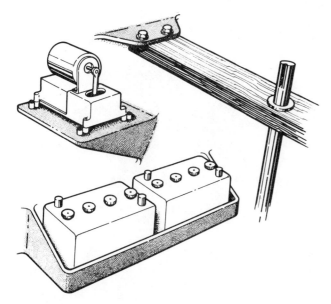

Fig. 242. A few ferro-cement machinery or accessory bracket variations.

Deadwood

As I have already discussed two variations of deadwood installations as a part of the mold construction (Figs. 146 & 147), we must now consider the method of fastening the mesh and rod along its trailing edge.

If the steel plate deadwood with a half-pipe trailing edge is to be used, all mesh and rod layers are brought fully against the forward side of the pipe, taking care that the finished armature will lie 1/8" inboard of the outside pipe radius. The reinforcing rod should be lightly tack-welded against the pipe. These welds are not to be considered structural upon curing the plastered hull but will serve only to hold the rod firmly as your construction continues.

Around the propeller aperture the rods will terminate approximately ¾" beyond the trailing edge of the plate and will be welded together on center. Where the rods reach the area of the propeller shaft, they will be cut just short of the finished aperture line and the ends tack-welded to the shaft pipe. This will form a swell around the shaft. Under no circumstances should the trailing edge of the propeller aperture be square or blunt as this will create unnecessary drag while under sail and can reduce propeller efficiency by 15% when powering.

At the base of the deadwood plate where the heel bearing is to be attached, you will have to use the prefabricated heel bearing weldment as a guide for properly fastening the rods and mesh so as to provide an adequate clearance on the hull for the installation of the bearing. One such method of shaping the armature is shown, but you will have to use your best judgment.

If your boat is to use a "T" bar deadwood weldment, your inner mesh layers should terminate at the forward side of the transverse iron. The reinforcing rod should come in contact with the same iron and is lightly tack-welded. Note that the outer rod layer is flush with the outer edges of the iron. When applying the outer mesh layers, cover the iron-work completely so that it will be fully encased in concrete. The base of the deadwood weldment may be fastened in the same manner as that previously described. It is wise, however, to consider the effects of drag and propeller efficiency when choosing "T" bar deadwood construction. (Plate 95)

The Heel Bearing

You will notice that I have indicated a "U"-shaped heel bearing in both deadwood examples and I will briefly describe the reasoning behind this type of design as it is one of the most important pieces of hardware on the boat. Consider first that as the rudder is turned from side to side it exerts hundreds of pounds of side force which, in part, is transferred directly to the heel bearing. Realize also that as the vessel gyrates in seaway, it creates tremendous additional side forces which tend to fatigue the bearing weld-

Plate 95. **Doreana's** upper deadwood and horn weldment after applying the first mesh layer to the armature. Note the shape of the rudder stock tube below its exit from the hull. The remainder of the armature will be faired slightly less than tangent to the rudder tube surfacing. A Bingham design.

ment and its fastenings. Also keep in mind that there will be the rising and falling weight of the rudder against the heel which will also take its toll on welds and fastenings. Last, but not least, is the problem of corrosion or galvanic action in that area of the hull which is the least accessible and, therefore, serviceability is an important factor.

Because the heel bearing attachment must be considered, and provisions for attachment built into the hull prior to plastering, the type of bearing weldment must be pre-determined. Keeping all of the foregoing factors in mind, let's establish some basic criteria for its construction. Sharp

these requirements, most being of a poorly fashioned and slipshod nature. (Plates 96—99) Most have been designed purely from the standpoint of easy fabrication. A rudder dropping out is as serious, if not more so, than breaking a spar. At least a dismasted vessel can be controlled to some degree.

The bearing which I have illustrated is of a galvanized mild or steel, stainless or Monel base plate bent to a "U" section which fits snugly under the keel. At the aft end of the baseplate, a short section of pipe is weld-attached. It must be of the same diameter as that used for the rudder tube

Plates 96—99. Heel bearings commonly seen on ferro-cement boats, more often than not, weldments which produce crevasses, sharp angles, and vertical bolt orientation. Many ultimately create stress risers and difficult maintenance problems.

corners (particularly welded) should be avoided if possible as these create "stress-risers." These corners are difficult areas in which to prevent rust or corrosion. After cutting or welding, all work should be ground smooth and edges rounded slightly, again to deter rusting. Attachment bolts must be either of Monel or highest schedule stainless steel, as even galvanized mild steel will eventually "freeze." For the most reliable attachment, bolts should fasten trans-versely, rather than vertically, into the base of the keel. In personally viewing hundreds of cement boats under con-struction, I have seen only a small number which meet

and in line with the stock. A rubber Cutless type bearing is inserted into this pipe section while the bottom end is closed to support the ball end of the rudder stock. In order to fair the finished weldment, common polyester putty or epoxy glazing compound may be used. This type of heel bearing is the most popular among fine custom yacht build-ers and is often fabricated as a solid casting.

To insure the proper fitting of the heel weldment, it must be prefitted on the armature then carefully tapped into position immediately upon applying the fresh mortar. Take great care to ensure the proper alignment between the heel bearing pipe and the deadwood half-pipe section. Leave the bearing in position until the mortar has taken a firm set. When you remove the bearing, some voids may be apparent, but they may be filled with epoxy grout. You will notice in the illustration that the amount of plaster applied in the heel area is such that it will create a rabbet for a flush mounting when the bearing weldment is installed. This will result in a smooth, professionally finished appearance. While the bearing weldment may be predrilled, the hull should not be bored for fastenings until after it has been fully cured. (Fig. 243)

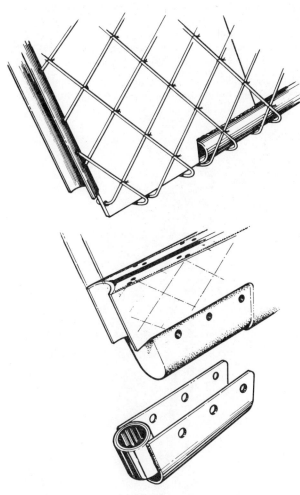

Fig. 243. A preferred rudder heel bearing installation.

Rudder Stock Reinforcement

As a means of distributing the upper portion of the rudder's bending loads at the stock entrance, a reinforced concrete boss should be constructed around the base of the rudder stock tube on the inside of the hull. This may take the form of reinforcing rod and mesh shaped as illustrated. The rods used for the boss should be worked well into the hull armature. (Fig. 244A & B)

Because of the weight of concrete involved, it is not recommended that this mass be poured until after the hull shell has been completely cured as this will prevent sagging and possible cracking. Prior to pouring, loose cement should be blown away and the area coated with an epoxy bonder employed to prevent runoff.

In addition to the concrete boss, I prefer to install a transverse oak timber between each side of the hull which is held in place by concrete brackets. The rudder stock will pass through a flange bearing set into this timber. The location of the timber should be as high and as close as possible to the rudder quadrant. The timber may also serve as a base for mounting a hydraulic steering ram or steering cable idlers.

Skegs

If your vessel is designed with an abrupt skeg forward of the rudder blade, such as found on many high-performance sailboats, it will deserve particular attention. While a long horizontal skeg is generally quite thick in section and does not require extra strengthening, the deep, thin skeg should be constructed with floors of either steel plate or a transverse rod webwork. These floors should be as close together as possible and should be built in such a manner so as to distribute the side forces well into the hull shell.

Plate 100. Steel plate skeg floors as seen from the interior of **Andromeda's** armature. Her original skeg was rather fine and drastic, hence the necessity for this engineering. These floors could have been constructed in web form. A Bingham design.

The design details of such a skeg will be carefully spelled out by your designer, but you should not discount the importance of careful work in this area. The trailing edge of the skeg is constructed essentially in the same manner as the normal deadwood. (Plate 100)

Concrete Tanks

I must point out at the outset that I can no longer approve of concrete tanks, except for those used as holding tanks. The reasons for my decision are quite varied and I suggest that you consider them carefully before going ahead with the installation of these units. During my travels I have run into quite a few vessels which have had problems with fuel bleeding through the concrete bulkheads. In spite of exasperating efforts to seal them thoroughly, it seems that hairline cracks, bubbles, or separations may have occurred, undetected, in the tank lining material. This may be impossible to avoid, particularly since it is extremely difficult to work within most concrete tanks. It should also be borne in mind that if a hull is damaged by grounding or impact, the tanks will be extremely vulnerable to penetration and seepage. The wiring up of tanks is a very tedious operation

237

one layer of 1/4" rod. For tank capacities over 150 gallons, two layers of 3/16" high tensile rod should be. The rod layers should be covered with two layers of 19-gauge square welded mesh on each side (total four layers) for small tanks and three layers each side for large tanks. Of course, an infinite variety of suitable combinations may be applied according to the specific tank shape and size. (Figs. 246 & 247)

Little thought to date has been given to ferro-cement construction without reinforcing rod, but in cases such as this where flat surfaces are encountered, a rod armature may be completely eliminated by substituting with an equal thickness of mesh. If the resultant mesh layup thus required is more than six laminates, a mortar penetration problem may exist. In cases such as this, the armature thickness would best be accomplished by using fewer layers of heavier mesh such as 16- or even 14-gauge square welded.

Because you will have to work inside of the tank and its baffled sections during fabrication, plastering, finishing and

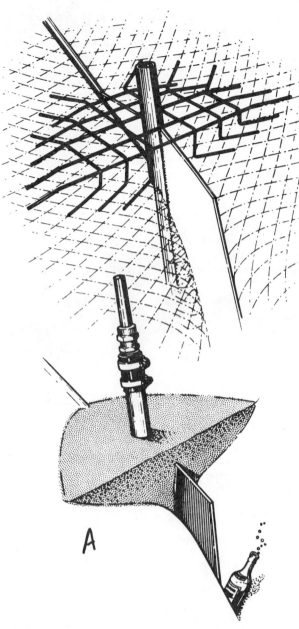

Fig. 244A. A typical rudder stock interior boss intended for the use of a self-aligning shaft stuffing box.

but not really as troublesome as the plastering of their interiors, unless the tank is very shallow and provided with an extremely large scuttle. The excess weight of concrete tank structures would be far more valuable if reserved for ballast, and here every pound counts on a ferro-cement yacht.

It is not my intention to override your own architect's engineering, so I would strongly consider your talking this over with him prior to your making the switch to steel or fiberglass tankage. Knowing full well, however, that many builders will not be swayed, I will attempt to describe ferro-cement tank construction in order that it may be executed properly.

These units are usually constructed in much the same manner as cement bulkheads with all of the reinforcing rods extended well into the hull armature. Using this conventional construction, tanks under 100-gallon capacity may be built using either two layers of 1/8" high tensile rod or

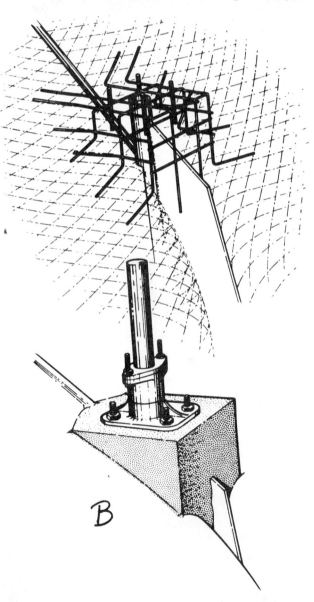

Fig. 244B. A typical rudder interior boss intended for the use of a "flange" type non-aligning shaft stuffing box.

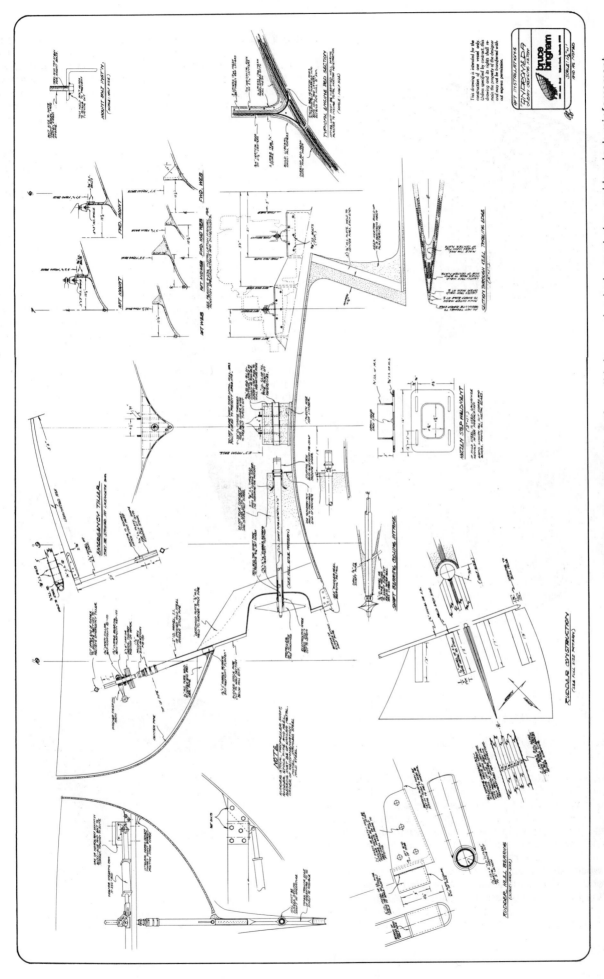

Fig. 245. A typical aft installations drawing for the **Andromeda** showing the construction of a solid skeg, engine beds, mast step, rudder stock reinforcement, deadwood and many other elements designed into the ferro-cement hull. A Bingham design.

Tank Scuttles

These provide a means of entering the tank for inspection, cleaning and repair and should be provided with a substantial steel bolt-on cover. The size of the scuttle should be at least 14" x 22" as this will permit admittance of a normal sized person. One scuttle should be located on the tank top of each baffled compartment.

The scuttle will require the construction of a ¼" x 1½" or 2" mild steel rim. This rim will be drilled and tapped on approximately 3½" to 3" centers to receive 3/8" x 1½" coarse threaded galvanized or stainless bolts. The undersides of each stud should be lightly tack-welded to prevent its turning out. Use the precut steel plate scuttle cover as a bolt locating pattern.

Before installing the inside tank mesh layers, install the rim from under the tank top, if two rod layers are used, by pushing the bolts fully through the armature. Weld the perimeter of the rim to each tank rod. Now install the inner tank mesh layers, taking care to completely cover the underside of the scuttle rim. Before plastering the tank, wrap all of the exposed scuttle bolts with masking tape to keep their threads clean.

Finishing the scuttle covers will require that all cut scores be ground absolutely smooth to retard rusting. Slightly round all plate edges and bolt holes. The water tank cover may be galvanized or epoxy coated. The scuttle cover may

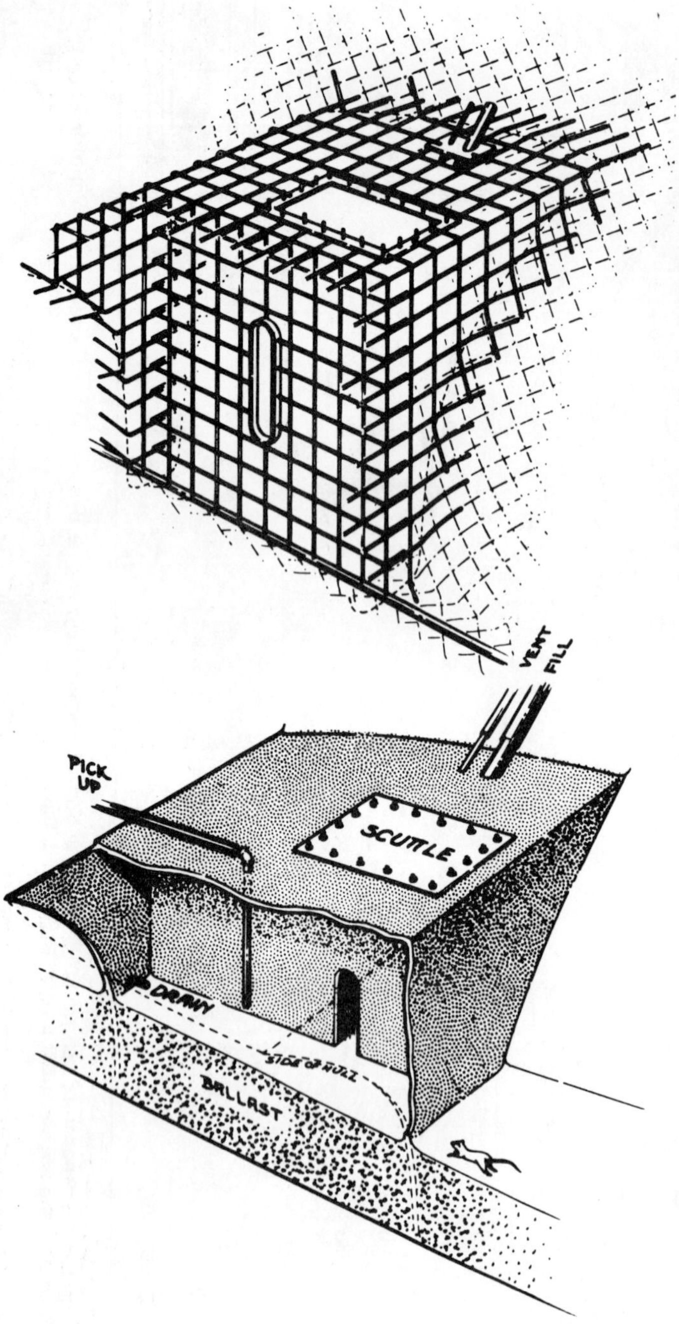

Fig. 246. Typical center tank construction.

occasional cleaning during service, all individual tank compartments must be fitted with scuttles.

Under no circumstances whatsoever should the ends of the fuel and water tanks share a common divider. If hairline cracks develop in the tank ends for any reason, it is possible that fuel and water will intermix through a slow but steady seepage. Over many years of service, fuel oil also has a tendency to become absorbed in concrete surfaces if the tank sealant is penetrated. I highly recommend, therefore, that when constructing fuel and water tanks end to end a plywood or foam cofferdam be installed between, thus building two completely separate units. (Fig. 248)

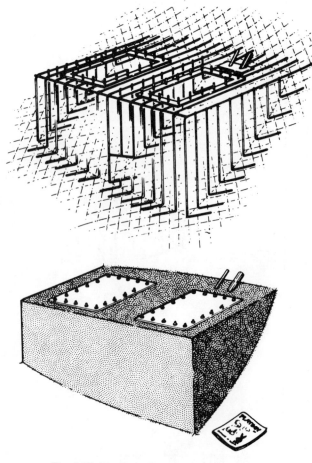

Fig. 247. Typical wing tank construction.

240

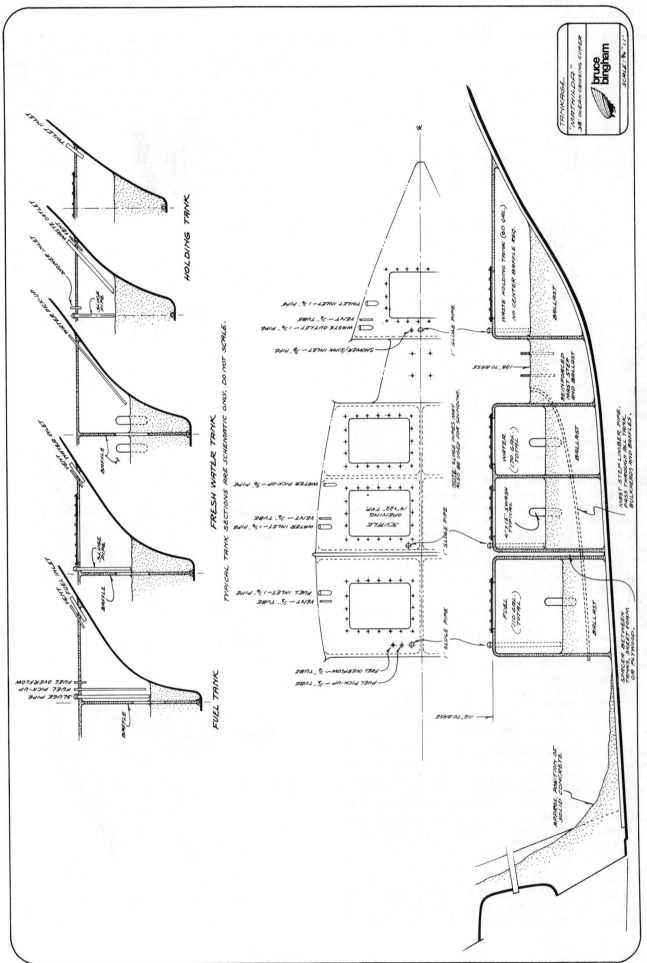

Fig. 248. A tankage diagram showing piping, scuttles, baffles and cofferdam between water and fuel tank.

be left as bare steel or epoxy coated, but do not galvanize. (Fig. 249)

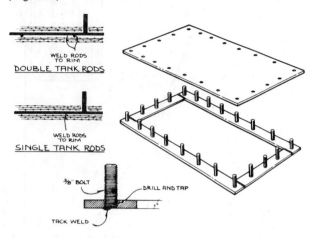

Fig. 249. Ferro-cement tank scuttle rim and cover.

Tank Piping

The actual installation of the tank fill connectors, vents, sludge pipe and fills will not actually take place during the construction of the tank armature, as it is far better to pack these pipe units with an epoxy grout rather than depending solely upon concrete to form a permanently sealed fixture, unless pipe flanges are used. It is required, however, that you install circular wood blanks into the tank armature to provide a hole for the pipe passage through the finished tank or to weld a flat plate bearing a connector pipe into the tank top rodwork. These blanks may be cut from ply or solid wood to a diameter 1/16" larger than the required pipe. The wood blanks should be sanded and plywood edge voids filled to prevent the concrete from "keying" to them. I always recommend that blanks be given a coating of melted wax prior to their insertion in the armature. You may prefer to make your blanks out of short sections of PVC pipe. This works very well, but I must caution you against leaving the PVC in the tank as a permanent fixture after plastering. (Fig. 250)

The pipe positions are critical. The sludge pipe should be located as close to the tank bottom as possible while still allowing accessibility and room for installing a valve. If this is not possible, a sludge pump pipe may be provided through the tank top. Otherwise, all purging must be accomplished through the scuttles. Fuel and water pickups may be located through the top or the side of the tank but never at the bottom. The lower ends of the pickup pipes should always be several inches from the bottom of the tank. Vent piping should always enter from the tank top.

Never use galvanized pipe in a fuel tank. It is far better to use untreated iron, steel or copper. PVC piping should also be avoided as it does not withstand vibration and may become brittle when extremely cold. Galvanized hardware is suitable for water tanks but some provision should be made for their eventual replacement. Coating tank piping with epoxy before its installation is a wise idea as this finish is most permanent. (Fig. 248)

Bolt-Holes

From time to time I have made references to prelocating the bolt-holes using waxed wood dowels. Their insertion into the armature does not usually require the cutting of a full hole in the mesh as such, but it will be necessary to snip the mesh wires at the determined location. The mesh wires are then bent out of the way and the dowel placed onto the armature. Now bend the mesh wires back lightly against the plug. The length of the plug should correspond exactly to the thickness of the finished hull and the ends of the plug should not protrude more than 1/16" beyond the outer mesh layers as this will impede the plastering and cause a slight bump on the surface of the concrete. On plastering day it is important that some person be delegated the responsibility of cleaning the plaster from the plug ends of concrete as this will provide for an easy removal without chipping the hull. After the hull has completely cured, tapping the bolt plugs with a centerpunch will cause their popping out quite readily. (Plate 101)

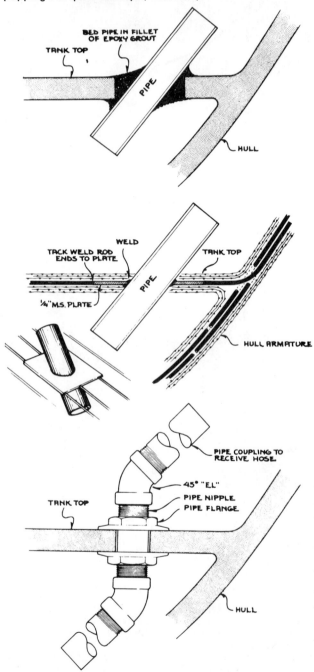

Fig. 250. Ferro-cement tank pipe coupling.

Plate 101. A waxed hardwood dowel inserted through the armature provides for prelocated bolt-holes before plastering.

Thru-Hulls

There is absolutely nothing more disgusting than to see a ferro-cement boat with holes for exhaust pipes, drains and seacocks that are chipped, broken, cracked and mangled, or just plain unhealthy. Stresses on a vessel have an uncanny knack of finding their way to these openings, and needless to say, the continual working of the boat during her length

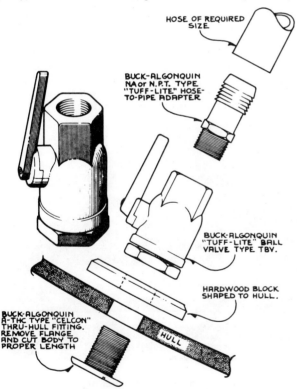

HOSE OF REQUIRED SIZE

BUCK-ALGONQUIN NA or N.P.T. TYPE "TUFF-LITE" HOSE-TO-PIPE ADAPTER

BUCK-ALGONQUIN "TUFF-LITE" BALL VALVE TYPE TBV.

HARDWOOD BLOCK SHAPED TO HULL.

BUCK-ALGONQUIN A-THC TYPE "CELCON" THRU-HULL FITTING. REMOVE FLANGE AND CUT BODY TO PROPER LENGTH

HULL

Fig. 251. Ferro-cement builders owe the Dickson Valve Division of Buck-Algonquin a debt of gratitude for producing nylon valves which can be coupled together with other synthetic parts to form seacocks. While other manufacturers of nylon seacocks may be available, they are unknown to the author. This one part, in itself, can solve a great portion of the ferro-cement electrolytic decomposition problems.

of service will eventually worsen the condition of the shell around these perimeters. It is hoped that future ferro-cement builders will learn a good lesson from these examples, drawing upon the experience of these failures. Your best bet is to make it a point not to repeat the same old mistakes you have already seen so often! (Plate 102)

It is absolutely necessary to the safety of the vessel that all piping that completely circulates below the waterline, whether in the inverted or upright position, be fitted with marine seacocks. Domestic garden hose valves will not do and will be refused by any competent surveyor because they are generally castings of brass or very low grades of bronze. Until recently, seacocks were manufactured only as bronze castings but are now available in molded nylon.

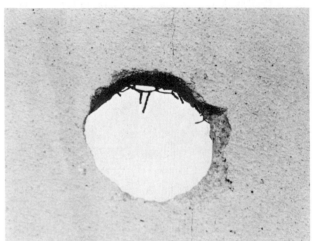

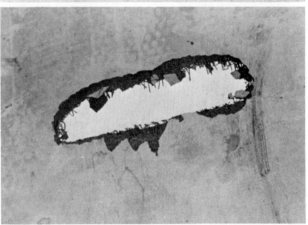

Plate 102. Severely broken thru-hull holes are the result of a badly "keyed" wooden blank. Notice the hairline crack passing through the hole as a result of having to hammer the blanks loose.

(Fig. 251) The nylon hardware has the obvious advantage of being unaffected by electrolysis. If you choose to use the bronze casting, a simple preventive measure can help retard galvanic effects and that is to form the true hole with a short section of PVC pipe. This will not only create a passage through the hull for the hardware but will also act somewhat as an insulator. The seacock installation should never be dependent upon simply tightening up against the pipe flanges for fastening but should always be bolted through the flanges whenever possible. Many of the seacocks on the market have not been drilled for bolts, so I recommend doing so before bedding this hardware.

The piping systems which do not circulate completely below the waterline, such as cockpit and sink drains, will only require simple thru-hulls without valves. These are also available in molded nylon and are recommended to eliminate galvanic action. Because these piping systems require many small hose sizes, it is not usual to bolt them to the hull. Provisions for hole blanks in the armature may also be accomplished with short sections of PVC pipe. The PVC may be left in the hull after plastering.

The most common method for installing blanks for holes is that of cutting the desired diameter or shape out of ¾″ plywood. (Fig. 252) When building to the cedar or closed-mold systems, the blank is accurately located and nailed to the mold. If you are building to an open-mold system, the hold is first marked onto the armature upon completing all mesh and rod layers. Use the wood blank as an accurate template. Now carefully cut away all mesh and rod in the

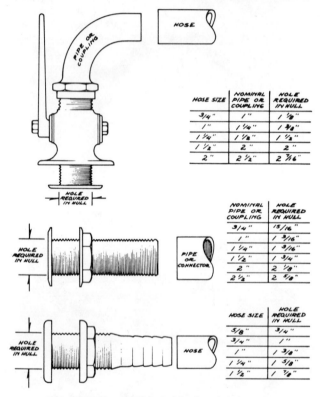

HOSE SIZE	NOMINAL PIPE OR COUPLING	HOLE REQUIRED IN HULL
3/4″	1″	1 1/8″
1″	1 1/4″	1 3/8″
1 1/4″	1 1/2″	1 1/2″
1 1/2″	2″	2″
2″	2 1/2″	2 7/16″

NOMINAL PIPE OR COUPLING	HOLE REQUIRED IN HULL
3/4″	15/16″
1″	1 3/16″
1 1/4″	1 9/16″
1 1/2″	1 3/4″
2″	2 1/8″
2 1/2″	2 5/8″

HOSE SIZE	HOLE REQUIRED IN HULL
5/8″	3/4″
3/4″	1″
1″	1 3/8″
1 1/4″	1 5/8″
1 1/2″	1 7/8″

Fig. 252. Thru-hull hole blank requirements.

area of the hole. The blank may now be inserted into the armature and held in place with tie wire passing through the blank and the armature. (Plate 103) Another method for installing the wood hole blank is that of slightly undercutting the mesh around the perimeter of the hole then bending the wire ends back slightly against the plug. The blank is then inserted into its hole and the mesh ends bent back against the blank.

Porthole Blanks

The actual fabrication of the wood blank has been grossly oversimplified and has been responsible for chipping and cracking the hole perimeters.

When the vessel is plastered, wet concrete will separate into the grain and voids of the plywood, therefore keying the blank firmly into the finished hull. (Plate 104) If this situation is allowed to exist, it will take a fair amount of ham-

mering to remove the blank. This may be solved, however, by first cutting the blank edges at a slight angle to create "draw" before installation. Secondly, the edges of the wood blank should be glazed and sanded to create a smooth bearing surface. Finally, coat the plywood blank with melted wax as this will allow for easy separation from the hardened concrete with only light tapping being necessary.

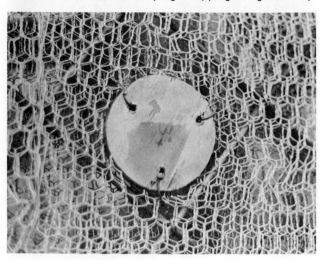

Plate 103. A plywood thru-hull blank wired into the armature.

The thickness of the thru-hull blanks should be exactly that of the finished hull which may necessitate a combination of different plywood thicknesses to achieve. When the vessel is plastered, it is imperative that no concrete be allowed to encroach onto the blank as this will key the plywood into the hull. (Plate 105) Some person should be delegated the responsibility of removing excess mortar from all the thru-hull blanks before the concrete is allowed to harden. This can be done with an ordinary butter knife.

Plate 104. The void in the middle layer of the plywood will cause "keying" of the thru-hull blank into the concrete. Eventual difficulty with its removal will result.

If you have allowed the thru-hull blanks to be constructed to a greater thickness than the finished shell, it will cause a ridge to be formed around the perimeter of the hole which cannot be remedied without considerable grinding of the finished hull.

Some designers have suggested that a polyethylene film be wrapped around the through-hole blanks to provide easy concrete separation. This problem here, however, is that wrinkles, overlaps or folds in the polyethylene will be reproduced in the concrete. (Plate 106) This will always necessitate grouting of the finished hole for a neater appearance.

Plate 105. Allowing the mortar to lap onto the surface of thru-hull blanks will make the removal of the blank virtually impossible without damage to the hull.

Ports and Windows

The holes in the hull for receiving these units may be blanked out with plywood plugs in exactly the same manner as the thru-hulls. Once again I urge you to glaze and sand the edges of the plywood and coat the blank with wax to prevent it from keying to the plaster. (Fig. 253) One significant point to bear in mind is that windows and ports

Plate 106. Wrapping thru-hull blanks with polyethylene will create a clean and easy separation of the concrete from the mold, but wrinkles in the polyethylene will be reproduced in the hull.

should **never** have square corners as these may become stress risers and a cause for eventual cracking of the hole. If you are using squarish windows, apply generous radii to eliminate this possibility.

Some designers and builders suggest that a flat bar screed be used around the perimeters of ports and windows. I have no objections to this, providing the shape can be accurately

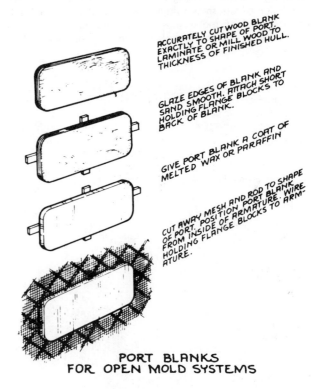

ACCURATELY CUT WOOD BLANK EXACTLY TO SHAPE OF PORT. LAMINATE OR MILL WOOD TO THICKNESS OF FINISHED HULL.

GLAZE EDGES OF BLANK AND SAND SMOOTH. ATTACH SHORT HOLDING FLANGE BLOCKS TO BACK OF BLANK.

GIVE PORT BLANK A COAT OF MELTED WAX OR PARAFFIN

CUT AWAY MESH AND ROD TO SHAPE OF PORT. POSITION PORT BLANK FROM INSIDE OF ARMATURE. WIRE HOLDING FLANGE BLOCKS TO ARMATURE.

**PORT BLANKS
FOR OPEN MOLD SYSTEMS**

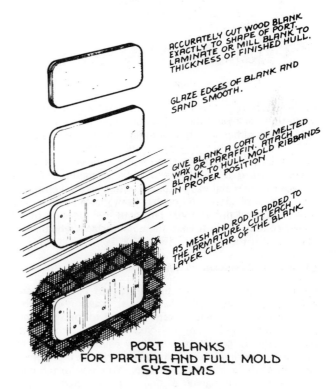

ACCURATELY CUT WOOD BLANK EXACTLY TO SHAPE OF PORT. LAMINATE OR MILL BLANK TO THICKNESS OF FINISHED HULL.

GLAZE EDGES OF BLANK AND SAND SMOOTH.

GIVE BLANK A COAT OF MELTED WAX OR PARAFFIN. ATTACH BLANK TO HULL MOLD RIBBANDS IN PROPER POSITION

AS MESH AND ROD IS ADDED TO THE ARMATURE, CUT EACH LAYER CLEAR OF THE BLANK.

**PORT BLANKS
FOR PARTIAL AND FULL MOLD
SYSTEMS**

Figure 253.

formed before its installation. This is not as easy as it sounds as I have seen many potentially good boats visually ruined by "not quite" round or square openings. I would suggest that the open screed be carefully formed around the edge of a plywood mold. Once the screed has been welded, remove it from the mold and readjust its shape accurately. Always grind this welding to a smooth surface on the exposed side of the screed then wire-brush or sandblast it to achieve a bright metal finish. Now carefully cut out the required hole and insert the screed into the armature. Bend

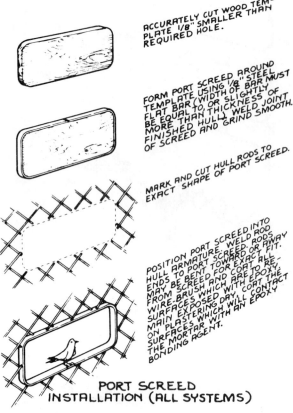

FORM PORT SCREED AROUND TEMPLATE USING 1/8" STEEL FLAT BAR (WIDTH OF BAR MUST BE EQUAL TO, OR SLIGHTLY MORE THAN THICKNESS OF FINISHED HULL). WELD JOINT OF SCREED AND GRIND SMOOTH.

MARK AND CUT HULL RODS TO EXACT SHAPE OF PORT SCREED.

POSITION PORT SCREED INTO HULL ARMATURE. WELD RODS ENDS TO PORT SCREED. RODS MAY BE BENT TOWARD OR AWAY FROM SCREED FOR EXACT FIT. WIRE-BRUSH AND COAT ALL SURFACES WHICH ARE TO RE-MAIN EXPOSED WITH EPOXY. ON PLASTERING DRY, COAT CONTACT SURFACES WHICH WILL CONTACT THE MORTAR WITH AN EPOXY BONDING AGENT.

**PORT SCREED
INSTALLATION (ALL SYSTEMS)**

Figure 254.

the armature rods toward the screed until a firm contact is made then tack-weld the screed for permanent position. If you would rather attach additional rods to the screed before installation, marrying these rods to those of the armature, it is perfectly all right and works very well. After the opening screed is installed within the armature, coat the surface of the metal to be exposed with a layer of epoxy. If you wait until after the hole has been plastered, they will surely rust and this will interfere with the adhesion of the epoxy coating. Do not coat the metal surfaces which are in contact with the armature. (Fig. 254)

Chocks and Hawser Holes

I cannot recommend that hawser chocks be constructed with sharp edges. Such sharp edges often result when wooden blanks or metal screeds are used to form the holes in the finished hull. (Plate 107) Sharp hawser hole edges are particularly dangerous because they cause chafing of the anchor or docking lines which can easily cause a 1" rope to wear through in a matter of hours. You may install commercial chock and hawser castings in the finished hull after the removal of a hole blank if you wish. These castings are available from such manufacturers as Perkins Marine Lamp and Hardware Corporation or Buck-Algonquin. Unfortunately, these castings require through-bolts and are not

engineered for the thickness of hull generally encountered in ferro-cement construction. Because the spigot lengths of these fittings may exceed the thickness of the finished hull, they should either be ground down to the proper dimension or may be shimmed on the inside of the hull with hardwood fashion plates. Be sure to install waxed dowels in their proper positions for receiving the chock fastening.

Some discriminating builders have gone so far as to have custom chock or port castings made by a local foundry. This may be very expensive but you are sure to secure hardware specifically engineered for use with ferro-cement. If custom castings are out of the question, you may elect to have hawser holes constructed of plate. They are very functional, present a good appearance, and may be fabricated by any local metalsmith. I have included working drawings for their construction. (Figs. 255 & 256)

Cement Deck Hatch Coaming
(Inverted Construction)

If the vessel is being built in the inverted position and you have constructed a concave deck mold as prescribed, your hatch position should be carefully located prior to your laying the first deck mesh layer. Accurately cut the opening through the deck mold and attach a square frame to the underside of the mold. The height dimensions of the hatch frame should be approximately 3", taking care to shape its transverse members accurately to the curve of the deck crown.

When applying the first layers of mesh on the deck mold, carefully avoid locating the coaming corner mesh joints at exactly the same place as this will cause a severe weakening.

When laying reinforcing rod onto the deck, I feel that it is more accurate to prebend and cut the rods to form the hatch coaming prior to stapling them in place. If your design calls for only one deck rod layer, it will be necessary to form additional short rods and attach them along the hatch coaming where the deck rods would not normally fall. The outside mesh layers are applied as before. To achieve a clean, sharp edge along the upper sides of the finished coaming, an angle iron screed may be attached to the ends of the hatch rods, the mesh being brought to the upper

Plate 107. Steel screeds may be used around ports and windows but they should be avoided at hawser chocks because the resultant sharp edge can chafe through dock and anchor lines.

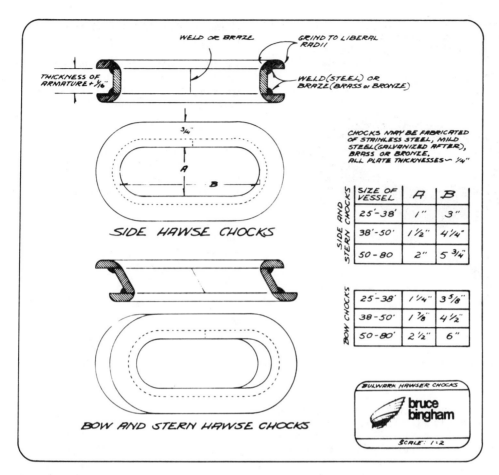

Fig. 255. Hawser bulwark weldments specifically designed for ferro-cement construction.

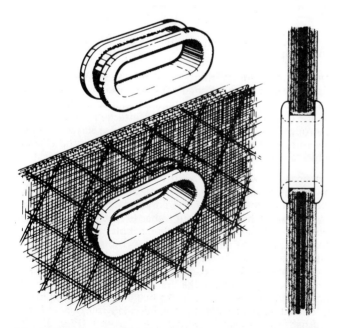

Fig. 256. The installation of bulwark hawser chock weldments requires their positioning as the mesh and rodwork are added to the hull in order to "key" them in permanently.

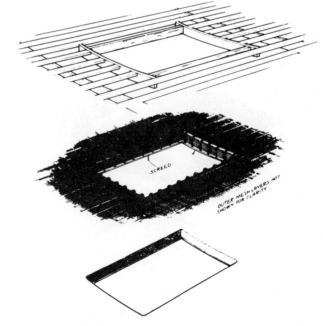

Fig. 257. A typical inverted hatch coaming construction in a ferro-cement deck.

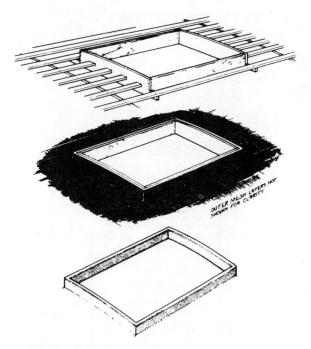

Fig. 258. Typical upright hatch coaming construction in a ferro-cement deck.

surface of the iron. Prior to installing the screed, I suggest wire-brushing the steel to a bright surface. After you have attached this screed to the armature, coat those surfaces with epoxy which will remain exposed after plastering. To do this after plastering will surely cause bleeding of rust. (Fig. 257)

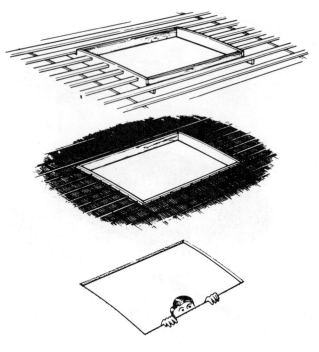

Fig. 259. Typical flush hatch opening construction in ferro-cement deck. Note that the opening screed height is equal to the thickness of the finished deck. (Outer mesh not shown for clarity.)

Cement Deck Hatch Coaming (Upright Construction)

Accurately mark and cut the hatch opening to the deck mold. Carefully fashion a wood frame to the shape of the coaming, taking care that the transverse members match the curve of the deck camber. The height of the hatch coaming should be approximately 3''. Attach the hatch coaming to the mold.

When applying the first deck mesh layers, simply run the mesh up the side of the coaming frames. Be sure that you do not locate successive corner mesh joints at exactly the same place each time, but stagger them as best you can. Cut the mesh along the top of the frame.

When laying rod, prebend and cut the ends then position the rod firmly against the coaming before stapling the length of the rod to the mold. In those areas where the deck rod would not normally come in contact with the hatch coaming, it will be necessary to fabricate supple-

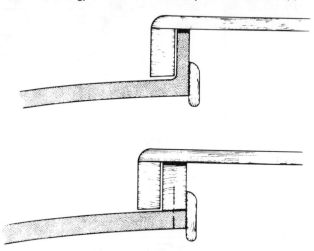

Fig. 260. Typical hatch coamings used with ferro-cement deck construction.

mentary rods for the purpose of reinforcing the coaming. For the clean sharp finished edge, an angle iron screed may be welded to the tops of the reinforcing rods around the lip of the coaming. After this screed is attached, wire-brush it thoroughly for a bright finish and coat those surfaces which are to remain exposed with epoxy. Do not coat those areas which are in contact with the armature. (Fig. 258)

Personally, I would not build a ferro-cement deck on my own boat unless it were my intention to use **FER-A-LITE** mortar. In either event, it would be my choice not to use an integral hatch coaming, but rather to install a brightwood timber frame around a flush hatch hole. While the guarantee of absolute watertightness is far less than with the integral coaming, it is much better looking and more yachtsmanlike. The construction of a flush ferro-cement hatch opening is exactly like that of the hatch opening with integral coaming except that the opening in the wood mold is carried upward only to the intended thickness of the finished deck where it will serve as a screed. (Figs. 259 & 260)

Cement Cabin Trunk Carlins

The purpose of the carlin is to provide a member onto

which a wood cabin trunk is attached. I will be describing several timber carlin systems later in this volume, but if you prefer, a reinforced cement carlin may be built in exactly the same manner as the integral hatch coamings. (Fig. 261)

Ferro-Cement Cockpit (Upright Construction)

Carefully locate the perimeter of the cockpit at the deck's edge and cut away that portion of the deck mold. It will be necessary to completely construct a temporary cockpit area, including seats. It is not necessary that the cockpit mold be of any substantial strength, but you must make provisions for cockpit drains, seat hatch covers, control wiring and piping, and through-holes. When completed, line the cockpit mold with polypropylene film and then apply mesh and rod in the normal manner. (Fig. 262)

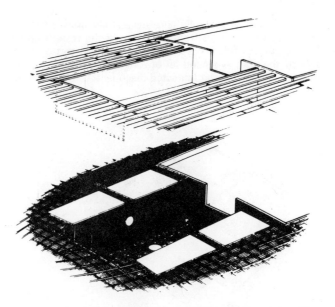

Fig. 262. Typical upright ferro-cement cockpit construction showing screeds and opening blanks. (Outer mesh not shown for clarity.)

Suspending the Armature for Upright Open-Mold Plastering

So that we may achieve total access to the armature during plastering, the time has almost come to remove the station molds and ribbands from the completely wired hull. Quite obviously, it is necessary to support the armature in some manner as we will no longer have the benefit of the mold hangers.

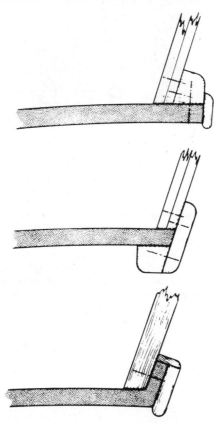

Fig. 261. Typical cabin trunk attachments to ferro-cement decking.

Personally, I do not like concrete cockpits as they are lacking in personality and seem to become colder and harder the longer one sits. The cockpit weight factor is considerable unless the cockpit is small in size or troweled with **FER-A-LITE** mortar. This is an area where the lightest possible mesh rod combination should be considered.

If you are building your vessel upside down and intend to employ a cement cockpit, this will require virtually building the cockpit mold inside out. Keep in mind that you will have to install blanks in your mold for seat hatches, drain holes, and control line passages. It will also be necessary to install wooden screeds wherever a raw concrete edge is to be formed. Prior to applying mesh and rod, it will be necessary to cover the cockpit mold with polypropylene film to ensure clean and easy separation. (Fig. 263 & 264)

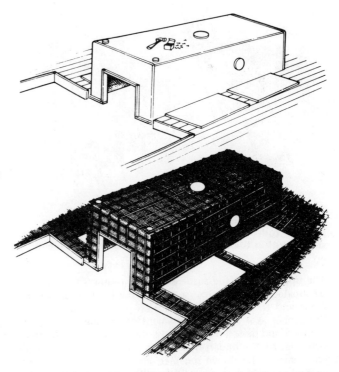

Fig. 263. A typical inverted ferro-cement cockpit construction showing screeds and opening blanks. (Outer mesh layers not shown for clarity.)

The system most generally and successfully used is that of attaching vertical rods from the sheer of the vessel's armature to the scaffold cross-span at each station and on each side of the hull. A second set of rods is attached to the flat portion of the bilge and directed diagonally to the scaffold.

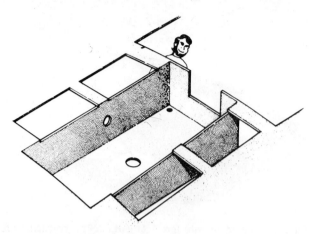

Fig. 264. A typical ferro-cement cockpit with all molds and screeds removed.

The attachment of the suspension rods to the hull is simply done by bending an "L" into the rod, inserting it into the armature and wiring it securely into position. An alternative to this is to tack-weld the suspension rods to the armature. (Fig. 265)

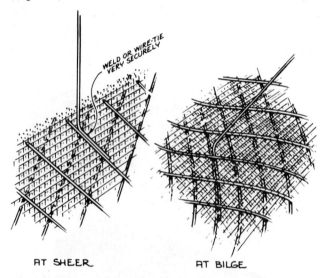

Fig. 265. Methods of attaching the upright open-mold suspension rods to the armature. These rods should be installed within the armature prior to attaching outside mesh layers.

The upper end of the suspension rods requires that wedges be located atop the cross-span to allow for tension adjustment and to prevent the suspension rod from sinking into the timber as it assumes the weight of the shell. The rod is crimped and passed over the cross-span and down the opposite side. At the underside of the cross-span, the suspension wires should be drawn together and welded, "U"-clamped or tied securely with wire. (Plate 108) As the diagonal wires will have a tendency to slide inboard, several preventative spikes should be driven against them and into the timber. You may use as many suspension rods as you wish. The

Plate 108. Wedging used for tuning the armature suspension rods.

more, the better. You should have a minimum of four at each station as well as plenty at the ends of the vessel.

In order to prevent possible swaying of the hull or springing of the sheer when the mold is removed before plastering, I suggest that you pass rods transversely over the armature from the scaffold uprights on each side. These transverse rods will be attached to the scaffold uprights, pulled tight and wired to the intersecting suspension rods. In order to help bear the weight of the hull during plastering, the keel should be blocked securely before removing mold. (Fig. 266)

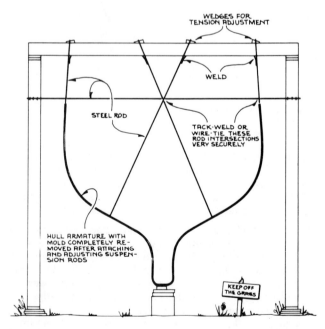

Fig. 266. The complete upright open-mold rod suspension system.

Plate 109. The sheer suspension rods and screed on an open-mold 46' **Andromeda** ketch.

Most builders are deathly afraid of no-mold suspension of their vessel's armatures, but I have personally seen it work beautifully time and time again on vessels over 45' L.O.A. with well over 3½' between suspension points. One thing to remember is that all of your armature rod intersections must be securely wire-tied before removing the mold. Once done, the armature will hold its shape perfectly and will easily bear the weight of several plasterers walking on the inside of the hull. (Plate 109)

Alternative to Open-Mold Armature Suspension

If you are skeptical about suspending an unsupported armature from wires and rods, it is understandable. I certainly cannot influence your apprehension. But it need not result in your decision to leave the mold in your hull as a visible and mechanical barrier to mortar penetration.

The alternative is to cut the mold ribbands along the station molds with a circular saw from the inside of the hull. The ribbands can then be removed from between the station molds. Short sections of ribbands will remain along the edges of the station molds and the armature will still be attached in these areas. (Plate 110) When plastering, voids may be created in the areas of these ribband sections but the effect will be minor as these discrepancies may be grouted after curing the hull. (Plate 111)

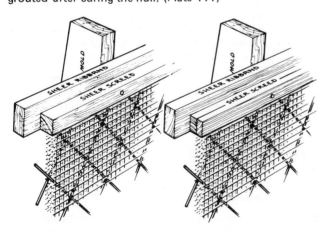

Fig. 267. Sheer screed variations for upright or inverted wood mold construction.

Plate 110. The partial open armature suspension system showing the remainder of a station mold and ribbands in the hull after plastering.

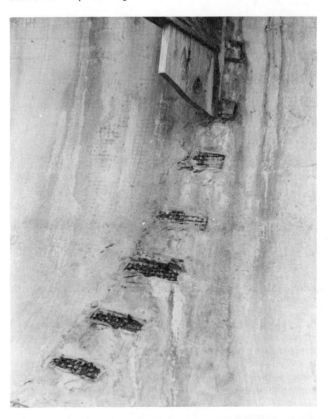

Plate 111. Voids left in the hull as a result of partial open-mold frame suspension during plastering. They must be ground-flushed and back-plastered with an epoxy grout to assure strength in this area.

Removing the Mold for Open-Mold Plastering

This is not as grueling a process as you may think. Once the armature has been suspended and the suspension rods tuned to proper tension, you may begin pulling the staples out of the armature and mold. If you have painted the staple strips before driving, you will have little trouble seeing them. This project can go very quickly if you make a party of it by inviting your friends and having some cases of beer on hand. The only equipment necessary will be several screw-

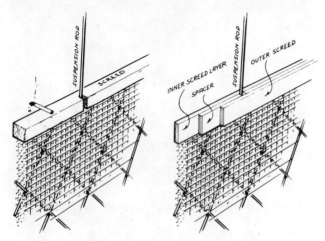

Fig. 268. Suggested sheer screed variations for upright suspended open-mold plastering.

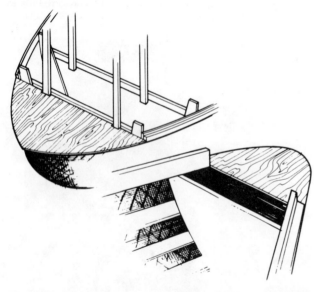

Fig. 269. Flat plywood screeds may be used when it is impossible to bend a timber screed to the tight curves of the sheer.

drivers and needle-nose pliers. If you miss an occasional staple, don't worry about it. I've seen six people remove thousands of staples and remove the mold from a 46-footer in only two days. They were able to resell the mold later by being careful not to tear up the lumber and by number-coding the parts as they were pulled out of the boat.

Sheer Screeds

If you are building to the partial upright or inverted construction system, the cedar or closed-mold method or the

open-mold with stations left in the armature, you may define your sheer line for plastering in the following manner. It will be necessary to attach a wood screed to your station molds around the entire hull so that the surface toward the armature represents the finished sheer line. The screed may be two vertically laminated strips of pine or fir equaling the thickness of the finished hull. If you prefer, you may use a square cut timber. (Fig. 267) The screed lumber should be of grade "A" fir or pine and scarf-joined for continuous length. Before raising to the mold, give it a coat of melted wax to ensure a clean, easy separation when removing it from the finished hull. When attaching the screed, it is intended that it fall directly on the sheer marks, but if you eye it very critically, you may detect slight unfairness, say less than 1/8" up or down. Errors such as this may be corrected with the screed and rightly so.

If you are building to the open (suspended) mold system, attaching the screed is a little more complicated because it will not only define the thickness and smoothness of the hull at the sheer, but it will also lessen the tendency of armature distortion before and during plastering. Obviously, when the mold is removed, fastening must depend upon the suspension rods extending from the sheer, in order that the screed may be made for passing the screed around the suspension rods. This can be accomplished in several ways: 1) Spring a continuous ¼" thick batten along the inner side of the suspension rods. Then glue ¼" short strips to the first batten between the suspension rods. Now spring an outer ¼" continuous batten and glue it to the center spacers. (Fig. 268; Plate 109) 2) Cut a batten measuring, say, ¾ x 1". Spring this batten along the sheer against the suspension rods and mark the suspension rod positions. Remove the batten at the sheer, forcing the notches over the suspension rods. If you have not cut the notches accurately, you will be able to coax a little rod movement. To prevent cement from oozing up into the notches in the screed, they may be covered from underneath with common cellophane tape. (Fig. 268) As with all other screeds and blanks, the screed should be coated with melted wax before its attachment.

A variation to the screed, which must be applied in areas of very tight curves such as canoe sterns or barrel bows, is that of using a ¼" sheet of plywood laid horizontally across the sheers. (Fig. 269)

Surveying

At this juncture you may have spent as many as 3,000 hours of tedious work in preparation for the big day, plastering. But there still may be a lot of loose ends which will have to be taken care of before hitting your armature with the trowel.

In order to establish a permanent record of your vessel's construction, which may ultimately affect insurability and resale value, photograph every conceivable detail of her armature. This photographic record should include typical wire ties, mesh butt joints, thru-hull blank detail, all integral structures, the mold detail on the inside of the hull, if it is to remain during plastering, and whatever other revealing views you may determine as being important for evaluating the quality of your boat.

If you have not already done so, construct several test panels, reflecting the exact armature layup of your hull. These panels should be, say, 1'6" x 3'. They will be very helpful to your surveyor in establishing strength factors necessary for his reports.

Hiring a surveyor, by the way, is an absolute necessity. A certificate of survey will most assuredly be required by your marine insurers as well as any prudent prospective buyer if the future selling of your vessel is intended. The survey certificate will also help your peace of mind when it comes to planning long-range cruises. I cannot over-emphasize this critical point. As previously mentioned, the photographs that you have taken of your armature will be of great value to him and will undoubtedly become included in his survey summary. All of these records should be carefully preserved and kept readily available throughout the life of the vessel. At this writing there are only a handful of surveyors fully qualified in ferro-cement construction, so be sure you ask around before choosing the man to do the job.

As long as I am discussing surveys, I should mention Lloyd's and A.B.S. requirements. These organizations have long been respected for their knowledge and competence in establishing construction parameters for ships and yachts. Most yachtsmen consider that the very finest boats maintain Lloyd's and A.B.S. A-1 certificates. In the area of ferro-cement, however, these organizations have failed miserably in research and regulations. In my opinion, their regulations are nebulous, unspecific and totally useless in aiding the builder in constructing a better vessel. Their specifications include very little in the way of critical procedures and even less on choosing materials. Lloyd's and A.B.S. scantlings for ferro-cement boats are, at this time, nonexistent. In short, an A-1 certificate will mean very little in the long run.

Obtaining a Lloyd's certificate is an extremely expensive business, as it will generally mean that several separate surveys of your vessel will have to be conducted. The surveyor will have to be flown from his point of origin to your construction site, provided with food, lodging and transportation, then returned to his office at your expense. The surveying, itself, will cost many hundreds of dollars per shot. The Lloyd's survey will not only include the propriety of the hull, but will also include the methods and materials of your interior structures, the engine and its installation, your electric circuitry, plumbing systems, masts and rigging, sails, ground tackle, and virtually every other conceivable aspect which could affect the longevity and safety of your boat. The Lloyd's specifications for these factors are very precise and complicated. If any one of these factors is amiss, you will have failed in securing the A-1 classification. If you do succeed, however, it is important to point out that your vessel may require additional annual or biannual surveys in order to maintain your classification.

In the United States there is an organization which pulls more weight in terms of its respect and influence on marine insurers. This is the American Boat and Yacht Council. Their regulations and specifications are precise and simple. These specifications may be found in the publication, *Safety Standards for Small Craft*, and is available from the American Boat and Yacht Council Inc., 15 East 26th Street, New York, New York, 10010.

Needless to say, the finishing of your vessel will also have to comply with all applicable Coast Guard regulations as well as any local ordinances. You might as well begin thinking ahead in this regard because it will affect future procedures.

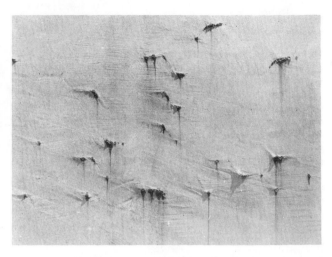

Plate 113. The cost of inexperience.

Plate 112. This is the result of a complacent attitude and sloppy workmanship on the part of untrained plasterers.

Plate 114. An excellent, if not elaborate, staging will add greatly to plastering efficiency.

plastering

Introduction

I have followed ferro-cement construction for years now, and have made it a point to be on hand for as many plastering jobs as possible. On every occasion, without exception, I have gone away feeling that I had learned a great deal or had seen details in procedure which could have been improved. I firmly believe that if I were to plaster a hundred boats, the next one would still be better.

Anyone who has ever lived through a plastering day knows what I mean. Those who have troweled only one boat (their own) wish they could do it over again in order to correct their first mistakes and to draw upon the invaluable experience gained through scores of unforeseen conditions.

I make no claim of being an expert on the handling of fresh mortar and I would surely lead you astray if I were to attempt to explain the idiosyncrasies of the material. I have, however, been closely associated with the one person I respect most in this delicate trade and have followed him far and wide to benefit from his background and judgment. I have asked master plasterer John Daniel to explain, firsthand, the fine art of successfully finishing the ferro-cement hull.

At this writing, Mr. Daniel has completed over 150 yachts and has gathered together an excellent team of professional masons. His reputation now ranges the entire length of the west coast of North America and he has been called in from as far as New England. He is truly a pioneer in ferro-cement construction and his finishing touches have done much to further the reputation and respect of ferro-cement.

Warning

Ferro-cement has a lot in common with marriage! Both are meant to last a lifetime; both require careful preparation; both must be practiced to be successful; both can fail miserably if one is not aware of the hidden pitfalls.

There is one basic difference, however; if your wife becomes fat or cranky or costs too much to keep or demands your constant attention, you can dump her! All you lose is time and money. But you can't dump a boat quite that easily and you're never sure exactly when she's liable to turn sour. If a marriage begins to break up, there may be a chance for remedy but if your boat is bad from the outset, there is little corrective recourse.

I have read scores of books on plastering; there are few good ones, but none of them can make a professional out of a novice. In the final analysis, experience is the most valuable commodity on plastering day. Unlike the mesh layup, plastering is usually a one-day affair and when it's over, the quality of your hull is fully committed. The finality of concrete is staggering and it can make or break the potentially fine yacht. There's no second chance! If your plastering is successful, your dreams, labor and financial investment have the best chances of paying off. But if you botch the job or make one critical mistake, your boat may be labeled worthless.

The evidence of the high percentage of failures may be seen in backyards throughout the country. They manifest themselves as junky hulks disfigured with bumps, cracks or inconsistent textures. Some boats become rust-stained due to an improper mortar thickness, while others are "holed" by their own cradles (evidence of the lousy concrete penetration). These boats are sad reminders of those who plunged ahead without adequate preparation or who thought they knew all about it. (Plate 112)

I must sadly admit that the failures presently outnumber the successes among those boats which have been hand-plastered by amateurs. My heart goes out to those well-intended builders who specifically avoided professional help. There are two basic reasons for this. First, it costs a good deal of money to bring in a highly trained crew with the required experience at fairing a hull. The plastering rates vary widely but the average generally runs about $600 to $700 for a 45' boat. Secondly, many builders are driven unreasonably by their quest for accomplishment and ego gratification. These people simply refuse to take a word of advice regardless to the credentials of the advisor. (Plate 113)

These philosophies are extremely dangerous. Granted, a builder may save a few hundred dollars by attempting the plastering himself. But the cost saving counts only if the builder never sells the boat or if he refuses to carry adequate insurance. If, however, he chooses to cash in or liquidate at some later time, a bad plastering job will take its toll in a lower resale value. The chances are that the owner will also be forced to pay a higher coverage premium throughout the boat's years of service. A poorly troweled yacht will also affect the owner's sense of pride, especially when he docks 'longside a well-built production yacht or custom vessel. His ego will surely suffer markedly.

In summary, my only recommendation is not to "go it alone" without professional help. It takes a long time to fully grasp the knack of "pushing mud" and the yacht's fine finish is best created by an experienced person. Admittedly, I have something to gain by this recommendation as

255

it is financially imperative for my own business. But if badly plastered boats continue to appear, the entire field of ferro-cement may be doomed to extinction. Concrete is a fantastic material . . . in the right hands.

In Preparation

The term "plastering day" is a very deceiving term. It has caused many builders to become blinded by complacency only to be rudely awakened after it is too late. If you think it's going to be easy, you never should have started.

All of my explanations will not replace the value of first-hand knowledge; my only function at this time is to attempt to convey to you some of the tricks of the trade and to point out some of the problems which are not readily apparent. The real name of this game is "look before you leap." Track down as many ferro-cement boats as you can well in advance of your own plastering. Look at the bumps; examine the visible problems; develop workable solutions. Think in terms of "what could have made the other boats better."

Ask the other man any questions which bother you or, better yet, force him into a bull session. Keep in mind, however, that he is also an amateur with limited experience and quite often "bum dope" may be transmitted in this way. "Instant experts" are plentiful so be on your guard for faulty information and irrational statements. Get the opinions of as many successful builders as possible, keeping notes as you go; arrange your notes in sequence and review them over and over again, continually filling in questionable areas as you go. Remember that one person's suggestions may not always apply to your own construction method or personal capability; keep an open mind. Do not commit your cementing procedure until you are absolutely sure of every detail. Avoid staggering along blindly, repeating the same old errors previously made by others.

It is essential that you choose your helpers very carefully and well enough in advance so they can adjust their schedules for the big day. Don't let yourself be stuck with the third or fourth draft choice but shoot for a winning team. Invite your people over several times so that you can point out the problem areas; delegate responsibilities and encourage their suggestions which might help in streamlining the plastering operation. Try not to bore them to death or scare them off but impress them with the importance of their work. Always include a few extra hands on your roster; some may not return.

The Scaffold

Most amateur builders grossly underrate this aspect of their upright hull. "After all," they say "it's only going to be torn down anyway." As a result of this faulty logic, many supportive systems are constructed only as they are immediately needed, using cast-off material and little forethought.

It is essential to keep in mind that the scaffold's purpose is not only that of hanging the mold system and armature but it must also be of such substance as to hold the tremendous weight of the fresh mortar. Remember that the armature build-up grows slowly so that discrepancies within the scaffold may not become readily apparent. The plaster, on the other hand, is applied very quickly (usually in a few hours) and will reveal sagging, warping or twisting if not adequately engineered.

You may spend your waiting time very profitably by concentrating on repair, additional shoring, bracing or otherwise strengthening the scaffolding. If there is any last-minute realignment to be made, do it now. The crucial moment is close at hand.

The Stage

I have purposely separated this terminology into two specific terms. The scaffold supports the hull but the stage supports people. Its sole purpose is that of providing a safe, stable walkway on which to work. It must be well planned so as to be positioned for a convenient reach to all areas of the boat but without interfering with the access to the hull.

The catwalks (horizontal planking) of the stage must be no less than 12" wide, but preferably laid with two 10" or 12" planks. The latter will assure that your crew will have both hands free for their work without performing a balancing act.

No part of the stage should touch the hull and a space of approximately 5" to 7" is considered ideal between the ledgers (the supporting beams for the catwalk planking) and the hull. The space will allow easy handling of the trowels in the catwalk area. You may cut the inboard ends of the ledgers to the angle of the hull for maximum efficiency.

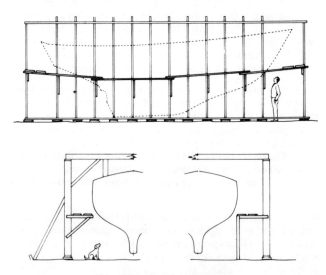

Fig. 270. Scaffold staging.

For upright construction, the ledgers may take the form of triangular cantilevers using two 2" x 4"s. These ledgers should be attached to each scaffold upright using 4" spikes. Catwalk planks are then laid onto the ledgers so that each end of every plank is supported fully. I honestly saw the wife of a client walk onto the unsupported end of a catwalk plank only to receive a broken neck when the plank teetered her to the ground. (Fig. 270; Plate 114)

If you are constructing your hull to an inverted method, an A-frame stage has been the most successful. This is just as it sounds. Long timbers are cut or laminated to a length which will allow their upper ends to meet when raised from each side of the hull. The bases of the frame timbers should be staked to the ground to prevent any possibility of slipping. Packing blocks may also be used to eliminate excessive settling.

The complete stage will require the building of several A-

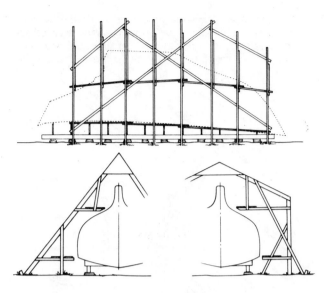

Fig. 271. "A"-frame staging.

Plate 116. A practical variation of the normal "A"-frame staging.

frames which will straddle the hull at intervals dictated by the catwalk planks. Each pair of A-frames must be liberally interconnected with diagonal braces to prevent the fore-'n-aft swaying of the stage. When determining the length of the frame legs, be sure to allow plenty of headroom over the hull so your men will not have to stoop or crawl along the catwalks.

Remember that the stage may have to support as many as 20 workers when plastering a large hull and there will be a lot of climbing about with buckets and equipment. Keep as much of the stage structure out of their way as possible.

The ledgers for the A-frame are "L"-shaped hangers, formed of 2" x 6" timbers. One ledger is attached to each leg of every A-frame. Their position relative to the hull is very important as it can determine the success of the plastering operation. On most inverted hulls, the lower corners of the ledgers should roughly follow the turn of the bilge. This will allow an easy reach to the top of the keel but will not be so far outboard as to require the workman to lean against the boat. The catwalk will usually be horizontal at the midship sections, then slant downward at the bow and stern.

If you do not have a sufficient reach to work the lower portion of the hull after installing the catwalk, a second set of ledgers must be installed at (or slightly above) the sheer.

Neither the ledgers nor the catwalk should touch the armature at any point. A clearance of 5" to 6" is considered ideal for accessibility to the hull under the catwalks. (Fig. 271; Plates 115 & 116)

If you are plastering your vessel to any of the upright open-mold systems, it will be necessary to erect staging inside of the vessel as most of your plastering work will occur on the interior side of the armature. This staging must be very stout and well braced to prevent its swaying in any direction. It should be arranged in such a way as to create a convenient reach to all parts of the hull, but the staging should not touch the armature at any point. A distance of about 6" is considered ideal between the ledgers and the meshwork, as this will allow accessibility to all parts of the hull with the trowels.

There are many possibilities as to the staging construction. Here, you will have to use your own ingenuity according to the particular hull type or circumstances. (Plate 117)

Plate 115. A typical "A"-frame stage constructed over a 60-footer prior to plastering day.

Plate 117. The internal staging within the open armature of a 46' **Andromeda** ketch.

257

The Tools

The right supplies in the correct amount are prerequisites for getting the best day's work out of any plastering crew. You may be able to hire them from the same firm who supplies your cement or they may be available from most "rent-all" companies.

The common rotary drum mixer should be avoided as it overworks the mortar. As this drum revolves, the fresh concrete makes a "slop, slop, slop" sound, and this sound indicates excessive agitation. Excessive agitation traps bubbles of air within the mortar while breaking down the formation of delicate cement crystals (the very strength element of concrete). Ideally, your mixer should be of the gas-powered, horizontal-screw blade type. (Plate 118) This provides the most consistent mixture with the least motion. Because mixing and concrete application equipment has been known to fail in the middle of the job, be sure you have a source for immediate replacement. You may even have to slip the rent-all people a few extra dollars to have them "freeze" the extra equipment for you. A broken mixer or punctured hose has caused the complete ruin of many potentially fine yachts. Whether you do your own plastering or use a professional team, be sure you have at least one man with the know-how to keep the gear running and in good order. This preparation alone may "save the day" from possible disaster should anything go wrong.

One of the secrets of good plastering is to make the concrete work for you. Because the "set" of mortar will try to race you to the finish, you must control time factors to your own benefit. One way to do this is to begin pushing mud at or before sunrise (when total visibility is not crucial). This keeps the mortar cool and retards the set allowing you more time to complete each plastering stage properly. There may be times, however, when it is to your

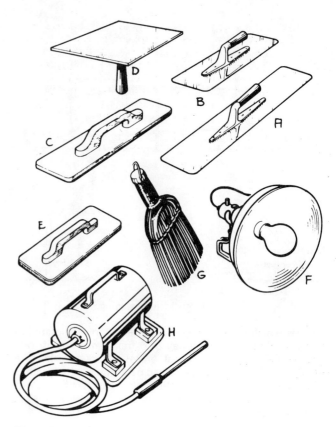

Fig. 272. Ferro-cement plastering tools. A) 24" pool trowel, B) 12" steel trowel, C) 18" wood float, D) hawk, E) 9" sponge float, F) photoflood reflectors, G) small whisk broom, H) electric pencil vibrator.

advantage to hasten hardening in given areas. This may be accomplished with portable kerosene blast heaters positioned at a considerable distance from the hull. I strongly urge you to have two of these heaters on hand for those unforeseen circumstances, even though you may not need them at all. It is well to be prepared.

Each man on the hull team should be equipped with a steel "pool" trowel (about 18") (Fig. 272A) and a shorter 12" plasterer's trowel (Fig. 272B). Be sure that these trowels are not bent in any way and are highly polished and free of rust. The steel trowels may be used not only for finishing but for "pushing" the mud, as well. Some builders prefer a wooden trowel for the latter purpose. (Fig. 272C)

The "hawk" is used by the mason to hold small quantities of mortar nearby and is simply a hand-held platform. In boat plastering it is particularly useful when applying the finish coat, as this is added to the hull little by little. (Fig. 272D)

The "sponge float" is a small steel trowel (about 10" long) with a ½" layer of sponge attached to the underside. In use, the sponge float is dampened (not wet) then drawn lightly over the finished hull. This should occur after the last steel-troweling when the hull has set hard but is still moist. The purpose of the sponge float has the visible effect of knocking off the sheen of the hull, and this is most desirable. (Fig. 272E)

If your boat is being constructed by one of the open-mold systems (mold removed entirely or partially before plastering), adequate lighting of the hull interior is absolutely

Plate 118. John Daniel thoroughly cleans his horizontal screw type mixer before mixing the first batch.

essential. Without good lighting sources, it is impossible to see the mortar penetration or to detect voids in the shell. Back-plastering is a miserable job if the men are hampered by poor lighting and the condition worsens as the hull is progressively closed in. You might look into the use of metal photoflood reflectors for use with common 150-watt bulbs as they are inexpensive and available with spring-action clamp handles for easy mounting. (Fig. 272F)

If you are plastering your vessel to one of the upright open-mold systems, it is desirable to create a rough, sandy finish on the inside of the hull as this will aid epoxy attachments. This rough surfacing may be achieved by brushing the plastered surface lightly with a small whisk broom. (Fig. 272G) This is called "brush" finishing.

Vibrating

Many types of commercial vibrators are available from your cement supplier or local rent-all. The most popular and versatile is the "pencil" vibrator. It consists of an electric power unit, a flexible hose-covered drive shaft and a 12" steel wand. The wand may be held flat, perpendicularly, or angularly against the hull. Individual circumstances best dictate its use. (Fig. 272H)

Some builders have made vibrators by attaching a long piece of plywood to a small electric sander. These may work quite well but I feel that they are dangerous. Should water or fresh mortar get into the workings of the sander, the workman may become a short circuit. The saving of a few dollars is not worth the risk.

Vibrating is an extremely effective way of assuring full mortar penetration in hard-to-get-at areas or where a large bulk of concrete is required, such as for solid mast steps. Vibrating is usually needed in the deadwood area, at the intersections of bulkhead webs or around the sheer where there may be a high concentration of rerod. It is imperative that the vibrator be applied to the length of the center-pipe and stem. You may also run the vibrator along the length of the mold battens or against large steel weldments as these are always trouble spots which usually yield massive voids after you remove the form. This is when you stop!

Vibrating may be good to a point but it can produce some disastrous effects. At the very moment you first touch the wand to the fresh mortar, water will suddenly appear on the surface. This is a visible indication of the strange phenomenon of material separation. If you persist with the wand, the aggregate will also depart from the cement. If the outward effects of vibrating are so apparent, you can imagine the extent of breakdown deep inside the shell. Vibrating also destroys the delicate crystal formation of the fresh mortar that is so necessary to its ultimate strength.

You should also realize that if you apply the wand at the aft end of the hull, the modulation will travel full length of the boat to the bow. In essence, you do vibrate the whole hull while trying to concentrate on a given area.

It is imperative that you refrain from vibrating any mortar which has "taken up" (begun to harden). It does not correct the slightest void but causes the mortar to loosen from the armature. If you are aware of a void in the hull but the mortar has set, break away as much of the plaster as you can and redo the area immediately. Of course, this drastic measure will be unnecessary if the void extends to the opposite side of the shell as it may then be successfully back-plastered.

It is far better to use your gloved hand whenever possible to achieve total mortar penetration. But if it is obvious that vibration is necessary, do it first (sparingly) then follow with the handwork. Moderation is the key to a successful outcome.

The Pump Gun

To this point I have assumed that you will be carrying fresh mortar to the hull in buckets. This is backbreaking work. My own professional team uses a plaster pump and gun which not only transports but forces it into the armature under tremendous pressure. (Plate 119) Another benefit of using the pump and gun is that the mortar entering the hose is always fresh and is forced along its route by even newer mud. The plaster never has a chance to "warm up" while

Plate 119. Here John Daniel uses a pump gun to ensure even spreading and total penetration of fresh plaster onto an inverted armature.

waiting to hit the hull. This is extremely important for strength because the cement crystals have had the least opportunity to form before being subjected to movement. (As mentioned previously, excessive movement breaks the delicate initial crystal formation.)

When applying mortar by hand, there is always the risk of misplacing the sequence of the buckets of concrete. Since concrete begins to set the very instant it becomes damp, the effect of pouring old mud is very detrimental. You not only break down the crystals as you work it in, but you cut down the working-time factor in that given area.

Using the plaster gun is not an easy operation and requires a tremendous amount of "feel" which is gained only through experience. A good gunner can dictate the mix ratio to his backup men as he determines by the rebound of the mortar as it hits the armature. The gunner also has control of the nozzle air pressure, mortar quantity and the size of the blast area. All of these variable elements, proportioned properly, give him the correct shot. It is an operation which requires agility and experience. I don't believe an amateur should try it for himself although it is one of the fastest applications, yielding superior finished strength.

Gunite

The best mortar is made with a minimum amount of water and its strength is determined by the rapidity with which it is applied to the hull after mixing. Gunite is a system for accomplishing this. The mixture is quite dry, almost powdery, but we know that cement need only be damp to begin the crystal growth.

Gunite is simply a trade name that was applied by the originators in 1910 for its specific type of mortar pump. It consists of a pressure chamber where the cement and sand are proportioned; a hose for transporting the dry cement and sand; and the nozzle where a small amount of water is injected into the mix.

The cement and aggregate are proportioned by the men at the pump but the finished mix is controlled by the gunner who judges the proper amount of water and nozzle pressure by the way the mortar "rebounds." This simply means that if the mix is properly hydrated, it will stick to the armature, but if the mortar is too dry, it will bounce off of the dry steel. If the gunner maintains a nozzle pressure of about 50—55 psi, penetration of the armature is of little concern.

Many people ask me, "Why don't you use gunite instead of the pump-gun, if gunite yields such superior strength?" Well, achieving ultimate strength is not the plasterer's only function. He also has to consider the workability of the mortar in order to produce a fine finish. The gunite mortar is so dry that the "set" time is extremely short and sculpturing for detail is almost impossible. Fairing a hull with a batten or wood screed really doesn't make it in my book, but it's about the only way to handle the gunite mixture. The amateur builder will not find gunite equipment on every corner and it is very expensive to purchase. Its use can be extremely beneficial but only in the hands of a well-trained or experienced plasterer.

Odds and Ends

While it might seem pointless to mention such obvious items as hose and shovels, frequently these items are in short supply. Your shovels will not only be used for mixing mortar but also for clearing the stages of splattered mud and for removing the piles of plaster which may accumulate on the ground. Wallowing around knee-deep in muck is extremely unpleasant, especially while balancing on the stage.

Make sure that your people have work gloves, an extra jacket (it may be very chilly in the early morning) and that their clothes are worthy of destruction. You are asking your friends to make a tremendous sacrifice on your behalf so you should assume these expenses for them. Having a stack of coveralls on hand is suggested. Do everything in your power to make your team as comfortable and contented as possible under the circumstances.

Providing each man on the hull with a whistle on a neck string is a good idea. As you can imagine, there will be a lot of yelling and shouting throughout the entire operation and orders have a way of getting muddled and mingled in these circumstances. The most important requirement of the day, however, is to keep the pushers well supplied with mortar and they only need to peep their whistles to make their demands known. This should immediately bring a runner with fresh mud.

The Water

Because the formulation of concrete is not a haphazard order, it is vital to strength that your water be accurately premeasured. Tap water will suffice but never use sea or pond water, extremely hard water, or water high in alkalis. Using five-gallon cans, set aside the required amount in 32-lb. lots. This will provide for a .35 to 1 water/cement ratio based on a one bag (94 lbs.) mix. This ratio is considered ideal for the highest strength and workability. Remember to deduct the weight of the cans when measuring your water. Keep your water covered to prevent the intrusion of foreign material.

Chromium Trioxide

To understand the importance of this additive you must first consider the nature of the electrochemical interreaction of dissimilar metals (zinc, galvanized mesh and steel reinforcing rod) when immersed in an electrolyte (wet mortar). Little has been said about this phenomenon but hundreds of boats have suffered tremendous losses in strength because of it.

When the fresh mortar is applied to the hull, it provides a complete galvanic circuit, which allows the flow of electrons from the zinc coatings to the re-rod. This electrical current can actually be measured. Where hydrogen ions acquire the electrons, hydrogen atoms are formed as a free gas along the surface of the steel. The gas layer breaks the bond between the mortar and the steel and may even collect as bubbles within the shell. On horizontal surfaces these bubbles may be readily visible but within vertical surfaces they may not be detected. Bubbles should be thought of as voids.

Another factor must be considered. When hydrogen comes in contact with steel (particularly high-tensile types) it may cause the steel to become brittle. This could lead to the eventual weakening of the armature.

One of the most notable attributes of ferro-cement is the anticorrosive protection of steel surfaces provided by the mortar. The hydrogen gas layer, if allowed to form, breaks this bond and, hence, invites internal chemical corrosion. If corrosion occurs within the hardened mortar, the growth of rust on the steel surfaces actually increases the steel wire or rod diameters. This solid expansion may eventually crack the plaster (even when standing) and will certainly make a hull highly susceptible to impact damage. You may have seen brand new boats with visible hairline cracks in the shell. This may be partly due to mortar shrinkage as well as to internal corrosion during curing. Cracking due to corrosion may continue to occur throughout the lifetime of the vessel.

One way to eliminate the foregoing problems would be to avoid dissimilar metals within the armature by using all nongalvanized steel. But employing a nongalvanized mesh would create other serious corrosion problems from other sources. Galvanizing re-bar is expensive and it softens the steel markedly.

There is an extremely simple solution to galvanic action and its related effects during plastering. If a small amount of chromium trioxide powder is added to the mortar (it is always premixed with the water), the chromium ions will inhibit the zinc and retard the electron flow. Extensive testing and research on this problem has been conducted by many leading institutions, notably the University of California, and the numbers of successful hulls prove the consistently favorable results.

Chromium trioxide (CrO_3) is purchased as a powder and need only be used very sparingly to be absolutely effective. Normally used in diluted concentrations of 100 to 300 parts per million (by weight to that of water) you should break its use down to approximately 5 to 13 grams or a level teaspoon of powder for every 100 pounds of water. This will turn the water green but it will not stain the hull or have any detrimental effects.

Cement

Many people completely misunderstand the nature of cement believing that it hardens by drying in much the same way as clay. But as explained earlier, this is not true. It is an interlocking crystalline formation which creates the binding effect and strength of cement and, as you know, these crystals cannot grow without hydration. If too much water is added to the cement, the crystals tend to become spindly and easily broken; hence the recommendation of notable authorities to use a minimum amount of water for workability. The molecular structure of the crystal itself is derived from the chemical elements of the cement so that its strength, in part, depends upon the cement makeup. There are many types of cement available and they have been specifically formulated for a given purpose: sulfate resistance, curing time, heat buildup, and shrinkage.

The standard Portland No. 1 is a general purpose cement most widely used in civil contracting. It is the most inexpensive cement available and its strength and setting time are moderate. Its use may be recommended for internal structures which will not be continually subjected to salt water or other sulfate actions.

Portland No. 2 will be slightly more expensive than the No. 1 and is considerably more resistant to sulfate attack. It holds up well in salt water and because of its moderate setting time, it can be considered excellent for amateur use. The Portland No. 2 is also excellent for use in fuel, water and holding tanks. This cement should be your second choice if Portland No. 5 is not readily available.

Portland No. 3 cement sets very fast and as a consequence is very difficult to work in the hands of an inexperienced plasterer. As I have mentioned previously, it is much to your benefit to be able to control the time factor. By using the type No. 3, time will not be in your favor. Otherwise its characteristics for ferro-cement are excellent.

Portland No. 5 should be your first choice, though it is quite expensive. Its characteristics are a slow set but high cured strength. Of all the cements, it is the most resistant to sulfate attack and is the least permeable by fuels, sewage and salt water. Because its use is so specialized, such as for oil rigs, etc., it may not be available in your area. I suggest, therefore, that you order this cement well in advance. Your supplier may not be inclined to stock this for you unless he is able to ship a substantial quantity, so it would be wise to combine your order with those of other ferro-cement builders.

Chem-Comp is a cement produced by Kaiser and is to be considered comparable to Portland No. 5. It has become extremely popular on the west coast of the United States where it is readily available. It is considered that Chem-Comp may be mixed with slightly more water than any other Portlands and hence, produces a more workable and creamy mortar. Chem-Comp is particularly suited for use with a pump gun, where a drier mix may tend to load or clog the equipment. The wetter mix makes for excellent armature penetration, especially in those areas which are not easily accessible. (Fig. 273)

Packaged Sand

Both crushed rock and beach sand are available in 94-lb bags. Before packaging, it is thoroughly washed and dried and is available in many grades and combinations. You may safely purchase bagged sand for ferro-cement construction although it costs a little more than a bulk delivery. There are advantages to this, however, which are well worth the extra cost:

1. You can be absolutely sure of proper grading so that you will not have to go through the tedious sifting and sorting process.
2. Because the sand is carefully air- or kiln-dried, you do not have to make adjustments in the mix formula to account for moisture in the sand.
3. Because the sand is preweighed, your batching of mortar is greatly simplified.
4. By opening your sandbags only as they are required, you do not have to worry about the intrusion of undesirable foreign elements or moisture. Bags also provide easy storage, especially if you are working in a windswept area.

The ideal combination of bagged sand seems to be a mixture of equal parts No. 16, No. 20 and No. 30 grades.

Just before the sand is required by the batching crew, dump the proper contents into a bin and thoroughly blend them together. It may also be possible to purchase sand in premixed combinations. It is also recommended that you order your sand well in advance to avoid possible hang-ups, and be sure to purchase considerably more than you will need. It is disastrous to run short and you can sell your overstock to another builder. (Fig. 274)

Bulk Sand

Having sand delivered in tonnage quantities by truck is less expensive and easier to procure but the disadvantages previously mentioned should be considered. Bulk sand may be graded for your specific order by the supplier, but you might as well purchase the packaged variety.

Grading bulk sand is an important procedure that is often overlooked by amateur builders. These are frequently the same people who made the most ruckus about strength. The particle size range of the aggregate has a great deal to do with tension and compression strengths as well as the effect on the hull under vibration. (Plate 120) The gradation also contributes to the rate of concrete shrinkage that must be considered. Remember that the small particles must fully surround the large ones and consequently the amount of small (fines) must be far greater in proportion.

CHEM-COMP is a registered trademark (1966) of Chemically Prestressed Concrete Corp., 14656 Oxnard St., Van Nuys, Ca. 91401.

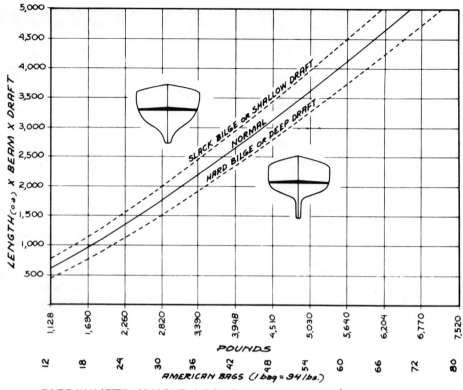

Fig. 273. Estimating cement quantity requirements.

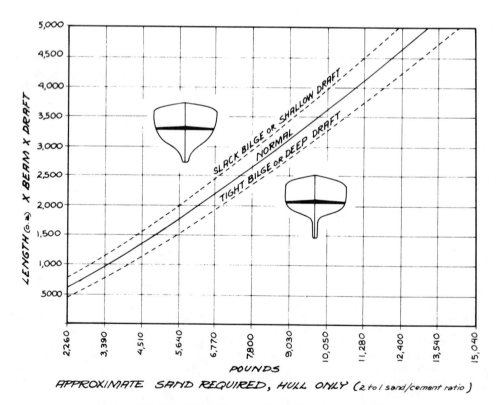

Fig. 274. Estimating sand quantity requirements.

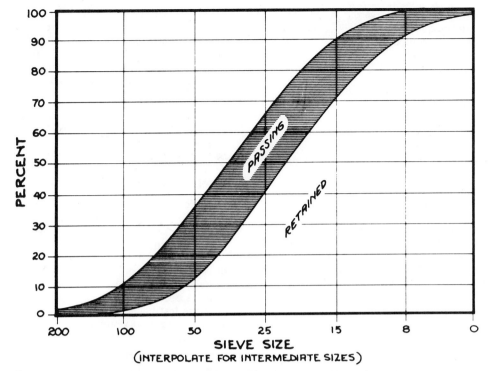

Fig. 275. Sand particle gradient chart for the proper mixing of particle sizes in their respective proportions when using bulk sand.

The more consistent the gradient between fine to coarse particles, the stronger the mortar will be. (Fig. 275) As some builders prefer not to go through the trouble of sifting, you may sacrifice some strength by purchasing a bulk combination of equal parts of No. 16, No. 20 and No. 30 grades.

Measuring Bulk Sand by Weight

Although this may seem like a menial task, it requires the greatest care. If you have ever seen a beach on a foggy day, you may have noticed how the sand appeared to be fluffed up. This fluffing or bulking is an indication of moisture present in the sand, although it may not always be readily visible.

There is always some moisture contained in bulk sand and you can often tell by the color or feel. This seemingly negligible water content counts for over 10% of the actual sand weight and has a tremendous effect on the proper cement/water ratio.

Because the moisture in the sand can actually increase the sand's normal dry volume by up to 25%, I cannot recommend measurement simply by filling a given size of container. This will not only throw off your proper sand/cement ratio, but the water/cement ratio as well. Some books give complicated formulas for the ratio adjustments, but I consider them to be extremely dangerous. It takes only an extra cup of water or a few pounds of sand to completely ruin good mortar.

The most accurate method of determining proper batching is by weight, and consistent accuracy is essential. First you must tumble the sand pile to ensure that dry sand is not concentrated on the surface, then remove several sand samples in five-pound quantities. Dry the samples thoroughly in an oven (not in the open air) and weigh them on an accurate scale.

For example: 1) 5.0 lbs.*   4.86 lbs.**
                    * damp sand weight
                    **dry sand weight

Now divide the damp weight by the dry weight. The result will be the ratio of damp weight increase required to equal the dry weight.

2) 5 lbs. ÷ 4.86 = 1.04 (104%)
3) 94 lbs.* x 1.04 = 97.76**
* dry sand required for a one-bag cement mix
**actual weight to be used for a one-bag cement mix

Plate 120. Inspection of this photograph will show a far greater proportion of small sand particles than large ones (note the dime for comparison).

263

Once you have established the proper damp sand weight for a one-bag cement mix, you must deduct the moisture content from your premeasured water.

$$
\begin{array}{r}
4) \quad 97.76^{**} \\
-94.00^{*} \\
\hline
3.76 \ \bullet \\
\end{array}
$$

$$
\begin{array}{r}
5) \quad 32.00 \ \bullet\bullet \\
-3.76 \ \bullet \\
\hline
28.24 \text{ lbs.} \ \bullet\bullet\bullet \\
\end{array}
$$

- • water content in damp sand
- •• water requirement for a one-bag cement mix
- ••• actual water to be used

Admixtures

Pozzolan is an extremely fine powdered aggregate which is used to increase the density of the concrete. This does add slightly to the strength of the concrete, and also reduces water absorption. The most common Pozzolan is diatomaceous earth and is popular because a 1% to 2% mixture by weight of cement has the same effect as 15% of most other Pozzolans.

Increasing the cement density has some very beneficial results as it reduces the amount of water required without affecting the workability of the mortar. The density also retards the takeup of the mortar, once again placing the time factor in your favor. Pozzolans also trap air within the mortar, therefore reducing detrimental effects within the concrete because of cold weather.

There are Pozzolans specifically formulated for the purpose of air entrainment. This additive greatly enhances penetration while helping to produce a slick, less porous and somewhat lighter outer skin coat. These air entrainment Pozzolans may reduce bulk weight from 3% to 5%, but they should be used very sparingly as structural strength may be decreased. The manufacturer's recommendations must be followed exactly. Pozzolith may be used specifically to reduce the required water content by as much as 12% while retarding the set of the mortar as much as 45 minutes. This retardation may be extremely helpful when plastering in hot weather. Pozzolith 8A is essentially the same product as Pozzolith 8 except that it also contains an air entrainment ingredient. The Master Builders Company (outlets worldwide) are the prime manufacturers of the foregoing products.

There are many other admixtures that may be used for various reasons, but I shall not go into their use as their application is extremely delicate and should be attempted only by professionals who are acquainted with them.

By using a small plastic bag, you may premeasure your admixtures. This will save a great deal of confusion on plastering day. Use the following figures for determining the proper amount for a one-bag mix.

EXTRA FINES

Pozzolans—14 lbs. or Diatomaceous earth— 1.4 lbs.	Pozzolith 8—4 oz. or Pozzolith 8A—4 oz. or Vinsol Resin Air (Entrainment Agent) —3.75 cc. (about one heaping teaspoon)

Testing and Surveying

This is one of the most vital elements in the life of your boat but few builders have pursued the proper course of action on this score. Because the armature construction and plastering process is the most crucial stage of hull development, it must be fully inspected and observed before and during plastering. Make arrangements with a reputable surveyor who is knowledgable in ferro-cement well in advance of plastering day so he can make his tour a day or two ahead.

Prior to his arrival, photograph every conceivable construction detail (stem, chain plates, center-pipe, keel, ports, bulkhead webs, etc.) and have an extra copy made for his certification of authenticity. These photos will become a permanent part of his report which should remain with the vessel throughout her life.

Don't forget to wire up a test panel armature. This must use the same mesh and rod as that of the hull, maintaining identical rod centers, wire ties and nesting. The panel should measure 18" x 36" although the test laboratory may elect to cut it down to a different size. If your surveyor does not intend to test the panel himself, you may have to make the arrangements. Certification of strength is imperative in order to guarantee insurability. On plastering day, have the test panel troweled in the same manner as the hull and have your surveyor initial it for authenticity. Cure it under the same conditions as the hull.

As the hull is being plastered, steal a handful or two of fresh mortar and force it into a 2" mailing tube. Rod it and vibrate slightly to ensure against voids and air pockets. This cement cylinder should also be cured in the same way as the hull and delivered to the test lab for compression, sheer and tensile calculations.

Because the makeup of the test cylinder and panel are difficult to ascertain once cured, save separate jars full of cement, sand, admixture and water samples. Small amounts may be analyzed from time to time during the ship's service so plan on storing it permanently so it may be retrieved as much as 30 years later. I cannot overemphasize the importance of this test information and certificate of survey. It will have a tremendous impact on the value of your vessel as well as your own peace of mind.

Rusted Metal

Have you ever seen a light rain sprinkle on powdered clay? When the droplets hit the ground, they simply form small balls of water without being absorbed into the soil. Fresh plaster will do exactly the same thing if it is applied to loose or powdered rust. A light surface rust is actually quite desirable as it forms a "tooth" in the steel, thus assuring an excellent bond. But if you are able to wipe the rust off of the armature you may be in trouble.

I strongly suggest blasting the entire armature with an air gun to blow off the rust within the rod and mesh layers then go over the hull with a wide wire brush. It is also possible to wash the rust away with a solution of oxalic acid and water. The acid is very mild and will not hurt your hands. Realizing that most weldments are covered by the armature, you should be most concerned with those which are to be exposed to water immersion or weather. Reach all the steelwork you can and attempt to produce a bright steel finish. On plastering morning, coat these same surfaces with an epoxy bonder to produce the best adhesion. Put your entire hull crew on this job, if necessary, even before you begin mixing the mortar. This is a lot of work but if quality and strength are your ultimate aim, it is well worth the additional effort.

Cold Joints

If you have preplastered any part of your boat (such as the deck) prior to troweling the hull, it is imperative that you clean away all loose, dry mortar or aggregate. This is best done with an air gun but a vacuum cleaner would suffice. Once cleaned, etch the edge joints with a 2% to 5% solution of muriatic acid. Two hours after applying the acid, neutralize with soda and rinse well with plenty of fresh water. On the morning of your plastering day it is vitally important that joint edges of the cured mortar be well coated with an epoxy bonder. If you fail to take this precaution, a weak "cold" joint will result which may eventually crack, rust or fail.

Plastering Day

Hand plastering is a fatiguing exercise that demands strong backs and arms. Plan on using your "heavies" on the bucket brigade, with the "light weights" positioned for back-plastering, vibrating, hard-troweling, detailing, etc. If you can recruit other amateurs who've done it before, so much the better, even if it means a free exchange of mutual services. Because plastering day will begin long before sunrise, make provisions for a hearty breakfast. Make plenty of coffee, even if it requires renting a large urn (it will contribute to the general morale throughout the day). This breakfast is extremely important as it will give your crew a chance to fully awaken, to "energize" and to review the modus operandi.

There may be times when it will be necessary to slow down, allowing the mortar to "set" a bit before proceeding again. Have some provisions ready for these occasions—cookies, soft drinks, chips, coffee—but keep the alcohol away from the operation. Getting your friends plastered is not the desired result but a quality yacht is. I've heard and seen much chaos and confusion resulting from such good-natured plastering "parties."

So far I haven't spoken of the important role of the women, but indeed it is! Physically they may not be up to "pushing mud" (sorry, women-libbers) although they should certainly lend a hand if back-plastering is required. They could also help with the sponge float as this takes a light hand and often the gals are more adept for the delicate strokes.

It is impossible to have too many people on hand as the size of the crew should be large enough to allow liberal rest periods without causing dissension among the ranks. Remember at all times that you are the captain of your ship and bear the entire responsibility here. It is important, however, that you arrange for "squad leaders" to help you keep the specific operations moving smoothly and interspace your professionals between the neophytes so they can take advantage of on-the-spot advice.

It is extremely difficult to efficiently organize a professional crew prior to taking on different vessels on different days as every situation is entirely unique. The professional crew does have one basic overriding advantage, however; they've been through it before and can anticipate a multitude of problems. I have never seen a well tuned amateur boat-plastering crew and I believe one of the basic errors is that they try to work too fast without any sense of pacing. The key phrase for plastering day is "stay loose . . . but together."

Mixing

It is difficult to say that one aspect of plastering is more crucial than another (mixing, penetration, finish, sculptured detail) as they are closely interrelated to produce the finest yacht. Careful mixing of the proper ingredients, however, is the basis upon which the other procedures will depend. (Plate 121) It dictates the workability of the mortar, shrinkage upon cure, smoothness of the surface and (in part) the strength of the hull.

It is imperative that the last batch be as close in formula as the first in every respect. To slow down the mix with extra water as the day warms is very unwise but I've seen it done many times. This is where premeasuring is important. On plastering day it is impossible to continually stop the mixing in order to weigh ingredients accurately. Have your sand, cement and additives laid out in order of use before firing the machinery. It will eliminate errors caused by the last-minute confusion.

Never mix more mortar than can be used conveniently. If your batching crew gets too far ahead of the hull crew's requirements, the mortar will simply "take up" in the buckets before it is used. Never trowel any mortar which is more than a few minutes old and discard that which has been allowed to thicken. Never mix old mortar into a new batch. Never thin old mortar with water to make it workable as it will lose much of its finished strength. Once the ingredients have been completely and evenly mixed (usually four to five minutes), turn the machinery off. Overmixing will only break down the cement crystals and entrap large quantities of air.

Plate 121. The day begins early with the mixing of the first batch.

The proper sequence of mixing should be followed strictly. Deviations may cause lumping of the mortar or concentrations of ingredients.

1. Pour about 80% of the water into the mixer.
2. Add premeasured amounts of Pozzolan and admixtures.
3. Pour in one bag of cement (94 pounds).
4. Add two bags sand (188 pounds dry).
5. Begin pouring the remaining water but only until the mortar becomes workable. If you use less than you have premeasured, the result will be a stronger hull. The less the better.

As soon as the batch seems ready, pick out a handful and place it on a board, then draw your finger across the surface of the mortar. This should leave a distinct impression of your finger without causing the mortar to crumble or flake.

(Plate 122) If crumbling does occur, the mix is just a bit too dry and would be difficult to work. (Plate 123) This requires adding a small amount of water; use only a cup or two at a time. (Plate 124)

If, on the other hand, the mortar tends to fill in behind your finger, you have used too much water. Of course, you don't have to throw this batch out because it will be good for the less structural areas of the hull or where penetration may be especially difficult. Remember that a mortar which is too wet will cause excessive shrinkage and possible hairline cracking.

Keep an accurate record of all mortar batches as well as notes on workability and viscosity. If any changes are required from the original batch, they should only be very minute and your notes will serve as a guide.

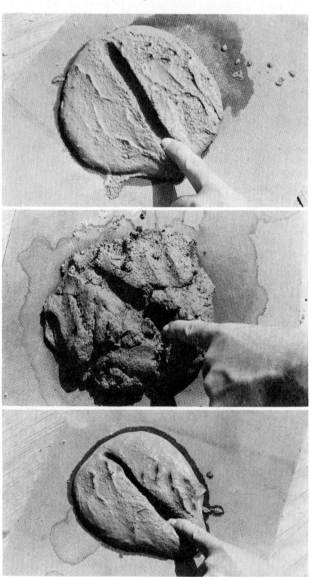

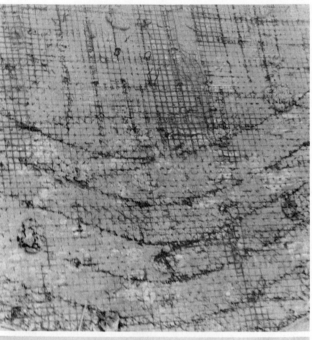

Plate 122. (Top) The proper consistency of mortar leaves a distinct finger impression without filling in or crumbling.

Plate 123. (Center) Upon drawing your finger through a mortar sample, dryness will be indicated by crumbling and chunking.

Plate 124. (Bottom) When the mortar is too wet, it will fill in behind your finger as it is drawn through the mix.

Plate 125. (Top) Massive surface voids left in the hull as a result of plastering against a solid mold. This example is typical of most cedar mold constructed vessels the author has seen.

Plate 126. (Bottom) Any backing against the armature will invariably cause voids unless extreme care is exercised. Vibrating may be applied with moderation. In this photo voids are evident as a result of a 2'' mold ribband.

266

Another method for determining a proper mix is called the "slump test." This consists of forming fresh mortar into a slump cone (available from your cement supplier) and then packing it with a rod or a stick to ensure settling and escapement of trapped air. Once the mortar is fully packed, the cone is carefully removed. After about 30 seconds, measure the degree of slump (drop) from its original height. This should not be less than 2" (dry) and not more than 4" (wet).

All of these precautions are not meant to worry you at a time when you have a great deal on your mind, but only to emphasize the importance of this crucial stage of plastering. I have seen too many broken hearts and wasted efforts caused by the careless and arbitrary use of cement, sand and water.

Penetrating Difficult Areas

Obstacles within or behind the armature always create the possibility of voids in the finished hull. The areas of the center-pipe, chain plates, corners of bulkhead webs and tanks, machinery or sole brackets, stem reinforcement or where extra rod and mesh has been used, deserve particularly careful mortar application. Plastering these trouble spots should precede the filling-in of the remainder of the hull but not so far ahead of the pushers as would cause this initial mortar to dry out.

If you are plastering to a partial or cedar wood mold system, it is vital that you completely penetrate the hull in the areas of ribbands or solid molds when they are relatively

Plate 128. (Top) There is no job more backbreaking in ferro-cement construction than the plastering of an upright hull by hand.
Plate 129. (Bottom) Steve Daniel assures complete mortar penetration in a difficult area by pushing mud by hand.

Plate 127. Steve Daniel reverts to the "bucket brigade" method of mortar application as a result of pump equipment failure. This quick decision must be based on knowing that your men are capable of rising to the task.

visible. Massive voids will result if you don't concentrate on these areas first. (Plates 125 & 126)

Begin by applying a sufficient quantity of mortar to the hull with a bucket (Plate 127) or trowel (Plate 128) then force it into the armature with a gloved hand (Plate 129) from both sides if accessible. You will be surprised at how much mortar these areas can absorb so continue to add mud until it appears to back out through the mesh. Do not attempt to cover completely at this time but only to a point where the mesh pattern is somewhat visible. You may vibrate these areas slightly but use restraint.

The deadwood always creates problems, especially if there is a flat plate built into the hull. It may be impossible to work this area from the inside of the armature because of

Plate 130. "Push, push, push" is the action necessary with the trowel to ensure total mortar penetration. This section will leave a distinct shingled appearance on the hull.

Plate 131. When the mortar seems to be backing out through the mesh as seen here, you know that the armature has been completely penetrated.

the lack of room but do have someone positioned for observation of penetration (open-mold systems). The deadwood, usually designed to be solid concrete, will require many gallons of mortar and because of this quantity it is best applied with a trowel or gun. The technique is first to lay a 1" thickness over the armature then push-push-push until the mortar is completely absorbed into the hull. Hold your trowel at a slight angle to the armature and move it upward a few inches with each thrust, resulting in a distinct shingle pattern. You may have to repeat several layers of mortar to ensure penetration but do not attempt to cover the mesh completely. Stop when the mud appears to be backing out through the armature, leaving the mesh pattern faintly visible. You may vibrate the deadwood area slightly but use your gloved hand whenever possible. Strive to achieve a good bond against the rudder tube or weldments.

Pushing Mortar

Once the difficult areas are completed, the remaining exposed armature must be penetrated. Pour about three gallons of mud at a time directly onto the hull and immediately spread it to a 1" thickness. Holding the trowel at a slight angle, push-push-push the mortar into the armature, moving the trowel up a few inches at a time. Once again, a distinct shingle pattern will result. (Plate 130) After pushing the area once, repeat the process as many times as necessary to force all of the mortar into the armature. Additional mortar should be applied until it tends to back out through the mesh surface. (Plate 131) Do not attempt to cover the mesh completely but only to achieve a faint mesh pattern.

Back-Plastering

If you are plastering to an open-mold system (mold removed), your penetration is assured when the mortar oozes through the opposite side of the armature. A back-plasterer positioned on this side should be calling signals to the pusher so as to inform him of the quantity of mud required and degree of penetration. (Plate 132) When the oozing has consistently passed through the armature and forms nipples (about 3/8" at each mesh square) the pusher should move to the next area. As soon as the mortar has taken a slight set, the excess mortar should be pushed back into the armature by the back-plasterer. It is not sufficient to simply skim-trowel when back-plastering as this will surely trap air under the surface. (Plate 133) It requires the same pushing technique as previously described without attempting to cover the mesh completely. The back-plasterer may apply additional mud as required to achieve the same faint mesh pattern as that of the original side of the hull.

Penetration Observation

When plastering to the inverted system, it has become popular to cover the mold with a polypropylene sheeting. This sheeting precludes any possibility of back-plastering but men should be stationed inside of the mold to ensure that the mortar is completely filling-in behind the mesh. Pushing mud will require additional care and it may be necessary to vibrate each area slightly as you go. Do not move on until all air pockets against the polypropylene are completely filled. To aid this filling, the observer may occasionally cut a small slit into the barrier as this will allow air escapement which might otherwise be impossible.

Positioning Your Men

Assuming that your entire team is comprised of amateur labor, and that you are not using a pump gun or gunite to achieve penetration, place your pushers (and back-plasterers) so they are not required to finish a larger area than 100 square feet at a time. To attempt a larger surface would surely cause the hull mortar to set up before you are ready for the skim coat.

It is best to locate an upper and lower team (except on very small hulls) so they do not have to scale the stage or change position. A lot of time and energy can be wasted in unnecessary climbing about. Set your teams so they overlap their areas and make sure that they can reach onto adjacent territories for the purpose of blending the boundaries harmoniously.

Of course, for open-mold plastering, one back-plasterer accompanies each of the pushers and they work in pairs throughout the entire day. For full or cedar mold construction, the entire team works from the outside of the hull.

Bulkheads, Tanks, Brackets, Web Beds

Some builders may elect to plaster these units on the same day as the hull and there is great merit in this as the cold-joint bonding problems will be eliminated. These structures do take very specialized attention, however, and should be provided with plastering teams totally separate from those on the hull. Staging must be made so that these men are required to put their weight on the armature as their movements could break the bond of the fresh hull mortar. If same-day plastering is planned for these units, this penetration should precede that of the hull to avoid a lighting problem. It is also possible to see the penetration of corners at the hull from the outside.

When plastering the tanks on the same day as the hull, proceed with the sides and ends, then the top. It is best to work from the outside in. Do not plaster the sides which are a part of the hull as they are done with the hull itself. Be particularly careful to penetrate the mortar around the corners and scuttles.

If you decide to plaster these units after the hull has cured (necessary with closed or cedar mold construction), the procedure must be changed slightly. Corner joints must be thoroughly cleaned of any loose concrete and this is best accomplished with an air gun. The joint areas must be coated with an epoxy bonder. Pushing for penetration and skim coat are handled in the same manner as the hull.

The advantage in plastering integral structural units after the hull cure is that you do not have to coordinate your plastering time factor as closely. The hull by then will be of sufficient strength to place full weight on the shell, thus increasing mobility and workability. Needless to say, you can reduce your original plastering team members by about 30%. Unfortunately, two separate curing cycles will be required for the maximum strength and this should be adhered to, to the letter.

Skim Coat

After you have achieved total penetration over all areas of the hull, the skim coat follows. This should not be done immediately, because the mortar on the hull will probably

Plate 132. Penetration of the armature can be visually checked using the open-mold systems. The proper mortar quantity may be judged when approximately 3/8" nipples of mortar protrude through the outer layer of mesh.

be too slack to withstand the additional trowel pressure without affecting its bond to the armature. You must wait until the hull has taken a rather firm set. The timing here is quite critical but you can tell the proper moment with this easy test. First, place the palm of your hand against the hull and press gently. Now lift your hand. If you have left a palm print on the hull, the mortar should set a little longer. If no print appears, pull your hand over the hull surface. If this leaves a roughened track, your hull is ready to skim coat. (Plate 134) If, when dragging your hand, no track is

Plate 133. Simply skimming over the armature when back-plastering may trap air within the hull. In this photograph you can actually see voids behind the skim coat.

left, you may be too late, so continually check the mortar take-up until the set is just right. This may be as much as several hours after completing your initial penetration so it may be a good opportunity to break for lunch. However, consider also the time difference between your first penetration area and your last.

The skim coat requires the use of the hawk to carry small amounts of mortar and to scrape the excess off your trowel. Some masons prefer to "flick" the mortar by hand evenly onto the hull while others may simply begin by spreading an even, thin layer over the surface. The trick is to achieve a consistent thickness of about 3/32", but if you feel a divot as the trowel glides along, add a little bit of

Plate 134. The palm test is used to determine the proper timing for applying the skim coat.

mortar. (Plate 135) If this requires more than 1/8" thickness, do not attempt to fill the divot completely as the skim coat will become very weak in these areas. You will fair them over with grout on another day. Many good hulls have been ruined by overplastering for the sake of airiness when the correction should have been accomplished by other methods.

When skim coating do **not** attempt to achieve a perfectly smooth finish at the outset. Don't worry about light trowel marks or occasional overlaps providing they are not excessively deep. You should be concerned only with covering the mesh adequately and uniformly. (Plate 136) If you are plastering to an open-mold system, skim coating will take place on both sides of the hull simultaneously.

Fairing Battens

Some notable builders recommended the use of an 8' plywood batten (1-3/8" x 5/16") to be used for fairing the

hull while the mortar is still somewhat creamy. Indeed, this may work quite well from time to time when in the proper hands. The batten is held against the hull with a man at each end and moved from side to side, while being moved upward little by little at the same time. This batten removes surplus mortar caused by bumps in the hull while completely missing the surface in the hollows.

I personally cannot recommend using the fairing batten in this way as it is not a very sensitive method. Where the hull reverses its contour, the batten rarely is able to follow the lines truthfully but often novices are duped into relying upon such gimmicks. In my opinion, a critical eye and delicate touch will betray unfairness much more readily. You

Plate 135. Applying the skim coat requires only enough mortar to lend a 3/32" thickness over the penetration coat.

must remember that a fairing batten may only be used in one direction at a time.

Pushing and troweling are absolutely necessary to obtain proper penetration. The batten may not be counted upon to accomplish this and often results in a thicker skim coat than necessary for strength and fairness. Don't fall into this trap or consider it to be "the easy way."

A fairing batten is not completely without merit; I have seen it used successfully but in a slightly different manner. If you desire its use for the purpose of detecting unfairness or hollows in the surface, I would suggest having a lumberyard cut strips 1/4" square by 5'. These strips will be extremely flexible and will adapt to almost any type of curve. Each man may have his own batten. After the skim coat has taken a slight set, bend the batten against the hull using one hand at each end. By doing so, slight divots will become apparent. This, of course, means that the area requires slightly more mortar. The batten should be used at differ-

Plate 136. The first passes over the fresh skim coat mortar should only be to spread it evenly and to remove the heavy trowel marks.

ent angles and in different areas in much the same way as it was used when fairing your mesh and rod layers; as you can see, the batten is not actually employed as a mortar-working tool.

Fishhooks

Though you may have been extremely careful during the days prior to plastering, you will surely discover an occasional loose end of mesh or wire tie exposed above the armature surface as you penetrate the hull. Little correction can be done at that time because of the lack of rigidity of the mortar. At the time you apply the skim coat, however, the inner mortar will have taken-up sufficiently to hold these fishhooks once they are pushed into the hull. Troweling during the skim coat will reveal these fishhooks once again and they may be hidden by forcing them into the armature with the flat corner of your trowel. (Plate 137) Do not attempt to hammer them. If this doesn't do the job, you have to leave them until it's time to grind the hull. But you will find that in the majority of cases, this action is not necessary.

Initial Sculpturing

It is during the application of the skim coat that you should begin shaping corners, edges and small, solid units. Inside radii of tanks, bulkheads, etc., may be nicely formed with a large salad or serving spoon. Putty knives may be used for getting into tight spots or to cut away mortar which has lapped onto the through-hull blanks, to clear mortar from metal parts which are to be left exposed.

If your hull is designed to receive the "U"-shaped heel bearing plate (Fig. 243), it must be fitted at this time. First apply a liberal amount of mortar to the base of the dead-wood in the area of the bearing. Now, carefully tap the bearing plate into its proper position, making sure that it aligns with the rudder tube. Hold a vibrator against the bearing plate briefly to ensure against air pockets, then scrape away any mortar that encroaches onto the exposed plate. Fair the hull in the area of the bearing so as to create a continuously smooth surface across the mortar/metal joint. Leave the bearing in position for several days after plastering, then remove and allow the hull to cure fully in this area.

I don't believe that the transom corners or bow should be fashioned to a sharp edge as they are extremely vulnerable to chipping throughout the service of the boat. Naval architects believe that a razor-sharp bow is actually detrimental to the yacht's performance. Some boats are designed with a large radius at the stem and no attempt should be made to change this. The corners of the transom should bear approximately a 1" radius as well as those of the cabin structures and cockpit perimeter (as applicable).

The upper edges of the billet head (clipper bows only) should be as well defined as possible. A very small radius is permitted to prevent chipping but an obviously rounded edge should be avoided at all costs. This, of course, depends partly upon the correctness of the armature but it is both improper and unsightly if allowed to become "soft" in appearance.

If your vessel is to have ferro-cement decks, the sheer line must not be allowed to form a sharp edge. This is the area on the boat most vulnerable to eventual chipping. A 1" sheer radius also reduces chafing of dock and fender lines. If the boat is to have wood decks, the edge of the hull will be dictated by the sheer screed.

Any surface which is to receive flat hardware or flanges (some types of stuffing boxes, cleats through-hull fittings, etc.), must be finished perfectly smooth. The strength or watertightness may depend on this detail.

Plate 137. Occasional "fishhooks" or loose wires may often be pushed back into the armature by using the corner of your trowel.

If any ferro-cement webs, flanges or other sculptured details are to be left exposed within the hull, do your very best to make them attractive, clean, and free of bumps. It is wise to give an artistic friend this task as a heavier, less sensitive hand may produce a totally unprofessional appearance. Keep in mind that grinding cured concrete is an exhausting job which can easily involve weeks of backbreaking work. Save yourself this punishment by looking ahead to this chore and correcting the imperfections at this time.

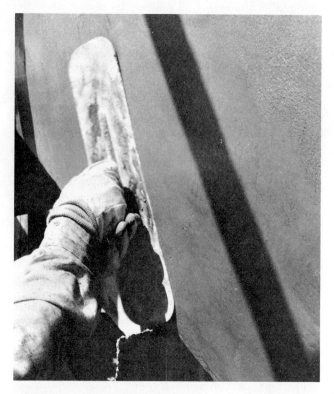

Plate 138. The finishing strokes of the skim coat should begin first with the scraping action of the angled trowel, then lightly with its flat surface.

The Hard Trowel (Finishing)

It is during this stage of plastering that you will see your hopes and efforts pay off . . . if all foregoing construction procedures have been observed religiously . . . all the way back to hanging the molds. If your hull was not properly aligned, if your meshing was done sloppily, if you allowed your boat to be plastered without first striving for a perfectly smooth armature, don't count on the trowel to cover up your errors. This stage will be the coup de grace and must be attempted only by your most experienced hands.

Once again the timing of the set is extremely critical and the palm test (previously described) should be used to determine the proper moment to begin. The purpose of hard troweling is twofold: to "knock off" any trowel marks which have been left in the skim coat and to create a satin-smooth surface. The first effect is accomplished by holding the steel float at a slight angle to the surface and scraping it across the hull with a moderate pressure. (Plate 138) When the trowel marks have been removed, hold the float flat against the hull and glide it back and forth until the surface begins to shine. Use long, graceful strokes. Do not be deliberate with your motion but sway to and fro, with rhythm and finesse.

It is said that the trowels of my boys sing a song of happiness as the glow of the mortar begins to appear. The attitude of the crew often changes from dead seriousness to exuberance and joy at this point. They have seen everything work perfectly while understanding the owner's worry and expectation. There is a distinct creative sensation in the hand as the once piles of used lumber and rolls of wire are transformed into the person of a fine lady. (Plate 139)

The Finish

The purpose of the sponge float is to roughen the hull slightly in order to provide a firm grip for the epoxy sealer. After dampening the sponge float slightly, pass it gently over the entire hull. Do not use pressure. This raises the aggregate on the surface leaving a consistently sandy texture while knocking off the visually pleasant sheen. I believe that a shiny concrete finish is detrimental to the effectiveness of the sealers and I recommend this procedure highly. Sponge floating takes neither skill, muscle or time and can be done in but a few minutes on the larger hulls. This is a good time to make preparations for your celebration as it will soon be all over. (Plate 140)

Hand rubbing produces a finish not unlike satin varnish. I have seen hulls so smooth that one could see distinct reflections along the topsides. While the appearance of a hand-rubbed hull is extremely impressive, this does not yield a proper bonding surface for the epoxy sealer. The hull requires sandblasting or liberal etching with acid after being fully cured and this, of course, eats away the beautiful gloss achieved by rubbing.

Plate 139. The sheen is already apparent by the subtle reflections in the bottom of this **Andromeda** hull.

The rubbing is accomplished with the palm of your hand swung in a circular motion. The hull surface must be almost dry and fully hard for the best results and the indication of the proper rubbing time may only be determined through trial and error. Your hands will pick up a smooth coating of cement, and this thin film will actually work its way into the pores of the hull. Once again, this procedure results in a glossy hull but will complicate sealing at a later date.

The Sayers Method
(Or Two-Stage Plastering)

Some notable builders recommend that amateurs should not attempt to completely finish their hull in one day. I believe this to have merit when the number of available personnel is not adequate and it does place the mortar setting time factor on your side. Two-phase plastering, however, may only be considered if you are constructing to one of the open mold systems. I do not personally recommend this method. Two-phase plastering not only requires rounding up and organizing of crew on two separate occasions, but also necessitates a full curing cycle before and after the second mortar application. Two-phase plastering is also complicated by the fact that you will be creating a continuous cold joint throughout the vessel's shell.

On the first plastering day, the mortar is mixed in the normal way. Do not penetrate the webs or integral structures of the hull but proceed directly to the outer surfaces of the armature. Impregnate slightly more than one half of the thickness of the armature using the pushing technique previously described. Do not attempt to cover the mesh completely but let a faint armature pattern remain slightly visible.

Once the plaster within the armature has taken a set, apply the skim coat to the outside of the hull in exactly the same way as for any other plastering method. Using the palm test as an indication of proper timing, hard-trowel the skim coat to the desired sheen then go over the entire hull lightly with the sponge float.

Remember that the same attention must be paid to correcting fishhooks and the sculpturing of detail as with any other plastering system. Remember also that the first phase of two-stage plastering will require observers inside the hull to help the pusher ascertain the proper depth of penetration.

Upon the completion of plastering of the outer surface of the hull, it must be steam- or wet-cured. Some builders recommend only a 75% cure cycle and this would surely be sufficient if no epoxy bonding were to be applied to the inner side of the hull prior to the second layer of concrete. I do not believe that it is wise to avoid a full cycle cure as the use of epoxy bonder normally requires its application only to bone-dry concrete. This necessary drying, of course, would disrupt the benefits of the full cure cycle.

Upon completion of the cure and drying of the hull, all loose aggregate or concrete must be purged from the exposed armature on the inside of the hull. Vibrating the hull may help, as well as brushing the armature with stiff, long-bristled brushes. It is necessary to vacuum or air-blast the armature to remove any free sand or concrete.

Before you apply the mortar to the inside of the hull, the exposed armature and concrete surface should be etched with muriatic (hydrochloric) acid, dried, then thoroughly coated with an epoxy bonder. It is best to use a spray gun although stiff-bristled brushes may serve adequately.

You may now proceed with the mortar penetration from the inside of the hull. You must use a slightly wetter mix than normal but be careful not to overdo it. It may require merely one additional gallon of water. Again, this is done with the pushing technique, taking extreme care so as not to simply skim the mortar over the armature. This, of course, would have the drastic result of trapping air be-

tween the two layers of concrete, therefore creating a continuous massive void. As mentioned previously, penetration of mortar against a solid surface is almost impossible to achieve by hand as evidenced by the examples in Plates 125 & 126. Because the pushers will have no visual assurance of total penetration, it must become my prime objection to two-stage plastering.

As the inner penetration has been completed and has allowed to take up, apply the skim coat in the normal manner. When the skim coat has set, follow with hard troweling, then the sponge float.

The entire hull shell has now been completed and will require the second full cycle wet- or steam-cure.

Super-Mortar

I have used this term for the lack of a better word, but, in essence, I am referring to formulations being developed for extremely high degrees of flexural strength, adhesion, or light finished weights. There are already an infinite variety of super-mortars being tested but none of them are commercially available at this writing except **FER-A-LITE**.

FER-A-LITE is a brand name for a mortar based upon a thermosetting polyester resin specifically tailored for the requirements of boat hulls constructed with steel reinforcement. It is the result of years of experimentation by the Aladdin Products Company, Westport, Maine, and although still being tested and improved, is available to the general public on a special order basis.

FER-A-LITE has many amazing attributes which are well worth considering and will be fully discussed in the next chapter. Its light weight (only 60 lbs. per cu. ft.) releases the naval architect from the burden of having to compromise hull forms often necessary when using heavier Portland cement (144 lbs. per. cu. ft.). **FER-A-LITE** produces many of the strength characteristics of aluminum while maintaining about the same weight range. Curing of **FER-A-LITE** mortar does not require hydration as it hardens through the use of a catalytic additive.

As I pointed out earlier in this chapter, one of the most difficult skills required in ferro-cement work is the plaster finishing and achieving full penetration. These are problems easily solved with **FER-A-LITE**. Voids within the troweled

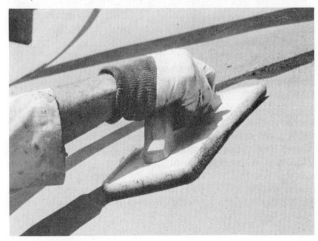

Plate 140. Applying the sponge float to the freshly plastered vessel roughens the surface, providing a "tooth" for the finishing epoxy.

mortar are almost unheard of because of the viscosity and cohesiveness of the material. The concern over the strength of cold joints may also be dismissed, thus allowing the plastering to be continued over several days' time. This, of course, eliminates the terrible pressure of the "panic party," so common when applying common Portland mortar. One method of creating a smooth finish with **FER-A-LITE** is that of first fiberglassing the outer surface of the steel armature. When the fiberglass has cured it virtually becomes an integral mold into which the mortar is troweled. This technique may only be used with "open mold" construction systems. Another trick, which seems far more practical, is that of applying sheets of newspaper onto the wet mortar. The newspaper makes it very easy to develop a smooth fair contour while troweling on a dry surface.

The addition of varying quantities of vinyl-latex binder to a common Portland mortar is another way of developing a super mix. Because the additive displaces a portion of the water normally used, the water content may generally be reduced. Because of a lack of available information concerning test results, I cannot make specific recommendations as to its actual use.

The vinyl-latex additive is reported to serve several important functions. It develops a creamier mix for easier penetration, it provides a far greater degree of mortar adhesion to all metal surfaces, as well as to itself it increases "cold-joint" strength markably, it serves to seal the armature metal more efficiently than a common mortar, it increases the density of the mortar for greater impermeability and it results in much higher tensile and bending strengths than that of a normal Portland mortar. You will surely hear more about these variations in the near future.

Another area which deserves future consideration and extensive testing programs is the use of light weight aggregates which may be employed to cut down high specific weight normally encountered with common Portland mortar. Many experiments are now being conducted by numerous universities, government research groups and independent test labs. It seems at this time that prelight and basalt used as a substitute for a portion of the normal sand aggregates holds the most promise. Once again, however, it is impossible for me to make specific recommendations as to its actual use.

In a recent report published by the United States Department of the Interior, Division of General Research (REC-ERC-72-10 and BNL 50328) polymer impregnation was reported to have almost doubled the modulus of elasticity while reducing water absorption by 96%. Results of compressive tests were the most remarkable showing an average of about 8,000 psi for common Portland mortar and an increase to 27,000 psi with the addition of 7.3% of polymer. This represents an improvement of 246% in compressive strength. Tests for flexure showed an improvement of 193% over common Portland mortar. The polymer used during this testing program was (60-40) styrene-TMPTMA. Because this research continues at this writing, I cannot make any specific recommendations regarding the use or its handling. But once again, the developments such as these will surely hold the future of ferro-cement.

Ballast

I am including my discussion on ballast at this time only because it relates to cement, although the installation of ballast will not occur until after the hull has been cured and is in the upright position.

The most common ballast used by amateur builders has been steel foundry balls (purchased by the keg) mixed into a common Portland mortar. When so mixed in its highest practical density, this type of ballast can achieve a weight of 330 lbs. per cu. ft. This will be poured directly into the hull in its estimated location.

A slightly more expensive, but efficient, alternative to steel balls is the use of 5 lb. lead pigs, which may be purchased directly from any major plumbing supplier. These lead pigs are fitted into the bottom of the hull layer by layer. As each layer is installed, it's covered with a soupy Portland mortar.

I do not recommend using scrap iron, steel or lead, as the irregularities of metal shapes decrease the ballast density drastically.

The most efficient ballast, of course, is solid lead. This material can reach a weight as high as 450 lbs. per cu. ft. and is highly recommended on vessels employing ferro-cement decks, as they have a history of being somewhat tender. The use of solid lead, of course, has the effect of lowering the center of gravity of the ballast. The installation of solid lead ballast requires first that the ballast area be lined with sheet lead in order to help insulate the hull from the direct heat of the molten metal. The melting of the lead should be done inside the hull where you can set up a coal or charcoal firebox upon which to set your iron caldron. Begin by melting a small quantity of lead, say, 15 lbs. Once this is in a molten state, you may add 5 lbs. at a time until you have reached the capacity of the caldron.

I cannot suggest pouring the lead into the hull from the caldron, as the splattering of molten metal is very dangerous, and you will not have complete control of its positioning. My recommendation is that the molten lead be drawn from the caldron with a ladle and poured carefully into the hull 5 lbs. at a time. As each ladle is withdrawn, 5 more pounds of solid lead are added to the caldron. Forming the shaping of the lead at its ends may be accomplished with a temporary plywood mold lined on one side with asbestos sheet. Installing lead ballast in this manner is a very lengthy process and may take a day and a half or two of 'round-the-clock work, but once completed, its benefits will surely pay off.

Regardless of the type of ballast you use, you should only install 80% of it at this time. I have never seen two ferro-cement boats of the same design built in exactly the same manner nor fitted out or stored identically. The result has been that no two vessels float or trim to the same plane. This problem has created a perplexing situation for ferro-cement designers. It has become absolutely impossible for us to pinpoint the exact weights and locations for final ballast. Because of this, the remaining 20% of ballast should be installed only after the full commissioning of the vessel with all rigging, hardware, stores, personal gear, electronic, and 75% of fuel and water are on board. At the time of installing the trimming ballast, attempt to occupy the lowest and most central part of the vessel if possible. Obviously, the trim of the hull will dictate the trim ballast position. It is imperative, however, that you do not simply insert additional weights at each end of the vessel or in the most convenient positions.

20

fer-a-lite

FER-A-LITE Brand Synthetic Mortar

FER-A-LITE brand mortar is a new material designed to replace Portland cement in ferro-cement boatbuilding. It is lighter, stronger, and reportedly easier to apply to boat hulls constructed with steel reinforcement than the mortar heretofore used in the common ferro-cement construction. It comes in two parts (a liquid polyester resin and an aggregate material called **REINFOR-CEMENT**) which are applied to the hull as ordinary mortar, except for some important advantages. **FER-A-LITE** brand mortar is much easier to mix and handle when applying it to the hull. Its light weight reduces the material-handling labor, compared with hauling heavy pails of cement up onto a scaffold when hand plastering. The premeasured batches in which **FER-A-LITE** is supplied simplify mixing. It penetrates the wire mesh easily, without voids, since it has a cohesive, easy-flowing consistency very much like peanut butter. The material is troweled on in batches sized for the convenience, mixing facilities, and pace desired by the builder. Work can be stopped at any time, since new mortar applied to previously cured material will bond with almost 100% strength. Depending on temperature, the mortar gels in three to four hours and gains 90% of its ultimate strength in one week. There is no need to keep it wet or apply steam or perform any other curing treatments. These advantages of great strength, light weight, easy mixing and application, and uncomplicated curing make this mortar an ideal boat construction material.

When cured, **FER-A-LITE** brand mortar has five times the bonding strength of Portland cement. It is also resilient and possesses greater impact strength. It weighs only 60 lbs. per cubic foot, compared with 140 to 160 lbs. for cement. This weight factor is very important since many ferro-cement boats are compromise designs with major consideration given to the penalty carried by excessive weight. Some ferro-cement boats have insufficient righting movement to carry sail efficiently, others may float below their design waterline.

Composite Construction—WIRE PLANK and FER-A-LITE

Use of the **WIRE PLANK** system enables packing a greater amount of steel reinforcement into the hull shell than when using ordinary mesh. This, in turn, permits construction of a thinner, hence lighter, hull. Use of both **WIRE PLANK** and **FER-A-LITE** in composite construction offers still greater advantage providing lighter and more resilient hulls having strength and weight comparable to aluminum.

Properties of FER-A-LITE

In an effort to provide the ferro-cement boatbuilder with the strongest, lightest, easiest-handling materials for the lowest cost, literally thousands of tests were made during the last six years using different resins, aggregate components and combinations thereof in the formulation of the synthetic mortar.

The following characteristics were desired:
　　High tensile and impact strength
　　High resiliency
　　Good secondary bonding to permit plastering in stages
　　Ease of mixing, handling and troweling
　　Good penetration and capture
　　Adequate working time

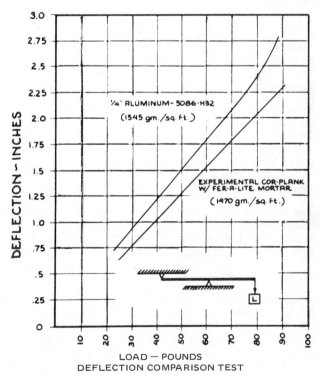

Fig. 276. Early tests comparing **FER-A-LITE** with aluminum.

This chapter is condensed from *A Revolution in Ferro Construction*, by Platt Monfort. Edited by Bruce Bingham; copyright 1973, by Aladdin Products Inc., Wiscasset, Maine.
FER-A-LITE and REINFOR-CEMENT are registered trademarks of Aladdin Products, Inc., R.F.D. 2, Wiscasset, Me. 04578.

Simple, uncomplicated curing procedure in a reasonable
working environment
Easy finishing and fairing
Reasonable cost

Many specimens were constructed and tested for these various characteristics. The result is a product which has five times the breaking strength (modulus of rupture), far greater resiliency and impact strength, and only 40% of the weight of Portland cement. When reinforced with **WIRE PLANK** brand mesh, specimens were obtained with bending strength and rigidity equal to or better than aluminum plate of comparable weight.

Specimens were set up for comparison of bending strength of aluminum and a composite of **WIRE PLANK** and **FER-A-LITE**. The specimens were supported on a bench over a fulcrum and weights were suspended from the ends to cause bending. The deflection of each specimen at the load end was measured and is shown in the graph. (Fig. 276) In the experiment, the deflection of the composite material was about 12% less than aluminum.

When comparing the properties of unreinforced Portland mortar (28-day wet cure) with **FER-A-LITE**, the differences are even more startling. The weight difference alone may lead to the development of light displacement racing boats.

	Portland cement	FER-A-LITE
Bending strength (Modulus of Rupture)	700 psi	3600 plus psi
Compression strength	5–10,000 psi	8000 psi
Weight (Pounds per cu.ft.)	144	60

Porosity

It is not porous. A very small percentage of water (difficult to measure) is absorbed in the **FER-A-LITE**. A sealer is normally applied primarily for cosmetic purposes and to fill small voids.

Sensitivity to Sunlight

Since **FER-A-LITE** is a polyester base material it will become chalky and dull after prolonged exposure to the sun, much like fiberglass. No significant effect on strength occurs. A coat of ordinary marine paint formulated for plastic hulls will suffice to protect the hull.

Smell

The characteristic smell of polyester resin will be strong during mixing and application of **FER-A-LITE** synthetic mortar. It decreases rapidly during the curing process and is not detectable after cure is complete.

Damage and Repairs

Overstress by severe impact would cause thin hairline cracking in the mortar, resulting in minor controlled seepage. The wire mesh prevents catastrophic leakage. Repairs are done by pulverizing the mortar in the damaged area with a hammer and a heavy weight on the other side. Loose particles are removed and the area replastered.

Electrolysis

FER-A-LITE is an excellent electrolyte insulator and will not contribute to any galvanic action. All metal thru-hull fittings should be installed to prevent contact with the steel

in the armature by providing a plaster layer of a minimum of 1/16" thickness.

Thermal Expansion

There is a slight difference between the thermal coefficient of expansion of **FER-A-LITE** and **WIRE PLANK** or other steel mesh. This will occasionally be evident as stress warpage on certain types of layup as in flat panels. The resultant stress is well within the elastic limit of both the **FER-A-LITE** and the steel. It will cause no problems whatever in a normal hull layup.

Shrinkage

Some shrinkage will occur during the curing process. This has not been measured precisely. It is evident by very slight depressions between the reinforcing wires if they are too close to the surface. The effect is not noticeable with a cover of 1/16" or more.

Flammability

FER-A-LITE will support combustion with a characteristic like an ordinary fiberglass boat. **FER-A-LITE** is not offered with fire retardant resin, as it is expensive and not generally available. For example, whereas the government has a specification for boatbuilding fire retardant polyester resin, there are no listed qualified manufacturers.

The requirement for fire retardancy in this application is far less than for ordinary fiberglass boats, because the interior of a home-built boat is generally built of wood, while a production fiberglass boat is completely molded of resin and glass.

Most fires start in the galley or in the engine room, so if you let them burn to the extent of the hull shell, you are already in serious trouble ... so, obviously, the answer is fire prevention! The use of fireproof paneling, such as cement asbestos board, stainless steel or even formica will be of much greater importance than a fireproof hull shell.

Marine Borers and Rot

There are no known marine organisms that can attack and destroy **FER-A-LITE**, and, of course, dry rot can be forgotten. As with any hull material, antifouling paint must be used to prevent the growth of weed and barnacles.

Adhesion to Portland Cement

In addition to adhering to itself in secondary bonds, **FER-A-LITE** will stick to cured, dry Portland cement with sufficient strength to pull the cement apart when tested. It is not recommended as a fairing compound for use on Portland cement hulls, however.

Many people with ferro-cement boats under construction are interested in completing their boats with **FER-A-LITE** decks. This, of course, is an excellent idea; it will contribute a great deal on the righting moment by keeping the center of gravity lower. (**See** Chapter 4.)

Color

FER-A-LITE is light gray, about the same color as ordinary concrete.

Installation of Ballast

Molten lead should not be poured into a **FER-A-LITE** shell directly. Ballast weight should be cast in place using ordinary cement mortar.

Insulation

FER-A-LITE has fairly good insulation properties compared with Portland cement; however, a wire-reinforced lay-up will take away from this insulation factor. Iceboxes should be insulated with urethane foam (not styrofoam).

Condensation

FER-A-LITE type mortar is a much better insulation than Portland cement and condensation is considerably less. However, if condensation does occur, ventilation and insulation will prevent it.

Effects on Styrofoam

Styrofoam will be dissolved by the polyester resin of the system. In cases where it is desired to lay up against a foam insulator, use urethane foam.

Fuel Tanks

FER-A-LITE may be used for construction of tanks; however, fuel tanks should be internally coated with **THIOKOL** compound.

In order to determine the quantity of **FER-A-LITE** required for your hull, you may refer to the hull area chart in Chapter 15. Divide the hull area by the proper square foot coverage to obtain the amount of mortar necessary to finish your vessel's shell. (Fig. 200)

	Coverage	
Thickness	Sq. Ft.	Lbs/Sq. Ft.
1"	12	5.0
3/4"	16	3.75
1/2"	24	2.5
3/8"	32	1.88
1/4"	48	1.25
1/8"	96	.63

Switching from Portland Cement to FER-A-LITE

Structurally speaking, there is no problem in converting a ferro-cement boat to **FER-A-LITE**. The drastic difference in hull weights is something to reckon with, however. At the outset, you would naturally think it would be beneficial to simply add more ballast to your yacht to achieve proper flotation, but it is not quite as easy as that. You must remember this conversion means adjusting the high specific weight of concrete (160 lbs. per ft.) to that of a material almost equal to fiberglass. The ultimate result will be almost identical to the weight expected of a normal fiberglass boat. This factor alone would increase the ballast by about 30%, which would create an intolerably stiff vessel. Not only would the boat's quick rolling characteristics cause discomfort to the passengers, but it would also place severe stress loads on the rigging and interior components.

There are three possible directions which your designer may elect to evolve a safe and serviceable solution (righting arm vs. heeling arm): 1) he may increase the vessel's rig; 2) he may build up the vessel's interior scantlings; 3) he may beef up the interior slightly while adding a little more sail area. Any one of the three solutions will reduce the ballast change requirement.

Regardless of the direction pursued, it should not be attempted by the amateur builder. The "simple switch" may involve stability calculations, rig and mast engineering, sail plan balance, and many other critical factors beyond the inexperienced person's capabilities. Don't try it yourself! Go to a professional for the sole security of knowing that the alteration will be done right. Better yet, choose a vessel designed specifically for **FER-A-LITE** right from the beginning, as it would be far less expensive in the long run.

Fiberglass Skinning (Inverted Construction, Small Boats Only)

This method has many advantages and is recommended for small boats where an open type of mold is used, facilitating access to the inside of the hull. The fiberglass skin can be very thin and therefore, inexpensive, as well as providing an excellent watertight barrier over the wires with an exact minimum thickness. Utilizing the following techniques, this skin is simple to apply and will be an invaluable help to simplify the problems of plastering day.

Fiberglass skinning may also be applied to the armatures constructed over temporary wood molds. In this instance, all staples and other rod fasteners must be removed to facilitate the removal of the hull upon completion of plastering. All rod welds and wire ties, of course, must be left intact.

Wash the mesh and rodwork thoroughly with solvent such as acetone, **MEK,** or similar cleaning fluid to remove any mill oil that may remain on the wire. This is important in order to insure a good bond with the fiberglass and also the **FER-A-LITE** mortar. If the armature rusts slightly, this will cause no problem with the bonding.

The important thing is to remove any oil or grease. Most solvents are very volatile and flammable. They should be used only in a well ventilated space with no smoking or open flame in the area. Beware of the fire hazard.

The fiberglass skin consists of one layer of ¾ ounce mat with no overlaps. This is covered with a thin, inexpensive fine-weave cotton fabric from a dry goods or department store. It is supplied in 36" widths, which will conform to the compound curvature of the hull with less wrinkling. Allow several inches at each end for clamping. Prepare the fiberglass mat into strips about 2' wide. Instead of cutting the mat, tear it with a straightedge. When it is wet with resin, the ragged edges will blend together to make an invisible joint which will be smoother and stronger than a butt joint.

Fit the fiberglass mat to the hull diagonally. Tear the edges of adjacent strips so they fit together roughly in the shape required to cover the hull. When mat is dry it will not follow the compound curvature of the hull and will not fit smoothly to the adjacent strip. When it is wet with resin, however, it becomes limp and pliable enough to be easily worked to a smooth, rounded contour. Fit the strips of mat on the hull and lay the fabric on top of it as smoothly as possible. Apply the resin with a roller until the wires show through clearly. If too much resin is applied, droplets will be seen forming below the wire and these may flow down to the second layer of **WIRE PLANK** which will then also be visible through the mat. Avoid this.

When placing the next strip, fold back the cotton along the edge so that the two pieces of mat will blend together, then overlap the cotton and pull it smooth. When the entire surface of the hull is covered and wetted, pull the cotton

THIOKOL is a registered trademark of Thiokol Corporation, Box 27, Bristol, Pa., 19007.

fabric taut and smooth and clamp the ends until the resin cures.

The type of resin to use with the fiberglass and cotton skin should be thixotropic or "super" boatbuilding resin with low wax content, which can be obtained at most boatyards. Aladdin Products resin **AL-072**, which is used with **REINFOR-CEMENT** aggregate, is also recommended. It is important to have all the material needed for the job at hand before starting this task.

An estimate of the amount of resin required can be made by weighing all the fiberglass mat and cotton fabric needed to cover the hull. Multiply this by four; this is the amount, by weight, of resin needed to saturate the skin. Resin weighs approximately 8 pounds per gallon.

Prior to working with polyester resin on the hull with either fiberglass skinning or **FER-A-LITE**, one should consider the following: Best results are obtained if the work is done indoors away from direct sunlight in temperatures from 55 to 75 degrees F. Since the cure is inhibited by dampness, the work should be done under cover. Direct sunlight will accelerate the cure to the point that it may start to gel when the job is only half completed.

One quart is a good batch size. A large clean and dry fruit juice can may be used as a mixing container. The required amount of catalyst will vary according to the manufacturer's specifications (but is usually specified as being 1% by weight). Select the largest amount of catalyst within the range given because a thin skin is slow to cure and that is what we are dealing with.

Plastering with FER-A-LITE

FER-A-LITE brand mortar is troweled on, starting inside the hull at either end of the boat. It is worked into the mesh systematically filling the area between the transverse rods one section at a time. Start from the keel and work upwards. A small pointed trowel will be needed to work the mortar into the mesh in the narrow space around the stem rod. Disposable wooden paddles in various shapes will be helpful in coaxing the mortar into place. Properly mixed mortar will allow a working time of about two hours, more or less, depending upon the temperature and the thickness of application. Scoop up a large quantity of mortar with the trowel and apply it to the panels formed by the transverse and longitudinal rods. Mash it down so that it spreads out and fills the void between the fiberglass outer skin (small boats only) and the rodwork covering the wire mesh at the same time. Avoid making bubbles or cavities, if possible, by applying each trowel full of mortar on top of the last, near the edge, and work upwards toward the gunwales on each side of the boat. Only a few minutes of work in handling the mortar are necessary to get the "feel" of it. Press it down firmly and squash it out so the mortar penetrates the mesh and oozes along the hull and over the rodwork. Do not try to finish the entire boat in one session. Stop, rest, and clean up the equipment while the first trial application cures. As a double check on the correct amount of catalyst required, note the time it takes for the **FER-A-LITE** to cure and then, if necessary, adjust the amount of catalyst slightly in the next mix. If curing takes more than

about four hours, add an additional small amount of catalyst. If the cure occurs too rapidly, reduce the amount of catalyst slightly until the right balance is achieved between gel time and working time required before gel sets in. Temperature will have a marked effect on curing time with higher temperature resulting in a faster cure. Avoid plastering in direct sunlight or in temperatures below 50 degrees F.

Covering Interior with Newsprint

You will find that it is not possible to trowel the **FER-A-LITE** to an absolutely smooth finish because of the somewhat tacky consistency of the mortar. By placing strips of newspaper directly on the mortar between the longitudinal rods, a barrier is formed which permits smoothing the surface nicely. The newspaper is left on during the cure and becomes impregnated with resin from the system. It develops a tough hide similar to kitchen countertops and serves to provide a positive minimum cover thickness over all of the wire. The paper strips should be cut to fit between the rods so that they do not cover the junctions of the transverse rods (open rod grid system). Wrinkles are a problem here. A small wood block, about 1½" x 3" can be used as a smoothing, tamping tool to produce a finished surface. Do not worry about the ends of the paper strips where they overlap; the excess can be torn off and the edges sanded or wire-brushed smooth after the resin cures.

Applying Newsprint to the Exterior of the Hull

In some instances on larger boats where the fiberglass skinning technique is not used, the newspaper covering idea is helpful as an assist in troweling the surface smooth.

As the mesh is being troweled with mortar, apply an excess of about 1/8" thickness over the wires. Sheets of newspaper are laid onto the wet mortar. A dry flat trowel is used on top of the paper to press the mortar to the desired thickness over the wires—1/16" is ideal.

The troweling action is more like patting to press the paper smooth with a little rotational rubbing. The paper will remain dry on the outside for a short while and it is easier to get the paper smooth while it is dry.

The compound curvature, depending on the size of the boat, will prevent the paper from conforming to the shape of the hull. In some places it must be tailored with slits or cut to smaller pieces to fit in place.

The overlaps that are created in this process are of no consequence. After the mortar is cured, the excess paper is torn off and the edges feathered smooth with sandpaper.

Interior Sealing

When all the interior plastering has been completed, a sealing coat of epoxy or polyester resin is brushed on. Pigment can be added for color. This coating seals the mortar, fills tiny voids, covers bare spots in the wire, and provides a smooth, pleasing appearance to the inside of the boat.

21

curing

Many people have assumed that the hardening of concrete is simply a matter of its drying out much as clay does. Clay, however, gains its questionable strength through the interlocking of previously suspended and dissolved particles and may be softened or completely broken down by rewetting. Little other chemical reaction occurs.

Cement, on the other hand, hardens and gains its strength through an entirely different process known as hydration. This is the development of a crystalline formation or ion complex created by the restructuring of the elements within the mortar. This chemical reaction is activated and maintained only through the presence of water. The water, in effect, acts as a catalyst much in the same way as with those of polyester resins. It has been said that cement will continue to strengthen for about 30 years, providing some moisture is continually present, but in truth, no one really knows the exact limit of the curing time span. It is doubtful that ferro-cement ever becomes absolutely bone-dry, especially when subjected to the marine environment.

There is a serious fallacy to the "ever-strengthening" theory voiced by many designers and builders. They have conveniently forgotten (for promotional reasons) the wicked effects of fatigue caused by the working of the material under changing load weights and directions. To my knowledge, there isn't a material known to man which becomes stronger under bend, sheer, tension, or compression! So don't be misled by the false claims: "Your boat will be better in ten years than the day she was built." A lot of unsuspecting souls have fallen for this sales pitch and may be in for a great surprise when their chain plates begin to give way.

To put the longevity of ferro-cement in its proper perspective, however, it is a conclusively proven fact that the continuous curing process of cement does cancel out a large portion of the service fatigue effects. The obvious result is that the rate of deterioration of ferro-cement is far less than that of wood, fiberglass, or steel (assuming, of course, that the vessels have been properly built with good materials and that farsighted operation and maintenance have been judiciously practiced).

With your armature correctly built, mortar carefully mixed and hull fully penetrated, the next most important function to carry out is the hydration program. Because hydration occurs on a rapidly decelerating scale, it is during the initial life of the mortar that this process is most critical and during which she may achieve as much as 90% of her final strength (Portland type No. 1).

There are several basic factors that readily affect the rate and success of hydration. As I have mentioned, the continuous presence of water moisture is imperative and must not be taken lightly. If any portion of the plastered hull, however small, is allowed to dry out prior to the completion of the curing cycle, hydration will be almost completely arrested. Hydration cannot be reactivated and the strength of the concrete at the time of drying will be increased only slightly during the remainder of the vessel's life. Let's examine the vast difference in finished strengths between properly cured concrete and that of dry-cured concrete.

You will notice that I am not only comparing wet- and dry-cured compression strengths, but that of steam-cured as well. (Fig. 277) Heat has a tremendous effect not only on the rate of strengthening but on shrinkage, too. You can see that the time required to reach maximum compression yield may be reduced substantially by inducing a higher temperature, but this must be done very deliberately to prevent blistering and cracking. (Fig. 278) Conversely, extremely low temperatures slow down the hydration process and, at freezing temperatures, curing ceases and severe damage may result from ice crystal expansion. This can be alleviated by using several space heaters within the curing chamber.

If the hull is not protected from direct sunlight or wind, there could be serious drying in the exposed areas. This, of course, not only arrests hydration but may also cause hairline cracks to develop. Another detrimental effect must be considered while we are on the subject; direct sunlight or wind may cause one side of the hull to heat or cool at a

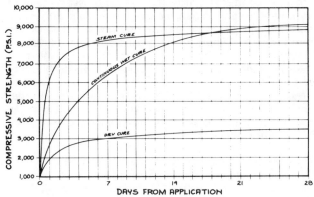

Fig. 277. Comparison of strengths achieved during equal times of dry, wet and steam cure.

different rate than the other. This produces a differential in the shell expansion which can break down the structure of the concrete, resulting in spalling or flaking of the mortar skim coats. This is a very common sight and is often the result of lackadaisical preparation and deaf ears to cautionary recommendations.

Polyethylene sheets are used as vapor barriers to cover the hull or the entire scaffold during curing and are available in translucent white and colors. While it is a little more expensive and less frequently available than clear poly, the translucent is well worth the time and extra effort to avert possible disaster. If you cannot immediately locate the translucent polyethylene at your local discount or surplus store, it should be special ordered well in advance to ensure that it is on hand when needed.

The polyethylene covering not only protects the hull from sun and wind but also serves to hold the warm, humid air around the vessel and to stabilize the curing temperature. The more airtight the polyethylene chamber, the better.

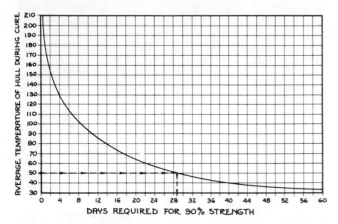

Fig. 278. For years it was generally accepted without question that wet-curing was normally a 28-day cycle. This chart shows, however, that temperatures can drastically affect the length of time required to reach 90% of strength.

The Wet Cure

Preparation of the curing system must be completed well in advance of plastering day and should be thoroughly tested before being actually put into service. Turning the system on after plastering should not be an exhausting or last-minute affair but one requiring the least amount of exertion, especially if you've put your trowels down at sunset.

The most popular and successful method of wet curing is to lay several lengths of garden soaker hoses along the vessel's sheer (upright construction) or keel center line (inverted construction). Because the water pressure is higher at the input end of the hoses, it is better to route the flow through manifolds rather than connecting them in a continuous series. The hoses may be held in position with weights or sheet metal brackets and an even water distribution is created by laying burlap strips or old blankets on the concrete surface. The important thing here is to keep every inch of the vessel wet.

The poly sheets may be preattached to enclose the entire scaffolding (upright construction) (Plate 141), or "A" frame staging (Plate 142). Once the poly is attached, roll it out of the way for plastering. The best fastening is by tacking pieces of scrap battens over the poly against the scaffold

Plate 141. A typical "hothouse" used for wet curing. This polyethylene-enclosed scaffold houses a 60-footer.

timbers. The longer the batten, the better, as this lessens the possibility of strong winds tearing the poly sheets. As soon as the last finishing strokes have caressed the new hull, roll the poly sheets to the closed position. Seal the seams with plenty of tape and pack sand or dirt around the base.

If the hull is being constructed on natural ground, excessive water collection may create a serious problem in that it may cause the vessel to settle into the muck. This rarely occurs evenly, thus straining and twisting the shell dangerously. Many builders have found it wise to rig a polypropylene sump under the boat to catch the drippings. A heavy-duty submersible bilge pump is then used to route this water back to the soaker hoses, thus devising a partially recirculating water system. Keep in mind that this pump must not only be strong enough to lift the water up to the soaker hose again but to pressurize the system as well.

Before allowing the fresh mortar to harden, poke a 1" hole fully through the side of the armature at the lowest portion of the keel (upright construction) and in any other area

Plate 142. Polyethylene sheets may be draped directly over an inverted hull. The soaker hoses are run directly along the center of the keel.

where standing water may collect. This prevents an excessive weight buildup, which could cause bulging of the hull and possible cracking. Do not depend on siphons or pumps to carry off this water, as they inevitably become jammed or clogged with concrete chips and loose aggregate. Once the hull has fully cured, the drain holes can be easily filled with an epoxy grout with no loss of hull strength.

The wet curing cycle should not actually begin until four to eight hours after the completion of plastering. To do so before this time would wash away the outer cement and dislodge the surface aggregate. If, however, hydration is delayed for more than 10 to 12 hours after plastering, you will run the risk of drying the hull and incurring hairline cracks.

Don't forget to cure your test panels. They may be wrapped in burlap and placed under the hull to take advantage of the drippings. Turn and reposition them from time to time to prevent erosion. Remember also to photograph every detail of your curing operation and keep accurate

Plate 143. The long vigil has begun! This workman is thoroughly wetting down the entire boat before draping it with burlap and soaker hoses. Notice the open-mold rod suspension system used on this 55' **Marco-Polo** hull.

records of your inspections and chamber temperatures. These documents will help to provide a graphic history of your boat's construction for future certification or possible buyers.

Because of the rapid early gain in strength, the first week of curing is the most vital. Do not walk on the hull or jar it in any way. Simply leave it alone. Keep the vessel totally soaked during this period, then gradually reduce the water flow to a uniform dampness for the last few days. This will prevent a sudden drying and change of the hull temperature. Make sure that no water has a tendency to constantly flow or drip onto any one concentrated area, as this may cause the eventual erosion of the "green" mortar. Remember that the hull will not become fully hard until it is finally dry, so don't erect the cradle or shoring until that time. For maximum strength, the hydration cycle should be maintained for at least the full period indicated.

While 28 days of wet curing have been commonly accepted as a general rule of thumb, it has been conclusively proven that extending this period to any degree increases both strength as well as the shell's protection against its corrosive environment.

Don't be concerned about stains and streaking on the sides of the vessel, as these are normal mineral deposits which will disappear when the hull is acid-etched before sealing.

Ideally, the hull should be inspected several times a day and periodic shifting of the hoses or burlap may be required from time to time. Spray down any area of the vessel which is not getting its full portion of the water flow. (Plate 143) Jay Benford suggests rigging a timing mechanism on the water supply, and I believe this is a great idea. His system is to salvage a solenoid valve actuated by a micro-switch from an automatic dishwasher. The micro-switch is operated by a cogged wheel attached to an old clock motor. The notches in the wheel are cut so that the water will be turned on for one minute out of every three. This has the effect of reducing the water bill as well as minimizing the collection of the water below the vessel.

Grinding and Grouting

I must say a few words about the finish of the vessel at this time because a great deal can be done toward smoothing the hull while it is still green. Grinding bumps from a fully cured hull is absolutely torturous and I strongly advise you to "put the block to her" during the first week of curing. Carborundum stones (available from any concrete supplier) are extremely effective in knocking down moderate high spots that may have eluded you on plastering day and will do the most work for you while the shell is still soft. When sanding with Carborundum, a mushy substance will develop on the surface and this will actually aid in sanding but be sure you rinse this stuff off as you move to a different area of the vessel, as it will set up, filling the microscopic pores of the mortar (this reduces the bond of the epoxy sealer to be applied later). Some fashioning of sculptured detail may also be accomplished in the same manner.

Fill shallow divots while the hull is wet, as a good bond is created at the earliest stage of cure. This filling is done with "grout" which is simply pure cement and water. It should be mixed on the basis of workability (a creamy paste) according to the particular circumstances, but I must advise caution to avoid building a grout thickness in excess of 1/8". Grout has very little strength in itself and may have a tendency to crack or flake if allowed to be built up to an unreasonable volume. A rubber squeegee or flexible wooden batten work beautifully as applicators. If grouting is done within the first week of hydration, it will then receive almost the full benefit of the cure cycle. More will be said about finishing in Chapter 23.

Steam Curing

As I noted earlier, the temperature of the water, air, and hull have a tremendous bearing on the rate of strengthening of concrete. In essence, the higher the temperatures, the faster the cure. Theoretically, submerging your vessel into boiling water would result in the most successful hydration but, unfortunately, theory and practice are often worlds apart.

The logistics of sinking and raising a hull are almost insurmountable for the amateur builder, not to mention problems involved in heating the required amount of water. More detrimental, however, would be the effect of thermal shock on the thin concrete shell, as it would be subjected to a sudden and extreme change in temperature. This thermal shock would cause an immediate expansion of the boundary surfaces while the inner material would remain relatively cool and undisturbed (however momentary). Concrete cannot withstand this bi-thermal expansion, as it results in massive spalling, cracking, and flaking.

The most successful alternative to submersion, then, is the use of steam which envelops the freshly plastered hull at normal atmospheric pressure.

The proper use of steam hydration may have the effect of producing up to 90% of final concrete strength within a period of only 24 hours. (Fig. 277) Compared with the time, labor, and mess of the 28-day wet cure, it is clear that there are many benefits to be derived, especially in the hands of the commercial boatbuilder. Steam curing also reduces unsightly mineral stains that have no structural effects but do have a visual influence on prospective buyers.

Plate 144. A typical commercial steam generator available from most large "rent-all" companies.

Steam curing has also been proven to have the marked effect of reducing the harmful electrolytic period required of wet curing as well as deterring corrosion through a more effective and dense crystal growth within the matrix.

The equipment used for steam curing varies widely. The most popular steam generator is the type used for cleaning car underbodies and may usually be rented from special machinery outlets. (Plate 144) Some builders have preferred to build their own generators using copper tubing and 50-gallon drums, firing the unit with bottled gas. I have included a construction drawing for just such a generator, if you prefer to exercise this option. (Fig. 279) Regardless of equipment used, it is imperative that you make provisions for a backup system capable of completing the cure in the event your primary system fails. This may even require shifting to a wet cure system right in the middle of the cycle after the hull has cooled sufficiently. Don't let yourself be caught short without recourse.

Covering the hull for steam curing differs widely from wet curing. Because polypropylene is affected by the high temperatures, canvas or heavy fabric serves much better. This chamber should be as airtight as possible and should be sealed as soon as the vessel has been plastered. It is not recommended, however, that you enclose the entire scaffold or "A" frame stage, as this would require steaming a tremendous volume, rarely done successfully. It is far better to construct a smaller framing with wood or steel hoops so as to encompass an area only slightly larger than the hull

itself. When building this frame, try to avoid creating large free spaces above the hull, as these areas could become "hot spots" of stagnant, hot air.

Routing of the steam exit piping must be planned to develop the most even temperature gradient around the hull. Because the steam pressure is higher nearer the generator, it is best to pass the steam through a manifold, then into various separate pipes rather than to connect them as a continuous series. When curing an inverted hull, two pipe routes must be laid to provide for steam both on the inside and outside of the shell. If you steam only under the vessel, the vapor will simply rise to the keel area where it becomes stagnant without benefiting the outside of the hull. This must be avoided at all costs, as it would cause an unevenness in the strength of the finished vessel.

As I have intimated, the uniformity of the temperature and humidity around the hull is vital for proper steam curing. Simply filling the chamber with hot vapors will not do at all and many potentially good boats have been ruined in this manner. My suggestion is to install several fans fitted with vapor-proof motors within the chamber and hull to keep the air in motion and well mixed. Among the many steam curing attempts I've seen, very few have given satisfactory results. Burners have run short of fuel in the middle of the night or the steam generator was too small to produce the high temperatures required. Some generators have even run out of water because no one bothered to check out the rate of consumption prior to its actual service. The answer to these problems is to conduct a complete test cycle some time before plastering. This involves additional time and expense but could avert a lot of heartache. Professional builders who intend to steam cure many successive vessels have the advantage of amortizing their test run expense, thus making it much more practical.

After the vessel has been plastered and the chamber closed, allow the mortar to set for three to four hours before begin-

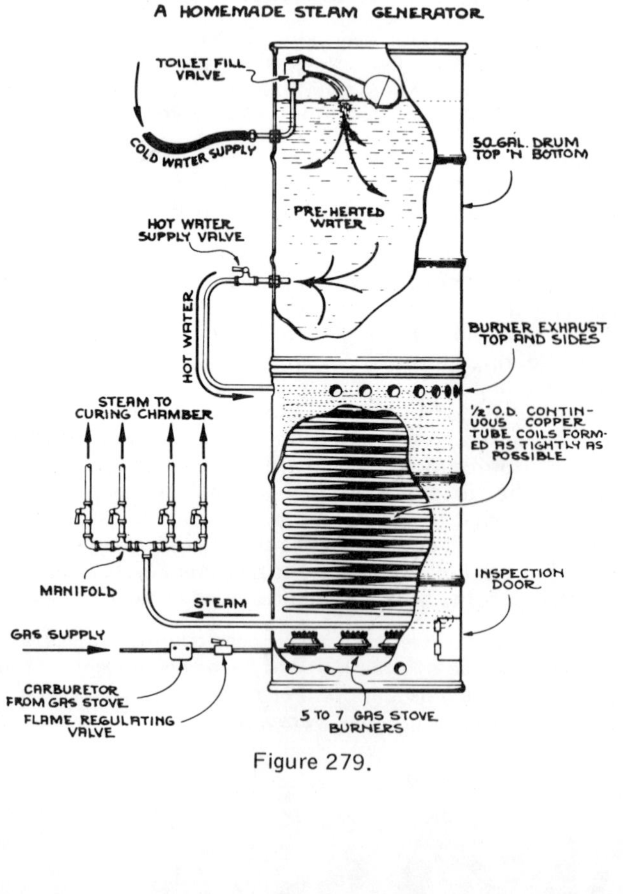

A HOMEMADE STEAM GENERATOR

TOILET FILL VALVE

COLD WATER SUPPLY

50 GAL. DRUM TOP 'N BOTTOM

PRE-HEATED WATER

HOT WATER SUPPLY VALVE

HOT WATER

BURNER EXHAUST TOP AND SIDES

½" O.D. CONTINUOUS COPPER TUBE COILS FORMED AS TIGHTLY AS POSSIBLE.

STEAM TO CURING CHAMBER

INSPECTION DOOR.

MANIFOLD

STEAM

GAS SUPPLY

CARBURETOR FROM GAS STOVE.

FLAME REGULATING VALVE

5 TO 7 GAS STOVE BURNERS

Figure 279.

ning the steaming process. Keeping in mind the effects of thermal shock, slowly raise the temperature of the chamber with steam to 150° F., if possible, over a six-hour period. Once this temperature is reached, it must be maintained for at least 12 hours but preferably a full 20 hours. Then gradually lower the chamber temperature over four to six hours until the ambient (outside) temperature is reached. Do not open the chamber immediately but give the hull plenty of time to cool and dry. If you are steam curing your hull under very cold ambient conditions, the heating and cooling periods should be lengthened.

If your system is unable to produce the recommended 150° buildup, all is not lost. It will only mean that your maximum heat period will have to be lengthened providing, of course, that the chamber air is water saturated. The chart will show variations in the steam cure cycle. (Fig. 280) Keep a very close watch as the hull cures and accurately record the hourly chamber temperature. Above all, don't assume that your system will tend itself because it will inevitably go on the blink the instant you drop your guard.

Leaching and Curing Agents

During a recent trip to Marina Del Rey on the California coast, I saw a ferro-cement boat belonging to a prominent American singer. This vessel has been "professionally" built by a well-known northern yard at a sizable expense to the owner. The vessel had weathered a storm during its trip south, which was sighted for its apparent damage and general disarray, but the owner's major complaint was that the hull paint had blistered and peeled severely during the short seven-day sail.

Upon my closer inspection of this vessel, I realized that the loss of paint had no connection with the storm at all, but rather was caused by the leaching of alkalis and lime from within the concrete shell. This leaching had loosened the bond of the epoxy hull sealant, thus dislodging large areas

of the finish. Unfortunately, the only recourse resulting from this preventable situation is to strip the hull to her humble beginnings and start again from scratch.

Sadly, I have seen many such happenings on ferro-cement hulls. Once again, the ultimate cause has been purely a matter of over-simplification by the early proponents of concrete hull construction. "Coat the hull with epoxy" was the common instruction called out by promoters as the "cure-all" to finishing maladies. Unfortunately, it is a little more complicated than that.

Once an untreated ferro-cement hull dries, you will be able to wipe off a fine, white surface powder with a stroke of your hand. A few weeks later the powder will have reappeared in the same place. You will be able to wipe the surface many, many times over, but still the powder will reappear. It is this powder building up under an epoxy sealer which may cause the ultimate failure of the finish.

This problem may be easily solved by swabbing the hull with **Crete-Seal** (Tri-Col Products, Bakersfield, California), allowing it to dry then washing away the residue with fresh water. This cycle should be repeated as many times as necessary until no residue appears upon drying.

Crete-Seal has the remarkable property of leaching entrapped lime and alkalis to the surface of the concrete, while sealing the concrete for increased resistance to water and oil absorption. These properties add greatly to the adhesion of the hull finish as well as tank linings. Once held as a "super secret process" by many self-seeking professionals, it is now available through a simple phone call.

While my mentioning **Crete-Seal** at this time may seem somewhat out of sequence as a hull finishing process, the material is far more effectively used while the hull is still damp. The moisture within the concrete as a result of curing will help to draw the **Crete-Seal** solution deeper into the concrete. I will discuss the attributes of this product in more detail in Chapter 23.

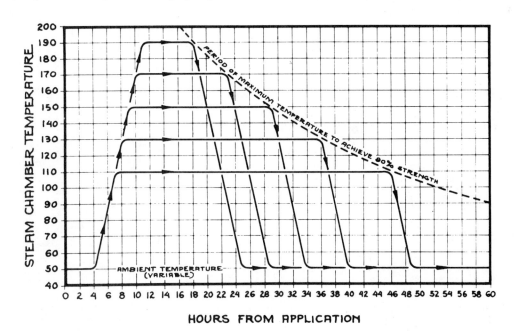

Fig. 280. The steam-curing cycle may be varied in length according to both ambient and chamber temperatures.

Crete-Seal is a registered trademark (1958) of Tri-Col, 1924 V Street, P.O. Box 971, Bakersfield, Ca., 93302.

section four

completing the boat

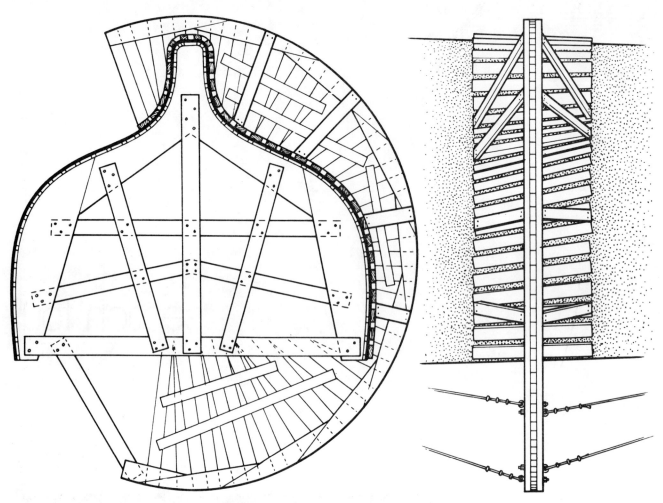

Fig. 281. A typical rolling wheel construction. Notice the heavy bracing within the station mold.

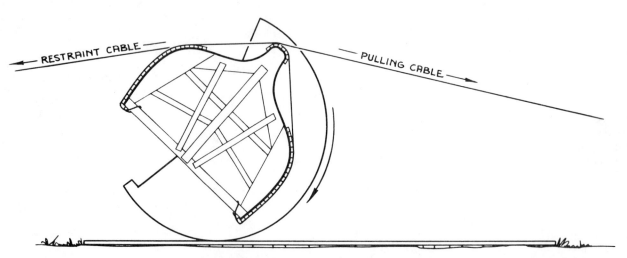

Fig. 282. A typical rolling operation. Note the bolstering between the cables and the hull to prevent chafing.

turning upright

While plastering ranks highest in terms of "builder's traumas," the second must surely be that of rolling the vessel over after constructing the hull in the inverted position. This one problem alone has kept builders and designers awake in thought for centuries. It has not been a challenge to the ferro-cement cult only, but it is a task that has faced wood boatbuilders since prehistoric times.

In light of rapid advancement in space technology, one would certainly think that rolling a boat would be child's play. One noted publication on ferro-cement even suggests that there is no need for worry. It is not a simple operation. A survey published in the *Ferro-Cement Times* reported the percentage of hull damage by turning running as high as 60%. The same source published a pictorial study of typical hull rolling failures emphasizing a score of unforeseen circumstances resulting in cracks and fractures. If you have chosen to build your boat to an inverted system for the sake of ease, the time has come to pay your dues in terms of both money and toil. There is no easy way out for you.

I don't mean to say that rolling over is an insurmountable problem. Hundreds of hulls have been flipped very successfully in a matter of minutes. In every successful case, however, the operation was carefully engineered and coordinated, reducing its dangers. The prudent builder has made it a point to be on hand for as many rollings as possible in order to gain the experience and feel of this operation and to judge the merit of the different methods available to him.

The Wooden Wheel

This system is exactly as it sounds. Literally, it requires the construction of several stout wheels built to fit the hull exactly at various station positions. The more wheels the better. The wheel patterns must be determined on the lofted body plan by first swinging a true circle around the entire hull. This circle should clear the keel and sheer at the largest station by a distance of six to eight inches. Note that each wheel will share both a common perimeter and center.

Now lay down 2 x 4's or 2 x 6's radiating from the center of the circle, trimming their outboard ends exactly to the circle perimeter while cutting their inboard ends 1-5/8" short of hull shape at the given station. (Remember that paper lofted patterns do not represent the outside of the hull but rather the station molds. You must add both the thickness of the mold and the hull to determine the proper hull shape.)

From 2 x 12's cut and attach timbers around the perimeter of the wheel. These may be first glued and nailed in the position, then bolted after lifting the wheel from the lofting. Before moving the wheel, interconnect the radial timbers liberally with diagonals, again gluing and nailing. Remember that the wheel may not be fully applied to the hull in one piece, so make allowance for its finished assembly on the hull. Do continue, however, with the fitting and assembly of its sections.

Before attaching the wheel to the hull, the inner bearing surface of the wheel must be beveled to conform to the vessel's lines. Then end-fasten bolsters of 2" x 4" x 4' timbers to the hull ends of several radials, spacing them about 4' apart. (Fig. 281)

Positioning the wheel on the hull will require the detachment of the hull from its strongback or skid and the careful jacking of the boat upward so that the wheel will clear the ground. When jacking, be sure that your equipment cannot shift or settle into the ground. Position at least two layers of timber sleepers horizontally under each jack so that they are firm and level. If possible, do not jack under the sheer but rather under the station mold cross-spalls as close to the hull as practical. Reinforce the station molds as necessary to bear the hull weight. Remember, the more jacks you use, the better. Have your men jack absolutely in unison, one click at a time. If there is any tendency for one jack to assume more weight than the others, let it pass a stroke or two but do this very judiciously.

As the hull begins to rise, have an additional team of men position packing blocks under the sheer or free cross-spalls. Packing blocks may be a combination of timbers, concrete blocks, railroad ties or what-have-you and are for the purpose of catching the hull in the event of its slipping off of the jacks. They will also serve to support the hull when the jacks are removed. Twenty or thirty big wedges will be needed, too, to equalize the load on the blocks when the jacks come down.

Never jack more than a few inches at a time. When the hull is lifted by this distance, lower it onto the packing blocks, drop the jacks down to the shortest lengths, and build up the sleepers to raise the jack positions. Now begin jacking again. Repeating this cycle will reduce the risk of capsizing the jacks.

Note: If it is impractical to lift the vessel by the mold cross-spalls, heavy transverse timbers may be passed under the sheer from side to side and used as jacking points. The wider the bearing surface against the sheer, the better. Now that the vessel is clear of the ground and the skid timbers hauled out of the way, the rolling wheels may be manhandled onto the hull. These are heavy so have plenty of people on hand for this job. With the wheel in place, adjust the bolsters so that they lie in perfect line with the hull surface. Now brace the wheel on each side with diagonal timbers from the wheel perimeter to the ends of the bolsters. Be sure to attach packing blocks at the ends of the diagonals to prevent their possible slipping and tearing loose.

Now that the wheel is firmly placed, pack the remaining open space between the hull and wheel with additional 2" x 4" packing blocks. "Toenail" the wheel radials to these blocks. If the packing blocks don't fit firmly into the space, the thickness must be built up to suit. If the backing blocks are too thick for placement, they must be milled or planed to fit. In short, attempt to provide a uniform bearing against the hull.

All rolling wheels are constructed and attached in the same manner. Next, the station molds at the wheel positions must be profusely reinforced to absorb compression strains on the hull. The lower portion of the wheels between the sheers must also be firmly attached to the mold system and thoroughly braced with longitudinal diagonal timbers. The lower wheel segment is probably the most vulnerable to crushing and capsizing so construct it accordingly.

It is vitally important that the wheels be absolutely parallel to each other. Take careful measurements to ensure this condition. Failure to do so will cause the wheels to veer in slightly different directions, resulting in the collapse of the entire wheel system. It would be wise to interconnect the wheels with cable or timbers to help absorb and transmit any tendency of fore-'n-aft swaying. (Plate 145)

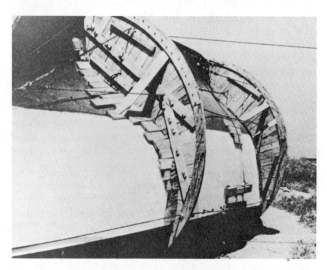

Plate 145. Typical rolling wheels attached to a **Marco-Polo** hull. Note the liberal use of fore and aft cables to prevent the wheels from collapsing. (A Joy Morrill and Pete Mock photo, the *Ferro-Cement Times*).

Before rolling, tracks or roadbeds must be built to ensure sufficient and uniform support of each wheel. These beds may be of timber or concrete and must be long enough and wide enough to allow for the complete 180° traverse of each wheel. Remember: If the wheels have not been attached to the hull perpendicularly to its center line, the rolling direction will be slightly diagonal. Construct your tracks or roadbed wide enough to accommodate this eventuality. (Fig. 282)

With possibly several weeks invested in your wheel construction, rolling day has finally arrived. Its successful conclusion now depends upon the procurement of proper equipment, and the materials list sounds more like a boneyard inventory. First, you should have at least two automobiles worthy of abuse (for boats under 35') or tractors. If cars are used, it is a good idea to fit them with snow chains for traction. You will also need several lengths of steel cable (size and length depends on the boat—check with a local rigging firm for their recommendation), cable fairlead blocks (as the circumstances of space dictate), and a pile of 2" x 4" x 2' timbers to be used as chafing gear and bolsters. You should also have your timbers on hand for erecting the vessel's cradle immediately upon completion of the inversion.

A pulling cable bridle should be fixed to the reinforced mold cross-spalls at the positions of the end wheels. The attachment of the bridle is by way of eyes formed in the wires being held firmly with at least two cable clamps each. The bridle is now passed under the sheer (the side opposite the pulling direction), up the side of the hull, and over the keel. A single pulling cable is then shackled to the bight of the bridle, thence to the chassis (not the bumper) of the pulling vehicle where it is looped and firmly cable-clamped. **Note:** If the running distance of the pulling vehicle is restricted by space, a fairlead block must be employed by attaching it to an additional stationary vehicle or structure.

A restraining bridle with cable is now fastened to the hull in the same manner as before, leading it under the sheer (the side toward the pull), over the keel, and to the restraining vehicle.

Before placing any strain on either bridle, place the 2" x 4" bolsters between the hull and cables at any position where the cables would normally contact the hull. Space the bolsters very closely together, paying particular attention to packing under and around the sheer. The bolsters should be attached to the bridle cables, to prevent their shifting, by partially driving spikes into the bolsters then bending them over the cables.

Now jack the entire vessel clear of its support blocking and lower it, bit by bit, onto the rolling wheels. Place a slight tension on the pulling and restraining cables to prevent any sudden movement of the hull and check out the entire mold and wheel structure for any evidence of failure. Don't begin rolling until you are satisfied that everything is holding together properly and that you have your signals coordinated between vehicles. To prevent a heavy lawsuit, get everyone out of the path of the vessel.

Warning: A hull of 45' L.O.A. can easily weigh up to 18,000 lbs. (including mold, concrete decks, bulkheads, etc.). Once this tremendous weight is in motion, it will be extremely difficult to stop, particularly as the hull's center of gravity begins to move away from its geometric balance. If the rolling becomes accelerated, the hull's momentum can easily pull the restraining vehicle right across the lot. Don't let the boat get away from you for a moment. I once saw a vessel roll herself fully beyond the upright position and completely off of her wheels. Needless to say, the builder was right back where he started, only more so!

Now, ever so carefully, place just enough strain on the pulling cable to cause the vessel's lateral movement, allowing the restraining vehicle to pick up the load. Until the hull lies on her bilge, little or no additional pulling will be necessary. *Slowly*, allow the boat to rotate until the restraint becomes slack. At this indication, she will be halfway there.

The major task is now up to the pulling vehicle. Placing a strain on the pulling cable, carefully raise the vessel to the upright position. As the hull nears the vertical, less effort will be required and the restraining vehicle must be prepared to prevent her over-rolling. As soon as the hull is vertical (check with a level placed upon a mold cross-spall), place temporary side shores and keel blocking. Maintain slight cable tensions until sufficient bracing is completed.

Once the wire bridles have been removed, you may begin the construction of your permanent cradle. The rolling wheels should be removed only when the entire weight of the vessel has been transferred from them.

The Rolling Sling

This system is considered by some as a more practical inversion method. Obviously, it does not require any complicated construction, as is necessary with the rolling wheel method, but you must be prepared to lay out several hundred dollars for hiring a crane and its operators. Most generally, the rolling sling operation will be handled entirely by a professional rigging team from start to finish (including the makeup of the sling itself). Very few experts in the field will trust their efforts or liability to equipment provided by their client, regardless of how well planned.

Needless to say, my recommendation for inverting your vessel is surely this system. At the same time, I must discourage any attempts to use the rolling sling method which is to be supported from a wooden or otherwise temporary scaffold or beam system. I grant that many commercial boatyards are now equipped with elaborate, expensive, and suitably strong rolling mechanisms, some of which are portable and able to handle most any size of vessel, but any attempt by the amateur to construct such structures for his own use will inevitably result in heartbreak.

The equipment list for the rolling sling inversion will include a crane of suitable capacity and footings, two rolling cable bridles, a hoist bridle fitted with a steel beam spreader, a pulling bridle and cable, a restraining bridle and cable, a restraining vehicle, and suitable cable fairlead blocks (if restricted space dictates).

It is also a wise idea for you to supply plenty of used tires or bales of hay to be used as fenders under the hull. Also, be prepared to shore and block the vessel's position as soon as she is placed upright, with properly spaced timber or blocks under her keel. Although the "pros" will probably do the job for you, I will describe the procedures for your edification.

On the day before rolling, the weight of the hull should be taken off of the construction skids or strongback. This may be done with a series of jacks placed under the mold cross-spalls. It is best not to jack directly against the sheet, but if you must, first place heavy timbers under the vessel from side to side. Be sure your jacks have a firm and level footing by positioning at least two layers of horizontal timber sleepers under the base of each jack. Once done, raise each jack until you feel a distinct pressure; go no further. Now begin cutting or otherwise detaching the hull mold from the skids until the vessel is completely free. Do not withdraw the skid or any part of the mold.

It is wise to reinforce two of the station molds so they may withstand the compression caused by the rolling bridles.

The two stations chosen should be approximately 1/3 the length of the vessel inward from her ends. Reinforcing should be in the form of several transverse timbers (preferably two layers of 2 x 6's or solid 4 x 6's) and should be interconnected vertically to reduce bending. All reinforcing should be through-bolted to the respective station molds.

Make sure that you can provide a large enough work area for the rolling crew and its equipment. Move out all extraneous lumber, dismantle your plastering "A" frame, rope off a no-parking territory, and remove any fencing which might get in the way.

The professional rollers will usually be comprised of one rigger, one crane operator, and a driver. They will rarely bring extra bodies unless you specifically request that they do so. Keep in mind that these men will probably be highly paid union employees working on an hourly basis, so it may be far better to arrange for your own "heavy helpers." The equipment is normally charged by the size of the gear times mileage to your site plus the duration of service.

At the outset, your "pros" and helpers should get together to work out the signals and modus operandi. The first task is the positioning of rolling slings. These are long, continuous cable loops which pass through blocks hung from the spreader. Wherever these slings are to touch the hull, they must be padded with 2" x 4" bolsters. The bolsters may be attached to the rolling slings by means of bent spikes or they may constitute a separate belt unit supplied by the riggers. Be sure the sheer receives particular attention to protect against damage.

Now the bridle spreader is hung from the crane hoist and a slight tension is placed upon the rolling slings. A final check is made for proper position. Several rope lines may span the two slings to prevent them from sliding toward the ends of the vessel.

A pulling wire is now attached with cable clamps to a reinforced mold cross-spall, passed up the side of the hull opposite the pulling vehicle, over the keel, then to the vehicle. This should be a truck with heavily treaded tires or may be several cars chained together. I have personally seen a four-wheel drive jeep pulled halfway across the boatyard

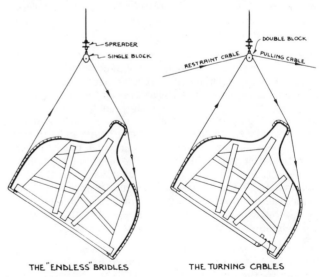

Fig. 283. The rigging required for inverting the vessel with a crane. Note the bolsters between the hull and the cables to prevent chafing.

by a vessel no larger than 40'. Prepare for this eventuality. A restraining wire is attached in the same manner but led from the other side of the hull and led in the opposite direction of the pulling wire over the keel. The restraint may be directed to a winch on the crane or through a fairlead anchored to prevent hull abrasion in the same manner as the rolling slings. To prevent the swaying of the vessel while suspended, guidelines should be led from the hull to some heavy helpers on the ground, with provisions for snubbing the lines, if possible. (Fig. 283)

With everything in readiness, the hull is raised clear of the remaining skid timbers. These are pulled out of the way immediately to clear the work area, then tires or hay bales are positioned under the hull to lessen damage in case of inadvertent dropping. The crane operator may elect to lower the vessel onto these fenders to help relieve the weight load and this is perfectly permissible as long as they are well distributed.

The pulling vehicle now places a steady strain on its wire while the restraint wire is kept only slightly taut. Unlike the rolling wheel system, the vessel will have considerably less tendency to "take off" so the puller must continue its strain throughout almost the entire procedure. There is the possibility with some hull types of a drastic change in the transverse position of the center of gravity, so watch closely for any unchecked momentum. Once the boat has reached its "bilge" position, the restraint wire may be required to increase its strain until she has reached the upright.

As soon as the vertical has been achieved, block and shore the vessel firmly. You may not have time to construct your permanent cradle (unless you can pay your riggers to stand around for a few days), so let your crew dismantle their gear once you are sure of the vessel's stability. Once complete, you may break out the libations for a well-deserved party. (Plates 146 & 147)

Other Rolling Methods

I have purposely gone into considerable detail to describe the rolling wheel and rolling sling inversion systems in an attempt to discourage experimentation with other systems. The described systems have been proven best through time and are the least marred by histories of failure. I suggest that you not use your own precious vessel as a proving ground for future methods.

One prominent firm suggests that your new hull may be inverted in the water by launching it upside down while still attached to its building skid. It is righted (theoretically) through the use of empty 50-gallon drums acting as floats attached to one side of the hull. Frankly, I think this system is highly questionable. An empty hull still has a center of gravity so high and center of buoyancy so low that the chances are she will only float on her bilge without first ballasting her with water. The new hull's stability problem is magnified by the additional weight of the skid still attached some distance above the deck line. I have never been on hand to see this rolling system applied, nor do I know anyone else who has witnessed it. It may work but I wouldn't try it with my own boat.

It has also been intimated by some that you can simply "dump" a small hull over onto bales of hay or tires, then raise it to the vertical with a forklift. My immediate reaction to this rolling system is absolute disgust at such folly. If the landing of the boat on her side doesn't wrench

the vessel severely, she is quite likely to be "forked" by the lift. In short, I recommend the rejection of this rolling system as totally absurd.

When reviewing the turning systems described, keep in mind the possibility of modification to suit your own particular circumstances but don't stray too drastically from these most accepted parameters. They may sound unnecessarily complicated, costly, and time-consuming, but it is far better to play it safe than to trust more dubious quantities.

Plate 146. All of the rigging components required for sling turning can be seen in this photo. Here, the operators are maintaining some of the weight of the vessel on the ground while packing the sheer with tires.

Plate 147. "We're halfway there!"

23

hull finishing

I can clearly remember my very earliest days in this crazy business of yacht design. I was making barely enough to live on and had to take on a second job painting after my regular working hours.

One day a 76' ketch pulled into the yard where I was employed as a laborer and it was a gruesome sight. She was more than 80 years old and terribly neglected. Broken planks, dry rot, deep gouges, and broken fastenings were in sad evidence everywhere and her desperate owner had sold her for salvage for a mere $5,000.

The work order posted on the office board was to restore her hull to a serviceable order, which looked impossible at the time. The rest of my cohorts were just as dismayed as I at the awful prospects of putting the old clunker back in shape. But the foreman, a salty codger with years on the warpath, set us to work with his words of wisdom and encouragement (mingled with other more forceful adjectives). We doubted that even he believed the job could be done but we went to it anyway for no other reason than to avoid his well-placed foot.

There were six men on the scaffolds burning off the quarter-inch of aged paint and it took an entire week to fully expose her planks. Then the timbers were replaced, followed by a complete refastening. At least six of her frames were renewed; then we set about driving hundreds of wedges to force the planking into some simile of fairness. The next order was caulking, then a careful planing off of 1/16" of her lumber to remove any traces of existing lumps.

Prior to applying the actual finish, she was lightly belt-sanded and given a coat of white prime. Each of us then took hold of long hand-sanding boards (about ¾" by 3') with handles at each end. On these boards was mounted a very coarse grit garnet paper. The use of the sanding boards meant pushing them back and forth in long, steady strokes, bearing down on each forward motion. As the sanding progressed day after day, large patches of bare wood began to appear while other areas of the hull remained untouched by the paper. The sanding boards, in effect, were betraying ripples on the surface which were not apparent to the sight or touch of even the most highly tuned craftsman.

The low spots where the prime coat remained after sanding were filled with a creamy glazing compound using rubber squeegees as trowels. When the glazing was thoroughly hardened, she was given a fresh coat of prime and the board sanding repeated.

We must have suffered through at least six full prime/sand cycles and the entire crew was on the verge of mutiny. But finally, on the last round, no bare wood appeared. This time our sanding produced a uniform, chalky smooth texture, completely free of irregular shadows. The old gal's hull looked more like velvet than wood. We were really pleased to witness the gratifying fruits of our otherwise thankless labors.

Before laying on her gloss finish coat, she was sanded several times with progressively finer grits until her hull felt like satin. The waterline was carefully rescored, then the painting areas were marked off. Three days were given to hand-brushing with the best paint and brushes in the yard.

Her launching was truly a glorious sight. Her topsides glistened like crystal and danced with sharp reflections. No flaws could be seen anywhere when reviewing her from various positions on the dock, and if her rig and style hadn't betrayed her age, one would think she had just been released from a mold.

I have seen the same finishing routine a hundred times since. It is almost identical to that of forming the "perfect" fiberglass mold "plug" or preparing a sophisticated racing "greyhound." Even plated steel vessels (usually a multitude of flat surfaces) are faired in basically the same fashion.

In ferro-cement construction, the same principles with only slight variations may be applied to smoothing the hull. The main ingredient for perfection, however, is your own patience. Without it, you will be condemning your boat to mediocrity or even dismal visual failure.

Make It Easy on Yourself

For some unknown reason, most amateur builders make a frustrating and exhausting project out of finishing their boats, without realizing that there are many laborsaving devices at their disposal. You don't have to be a genius to make it easy on yourself, as any professional can tell you. The name of the game is "conserving energy."

The staging constructed for the wiring of the armature and plastering may be of little value during the sanding, glazing, and painting. These jobs require its modification. Additional ledgers should be installed on the scaffold or "A" frame so that variations in the catwalk heights are possible. The catwalk planking may then be adjusted to provide a convenient working area measured from the shoulders to the knees of your workmen. This eliminates unnecessary reaching and stooping with brushes and heavy equipment.

During painting, the catwalks may be staggered to form several tiers. The workmen on the upper tier begin first, and when they have sufficiently covered their area, they move

291

their catwalk aft. At this time, the men on the second tier begin to paint their area just below the first. When completed, they also move their staging aft. The number of tiers and staggered starts required is dictated by the size of the vessel. Continue in this fashion completely around the hull. (Fig. 284)

In order to ease the physical weight of the power sander on your arms and shoulders, consider rigging a counterweight attached to a line passing through a block on the scaffold cross-span or "A" frame. This one measure, alone, can decrease your man-hours measurably.

There may be a hundred other less obvious ways of eliminating steps, wasted motion, and rework. While it is impossible for me to speak for your particular circumstances, tune yourself in to this way of thinking. Above all, don't try to "go it alone." Two men working together can accomplish as much as three or four working separately.

Leaching and Sealing Agents

As I mentioned in Chapter 21, Curing, it is imperative that the powdering of alkalis and lime on the surface of the concrete be completely arrested before applying epoxy sealants. Unless you do so, this powder may form under the finish of the vessel and cause its eventual blistering and peeling. To my knowledge, the best product available for this is **Crete-Seal** from Tri-Col Products, Bakersfield, California. If you have not applied such a product during your curing cycle, it may be applied at this time and most certainly before any etching, sandblasting, or painting begins.

Crete-Seal must be used directly from the can without dilution. When applying to the hull, literally saturate the shell surface. Allow this solution to stand for about four hours, then rinse it away with fresh water. It generally takes 12 to 14 hours for **Crete-Seal's** chemical reaction to occur and about six days for the solution to completely harden within the concrete. Once the hull has completely dried, a white powder will appear on the hull surface which is simply a deposit of lime which has been leached out of the interior of the shell. This residue is easily rinsed away with fresh water.

Successive applications of **Crete-Seal** are very desirable, as they will cumulatively add to the benefits of the first washing. When, at last, no lime dust remains upon drying, you may confidently proceed with the hull finishing process.

It has also been suggested to me that Thompson's **Water Seal** may be used in the same manner as **Crete-Seal**, although it is not a leaching agent. This is a petroleum product which smells like kerosene and is applied per their instructions in a nondiluted form. I have had no direct experience with this product but it has been on the market for many years, though not necessarily in a marine environment. You will find **Water Seal** at any concrete supply company.

Grinding and Blocking

Heavy grit Carborundum paper is available in the form of sanding belts and I highly recommend its use. I discourage the use of disc sanders for grinding down high spots on the hull, as they usually leave deep circular scores when improperly handled and will offer you little "feel" for unfair-

Fig. 284. The "staggered start" method of painting a large hull surface using two teams, each with brush and roller.

Consider the benefits of **Crete-Seal**:
1. It literally seeps into the concrete through capillary action, thus filling the microscopic voids of the mortar caused by entrapped air. This results in a denser concrete shell.
2. The binding properties of **Crete-Seal** enhance the internal adhesion of the mortar components, thus creating an increase in shell strength.
3. The increased density of the concrete skim coat caused by **Crete-Seal** enhances the waterproof properties of the shell, as well as its resistance to oil penetration and armature corrosion.
4. **Crete-Seal** retards the formation of "hot spots" caused by uneven drying of the cured hull, thus lessening the possibility of forming hairline cracks.
5. Some hairline cracks may be sealed and bonded by the saturation of **Crete-Seal**.
6. **Crete-Seal** neutralized and helps leach alkalis and lime which may exist within the cured shell. This one factor alone greatly aids the adhesion of epoxy sealers and paint.

ness. The belt sander, on the other hand, is fast and powerful and betrays irregularities more readily with less risk of causing gouges. The longer the sander, the better.

When belt sanding, concentrate only on the most obvious bumps. Don't try to smooth the hull completely, as this may expose mesh lying near the hull surface. Keep the sander in motion at all times; run it in all directions, not just back and forth; change belts as soon as they begin to clog. "Open coat" belts last much longer than the "closed coat" type and cut more effectively. Be sure to wear goggles and a painter's mask, for concrete dust can cause serious eye and lung damage.

Carborundum blocks (available at your cement supplier) may be used in the same way as the belt sander but should be restricted to sculpturing and knocking down the smallest imperfections. Hand blocking is very slow work and should be used with long pushing strokes, not with quick back and forth motions. If the pores of the block tend to fill with the cement powder, it may be cleaned with a wire brush.

Water Seal is a registered trademark of E. A. Thompson Company, Inc., 1333 Gough St., San Francisco, Ca.

Blasting and Etching

Regardless of the glazing or sealing material you intend to use, it is imperative that the concrete surface be porous or gritty to provide a "tooth" for the finish. Without this condition, the sealants and fillers will not become a permanent part of the hull and may dislodge during the service of the vessel.

If you sponge-floated on plastering day, you are a long way ahead of the game, but still, grinding and blocking have caused some areas to become smooth while filling the concrete pores with cement dust. By far the best abrading preparation is light sandblasting, as this leaves a consistently rough surface without using chemicals that could affect the curing of the glaze and sealant. Be sure to vacuum the hull after blasting to remove dust and aggregate. If you do not prefer this additional expense, the alternative to sandblasting is washing the hull with an acid bath. There are several acids which may be used but require different dilutions and working times.
1. Muriatic (hydrochloric) or acetic acid (powder): dilute 1:10, or per instructions, by weight and acid powder to water. Apply liberally with brooms and let set for two hours. If any areas of the hull tend to dry during this period, rewet with acid.
2. Muriatic or acetic acid (concentrated liquid): dilute 1:3 by volume of acid concentrate to water. Apply in the same manner as above.

During etching, the hull will foam and may even smoke; this is simply a normal indication that the acid is doing its job. Be sure to protect your eyes with goggles during this operation and protect your hands with rubber gloves.

If foaming persists after the prescribed etching time, the acid must be neutralized to arrest its action. This may be done by washing the hull with 5% solution of baking soda or a 1% solution of ammonia. To check neutralization, press strips of litmus paper against the hull in various areas. If the paper turns pink, continue the soda or ammonia wash until succeeding test papers remain unchanged in color.

After neutralization, rinse the hull thoroughly with high-pressure water. You cannot overrinse. No acid or salt (a product of neutralization) should remain, as it may affect the curing or bond of the sealant. Allow the hull to dry for several days after the final rinse before applying finish coatings. (Plate 148)

Initial Glazing

Most books will tell you to simply cover the hull with an epoxy sealer. There's a lot more to it than that if you want a professional appearance. Your hull surface will still bear shallow divots, even after your initial grinding and blocking. These divots should be corrected **now** . . . before proceeding with any finish coatings.

The best glazing compounds to be applied to untreated concrete are of an epoxy base, not polyester. Polyester putties, such as used on auto bodies, do not have the absorbent characteristic of epoxies and have only a fraction of the strength. Remember that the initial glazing must become an actual part of the hull, not simply a cosmetic coat. The epoxy to be used should be specifically formulated for bonding to concrete and is made by scores of companies. After the epoxy has been mixed as directed by the manufacturer, it may be thickened to the consistency of cake batter by adding talc, Cab-O-Sil, or micro-balloon powder.

Cab-O-Sil is a registered trademark (March 27, 1953) of Cabot Corporation, 125 High Street, Boston, Ma., 02110.

The glazing compound is now applied to the hull with a long, stiff rubber squeegee or a 2½' strip of ¼" plywood. Before using, wet the squeegee or plywood with soapy water, as this will retard the sticking of the glazing compound.

Apply a sufficient quantity of glazing to the divot area. Spread the compound by bending the squeegee to the fair shape of the hull and draw it slowly over the divot. If the divot does not fill on the first pass, draw the squeegee once

Plate 148. Any one of several factors may have caused this vessel to peel severely in the early stages of her construction. The shell may have been damp when applying the finish; the hull may have been too smoothly finished; the cement may have leached lime and alkali powders under the epoxy; the finish may not have been specifically formulated for concrete work. All of these factors add up to undue haste . . . and waste.

Plate 149. Close but not quite! The apparent irregular shadows along the topsides of this vessel betray slightly unfair areas. An additional week's work could have produced a remarkable finish.

again, but no more. Overworking of the compound will cause the soapy solution to impregnate the epoxy.

Don't expect a perfect surface on the first application. Allow the glazing to cure until it is hard enough to rasp or sand smooth. Knock off any peaks which lie above the fair line of the hull and reglaze all divot areas again. **Caution:** Never apply new epoxy to old without first abrading the original surface. The complete glazing and sanding cycle should be repeated over and over until all hollows in the hull surface are eliminated. The hull fairness may be easily checked at night by shining a flashlight at a low angle along the surface. Remaining hollows will be betrayed by dark shadows. Mark the hollows with a felt-tip pen so they may be located the next day. The final fairing work is best done just before sunset, as this will visibly exaggerate irregular shapes. You might also learn to find the bumps and hollows by quickly passing the palm of your hand over the surface. (Plate 149)

Exterior Sealing

After the initial glazing of the hull has been completed and sanded, it may be coated with a two-part epoxy specifically formulated for sealing concrete. This sealant will have a penetrating quality that will cause it to become deeply absorbed into the hull surface.

It has been suggested that the first epoxy application be diluted to almost a watery consistency in order to enhance its absorption into the hull. This reportedly creates a more effective, mechanical interface between the concrete and the hull finishing. While I have had no personal experience with this method, a thorough testing program has been conducted by Doctor Bernard Jacobs and results of his tests have been extremely encouraging. It should be noted here that several very thin epoxy coats applied with a decreasing dilution seem to be the most effective.

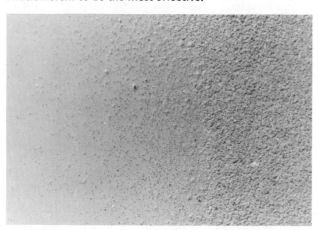

Plate 150. A closeup view of sand particles which have become exposed due to sponge-floating and acid-etching prior to the application of the epoxy (right side). The epoxy sealer, being applied by spray, has worked itself deeply into the surface aggregate and completely covers the cement texture (left side).

The epoxy's purpose is to make an absolutely watertight seal, to protect the hull from corrosion, abrasion, and chemical attack, and to provide a base for subsequent finishing coats. Epoxy sealants are available as clear or amber liquids and are of low viscosity. They generally have a pot life of about five hours at 75° F. and cure to tack approximately two hours after application. This cycle may

be repeated as many times as desired and each successive coat will fill in more of the surface irregularities. (Plate 150)

To enhance the glazing and filling qualities of the sealant, Cab-O-Sil or talc may be added to the epoxy, which works especially well with the squeegee applicator. For spraying, however, avoid building the viscosity to the point of clogging the equipment.

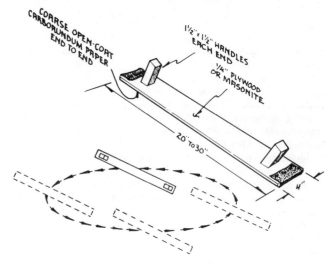

Fig. 285. The fairing and blocking board. Hold the board firmly against the hull so that it forms itself to the curves of the surface. Move it in long rhythmic ovals keeping your weight into it at all times. Do not simply push back and forth in a straight line but hold the board at about a 30° angle.

Sanding between coats should be first done with Carborundum paper attached to a 4" x 30" plywood strip with handles at each end. (Fig. 285) This will knock down remaining high spots. The low areas and divots must be done by hand.

Rub Rails

If your design calls for hull fenders or rub rails, they must be attached prior to any successive finishing sequence. The actual construction of these coamings is discussed in Chapter 24.

Some builders have suggested attaching rub rails and fenders to the hull by using an epoxy adhesive. I admit that this is a strong and permanent method, but they are sure to become damaged during the life of the vessel through collision, docking incidents, and pounding against pilings, which will dictate their eventual replacement. Unfortunately, an epoxy bond cannot be broken without destroying the surface concrete so its use must be restricted to other functions.

Galvanized or stainless through-bolts accommodate the only rub rail attachment I can recommend. Hopefully, you have provided for prelocated holes, but if not, you will have to drill the hull using a carbide or diamond-tip drill bit. Be sure to locate the holes accurately with a fairing spline (batten) while scaling the positions from your Lines Drawing or Outboard Profile. Before installing the rails, coat the undersides with bedding compound as well as the bolt shanks to prevent leakage. As the timbers are drawn up tight, the bedding should ooze from under and may be wiped clean with rags and turpentine. After attaching, glaze all visible gaps between the hull and rail and fill or plug over the bolt heads. The actual rub rail construction is discussed in Chapter 24.

Sanding Undercoat

So far your efforts have been aimed specifically at correcting unfairness by concentration on obvious bumps and hollows. Now that they have been brought in line with a smoothly shaped hull and no severe irregular shadows appear when viewing with a flashlight, you may begin your fine finish work.

Various white sanding undercoatings have been specifically manufactured by virtually all the marine paint companies for use on plastics and other synthetic materials, i.e., polyester, epoxies, vinyls, etc. Any of these products may be used successfully over your epoxy sealer. Do not use an undercoating formulated for wood or concrete, as its inhibiting qualities will be totally ineffective on the epoxy. The purpose of the undercoating is to fill any pores which exist in the hull surface and to provide the proper chemical bond for the finish paint. Because each paint manufacturer has devised his own sets of formulas for various finishes in order to ensure that his products are compatible, stay with one brand of paint through the remaining finish coats.

Most boat owners think in terms of painting with a brush alone and that is fine if time is not important. But I have found it saves a lot of time when one man lays the paint on with a roller while a second man immediately follows with the brush. He can then concentrate solely on smoothing the finish without wasting precious time dipping and spreading. This prevents unsightly overlaps from occurring, especially on hot days. Your first brush strokes should be short diagonals followed by long, slow, horizontal passes. This combination will ensure an even spread while preventing most runs. Do not "flood" the surface with too much undercoat at one time, as this will affect the drying and shrinking rate and cause runs. Allow at least a full day of set before sanding.

Finish Glazing

Until now, your sanding has been done with a coarse Carborundum. Now you should shift to aluminum oxide opencoat paper. After each undercoat, use a progressively finer grit starting with No. 100 and working down to No. 200 just prior to final painting. You should now work with a heavy-duty rotary or reciprocating electric vibrating sander. Don't even bother buying a magnetic vibrator, as it is totally useless for boatbuilders. Don't even touch the hull with a disk or belt from this point on, as it will completely undo your efforts in an instant. Once again, check your finish with a flashlight. This time, indicate any low spots with colored chalk, not a felt-tip pen (it will bleed through paint).

"Plastic" marine glazing compound is available from your chosen manufacturer and is formulated to be compatible with your undercoating. Glazing may be applied directly out of the can using a wide putty knife (a 6" blade works well) or a rubber squeegee. If you use a squeegee, the glaze must be thinned slightly to a creamy paste, accomplished by mixing in a small quantity of white uncoating (not solvents). Fill the hollows and divots as well as you can on one pass but avoid overbuilding. Now begin concentrating on the smaller, less obvious imperfections. After allowing the glaze to dry thoroughly, sand down any peaks. Give the entire hull another coat of prime. (Plate 151)

The prime-glaze-sand cycle should be repeated as many times as necessary to produce a shadow-free surface. Of course, all of these preparatory stages are physically exhausting and seemingly unrewarding; your efforts will pay off long after the last brush stroke, but unfortunately, this is the nature of all painting. A good finish cannot be rushed. It usually takes several weeks to develop a flawless gloss and your patience here is your most valuable asset. Bear in mind that smoothness of the hull is the result of a chain reaction: fair mold, fair armature, fair plastering, fair grinding, fair glazing. Every boat, regardless of her material, must pass through this sequence successfully to achieve a professional appearance, so don't blame ferro-cement for the shortcomings of a bumpy hull. (Plate 152)

Scribing the Lower Boot-Top Line

The boot-top is the color band which encircles the hull at a slight distance above the waterline. While the boot-top is often called the waterline, it should never be located at the plane of flotation. The reason is that a certain amount of

Plate 151. A closeup view of the vessel pictured in Plate 149 shows the raggedy texture of the concrete surface through the hull finish. Failure to correct the "orange peel" surface, as well as leaving the edges of the body putty fairing compound unfeathered, have undone the image of months of hard work.

Plate 152. A nicely faired hull at the Betts Marine yard in Alameda, California, awaits her outer hull finishing coats after being sealed, glazed and sanded many times over.

scum and grass will always accumulate above the waterline where the hull should afford some antifouling protection, and a hard finish in the boot-top permits frequent scrubbing. The boot-top position is also a matter of accepted marine tradition and to alter this tradition is to create an obvious amateurish visual impropriety which will be quickly noticed by experienced yachtsmen. If it is not a proper distance above the actual waterline, your ship will appear to be sinking.

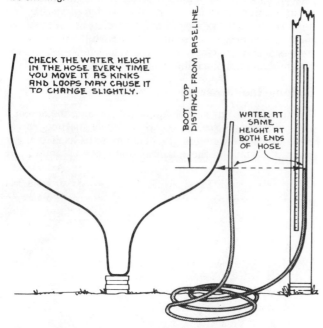

Fig. 286. Using a water level for locating the waterline along the hull.

The lower edge of the boot-top should be located about 5% of the minimum freeboard (distance from water to sheer) above the waterline. This may vary slightly according to the type of vessel, so carefully scale the proper distance from the Outboard Profile drawing. This dimension should be translated into "distance from baseline."

The actual transfer of the lower boot-top line may be accomplished in several ways. The "water level" (a long, flexible plastic tubing filled with water) may be attached to the scaffolding or other upright, then filled with water until it reaches the exact boot-top level. The free end may now be carried to any location along the hull. Place marks at the waterline on the vessel at approximately 18" intervals. Now she is ready for masking. (Fig. 286)

Another successful transfer method is that of stretching a tight wire from one end of the vessel to the other, including temporary level supports at each end so the wire stretches about a foot away from the hull. Now suspend several light bulbs from the scaffold cross-spans at equal distance. The light wire and the bulb centers should coincide exactly with the lower boot-top line. The shadow of wire as it falls on the hull will be your masking line. **Caution:** On large vessels, the tight wire will sag significantly between its supports so intermediate supports will be required for accuracy. (Fig. 287)

A long carpenter's spirit level may also be used for transferring the lower boot-top line. With this method, the boot-top position is marked on each scaffold upright; thus, the level is employed to project the mark onto the hull. Because the hull may be a considerable distance from the scaffold, a straightedge extension should be attached to the level.

A variation of the last method is to attach the carpenter's level perpendicular to a long 1" x 2" staff. The length of the staff to the level should be equal to the boot-top relative to the base line. Now hold the rig so that the end of the staff is against the bottom of the scaffold cross-spans (they usually represent the base line) while the end of the level touches the hull. Mark the hull at 18" intervals. (Fig. 288)

Scribing the Upper Boot-Top Line

Take a look at plans by reputable naval architects; you will notice that the upper edge of the boot-top is rarely a straight line but rather a long, gentle curve. The boot-top, when seen in profile, is usually about 30 to 40% wider at its ends than at the amidship section, the narrowest portion located approximately 65% of the waterline aft. The boot-top width at its narrowest point usually is about 7 to 8% of the freeboard.

Tapering of the boot-top creates several effects. Because of the twist of the planes of the hull at the waterline, an optical illusion occurs which causes the boot-top to visually diminish, bow and stern. By altering the boot-top width, this illusion is canceled. Tapering the boot-top also tends to accentuate the curve of the sheer. Boats which have a straight boot-top often look "down by the bow," lacking in spirit, while appearing very amateurish.

If your architect has not indicated the boot curvature on the Lines of Profile Drawing, you may do so using a ¼" x ¼" pine batten, holding its position with weights. Carefully scale the line position from the base line at each station, then transfer the dimensions to the hull using a spirit or water level. You are now ready for masking.

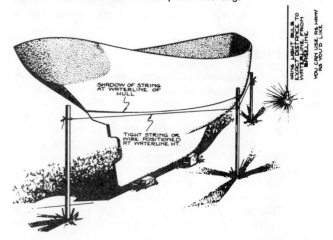

Fig. 287. An interesting method for striking the waterline. Both the string and the light bulb must be positioned at exactly the waterline height.

Caution: Most professional wood and fiberglass boatbuilders score the hull with a saw or router along the boot-top lines and cove stripe. This is done to aid masking, to prevent paint bleed, and to preserve the line for the life of the vessel. Scoring should **not** be done on ferro-cement hulls, however, as it can become an area for the concentration of bending stress resulting in severe cracks and possible hull failure. An alternative to scoring is to permanently and accurately attach an epoxy-soaked string to the hull along the boot-top edges. This will adequately serve the same purpose as scoring.

Masking

Proper masking should be a two-man job with one member of the team tightly pressing the tape to the hull while the other person stretches the tape out 10' or 15' ahead, adjusting its vertical position as he sights down the tape. The

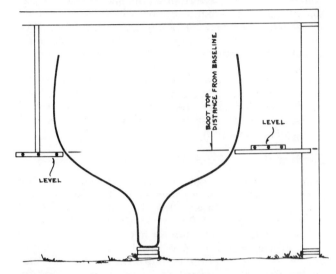

Fig. 288. Using a spirit level for locating the waterline on the hull.

roll of tape should be held away from the hull so that it will not make premature contact. Only when the stretcher is satisfied that the tape line is straight or fair will he allow the toucher to attach the tape. (Fig. 289)

The area of the stern counter and turn around the bow always creates problems because the tape will tend to veer off onto its own course. If the tape refuses to lie fairly, it may be partly cut through or applied in small pieces. After you have masked both edges of the waterline, it is well to add a second width simply to speed up painting. Be sure to rub the paint side of the masking well against the hull, or burnish it with your thumbnail, to ensure a firm contact.

Removal of masking should occur as soon as your paint has

set up. Don't wait for it to dry, as this could cause peeling or chipping. Using a slow, steady tension, pull the tape at an angle to the line, never straight up. Always pull the tape in a direction away from the fresh paint.

When masking over freshly applied paint, always allow two or three days to pass before applying the tape, as this will prevent pulling up the new finish.

Mixing Paint

Assuming you have chosen stock colors of a recognized marine finish, do not rely solely on the paint store to mix your paint adequately. If your paint has been on the shelf for any length of time, the pigment will surely have settled to the bottom of the can and may have become a solid mass. Upon opening the containers, stir the paint with a stick (a mixing bit in an electric drill also works well) until you can no longer feel any concentrated solids or paste. Now strain the paint through a piece of cheesecloth or nylon stocking to remove any remaining lumps. "Box" the paint back and forth between several clean cans or buckets, then pour off the quantity you intend to use. Don't paint out of the original container, as this will foul the sealing rim, thus rendering its future storage capability almost useless. Keep this can clean and covered at all times when not in use, however short a time. Thinning of new paint should not be done to any good paint, as the proper viscosity has been carefully programmed for the best results. Of course, the paint will thicken slightly as it is continuously exposed to air but follow the manufacturer's instructions to the letter, using only the recommended solvent.

Finish Coat

The boat need not be painted right away and I would prefer to wait until the completion of all construction. Hulls have a way of being battered with ladders and lumber while becoming generally soiled or scraped during building.

The day and time for painting must be chosen judiciously. Avoid high humidity, especially if it is foggy or misty, as this will not only produce poor paint adhesion but will also cause a potentially glossy surface to dry out flat. Do not begin painting near sunrise or sunset, as the daily humidity is always highest during these times. Painting in the direct sunlight or during extremely hot weather (50-70% is ideal) will accelerate drying, which will also intensify brush marks

Fig. 289. Masking.

and overlaps, as well as cause possible blistering of the finish. I strongly suggest covering the sunny side of your scaffolding with bed sheets, blankets, or tarps so that your work will be done in the shade. Never paint on a windy day for obvious reasons. Nothing can destroy a fine finish quicker than dust or sand particles. It's a prudent measure to spray the entire work area lightly (not the hull) with water just prior to beginning painting. Spraying should include the scaffold, if possible, as most debris will originate from overhead. If this measure is impractical, use the vacuum, then give it a light wiping with lint-free rags dampened with mineral spirits or an epoxy cleaning solvent (for epoxy only). This will help the paint absorb into the primer coat. If any of the preceding conditions cannot be met, postpone your painting until another day.

The darkest color of the hull is normally applied first. Again, I suggest at least a two-man team here. The first man lays the paint on with a roller while the second brushes out the finish. Do not flood the surface but keep the coating rather thin; two or more light applications are more effective than one. Do not let the roller get more than a few seconds ahead of the brusher. The first brush strokes should be short and diagonal followed by long, steady horizontal sweeps. Each finishing stroke should terminate at the end nearest the previously applied paint and never with excessive pressure. Just barely glide over the surface with the brush tip and keep your strokes to a minimum. Don't repeatedly rework the finish, as this will intensify brush marks and overlaps and may curdle the surface. Brush

Plate 153. The glasslike finish on Don Terry's **Sea Smoke** is striking evidence that a superior finish can be achieved with ferro-cement. It is, however, the result of hundreds of patient hours' work. A Samson Marine design.

marks will always be left in the immediate wake of the brush but if you have purchased and thoroughly mixed the best paint available, these marks will flow out into a flawless gloss.

Always sand between every coat, wiping clean each time. No. 160—No. 200 grit aluminum oxide open-coat paper is the best but be sure that each application is separated by several days of hardening.

The last coat of the bottom antifouling should never be applied until just before launching, say the day before. To do so before will only cause the toxides and poison additives to become stale and inactive. Ideally, your boat should be put into the water as soon as the bottom paint has dried and you have covered the "holidays" (bare spots caused by obstacles such as shoring and blocking). For the best sailing performance, several coats of bottom paint may be used (sanding between coats). The last application should be wet sanded to a velvet sheen free of brush marks. Do not wax the bottom. (Plate 153)

Brushes

Poor equipment will not produce a good paint job. Bad brushes, in fact, can do more harm than good. Unfortunately, a quality brush costs a lot of money and you should expect to pay $18 or more for a good 4" hog brush or one that is a combination of hog and horse hair. Most amateurs are only willing to spend a few dollars apiece, usually for synthetic brushes, but these are generally latex or like finishes. These will rarely produce a fine marine finish. A good brush can last 15 years when properly used and cared for and should be respected as highly as your treasured power tools. Don't begrudge your new boat the best or sacrifice your high principles solely on the basis of your pocketbook.

When buying brushes, look for those with long, natural bristles (hog and horse are best). Short bristles will not hold paint or have any "feel." Your brushes should feel "live" and springy, not limp or stiff. When looking straight down the bristles, the brush should appear solid, not simply a ring of bristles wrapped around the handle. The extreme ends of the bristles should be split into finer hairs and the brush should be slightly tapered for the best control when painting near edges or "cutting in."

A good painter will never dunk his brush to the full depth of the bristle, nor will he allow paint to collect and harden around the collar. Keeping your brushes tidy while in use will make painting a more pleasurable chore and ease the burden of cleaning up. The proper way to use a brush is first to fill your bucket (not the paint can) with only about 2" of paint. Dip your brush so that it just kisses the bottom of the pail, then withdraw it and pat it gently against the sides of the pail. Never scrape the brush over the edge of the pail, as this will not only turn your job into a messy fiasco and wear out your brush, but you may also pick up dirt or dried paint particles which will ultimately end up on the boat.

Cleaning brushes has caused a lot of needless suffering and is a term which is somewhat misleading. No brush is ever completely clean! No matter how much time you spend squeezing, dumping, and mashing your brush into an approved solvent, some paint will always remain. Certainly you should remove as much paint as you can, especially when using epoxy or urethane, but don't ever allow the brush to dry out. Once any brush "sets up" it will be of little use for fine work again. After rinsing the brush gently, fold a cloth envelope around the bristles (to preserve the shape) then hang the brush bristle-deep in a can of solvent. Never allow the bristles to touch the bottom of the can or to bear any of the weight of the brush. If a brush is allowed to curl in this manner, it may never be straightened out. Check the level of your soaking solvent from time to time and keep the can covered with foil to prevent excess evaporation.

I recommend purchasing three brushes for each color you use (including varnish). They should be among your most prized possessions (along with your sextant, compass, and life raft). For the life of these brushes, they will always be used for their original color and never (or rarely) changed. If you must wash brushes for color changes, be gentle but thorough. Brushes for white paint should be guarded against other uses with life and limb.

Never store your paints, solvents, or brushes inside the boat. Their fumes are volatile, even within the most efficiently ventilated space. If you must carry this gear on-board, build an attractive deck box provided with fiddles to prevent cans from shifting. This interior should be lined with foil just in case something leaks. Never fill soaking containers to the brim and keep their foil covers tightly sealed with rubber bands or string.

24

the woodwork

Observation

If you make it a habit (and you should) of walking the docks of your local marina or yacht club, you may be aware that certain boats immediately catch your eye while others go completely unnoticed. Sometimes your attention will be drawn to a vessel a dozen wells from you. Why this magnetism toward some and not others? Occasionally, it may be an unusual color scheme or a unique cabin shape but usually a yacht's quality is distinguished by her woodwork. Surely no other element can make or destroy a favorable impression of passersby or represent the owner's attitude more clearly than the quality of the ship's carpentry.

It is impossible for me to explain within these pages all of the techniques, variations or problems of handling your joinery, so I must immediately advise you to visit your library and check out every book you can on cabinetmaking, general carpentry and wood boatbuilding. A book list would be too extensive to include here but I might point out the best publications on woodwork are rarely written specifically for the boatbuilder.

Acquire the habit of looking at specific details when viewing a boat, not simply the vessel as a whole. Observe the way toe-rails are shaped and constructed; notice the variations in handrail and ventilator Dorade treatments; pay particular attention to hatch construction and window trim; examine cockpit coamings, winch and hardware pedestals, cabin moldings, tiller, bowsprits, pin and fife rails. If you have a camera, take close-up pictures of these various elements so you will be able to study them at a later date. Mount them in a scrapbook under their respective categories and they will be an invaluable reference.

The interior of a yacht will yield even more detail than her exterior. Take every opportunity to go aboard yachts for the purpose of inspecting doors, drawers, fiddles, moldings, jambs, bunk boards, shelves, hatch and cabin carlin facings, and the overall element of care of construction. In areas where it is too dark to use a camera make sketches of the things which have pleased you. Above all, tune your senses for observation. It will help you far more than anything else I can suggest.

Scrimping on Material

At the outset, you may have decided to build a ferro-cement boat because it seemed like an inexpensive way of obtaining your dream yacht. By now you will have realized that it is going to cost considerably more than you originally thought. Well, be prepared! The real shock of boat-

building has yet to come. Before you decide to scrimp on lumber, let me point out some things that could determine your success or failure.

Your ferro-cement hull may be extremely strong but, like vessels of all other materials, she will depend on her internal girdering to withstand the severe strains and shocks of the sea. During her years of service she will be grounded many times; she will bang into docks; she will be covered with green water and mercilessly pounded by storms.

Some day, you may wish to sell her to make way for a larger vessel. Her ultimate value may depend upon the materials in her and the way they have held up. Far more important than all of these factors, however, is your own sense of pride and the respect you have built for your own creation thus far.

The Tools

For years ferro-cement promoters have been claiming this type of boat construction does not require the investment in expensive tools as do other construction systems. Unfortunately, they have not considered the interior which must be constructed entirely of wood. Since the wood affects the finished quality of your boat, lacking the proper equipment can ruin the results.

I do not advise renting tools because it is as expensive as buying them. When buying tools, remember that you will probably be able to sell them later to other boatbuilders with little difficulty.

To avoid expensive milling charges, I believe a 10" table saw with heavy-duty motor will be your most important piece of gear. You will use it for making long, straight cuts when laying the sole, laminating bulkheads and deck, constructing joinery, building the toe-rail and masts. A suggested list of tools follows.

(1) 10" table saw
(1) variable speed saber (jig) saw
(1) 7" hand circular saw
(1) 3" x 21" belt sander
(1) variable speed reciprocating sand or orbital sander
(1) 3/8" variable-speed drill
(1) 1/4" single-speed drill
(1) hand power plane
(1) 6" jointer/planer
(1) belt/disk table sander
assortment of good, high-speed drill bits
assorted screw countersink bits
counterbore bits as required
flathead bolt countersinks as required

301

a variety of blade types for circular and table saws
(2) claw hammers
a variety of screwdrivers
a variety of chisels from 1/2" to 1-1/2"
(1) spirit level
(1) large steel square
(1) adjustable level gauge
(6) 9" spring clamps
as many "C" clamps as you can accumulate (judge the sizes as you proceed)
a variety of putty knives
(1) large and small nail set
a variety of plug cutters (cut plugs are available)
(1) portable electric rotor with table
a variety of rotor rabbet and corner radius bits
(1) good, hand crosscut saw
(1) good, hand ripsaw
a substantial woodworking bench
(1) good, woodworking vise
an 18" band saw (if you can manage it)

Solid Wood

I've seen a lot of potentially fine boats (by virtue of good hulls) turn into hunks of junk simply because the builder ran down to his local corner lumberyard and bought junky wood that was intended for the construction of mediocre tract houses. This is unwise if you really want to build a decent boat, so I offer a few guidelines that may help you reach your intended goal.

Air-dried lumber. There is no doubt that air-dried lumber is the best for boatbuilding, as it markedly reduces wood shrinkage after the boat has been finished. Shrinking of lumber often causes severe strains (even frequent breaking) of fastenings and glue and usually results in opening joints. Even the finest built hatch or cabin trunk can leak if "green" lumber is used and the joints of deck framing can fail.

Unfortunately, the cost of air-dried stock can be almost double that of green lumber and few amateurs are willing to pay the high price; besides, it has become almost impossible to locate, particularly in the grades and cuts required. The alternative, then, is to use only kiln-dried lumber rather than green stuff.

Granted, kiln-dried stock is still pretty wet stuff, not necessarily with water, but with sap and oils. Kiln drying is usually equal to 1-1/2 to 2 years of natural air drying and this is not long enough for the best results. It will help if you can add another year to that. Purchase as much of your lumber as far in advance as possible. Air drying necessitates laying lumber out flat on battens to allow the circulation of air. Set the battens close enough together to prevent sagging. Be sure the planks do not touch each other. For each layer of lumber being stacked, lay down another row of battens.

Before stacking the lumber for drying, give it a liberal coat of Cuprinol. This is a watery, green, wood preservative which can add ten years to the life of your boat. It also helps to prevent the checking of lumber while it is being dried, as well as repelling bugs, mites and worms. Apply a coat or two of paint to the ends of your lumber; this will also retard checking during drying.

Erect a tent or shelter over your drying lumber to shield it from rain, dew and direct sunlight. Drying should be slow and moderate. Every few weeks, rotate the drying lumber to

Cuprinol is a registered trademark (Sept. 26, 1933) of Darworth, Inc., Avon, Ct., 06001.

prevent bowing, and turn it over occasionally so that drying will be even.

Dried lumber will often become grey, but the color will have no effect on strength. When the dried lumber has been surfaced or sanded, it will return to its original color. Once dried, it may be difficult to tell what species it is; mark the wood well in advance of stacking.

The cut. If you were to call your lumberyard and ask them to send over a 1" x 2" or 2" x 4" or 1" x 6", they would probably deliver a ¾" x 1½", a 1½" x 3½", or a ¾" x 5½". The reason for this is that lumber sizes refer to the rough-milled size before surfacing. You will actually be paying for a larger piece of wood than you receive. In your plans, a 1" x 2" or 1" x 4" or 1"x 6" will ordinarily refer to the rough-cut size. It is really assumed that you are to use ¾" stock. All other sizes stated on your plans will refer to the actual surfaced dimensions of the lumber. (Fig. 290)

SIZE TO ORDER	ACTUAL SIZE SURFACED(DRY)	SIZE TO ORDER	ACTUAL SIZE SURFACED(DRY)
1 x 3	¾" x 2½"	2 x 3	1½" x 2½"
1 x 4	¾" x 3½"	2 x 4	1½" x 3½"
1 x 6	¾" x 5½"	2 x 6	1½" x 5½"
1 x 8	¾" x 7¼"	2 x 8	1½" x 7¼"
1 x 10	¾" x 9¼"	2 x 10	1½" x 9¼"
1 x 12	¾" x 11¼"	2 x 10	1½" x 11¼"

other thicknesses are represented by the widths of the lumber shown in tables above
STOCK FINISHED LUMBER SIZE

Figure 290.

You may be familiar with the terms "quarter-sawn" lumber and "edge-grain" lumber. They are interchangeable and refer to the position and direction of the plank relative to the trunk as it was taken out of the tree; i.e., the plank was on a line in the trunk which passed through the center of the tree. You will recognize quarter-sawn lumber by the fact that the grain runs up and down when the plank is on its side (not edge).

Quarter-sawn lumber resists bending more than other cuts; it shrinks more evenly and will not warp. For these reasons, most of the lumber purchased for your boat should be quarter-sawn if possible.

There are a few exceptions to this rule, most specifically dimensioned timbers for sole beams and cut deck beams. In these instances, the grain should run vertically when placed into the boat, although they may have been cut from quarter-sawn planks. (Fig. 291)

Hardwood or softwood. These terms do not refer to the actual hardness of the wood but rather to its gender. Hardwoods come from broadleaf trees (oak, birch, cedar, poplar, elm) while softwoods come from evergreens (fir, pine, spruce).

Hardwoods are expensive and heavy but take a fastening well. There are few places on a synthetic hull, however, where hardwoods are necessary. They are definitely too heavy for deck beaming (they do not resist bend as well as softwoods relative to comparative weights) but can be used for framing the sole where the weight will be low in the hull. Hardwoods are also preferred for timber engine beds, mast steps and mast partners, as well as knees and hooks.

Softwoods are excellent for the majority of lumber to be used in your boat. Spruce is the best. Since it resists bend

302

exceptionally well, it should be your first choice for deck beams. Fir is a fine wood, being quite stiff for its weight also. However, it doesn't resist rot or take a fastening as well as spruce. "Construction grade" fir is the most readily available of all American woods but it should not be used anywhere on your boat, as it is of very low grade. You will find that good spruce is as available and no more expensive on special order than an equal quality of fir.

Mahogany is sometimes classified as a hardwood but it really depends on the type. Honduras and African mahogany is harder than Philippine and is usually used for brightwork (toe-rails, fiddles, etc.) while Philippine mahogany is much lighter and is an excellent general construction wood. Because lumber prices have sharply increased lately, you may find it to be no more expensive to use Philippine mahogany throughout your boat (sheer clamp, beam shelf, sole beam, joiner cleats, as well as bright facings, drawer fronts, doors, fiddles, trim, toe-rails, etc.). The weight of Philippine mahogany is almost the same as that of yellow pine and it takes a fastening exceptionally well while being tough and resistant to rot. In short, Philippine mahogany is my first recommendation even though I often detail my own designs for fir or spruce.

Teak is a beautiful wood, resistant to rot, wears well and takes a great fastening. Its price, however, is almost prohibitive and it can seldom be purchased in a consistent color, cut or grade. Most purchases of teak must be made on a "random width and length" basis, which doesn't leave you with a very selective choice of what you get. I doubt if there are many amateur builders who are truly capable of building the high quality of joinery or decking which is deserving of teak lumber, so consider your own experience and equipment before you pour your hard-earned money into this fine lumber. In the area of decking, quarter-sawn fir, pine or spruce planking trimmed with mahogany margins will produce just as beautiful a piece of work as teak.

Fig. 291. Quarter-sawn lumber is the best for boatbuilding.

Grading of lumber. This is a nebulous area because different woods are graded in different ways. The easiest way to explain the choice of wood for boatbuilding is to advise you to ask for best quality. Your lumber, regardless of type, must be free of knots, checks, splits, shakes, sap pockets, and must be of extremely straight and regular grain. The grain should run as parallel with the edge of the plank or timber as possible and leave the plank as seldom as

practical. Many lumberyards will special order "boat-grade" spruce or fir and it is, indeed, excellent lumber. "Construction grade" absolutely does not belong on a boat; it is soft, irregular, knotty and sappy. When ordering fir or spruce, ask only for "C and better." When ordering pine, ask for "C and better" or "No. 1 clear." When ordering mahogany, ask for "No. 1" or "A."

Cutting and milling. This can easily increase your lumber bill by a third. It is advisable to invest in a large table saw (10" minimum) and a joiner/planer in order to save on milling charges. It will be possible to purchase larger timbers at a lower cost and cut them into the desired sizes on your own equipment. This is most practical because so many different sizes are required in the construction of a boat. When doing your own milling, draw out the timber section very carefully, showing the desired cuts, and make sure to allow 1/8" for the saw blade. If you are surfacing your "recut" lumber, also allow 1/16" per side when drawing your milling plan.

Plywood

The type and quality of plywood used for the construction of your boat is extremely important. The wrong choice could have disastrous consequences. There are several factors which will influence your choice: hardwood or softwood; surface grade; the glue used between the veneers; solid cores or open cores. Let us consider these factors one at a time.

1. *Hardwood* refers to woods from broadleaf trees such as *birch, maple, walnut, oak, teak,* etc. Hardwood plywoods are all very expensive if suitable for marine construction but, fortunately, their use is rather limited. I believe that birch is one of the toughest and is constructed of 1/16" veneers or less regardless of the thickness of the sheet. It is manufactured in Scandanavia and may require special order well in advance of its use. Birch (or oak) plywood is the best suited to rudder construction and the building of large laminates such as the breasthook. Most hardwood plywoods are confined to use as interior decorative paneling.

Softwood refers to wood from evergreen trees. The most notable of softwood plywoods is *Douglas fir* and it is readily available everywhere. Fir plywood is the most popular among boatbuilders because it is one of the most inexpensive. Douglas fir may be used virtually anywhere in the boat except for rudders or the building of heavy laminates such as the breasthook. *Luan Mahogany* (Philippine) is also available as a softwood plywood and is very tough but light. This is my first recommendation for the construction of bulkheads, major joinery, and may also be used for building rudders, cabin trunks, decking and fine joinery. Luan plywood is manufactured in the Orient and is very expensive but worth the added cost to ensure a rugged and lasting boat.

2. If the glue used is not waterproof, don't consider the plywood for use anywhere on your boat. Waterproof glues are used in the manufacturing of both marine plywood and exterior plywood and from this standpoint are exactly the same. (There are structural differences between marine and exterior plywood.)

3. The structural differences between marine and exterior plywood are that the joints in the middle veneers of marine plywood are butted firmly together or splined so that no voids exist anywhere within the plywood sheet. This is not true with exterior plywood. In the manufacturing of marine

plywood, the knotholes in the middle veneers are patched so that no voids remain. Again, this is not the case with exterior plywood.

Voids left in exterior plywood weaken it considerably and are in danger of becoming possible rot traps which will affect the longevity of your boat. Of course, there are areas where the strength of the boat will not be severely compromised, such as shelves, drawers, small cabinets, etc., and I agree that the cost of marine plywood here is a needless expense. The construction of major bulkheads, decks, berths, countertops, etc., must be of marine ply, as they are major structures which complement portions of the interior box girder. Don't sacrifice your wood quality here. Marine plywood is available in fir, Luan mahogany, birch, and many other species.

4. The surface quality grading is different for hardwood and softwood plywood. A grading table will be more useful than my attempting to describe them. (Fig. 292)

	HARDWOOD		SOFTWOOD
#1	Flawless, expensive, usually available by special order.	"N"	Flawless.
"A"	Well matched surface veneers, no other flaws.	"A"	Small defects repaired, otherwise flawless.
#2	Slight flaws of matched veneers and very small knots.	"B"	Repaired large and small defects.
#3	Large knots, irregular grain and color. Good if it is to be painted.	"C"	Large defects repaired; small defects left as is. May also have slight splits.
		"D"	Large knotholes and splits.
		"X"	Large knotholes, splits and very rough grain.

Figure 292.

You may purchase most plywood with different surfaces on each side. When constructing cabinets or counter fronts, for instance, the invisible side of the plywood may be "D" or "X", whereas the visible side should be "N", "A", or "B." A typical nomenclature for plywood, then, would be "1/2" A/C marine fir" or "3/4" 1/3 marine Luan."

Painting plywood deserves a brief mention here. Most amateurs tend to sand fir plywood before applying undercoat or final finish. This should never be done if a smooth painted surface is your ultimate goal. Sanding bare fir plywood will only accentuate the grain rather than make it smoother. So before sanding, be sure to apply two successive coats of sanding sealer. The results will be a remarkable velvety surface that will take paint like a sheet of Masonite.

Glue

Throughout the vessel's construction you will be called upon to glue wood joints, even though the joint may be supplemented with screws or bolts. You will also be required to construct laminates and many joints made with glue alone. It is imperative that you choose the proper glue for the job.

Although your woodwork will not be submerged in water, the humid atmosphere in the marine environment will affect everything on board. Glue joints are no exception. Wood joints will also be subjected to continual working, stress and sheer. A common household or general building glue cannot possibly hold up on a boat. So be sure to

purchase glues which are truly waterproof, not merely water-resistant. The life of your vessel (and perhaps yours) will depend on it.

Resin glue. I believe this is the best type of all. However, it should not be used on oak because it is not chemically compatible with the acids in the wood. It is a dry powder which is mixed with water into a creamy substance. Once mixed, it will have a pot life of about three or four hours at 70° F. but the sooner you use it, the better. If the temperature becomes cold (say 60° or less), the glue will still harden but may lose its strength completely. If you are working in cold climates or gluing at night where the temperature is likely to fall, position infrared lamps near the work but not so close that the work becomes hot. The infrared lamps are the same as those used in restaurants for keeping food warm. One of the prime benefits of resin glue is that it will not stain through paint. Resorcinol does, on the other hand, even after a half dozen coats of paint.

Resin glue is sold under several different names, i.e., *resin glue, plastic resin, phenolic resin,* and so on. Whatever brand you use, it will be specified with the word *resin.* The prime manufacturers are: Borden (Elmer's plastic resin cascamite glue) and U.S. Plywood Corporation (Weldwood plastic resin glue).

Resorcinol. This is a two-part mixture of a light brown powder and a reddish brown syrupy liquid. When mixed in proper proportion, it develops a creamy substance. Resorcinol is considerably more expensive than resin glue but provides a more watertight joint. It should definitely be used in cabin, deck and hatch construction.

The pot life of resorcinol is quite short, say 30 minutes at 70° F., but the curing time is longer than resin glue, about 10 hours at this temperature. As with resin glue, the temperature of the work should not be allowed to fall below 60° F. and infrared lamps should be brought into action if colder temperatures are expected. One critical factor that affects resorcinol drastically is the mixture ratio. I recommend weighing the mix components very accurately with at least a mailing scale. Don't simply trust your judgment of viscosity. As soon as the wood joints have been glued and clamped, wipe the excess glue away with a damp sponge and dry rag. If this is not done, the remaining glue will stain the paintwork for the life of the boat.

Epoxy. There will be times when your talents or equipment are unable to form a "perfect" fitting joint. Here's where epoxy fills a need like no other glue because glue thickness (unlike resin glue or resorcinol) is not a detriment to joint strength. The more "imperfect" the joint, the thicker the epoxy consistency should be and this may be accomplished by adding Cab-O-Sil powder to the mix.

Epoxies are two-part mixes whose proportions are extremely critical. The mix may vary slightly with epoxy types or manufacturers, so follow their instructions to the letter. Temperatures are not as critical as with the other glue types, however, and even if the work is allowed to become cold, it will eventually cure fully upon reheating.

One of the greatest attributes of epoxy is that it may be used to glue dissimilar materials together, i.e., wood to fiberglass, wood to concrete, steel to wood, aluminum to glass, etc. It is fine for gluing components to the inside of a ferro-cement hull which has been painted with an epoxy sealer, although the bond to raw cement is much more effective.

When using epoxies, the joint strength can be enhanced markably by providing "tooth," so sand both surfaces well to roughen otherwise smooth materials.

Contact cement. This is a one-part yellow liquid used directly from the can with little regard to temperatures and is available at most hardware stores. Its use in marine work, however, should be restricted to attaching Formica to wood counter tops, decorator vinyl to bulkheads and foam hull lining or foam insulation to the inside of the hull. Contact cement has no structural properties.

Applying contact cement requires a serrated towel (from any hardware store), as a thick surface coat is required on both mating materials. After application, it must dry for about 30 minutes or to a tacky state. If the cement has soaked into either of the materials, a second or third coat must be applied. Regardless of the material, I make it a practice to use three coats to increase the bonding effects. When applying contact cement to fiberglass (e.g., attaching foam sheets to the inside of the hull), you must roughen the surface quite deeply with coarse Carborundum paper in order to provide sufficient tooth.

Fastenings

If fastenings are to be in constant contact with water, i.e., heel bearing bolts, plank fastenings, etc., more than the strength of the fastenings must be considered because they will also be subjected to galvanic electrolysis. In these areas, the galvanic scale must be taken into account (the fastening should be as much like the surrounding materials as possible). In short, mild steel bolts in a hull with a mild steel re-bar hull should (theoretically) hold up the longest. This may not necessarily be the case, however, because the fastenings could also be in close proximity with a bronze propeller or shaft, etc. Because certain schedules of stainless steel, Monel and bronze (e.g., Everdure) are almost galvanically inert, it may be true that the best wooden boats were, in fact, fastened with these materials. But, fortunately, we are not facing galvanic problems when constructing decking and interior components.

Your concern when choosing fastenings, then, should center around two factors: the required strength at the joint and the chemical corrosion to which the fastening may be subjected. In regard to the first factor, I will discuss only general rules:

1. If you are joining major structural members which are to be subjected to great tension and where it is not necessary to completely conceal the fastening for the sake of appearance, use a hex-head bolt and nut with large washers.
2. If major structural members are to be joined where the fastening must be concealed on the finished side, use a flat-head bolt, countersunk and plugged (or glazed) with a matching washer under the nut.
3. When attaching metal parts to the hull or wood, where the appearance of the fastening is not critical, use a round head machine screw unless the metal has been countersunk.
4. If you are attaching minor structural wood members which are to be subjected to moderate tension or considerable sheer, use a flat-head screw or stove bolt. The head should always be plugged or glazed. When the appearance will be affected, the screw length should be as long as possible without protruding through the other side of the joining timber. The screw diameter should always be as large as possible as long as there is no risk of splitting the lumber. Every screw should always be preceded with a pilothole for both the screw threads and the shank. There

are no exceptions to this rule, whatsoever. There are screw pilot bits available for almost every screw length and diameter (except for the very largest screws) and they should be used prudently. Don't use a No. 14 pilot for a No. 12 screw or vice versa.
5. Serrated nails, namely Anchorfast nails, may be substituted for almost every screw on your vessel where the finished appearance of the head is not critical or where the head may be successfully glazed over for painting. Bronze Anchorfast nails are the most common, but they are also available in stainless steel. However, I feel that stainless steel screws or nails are not necessary anywhere on the boat because they will always pull out or destroy the wood long before the fastening is in danger of breaking.

Do not drill pilotholes for Anchorfast nails except when fastening oak, green heart or comparable extreme hardwoods. In these circumstances, the pilotholes should never exceed half the diameter of the nail.

Anchorfast nails may be used for joining decking to beams, joiner cleats to bulkheads, joiner tops and fronts to cleats (when joinery is to be painted), laminating the wooden sheer clamp and many other applications where you may have considered using a screw. Anchorfast fastening is strong enough to destory the wood around itself before pulling out the same manner as a screw.

6. Galvanized finishing nails may be used throughout the interior of your boat for attaching joiner cleats to bulkheads, attaching most joiner tops and fronts to cleats, fastening the sole to the beams and constructing small cabinetry such as drawers and small shelves. Finishing nails should never be used in areas which are to be subjected to large tension loads (large shelf bottoms, for instance) or large sheer loads (berth top cleats at bulkheads).

Finishing nails may be employed whether the nailed surface is to be painted or varnished. For painted surfaces, the nail head is set into the wood and simply glazed over. For varnished surfaces, the head may be invisibly covered with a matching color of Plastic Wood (it is available in light, medium and dark colors which are compatible with the color tones of pine, teak and mahogany). Galvanized finishing nails may be used successfully for attaching deck planking to plywood subdecking in this manner (my only recommended exception to galvanized fastenings topside).

Fastening material. Stainless steel fastenings are expensive and only necessary when the appearance of the fastening is of prime importance, especially when attaching white metal hardware (stainless steel, chrome-plated brass or aluminum). Below the waterline, you may also consider stainless bolts for attaching the heel bearing and thru-hull fittings but here Tobin bronze, Silicon bronze or Monel will be almost as galvanically inert and extremely strong.

On decks where the fastenings may become subjected to attack by salt water, rain and dew, I would not suggest using galvanized steel bolts and screws. They will eventually bleed rust, particularly if they are in contact with moving metal parts.

Bronze bolts and screws are long-lasting. They will never "freeze" tight like a galvanized fastening and are extremely strong. This type is the best buy for constructing a quality boat.

Galvanized screws, bolts and finishing nails may be used virtually everywhere below with little concern for longevity, bleeding or strength. For interior fastening, the invest-

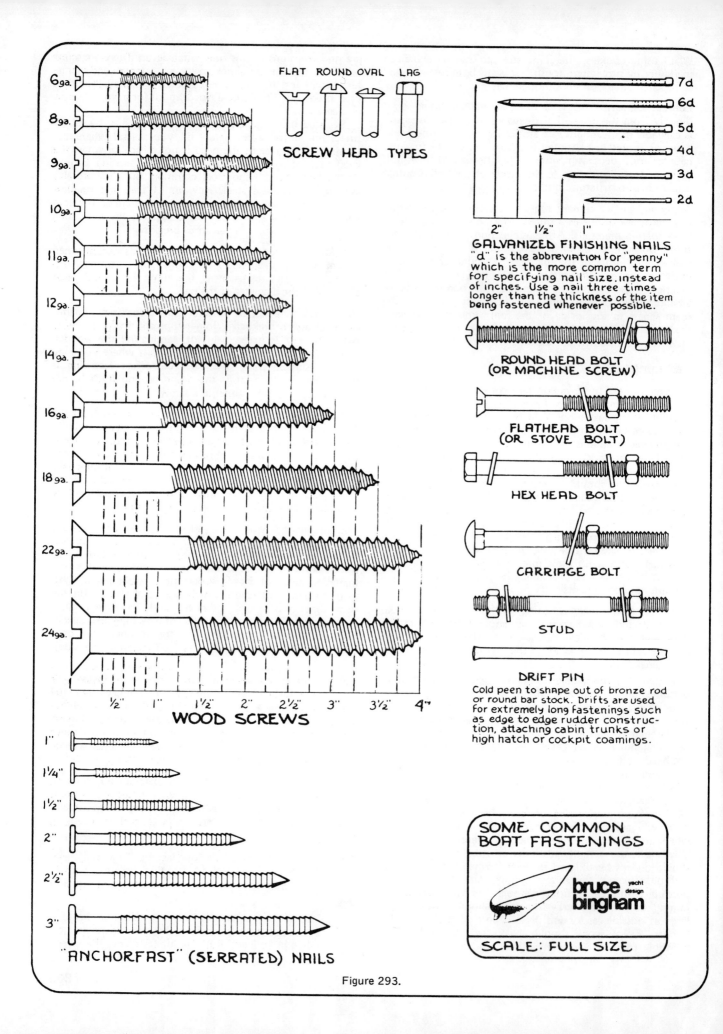

FLAT ROUND OVAL LAG
SCREW HEAD TYPES

WOOD SCREWS

6ga.
8ga.
9ga.
10ga.
11ga.
12ga.
14ga.
16ga.
18ga.
22ga.
24ga.

½" 1" 1½" 2" 2½" 3" 3½" 4"

"ANCHORFAST" (SERRATED) NAILS

1"
1¼"
1½"
2"
2½"
3"

GALVANIZED FINISHING NAILS

7d
6d
5d
4d
3d
2d

2" 1½" 1"

"d" is the abbreviation for "penny" which is the more common term for specifying nail size, instead of inches. Use a nail three times longer than the thickness of the item being fastened whenever possible.

ROUND HEAD BOLT
(OR MACHINE SCREW)

FLATHEAD BOLT
(OR STOVE BOLT)

HEX HEAD BOLT

CARRIAGE BOLT

STUD

DRIFT PIN

Cold peen to shape out of bronze rod or round bar stock. Drifts are used for extremely long fastenings such as edge to edge rudder construction, attaching cabin trunks or high hatch or cockpit coamings.

SOME COMMON
BOAT FASTENINGS

bruce
bingham
yacht
design

SCALE: FULL SIZE

Figure 293.

ment in bronze or stainless fastenings is somewhat questionable unless you are building a "gold plater." (Fig. 293)

Epoxy Attachments

While many builders object strongly to the use of epoxy fillets for attaching interior structures, I must point out that bulkheads and horizontal surfaces are primarily compressive units, not tension units. Rarely are there any forces tending to pull bulkheads inwardly away from the hull and fore and aft forces are almost nonexistent. The primary strains at a bulkhead joint are bending and sheer. In short, the case for overwhelming strength of attachment has been grossly overemphasized and misstated. Unfortunately, the more flagrant errors I have observed in yacht construction have been the result of greatly increased weight from "overbuilding."

I am not suggesting that epoxy fillet attachments are weak because they are not as strong as web attachments. What is intimated, however, is that web attachments are far, far stronger than would ever be necessary (even under the most adverse conditions) and that epoxy fastening is rejected more because it is a misunderstood concept.

What few people realize is that epoxy often exceeds the strength of mechanical fastening. To cite some examples: In California, the highway lane indicators are fastened to the concrete with only a few drops of epoxy and remain in place through years of battering and extreme weather conditions. An important note here is that the highway concrete is not formulated or cured with the precision or care of that used in boatbuilding. In aircraft and race car construction, many metal structural members are actually epoxied together rather than being riveted or welded. The famous Can-Am Champion Chaparrals, designed and built by driver Jim Hall, were entirely epoxy glued together.

In my own experiments with epoxies, I have completely destroyed 2" of white oak before being able to dislodge it from fully cured concrete (perpendicular tension in excess of 800 psi). In another test, I butt-joined two separate ferro-cement panels end to end, then bent the panels until the ferro-cement failed while the epoxy joint remained intact.

Epoxies are formulated for many different purposes but most conform to universal government specifications. It will make little difference as to the manufacturer, so you will be sure to find a reliable source in your locality. When ordering epoxy, tell your dealer exactly what you are using it for so he can judge your requirements. Most epoxies cure into a slightly pliable state and do not become "brittle" hard. This is a beneficial characteristic when attaching units to ferro-cement because the joints tend to "give" a little under bending loads (this movement is almost microscopic).

Epoxies do not require as careful temperature control as polyesters. If the temperature falls drastically during epoxy curing, the rate of curing will slow down but will re-accelerate when the temperatures advance again. In short, epoxy cure will not completely cease (as with polyesters) when subjected to cold conditions. When mixing epoxies, it is necessary to keep this in mind, because curing is caused by a very critical chemical balance. If the balance is altered (too much part A or too much part B), the curing will be seriously affected. You cannot speed up the epoxy cure by adding more of one part or the other; this will only slow the cure rate or stop it completely.

Let us apply the strength characteristics of epoxy to bulkhead attachments. If you were to create a 3" epoxy grout fillet on each side of the bulkhead, you would have a total of 6 square inches of interface per inch of linear attachment. The weakest part of the attachment is expected to be the cement surface with a tensile strength of about 800 psi. Thus the total strength obtainable would be 4,800 lbs. per linear inch of attachment (or 57,600 lbs. per linear foot). Now you can see why I'm so enthusiastic about synthetic materials. If these numbers boggle the mind, remember that sheer strength is several times higher.

Installation Sequence

In most designs, it is imperative that the engine, tanks, exhaust system, mast steps and under-sole piping and wiring are installed prior to the insertion of sole beams. On this point, however, there is one major structural difference between boats. Many architects prefer to extend bulkheads below the sole fully to the bottom of the hull. In this instance, the sole must remain unfinished until after the fastening of bulkheads. (Fig. 294A) If your designer specifies that the bulkheads terminate and are fastened to the sole, however, the beams and sole covering must be placed before the bulkheads. (Fig. 294B) Some designs may call for a combination of these two systems, in which case each circumstance must be handled accordingly.

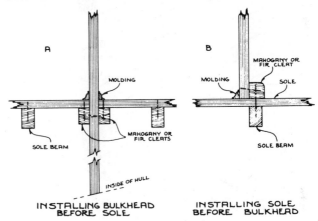

Fig. 294. The variation of installation sequence will depend upon whether bulkheads continue below the sole.

One argument I have always had against cement decks is that when they are in place before the interior installation proceeds (which is always the case), the deck will be a continual obstruction. This is particularly true when attempting to install large bulkheads, as the deck openings may preclude their passing below in one piece. Decking will not only be in the way of swinging large timbers and sheets of plywood but will create communications, ventilation, cleaning and mobility problems. Professional boatbuilders in wood, fiberglass or steel would never consider attaching the decking before completely finishing the interior of the vessel. Some shipwrights even complete most of the painting and varnishing before "boxing" the boat in. If you have already constructed a concrete deck (against my own recommendations), you must accept these severe handicaps.

Halved Joints

Halved joints will be required when two timbers in opposite directions intersect, provided their upper surfaces lie on the same plane. The most notable occurrences are at the intersections of carlins and deck or sole beams. The halved joint requires the cutting of notches in both timbers enabling them to fit snugly into each other to form a flush finished joint. The notches must be very accurately cut to prevent losing significant strength at the joint.

Notching of long timbers or those permanently fastened into the hull are done with the circular saw while short timbers may be done on the table saw. Begin by placing the timbers in their proper positions within the hull (one of the timbers will obviously rest on top of its perpendicular counterparts). With a very sharp pencil, draw lines on all timbers and intersecting timbers. (Fig. 295A) Now adjust

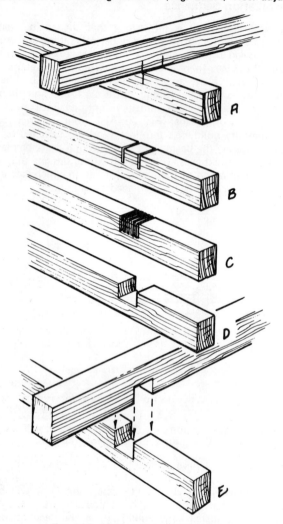

Fig. 295. Making the halved joint.

your saw depth to exactly half of the thickness of one of the timbers. Very carefully saw across the perpendicular timbers at the inside of the pencil marks. (Fig. 295B) Make a series of additional cuts between the first two to "riddle" the areas of the joints. (Fig. 295C) Knock out the riddles, they will leave a rough notch. Without changing the saw blade depth, run the saw slowly back and forth and side to side within the notches. This will clean out and finish notches perfectly. (Fig. 295D)

The notches to be cut in the perpendicular timbers are done exactly in the same way except that the saw blade depth should be set about 1/16" short of the actual required notch depth. After all of the second set of notches have

been cut, place all of the timbers' joints together to test them for accuracy. If the joint notches need more depth to create a flush intersection, readjust the saw and cut a little deeper or until the timbers fit perfectly.

Once fitted, glue the joints with resorcinol or resin glue. Drill, countersink and drive one screw or bolt only into each joint. (Fig. 295E)

The Sole

If plywood, shiplap pine or tongue-and-groove is used, the thickness will inevitably be 3/4". If plywood is chosen, you may construct accurate paper patterns using a variation of the "tick-stick" system (described later in this chapter). If a planked sole is preferred, lay off the hull shape with dividers on each board as you go. In any case, all transverse (plywood) or thwartship joints in the sole must occur on a timber. While sole beams are always well detailed, rarely will a construction drawing show longitudinal timbers except for engine space carlins or, occasionally, bilge access carlins. Longitudinal sole timbers are never specifically engineered to accommodate plywood planking joints so you must provide them in your construction.

As each plywood sheet is cut to final shape, it is laid upon the beam in proper position. If the longitudinal edge is not supported, mark the respective beams and notch the beams for a 1½" x 1½" longitudinal stringer, then glue and screw (or bolt) the stringer into place. Then bevel the upper edge of the plywood sole at the hull for an epoxy fillet attachment. Next, Anchorfast-nail the sole panel into place. If the sole is glued when fastening it will increase the torsional strength of the boat, but for some reason, most builders think they will have to remove the sole some day so they reject this practice without further consideration. (Fig. 296)

At the edge of the sole at the hull, a superior attachment may be constructed using epoxy grout formed into a generous fillet with a quart soda bottle. Occasionally dip the bottle in soapy water to prevent the epoxy from sticking to it. The fillet joint may be fiberglassed with epoxy resin if so desired. Two layers of 18 oz. roving and one layer of 1½ oz. mat should be sufficient to do a good job. (Fig. 297)

Bilge Access Openings

Removable panels must be provided over all tank scuttles, shaft couplings, exhaust mufflers, around mast steps, bilge pump pick-up, battery compartments (if in the bilge) and any spaces to be used for storage. In short, every inch of the bilge should be within a normal arm's reach through a bilge access opening. For the sake of safety, it is important to be able to repair piping, electrical circuits, or even the hull, if necessary. Rarely do designers specifically detail the positions for these openings since they consider it to be a standard construction procedure not needing special mention.

If possible, draw out your sole to accommodate these openings. The strongest access panels rest on sole beams at each end and on 1½" x 1½" carlins along the sides. These carlins should be halved, glued and screwed (or bolted) to the sole beams. When cutting the panel for the opening, undercut on all sides by 1/8", otherwise it will mysteriously jam at a time when the opening is most needed. Be sure to provide a finger hole or flush ring lift hardware in each panel. (Fig. 296)

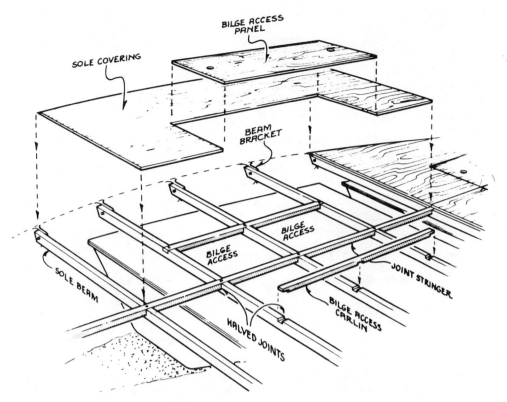

Fig. 296. Sole construction. Notice that longitudinal stringers are positioned under the plywood joints and edges of bilge access openings.

The Breasthook (for Wood Decks)

If the vessel is to have wood decking, it must be provided with a sheer clamp as means for attaching and sealing the deck to the hull. This should be installed before the bulkheads. Because of the drastic tapering of the hull at the stem, however, it is extremely difficult to shape the clamp members accurately enough to provide the necessary strength and seal required. For this reason a separate member, called a breasthook, is used here. It is a triangular block of wood fashioned to fit the inside of the hull perfectly before its attachment.

The breasthook may be solid wood or laminated of birch or oak plywood to sufficient thickness, which is usually slightly deeper than the sheer clamp. Shaping is accomplished with a plane or belt sander. When installing, its upper surface will extend a slight distance above the hull. This will allow its shaping to the deck camber prior to laying the deck.

After the breasthook has been fashioned, it may be installed by troweling the inside of the hull with epoxy grout, then carefully inserting the hook in its proper position. Clamp it securely until cured. Visible gaps between the timber and the hull may be filled later with an epoxy glaze. An epoxy grout fillet should be formed along the underside of the hook and the hull for additional strength. A common water glass or bottle makes an excellent trowel for this purpose. (Fig. 298 & Plate 154)

Note: Shaping the breasthook may require the cutting of a fore-'n-aft slot to accommodate the stem head structure.

Sheer Clamp (for Wood Decks)

The major complaints that ferro-cement proponents have

expressed against wood decking are: (1) it costs too much for the amateur's pocketbook, and (2) it is difficult to create a watertight joint along the sheer. First, I have seen dozens of inverted planked deck molds for ferro-cement that cost almost as much as a plywood deck covering. Second, I will describe a sheer clamp system that has been used by professional custom fiberglass yacht builders for over ten years which yields a strong, positive bond. The glass builder faces essentially the same problems that you do from this point except that different adhesives are used.

In order to insure proper bonding of the epoxy, the sheer clamp area must be of exposed, raw concrete. If the inside of the hull has been sealed, it must be well sanded and vacuumed clean of any cement dust. All deep hollows along the inside of the sheer must be built up fairly with epoxy grout or cement containing a vinyl-latex binder. Fairing in

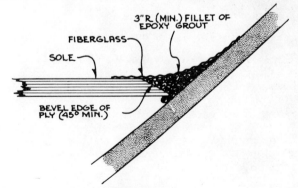

Fig. 297. The connection of the sole to the hull using epoxy grout. Notice that the plywood has been beveled in order to provide for an interface between the grout and all ply layers.

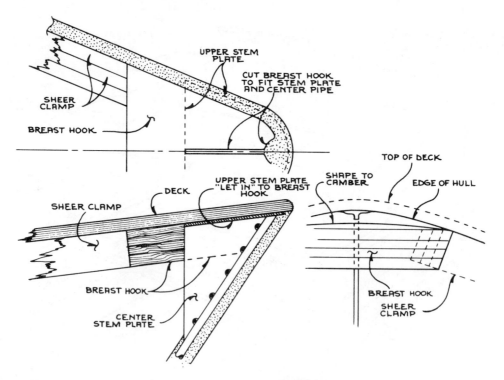

Fig. 298. The breasthook installation. The connection of the plywood deck covering in the horizontal plate area must be accomplished with small through-bolts.

this manner will eliminate most severe air pockets behind the finished sheer clamp timber.

Basically, the finished sheer clamp timber will be a laminate of two (small boats) or three (large boats) of ¾" mahogany, Douglas fir or pine. Each layer will be completely installed before proceeding to the next.

It is easy to bend the sheer planks to the plan view of the sheer but virtually impossible to bend it up and down so the lumber must be "spiled" (cut to curve). To accomplish this, clamp the plank to the inside of the sheer, then scribe a pencil line to the plank from the vessel's edge. (Fig. 299)

Plate 155. Sheer clamp timbers being spiled to the curvature to the hull prior to installation.

Plate 154. A laminated plywood breasthook being installed at the bow of a 20' **Flicka**. It will be shaped to the crown prior to laying the deck covering.

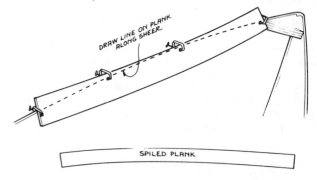

Fig. 299. Spiling a sheer clamp plank to the curvature of the hull.

Plate 156. The first layer of the sheer clamp laminate being sprung and clamped to the hull. Notice its relationship to the breasthook.

Remove the plank and lay off the sheer clamp depth with a pair of dividers. The divider measurement must be perpendicular to the upper curved line. Now the plank may be cut. Sand or plane its edges smooth, then square the plank ends while cutting to at least a 45° bevel where it is to butt against successive clamp planks. All other sheer planks will be shaped in the same manner. (Plate 155)

Mix an epoxy grout (epoxy and sand or ¼" chopped fiberglass strand) and trowel it liberally along the clamp area on the inside of the hull. Position the first plank, clamping it securely in place so that the epoxy oozes from between. Remove the excess epoxy with a putty knife and wipe clean with solvent. You are now ready to spile the next sheer plank.

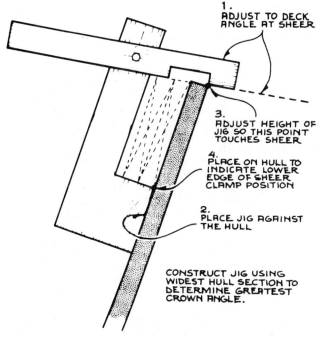

1.
ADJUST TO DECK ANGLE AT SHEER

3.
ADJUST HEIGHT OF JIG SO THIS POINT TOUCHES SHEER

4.
PLACE ON HULL TO INDICATE LOWER EDGE OF SHEER CLAMP POSITION

2.
PLACE JIG AGAINST THE HULL

CONSTRUCT JIG USING WIDEST HULL SECTION TO DETERMINE GREATEST CROWN ANGLE.

Fig. 300. Sheer clamp jig.

Each successive plank, then, is spiled, cut to shape, cut square at the ends, beveled to fit the preceding plank, and epoxied to the hull until the vessel is completely surrounded. (Plate 156) The clamp at the transom is accomplished in exactly the same way. If you are building a double-ender, however, you may have trouble bending the clamp planks to the abrupt curve of the sheer. In this case, you will have to substitute the ¾" thickness for two layers of 3/8".

The second sheer clamp layer is spiled in the same way as the first. Its attachment, however, will be by means of resorcinol or phenolic glue and 1¾" flat head screws or Anchorfast nails. Clamps will not be necessary. Once again, fasten the sheer clamp layer around the entire perimeter of the boat before proceeding to the next layer (if one is planned).

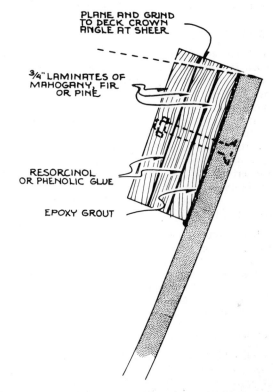

PLANE AND GRIND TO DECK CROWN ANGLE AT SHEER

¾" LAMINATES OF MAHOGANY, FIR OR PINE

RESORCINOL OR PHENOLIC GLUE

EPOXY GROUT

Fig. 301. A laminated sheer clamp completely installed and beveled to receive the plywood decking. While it may be attached with epoxy grout only, the through-bolt is shown as a builder's option. The holes for these through-bolts should be provided in advance by locating waxed wooden dowels in the hull armature.

A third sheer layer is attached in exactly the same manner as the second. **Note:** When positioning the sheer clamp, its vertical position is important, particularly if you plan to use deck beams instead of a laminated deck structure. I would suggest making up an adjustable wooden jig representing the clamp position and deck angle. This will help to pinpoint the correct height. (Fig. 300)

Once the entire sheer clamp has been installed, plane away most of the excess lumber above the hull at the approximate deck angle, leaving about 1/8". Now finish shaping the sheer with very rough Carborundum on a belt sander. This shaping must reflect the deck angle at the sheer and will not only knock down the excess wood but the concrete as well. (Fig. 301)

311

The sheer is now ready to receive the deck beam shelf (if used) or deck laminate, which will not be installed until after the majority of the interior structures.

I will briefly describe the attachment at the sheer now so you will understand the nature of the watertight seal. Later in this chapter section drawings of these finished joints are shown. As most plywood decks are built of two or more layers, it is the first fastening which is most important. Once the plywood sheets are accurately shaped, epoxy is troweled thickly along the upper edge of the hull and clamped. The plywood is immediately placed in position and driven home using Anchorfast nails closely spaced. When the entire deck has been laid, any open portions of the seam may be filled with epoxy glaze (not grout; it is too coarse for penetration into fine cracks). You now have a strong, trouble-free, watertight wood deck connection which will last indefinitely.

Bolting the Sheer Clamp

Some builders are fearful of epoxies and prefer to use mechanical fasteners as well as the epoxy. You may bolt through the hull and sheer clamp after the entire clamp has been installed but drilling ferro-cement is backbreaking and time-consuming. The answer to this is to install waxed wooden dowels in the armature prior to plastering. They are easy to tap out after the hull has cured. If bolting is your intention, countersink the holes for flat head bolts, then drill the sheer clamp from the outside of the hull. Before inserting the bolt, give the hole a liberal shot of bedding compound to insure a watertight seal. The head of the bolt may be covered neatly over with an epoxy or polyester glaze. (Fig. 301)

If you have thoughts about installing sheer clamp bolts within your hull armature before plastering, forget them. While it may sound good at the outset, you would find that (1) it is impossible to line up the drill holes in the sheer planking to the bolts in the hull, and (2) the "negative draw" or reverse angulation of bolt centerlines will not match that of the bolt holes in the lumber.

Beamed vs. Laminated Decks

Even though a beamed deck system usually employs a two-layer plywood covering, it is not considered a true laminate because very little additional strength is achieved over a single layer. Consequently, the deck beams are the prime structural units resisting compression, tension and flexure. Plywood decks laid on a base system are light and strong, particularly if laminated beams are used, but there is no doubt that beaming a hull requires a tremendous amount of time and is costly (costlier than ferro-cement decks). Laminating, attaching and aligning deck beams, in itself, is a challenging task. When completed, a beamed wood deck is visually rewarding, as there is little else inside the vessel quite as nautical. It is true that headroom does suffer somewhat and maintenance on the interior overhead (even if painted only biannually) is literally a pain in the neck.

A laminated deck does utilize a few deck beams but these are usually confined to those of the mast partners, upper edges of bulkheads and the cabin trunk carlins. Because of the general lack of beaming, headroom and maintenance are improved. Laminated deck construction is much simpler than beaming and requires an increase in deck thickness of about 30%. Strength and shape are achieved through the use of, say, six layers of ¼" plywood instead of two layers of ½" ply on beams. The weights of beamed decks and

laminate decks are almost identical so you may substitute a beamed deck construction with that of a laminate without fear of loss of strength or effect on flotation or stability.

The choice is really yours, even though your vessel may or may not have been detailed specifically for beams. No special engineering is required for the conversion, only a knowledge of laminate construction procedure. This will be described later.

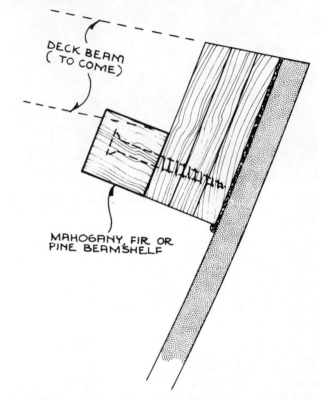

Fig. 302. A beam shelf must be attached to the first sheer clamp in order to provide for the attachment of deck beams if you do not intend to use the laminated decking system.

The Beam Shelf (for Beamed Decks)

If you have chosen the beamed deck system, you must provide a method of attaching the deck beams to the hull. While some builders have gone to the expense and trouble of individually designed and welded steel brackets, the easiest and least expensive method is by bolting the beams to a wooden beam shelf. This is simply a square or rectangular timber which is laminated to the lower portion of the sheer clamp. On small vessels (under 20,000 lbs. displacement), it will measure about 2" x 2" when constructed of fir. On larger vessels (say up to 40,000 lbs. displacement), it will be approximately 2½" x 2½" or 3" x 3". This will be detailed in your construction drawings. (Fig. 302).

Although the beam shelf may require some occasional notching when installing the beams, no special carpentry is needed at this time. The timber is cut to its proper section, scarfed at the ends of successive lengths and glued into its position against the sheer clamp. The horizontal position may be determined by using a jig similar to that shown in Fig. 300 or by holding a mock beam timber in its proper place as the beam shelf is clamped. During or after gluing, the beam shelf may be screw-fastened to the clamp.

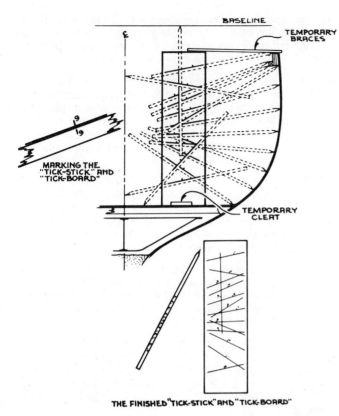

BASELINE

TEMPORARY BRACES

MARKING THE "TICK-STICK" AND "TICK-BOARD"

TEMPORARY CLEAT

THE FINISHED "TICK-STICK" AND "TICK-BOARD"

Fig. 303. Taking the shape off of the inside of the hull using a tick-stick and tick-board.

Pulling Patterns (the "Tick-Stick")

At one time I supplied bulkhead, berth and counter shapes within my full-size lofted pattern prints. I soon discovered that no two builders constructed or faired the hull molds in exactly the same fashion, thus causing slight variations in the hull shapes of the same design. Also, some vessels "sagged" a little more than others, or were slightly distorted during rolling or cradling. Obviously, these distortions and hull variances precluded the use of prelofted patterns so I can no longer recommend their use as accurate templates.

I don't feel that pulling patterns from the hull itself is, in any way, a major undertaking provided you know an old boatbuilding trick. I have seen so many amateurs slaving away with a pair of dividers in order to transfer the hull shape onto a piece of plywood that I have impulsively jumped in to lend them a hand. The system I have shown to hundreds for pattern making is so fast and so accurate that you will hardly believe the simplicity of it (after you have tried it. It is called "tick-sticking."

1. Cut a sharp point at one end of a long section of 1" x 1" pine or fir. This can be made out of any scrap wood you might have laying around. This will be the tick-stick.
2. Cut a sheet of plywood to a convenient size (say 3' x 5', for taking bulkhead patterns). This will be the tick-board.
3. Temporarily erect the tick-board in the position to be occupied by the bulkhead, berth or countertop. Light wooden braces may be used to hold it steady. Be sure it is aligned to the same plane as the structure to be patterned.
4. Hold the tick-stick against the tick-board so that the stick point touches the hull or critical cutting point of a structural timber.

5. Draw a line on the tick-board using the tick-stick as a straightedge. (Use either side of the stick.)
6. Before removing the stick from the tick-board, place a reference "tick" on both the stick and board, placing a small number on each.
7. Move the point of the stick to the next critical cutting point, draw a new line on the tick-board, tick and number the stick and board again.
8. Continue moving the stick to new positions, lining and ticking as you go, until you have ticked off the entire perimeter of the shape to be transferred. The tick-board will be crisscrossed with straight lines bearing the memory of the original tick-stick positions. The more ticks you have marked, the more accurate the pattern will be. If you are pulling a pattern for a bulkhead, you must also carefully transfer the centerline onto the tick-board from points taken from plumb bob string. It is important that you don't forget this. (Fig. 303)

Drawing the "Tick-Pattern"

9. It is best to lay off the pattern on meat wrapping paper first so that you can then adjust its position more strategically within the plywood sheet and avoid wasting a lot of valuable material. Lay the tick-board face up on the paper.
10. Place the tick-stick on the tick-board so that it is aligned with the No. 1 line and tick.
11. Place a pencil dot on the paper at the point of the stick.
12. Move the stick to successive tick positions on the board,

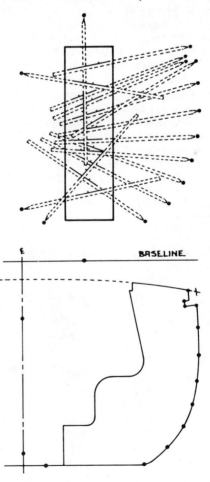

BASELINE

Fig. 304. Transferring the shape of the hull to plywood or a paper pattern using the tick-stick and tick-board.

313

making dots on the paper at the point of the stick as you go, until all of the tick references have been used.

13. Remove the board from the paper. Now connect all dots, using a flexible wood batten (or straightedge, as circumstances dictate).

14. Referring to your construction drawings, draw all remaining detail required to cut the finished piece accurately.

15. Position and transfer the completed pattern onto the wood using a tailor's wheel and chalk bag (described in Patterns, Chapter 9).

The tick-stick may be used for drawing any kind of pattern to the hull shape, whether the unit is large or small, vertical or horizontal, complicated or simple. The degree of accuracy will astound you, as the tolerance should fall within a pencil-line thickness every time. (Fig. 304) Remember, the more ticks used, the more accurate the pattern. Use tick points to pick up all hull irregularities, notches, protrusions and structural members. Don't miss a hitch!

Laminating Bulkheads

There is little likelihood that you will be able to construct your bulkheads out of a single sheet of plywood. The foremost bulkhead (the chain locker) may possibly fit a 4' x 8' sheet if your vessel is rather small. All other bulkheads will probably require laminating ply sheets to accommodate their larger sizes.

On boats over 20,000 lbs. displacement, you will probably be required to construct at least two of the bulkheads out of 1" plywood. Because it is difficult to obtain sizes over ¾" in most lumberyards, unless specially ordered, it may be necessary, once again, to laminate to proper thickness using "half-size" sheets.

After pulling the pattern from the hull and transferring it to meat wrapping paper, lay out sheets of plywood of one-half the finished thickness end to end and edge to edge. Use as many sheets as necessary to cover the bulkhead area, staggering the corner joints in all directions so that they do not form a cross but rather a tee. Lay the paper pattern onto the plywood and very roughly draw out the bulkhead perimeter and major openings. Don't draw too darkly because the lines will be only general guides. Number the plywood panels within the bulkhead perimeter for future refitting together. (Fig. 305A)

Remove the panels from the floor one by one and cut the

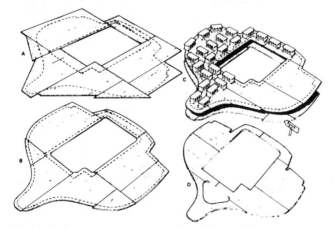

Fig. 305. Laminating a large bulkhead into one continuous piece using two layers of "half-size" plywood.

bulkhead perimeter, leaving about a 4" margin. Now bevel all of the matching edges of the panels to at least 45°; the greater the bevel, the stronger will be the finished joint to bending stresses. All of the bevels must match.

Lay the floor with waxed paper, then replace the panels back in the original relative positions leaving a workable gap between each. Coat the bevels with resorcinol or phenolic glue. Push the panels tightly together and temporarily nail them to each other with wood battens to prevent any possible movement during curing. This completes the first bulkhead layer. (Fig. 305B)

The next layer of the laminate will require arranging the second plywood panels with well staggered joints on top of the first bulkhead layer. Take care to separate the joints of the first and second layers as far apart as possible. Roughly mark the bulkhead perimeter, number the panels, cut to shape and bevel the matching edges.

Now mix a substantial amount of resorcinol or phenolic glue and roll it onto the surface of the first bulkhead layer. One by one, position the second bulkhead layer panels in their respective positions upon the first layer. As this is done, glue the beveled joints liberally.

To ensure a good bond between the bulkhead layers, stack concrete or cinder blocks on top of the laminate until the glue has fully cured (say, 10 hours at 70° F.). Once cured, you may remove the blocks and begin laying out the bulkhead for final cutting and attachment of joiner cleats. (Fig. 305C & D)

Bulkhead Layout (Attaching Joinery Framing)

After transferring the hull shape to the plywood, but before final cutting, it is advisable to draw in all joinery structures while the lumber is still on the floor. It is much easier to accomplish accurate measuring in this way rather than having to continually drop plumb bobs and tape measures from the overhead base line. When laying out these elements, you must completely understand your Accommodation Plan, Construction Plan and Joinery Sections, so study them carefully before proceeding. (Fig. 306) Cut the bulkhead as accurately as possible, allowing enough excess lumber for beveling for the deck angle, etc.

In my opinion, joiner cleats should be attached to the bulkhead prior to raising to the hull, as this will provide a solid surface into which to drive nails. Joiner cleats are usually 1" x 2" (¾" x 1-5/8" actual size). Mahogany or other moderately hard wood is best but extremely expensive. The cleats are precut to accurate length, glued, and nailed in position with Anchorfast nails or galvanized steel common nails. It is not necessary to screw-fasten the joinery cleats. (Fig. 307)

It is essential to understand the method of constructing corners and cabinet edges with the cleat system so that weights on the joinery will be properly distributed. If not so, the plywood joints could separate after lengthy service. All weight must be transferred to the sole or hull in a progressive fashion, i.e., the weight of the counter top and drawers to the front of the counter, then downward. To accomplish this, the cleats under the top surface of the unit just rest on top of the vertical side (or front) cleats.

Attaching Transverse Bulkheads (Web Method)

The most common method of attaching bulkheads to the

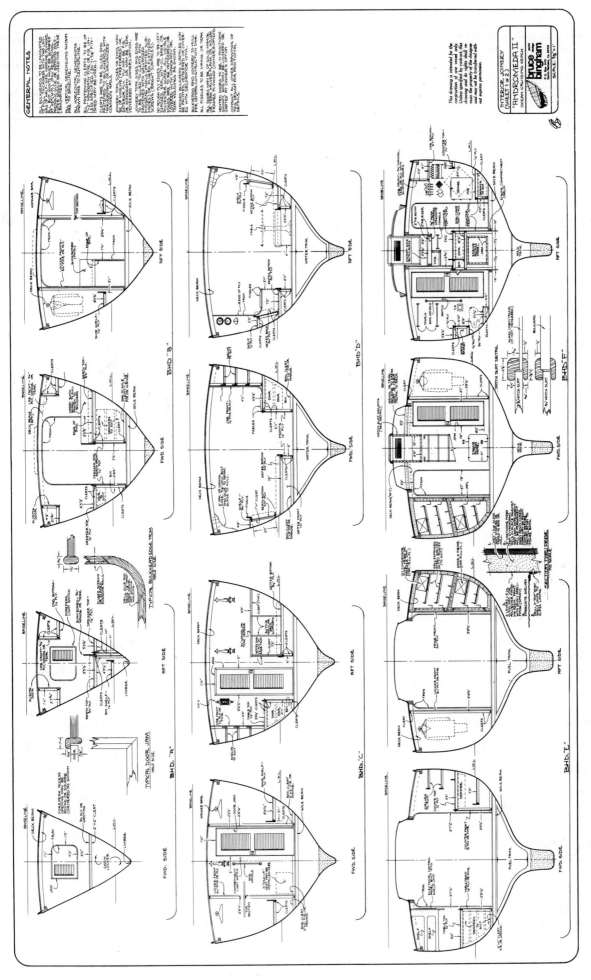

Fig. 306. A typical joinery detail sheet showing the location of all pertinent items at each bulkhead. The joiner cleats, all dimensioned, may be prelocated on the bulkheads prior to their installation. (**Andromeda**, a Bingham design.)

hull is by bolting to a ferro-cement frame or web. The frame or bulkhead web construction has been described in Chapters 14 and 18. If you have chosen this method, you should provide waxed wooden bolthole dowels in the frame armature to eliminate tedious drilling at the time of bulkhead installation.

Before installing the bulkhead, be sure to seal the plywood edges with epoxy or polyester resin, several coats of paint or a thin coat of glaze. This will deter water absorption and possible rot. The bulkhead, once shaped and fitted with joiner cleats, is lifted into the hull, but not placed against the web until the mating surfaces are coated with epoxy. During this positioning, it is imperative that the bulkhead position may be accomplished by inserting small wedges between the hull and bulkhead edge.

entire bulkhead. This is done for several important reasons: (1) all plys of the bulkhead will become exposed to the epoxy grout, not just the outer surface plys; (2) it will provide a wider base for the epoxy fillet at the hull; and (3) it will eliminate any possibility of voids in the epoxy being created between the plywood and hull. After leveling this way, lay out and attach the joinery cleats in the normal manner.

Before inserting the bulkhead, be sure that raw concrete is exposed along the bulkhead position. If you have sealed the interior of the hull, hit the area with a disk sander, then vacuum to remove all cement powder from the concrete pores. Etch the joint area with muriatic acid to help provide a tooth for adhesion of the epoxy fillet. Flush the area with fresh water and allow it to dry thoroughly.

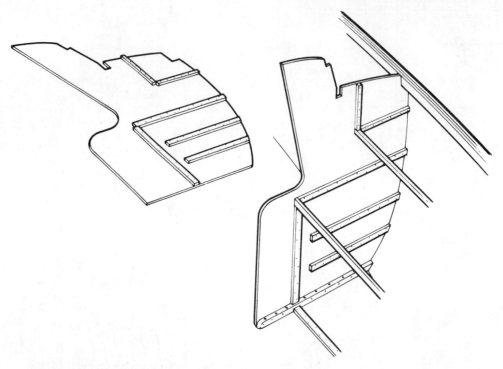

Fig. 307. Attaching joinery cleats prior to the installation of the bulkhead into the hull.

Once the bulkhead position is correct, drill through the bolthole plugs and plywood. Insert the bolt (cut to the exact length required) from the web side, placing a large washer and nut from the plywood side. Draw the bolt just tight enough to cause the epoxy to ooze from under the plywood. Before the epoxy cures, clean the joint with a putty knife and wipe the area with solvent.

If, after drawing up the bolts, a portion of a bolt protrudes beyond the nut, cut or grind it flush and smooth to prevent snagging of clothing in hanging lockers, etc.

There will undoubtedly be a small gap between the edge of the plywood and the hull upon completion of installation. It should be filled with an epoxy grout or filler. (Fig. 308A)

Attaching Transverse Bulkheads (Epoxy Fillet Method)

Once you have accurately shaped the bulkhead, but before attaching joiner cleats, plane or belt-sand the bulkhead perimeter to a long, distinct bevel from each side. The bevels should meet in the form of a pointed edge around the

Raise the bulkhead into the hull and align it accurately to baseline, centerline and bulkhead position. Be sure the bulkhead is absolutely vertical and perpendicular to the center line. Once done, you may insert small wedges into any existing spaces between the bulkhead to prevent movement.

The epoxy to be used must be specifically formulated for wood to concrete attachment. Once its two parts have been mixed, stir in ¼" chopped fiberglass strand and sand or Cab-O-Sil, in equal parts, until the epoxy has become a pliable mush. Trowel this grout liberally into the corners along the bulkhead/hull joint on both sides. The best trowel for finishing the fillet is the bottom edge of a quart soda bottle, as it creates a large, smooth, consistent radius. To prevent the epoxy from sticking to the bottle, occasionally dip it into soapy water. This will aid troweling markably but don't overtrowel the epoxy, as the soap will work its way in and decrease the effectiveness of the bond.

If you wish, you may fiberglass over the fillet with two layers of 1½-ounce mat, using epoxy resin. This will help to stiffen the joint somewhat but is primarily cosmetic. Glassing, however should be done while the fillet is still quite tacky. (Fig. 308B) (Plate 157)

Longitudinal Bulkheads

If the span of a longitudinal bulkhead (not including cabinet fronts) is less than, say 3', it is usually constructed of a single layer of ¾" plywood. The base of the bulkhead is cleat-fastened to the sole with a mahogany or fir timber located on the least visible side.

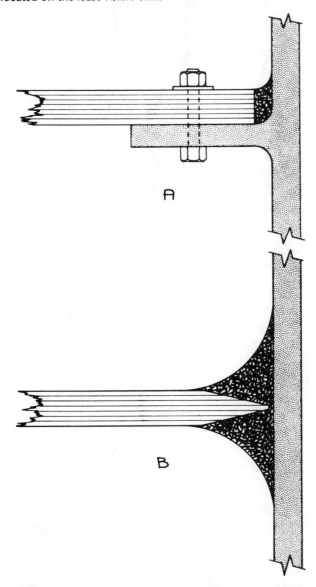

Fig. 308. Attaching bulkheads to the hull using an integral ferro-cement web and bolt (A). Attaching bulkheads to the hull using epoxy grout fillets. Note that the bulkhead edge has been tapered in order to provide an interface of the epoxy grout to all ply layers (B).

If the bulkhead is to span a longer distance, it must be provided with more stiffness. This is done by: (1) laminating the bulkhead into one full section of two layers of ½" ply; (2) fastening 1½" x 1½" mahogany or fir vertical studs on approximately 18"–20" centers to the least visible side of the bulkhead. (Fig. 309A) These studs may be covered with a cosmetic paneling. (Fig. 309B) (3) you may construct the bulkhead using two layers of ½" or 3/8" ply with

1½" x 1½" studding in between on approximately 18" centers. (Fig. 309C) The latter two systems yield a smooth, clean appearance but only the last is suitable for attaching shelves or cabinets directly to the inner, as well as outer, sides of the bulkhead.

There are many ways of attaching the longitudinal bulkheads to the transverse. In most instances, their intersections will form a "T", which poses the easiest construction. If the longitudinal bulkhead is of a single ¾" ply or a laminate of 1", it is simply cleated on the least visible side with a 1½" x 1½" mahogany or timber. If the bulkhead is studded (whether two layers or one), the end studs are attached directly to the bulkhead.

The attachment of all longitudinal bulkheads at the underside of the deck will require cleats in the same way as the base of the bulkhead. The upper end, however, should not be cut to final trim until actually attached (or at least carefully marked to coincide with the transverse bulkhead).

Corner Posts

When two bulkheads (or a cabinet or berth front and a bulkhead) meet at a corner which forms an "L", the joint between the two perpendiculars is easiest to construct by gluing and nailing a 1½" x 1½" mahogany or fir corner cleat. This construction, while certainly strong enough, leaves an exposed edge which looks cheap and unprofessional, even when covered with a molding or bright trim. It's all right for workboats but not fine yachts where the corner post construction of "L" bulkhead corners will give a better appearance.

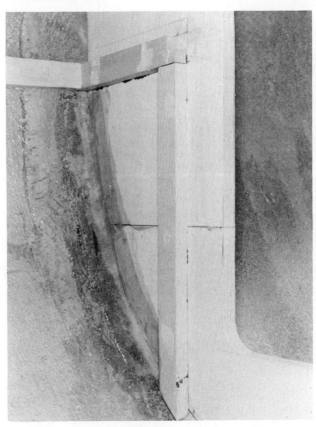

Plate 157. A set of joiner cleats attached to a small bulkhead. The attachment of the bulkhead to the hull has been accomplished by using epoxy grout, fiberglass covered.

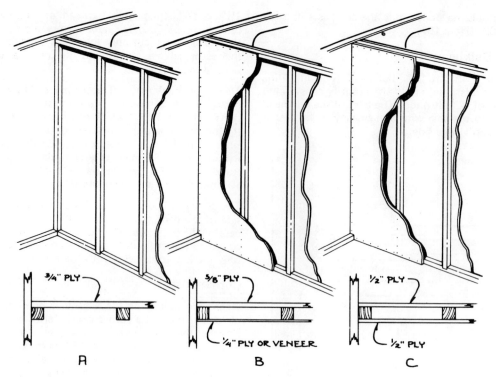

Fig. 309. Longitudinal bulkhead variations. The construction shown above is not intended to be used for small units such as hanging locker faces but will apply to units such as staterooms and other large compartments.

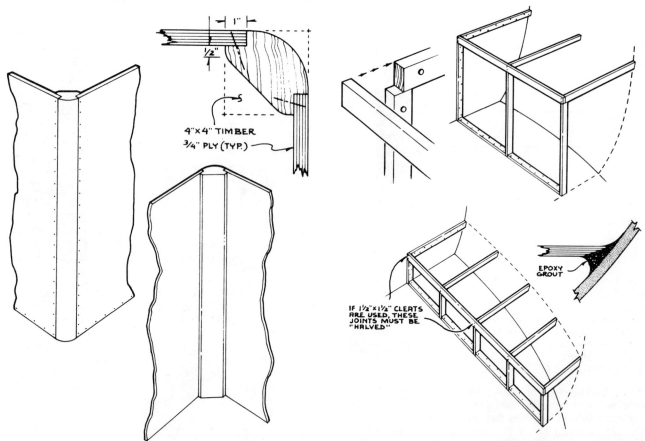

Fig. 310. Although your construction drawings may show cleated corners, the rounded, rabbeted corner post is always preferred as it is stronger and more attractive.

Fig. 311. Typical berth and counter framing showing the use of light transverse beams. The attachment of the plywood tops to the hull is accomplished through the use of epoxy grout.

The corner post must be fashioned from a large, square timber of mahogany, fir, teak, or other. This timber may be laminated to proper section if not readily available in the size required. It generally seems that the larger the corner post the more class is given to the yacht as a whole. I would suggest a minimum size of 3" x 3" but it could be as large as 6" x 6" if your equipment can handle it.

The square timber should be laid out full size on paper first in order to determine the proper saw angles and rabbet positions. Two rabbets will be cut into two adjacent sides of the post. They should be about 1" wide to accommodate the fastenings and glue for the plywood bulkheads. The depth of the rabbet will equal the plywood thickness. After cutting both rabbets, your next cut will be 45° across the inner corner of the post. This cut should not take off so much wood that fastenings will protrude. Leave enough timber behind the rabbet to accommodate 1" of fastening under the plywood. This will leave small exposed square corners on each side of the angular cut, but you can radius these with a router. Now, to finish the outer corner of the post. This will first require "roughing in" a radius with three consecutive saw cuts. Make sure you predetermine these cuts on your full-size pattern first. Once the radius has been roughed in, finish the radius smoothly by plane, sander, then hand.

The finished corner post is attached to the transverse bulkhead first by gluing and fastening the bulkhead edge into the post rabbet. The longitudinal bulkhead will then be fitted and attached to the post in the same manner.

When applying the corner post to joinery, follow the exact same procedure. The dimensions of the post, however, are usually somewhat smaller than found at bulkhead corners.

Whether constructing the corner post for bulkhead corners or joinery, do not cut the length of the post until both surfaces have been attached to it, then cut it flush. (Fig. 310)

Longitudinal Joinery Attachment

Obviously, the longitudinal joiner cleats cannot be attached until the bulkheads have been installed. They will be positioned into the notch provided at the upper corner of the counter or berth unit as well as another fastened to the sole. These cleats are usually 1 x 2s (¾" x 1-5/8" actual) of mahogany, fir or pine. They are nailed and glued to the sole in the same manner as the cleats on the bulkheads. The upper edge cleats, however, are set in the bulkhead cleat notches and glue/screw end-fastened.

In ferro-cement construction, it is not necessary to attach longitudinal cleats along the hull to support berths or counter tops. This support is accomplished through the use of epoxy grout "troweled" into the hull/ply corner using an ordinary water glass as a trowel. Once the grout has fully cured, the plywood attachment will be extremely strong and trouble-free. If you prefer, the joint may be covered with two narrow strips of 1½-ounce fiberglass mat. If possible, position the fillet on the least visible side of the plywood. Of course, a fillet on both ply sides is much stronger but rarely necessary.

Joinery Beams

When constructing a berth or counter top, the cleats attached to the bulkhead will not be sufficient to support the top alone. Intermediate cleats or joiner beams will have

to be installed between the upper longitudinal edge cleat and the hull. These intermediate cleats should be spaced apart by no more than 18" for berths and settees and 24" for long counters. The attachment of these cleats to the hull is by means of epoxy grout while the inboard ends must be glued and end-screwed to the upper edge longitudinal cleat. (Fig. 311)

Joinery Fronts

Upon reviewing your Accommodations Plan, you will clearly see that almost all cabinet, counter and berth fronts are primarily drawer and door openings. If you decide to finish the drawer and door fronts of brightwood, or of the overlapping plywood type, it would be an obvious waste of lumber to construct the fronts out of plywood. After all, the holes cut into these fronts can represent over 80% of the surfaces. The alternative, then, is to build these fronts out of solid pine, fir or mahogany, using "halved" joints throughout. While it will definitely take more time to accomplish, it will cost considerably less than plywood and will be much easier to finish with gloss, semi-gloss paint, or varnish.

Because most joinery fronts are specified as being ½" plywood over cleat framing, your decision to use solid lumber instead depends upon maintaining the ½" requirement. This will usually mean milling down ¾" lumber. To use a thicker timber could add hundreds of pounds to your vessel without any significant reward of strength. "Halving" the joints of cabinet, counter or berth fronts may be accomplished on the table saw or with a router. These joints are fastened only with glue. You may construct the front in position after mounting the attachment cleats, or you may build up the front on the floor. When installing the fronts, all construction dimensions must be maintained.

You will notice on your joinery sections that the hanging locker fronts have been specified for ¾" lumber. This is because they also serve as structural deck stanchions (in the same way as the bulkheads) as well as having to withstand the battering of slamming doors. These fronts should be constructed of ¾" halved lumber or ¾" ply. (Fig. 312)

Counter and Berth Tops

Joinery tops which will require bearing a lot of weight (berths, settees, toilet platform) should always be constructed of ¾" plywood. The counter tops may be of: (1) mahogany or teak veneer, (2) Formica covered or Micarta covered, (3) lightly planked over with brightwood. The latter makes the best looking counter. The covering of the ply should always precede the installation of the top.

The plywood edge of the counter or berth top should always extend over the edge of the front lumber except when a berth front is extended upward to serve as a bunkboard. This will prevent any possibility of a joint separating under weight. Because the top must be fastened to the hull with epoxy grout from underneath, however, it must be glued and nailed (or screwed) to the attachment cleats before the front. To reverse the sequence would make it almost impossible to reach the edge of the top at the hull for troweling the epoxy. Obviously, the counter or berth front must be very carefully pre-fitted and then set aside to ensure a tightly fitting joint.

If there are to be storage access hatches built into the top, they should be constructed prior to installing the top. This will require cutting an accurate, clean hole (the plywood removed will become the cover) then gluing and screwing ¾" x 1-5/8" pine or fir frame around the underside of the hole. The raw ply edges of the hole should be well sealed with glaze and paint or edged with solid wood strips. This frame must protrude beyond the edge of the hole by about ½" in order to support the cover. (Fig. 313) (Plate 158)

Saving the Centerpieces

If you have decided to construct the joinery frame out of plywood, you need not consider discarding the centerpiece which will be removed when cutting door and drawer openings. There are three basic ways to use these valuable pieces.

1. You may construct doors and drawer fronts of the overlapping type. This may be accomplished by adding a ½" x ¾" strip of solid wood around the edge of the plywood drawer front, thus enlarging its size. If the edging is well fit of mahogany or teak with nicely mitered corners, it may be varnished while painting the front within its border. This is one way of adding an exquisite touch of wood to an otherwise starkly painted interior.

2. You may construct all door and drawer fronts so that they will lie flush with the surface of cabinet work. This may be a waste of time because you should "edge" the raw plywood edge as a matter of course, so why not follow my first suggestion, anyway?

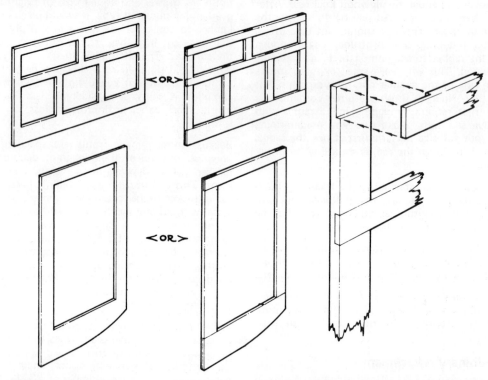

Fig. 312. Joinery faces may be constructed of plywood or a comparable thickness of solid lumber using halved joints throughout. The latter is always more attractive when well executed and considerably less expensive.

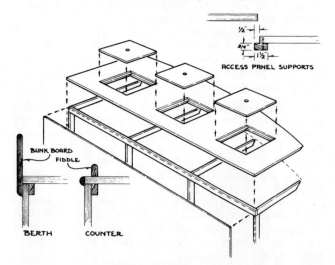

Fig. 313. Applying the top and fronts to a typical cabinet or berth. Notice the jamb which is applied to the underside of the top to support access panels.

Plate 158. A typical berth or counter top storage access opening.

320

3. Set aside the pieces of plywood remaining after the holes were cut to be used later in some other part of the boat. These pieces will far outnumber their practical usefulness and most amateurs create a great amount of waste in this way. (See Drawer Detail, Fig. 332)

Laminated Beams

A laminated beam should be constructed of as many horizontal layers of wood as practical. The more layers you use, the less springback there will be when removed from the laminating form and the stiffer the beams will become. Under no circumstances should you use less than four wood layers; I recommend six.

If you are constructing a constant deck camber, you may construct a beam mold. If you are constructing a variable radius camber deck, each beam will have to be drawn out full size and laminated individually. I'll explain each system.

Building a beam mold begins with the cutting of two identical crown patterns out of 2" lumber. Be sure to mark the crown center lines. Set the patterns on edge about one foot apart and interconnect with 1" x 2" staves spaced approximately 5" from one to the other. The ends of the staves should overhang the crown patterns by about 1½". Make a series of beam clamps, also cut 1" x 2"s to the same length as the crown mold staves. Drill matching holes through the beam clamps and staves to receive long bolts

Plate 159 (left). The interior of a 20' **Flicka** has been completely roughed in and is ready for the cabin trunk construction. You can clearly see berth-top access panels, galley stove counter-top recess, counter-top sink opening, cabinet door opening, sole and generously rounded bulkhead edges. Although **Flicka** is only 20' in length, Laura's height (5'6") emphasizes the vessel's generous accommodations. A Bingham design.

Plate 160 (right). **Flicka's** interior as seen from the foredeck. The opening on the starboard side of the cabin bulkhead is a hanging locker, while the access panel on the center line provides for engine servicing space. The aftermost opening in the galley top is her icebox. She is ready for her cabin trunk installation. A Bingham design.

The best wood for beaming is spruce because it has more bending resistance per pound than any other North American wood. If this is not readily available, you may use fir, but I recommend increasing the beam dimensions by about 20%. Don't use oak for beams, even if you are tempted by your quest for more strength, as oak is far too heavy for beaming and it does not take gluing well. An attractive touch, however, may be the use of alternating layers of spruce and mahogany, finished bright instead of painting. The added weight here will be insignificant.

with wing nuts. This will complete the beam mold. (Fig. 314A) After milling your lumber to proper thickness and say, four and a quarter beam widths wide, apply glue and set the laminates on top of each other while resting centered on the beam mold. (Fig. 314B) Place the center beam clamp over the laminate on the beam mold and draw the wing nuts up tight. Continue placing clamps from the center outward (never from the ends inward). It may help if you have one person at each end of the laminate bending the lumber downward as you place and tighten the clamps.

Leave the laminate on the form until the glue has sufficiently cured. (Fig. 314C) The individual beam width may now be cut on the table saw then planed or sanded for edge smoothness. (Fig. 314D & E) (Plate 163)

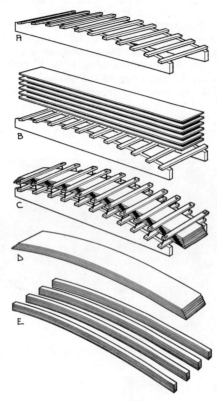

Fig. 314. Laminating deck beams using a constant camber jig. This system may be used only when the deck camber is constant throughout the length of the vessel.

If you are constructing a variable radius camber deck, draw the required beam arc on a sheet of ½" (minimum) plywood and cover the plywood with wax paper. (Fig. 315A) Temporarily nail a series of 2" x 4" blocks about six inches apart onto the plywood along the upper side of the crown line. (Fig. 315B) Mill your beam laminates to proper thickness and width (plus 1/16" for surfacing). Stack the laminates on top of each other, gluing in between. Now quickly flip the laminate on its edge and immediately place "C" clamps along the laminate from the center blocks outward, holding the laminate to the form blocks. (Fig. 315C) It will help if you have one person at each end for springing the laminate as you set the clamps. After the laminate glue is fully cured, remove the beam from the form and surface the edges smoothly. (Fig. 315D)

Cut Beams

Cut beams are not as strong as laminated beams. The Lloyd's, Herreshoff, Nevins and other construction system rules generally allow a significant reduction in beam sizes over cut or solid beams. Cut beams do have their uses, however, so I will explain their construction with the least amount of lumber waste.

Mill the beam lumber to the width of the beam (plus 1/16" for surfacing). The width of the beam lumber should not exceed that of twice the beam width, as it could otherwise lead to warpage. It is important here that the wood grain runs vertically when looking at the end of the beam in its

proper position. The grain, thus, will reduce bending and possible fracture under weight and compression loads. Lay the beam lumber, side down, on a flat surface and glue the planks together into the form of one large laminated board. When gluing, use plenty of pipe clamps. (Fig. 316 A & B)

When the glue has cured fully, trace your constant beam pattern onto the laminated board, progressively moving the pattern upward or downward for successive beams. (Fig. 316C) Now cut the individual beams out of the laminated board and surface them neatly. (Fig. 316D) You should not use this cut beam construction for variable radius camber, as it results in extreme wastage of lumber.

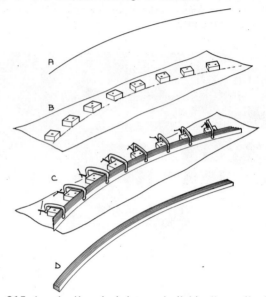

Fig. 315. Laminating deck beams individually on the floor. This system should be used when the vessel has been designed with a variable radius camber.

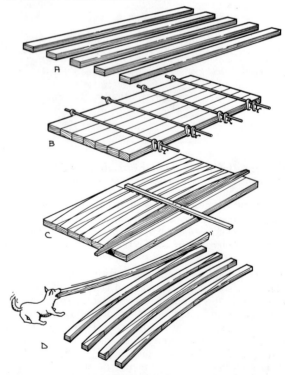

Fig. 316. A method for cutting radius deck beams so as to provide for the least amount of waste timber.

Deck Framing (Beamed Decks Only)

At this juncture you will have installed the major portion of your interior, machinery, sheer clamp, beam shelf (for a beamed deck construction) and breasthook. The time has come to "weather the boat in." With your deck beams shaped and numbered, and your hatch and cabin carlins milled to their respective shapes, you may begin the long but rewarding work ahead.

Every other deck beam is positioned on the sheer of the hull in its proper location. Be sure that they are absolutely perpendicular to the hull's center line. Now mark each beam very carefully to indicate the cut at the beam ends. You will have to account for the flare or tumble home of the hull as well as the bevel of the sheer in its plan view.

Cut the deck beam ends very carefully and set them, again, in their proper positions but resting upon the beam shelves along the sheer clamp. Because the deck beams will slant forward or aft according to the angle at the sheer, it will be necessary to correct this angle to the perpendicular. This is done by "chipping" the beam shelf or by inserting thin hardwood wedges under the beams. The solution chosen is determined by the height of the beam relative to the edge of the hull. If the top of the beam is higher than the hull, chip, if the beam is lower than the hull, wedge.

When chipping or wedging, the accurate shaping for a tight finished joint is very important. Obviously, the greater accuracy attained, the stronger will be the connection of the beam to the shelf and hull. Leave the beams in position once fitted but do not drill just yet. When inserting the wedges, however, they should be glued (one side only) to the shelf.

The longitudinal fairness of the deck crown must be checked and/or corrected before fastening the beams to the hull. This is done by placing a 2" x 4" strongback (small boats) or 4" x 4" (large boats) along the top of the beams at center. It is best if this strongback is scarfed to the full length of the vessel. Clamp each deck beam (including those on bulkheads) to the strongback. Some beams will be a little higher than others, but the bulkhead beams will be the controlling factor. Clamping may cause some of the beams to force the hull outward. This condition is undesirable and will require a slight shortening of the beam ends. If, however, some of the beams do not naturally reach fully up to the strongback, you will have to insert a thin wooden shim between the hull and beam end to help spring the beam upward. When inserting these shims, they should be epoxied (one side only) to the hull.

Now, one by one, remove each beam for gluing, replace onto the shelf, reclamp to the strongback and immediately drill, bolt and draw up tight. It is not necessary to counterbore the lower end of the bolts into the beam shelf but the heads must be sunk at least flush with the upper surface of the beams to allow for the laying of the deck.

Remember that each deck beam must be finally fastened while being simultaneously clamped to the strongback. All other beams must remain in place and firmly clamped to ensure finished fairness. You may now begin fitting and fastening the remaining beams in exactly the same manner as the first.

The half-beams (those which do not extend completely across the vessel) require being clamped to a temporary strongback at their inner ends, much in the same way as the full beams. This will provide for their support while they are being fitted and fastened to the beam shelf. The strongbacks supporting the half-beams should extend a considerable way onto the full deck to ensure proper fairness for the full distance of the carlin. These strongbacks should not be removed, at least until the carlins have been completely fastened.

Once all deck beams have been firmly fastened, the half-breadths of the cabin and hatch carlins must be marked off on the beams. Spring the carlin timber to check these marks for fairness and correct them where necessary. Before removing the carlin timbers, mark them also with each beam position.

Plate 161 (top). Sherry Mason stands outside the entrance to **Andromeda's** massive forepeak which will be a stateroom for her two boys. Bilge access openings, integral chain plates, mast partners and chain locker bulkhead are shown in this photograph. The dark line along the bulkhead and sole edges is a fiberglass tape placed over an epoxy attachment. A Bingham design.

Plate 162 (bottom). Clark Mason reveals **Andromeda's** huge scale as he looks forward into the main salon area from the starboard galley entrance. The port side opening (right) leads into the head, thence to the aft double stateroom. Major structures can be clearly seen including the companionway carlins, glass-covered epoxy attachments and bilge access openings. At this shooting the doors had not been cut to their complete height. A Bingham design.

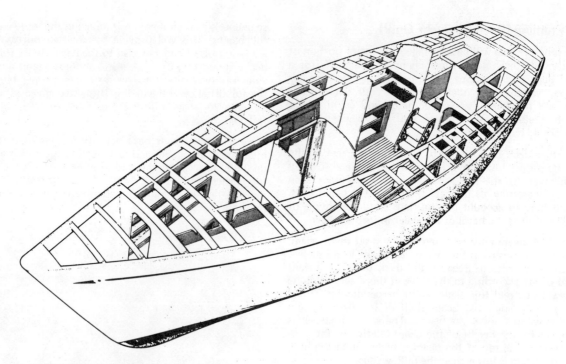

Fig. 317. **Regina's** complete deck framing as seen immediately prior to applying the deck covering. Notice that all major interior components have been installed at this time. A Bingham design.

Cut half joint notches into the beams and carlin timbers as marked. Note that the fore- and aftmost ends of the carlins must rest on top of the deck beam but all other beams will rest on top of the carlin. Do not cut the inboard ends of the half-beams to length at this time.

Respring each carlin timber and place the glued halved joints together, driving one carriage bolt or screw into each. I would suggest counterboring for both ends of these bolts, the lower end being set flush to prevent splitting open someone's head.

When the carlins have been completely fastened, you may cut the half-beams to their finished lengths and flush with the inboard side of the carlin. Before going on to laying the deck covering, you must bevel the upper edge of all frames to the angle of the deck. Because the deck at the center is not the same as that at the sheer, be sure to alter the plane angle accordingly. (Fig. 317) (Plates 164-166-167)

Mast Partners

Whether you are constructing a laminate deck or a beamed deck, it will be necessary to laminate or cut heavy deck beams in way of the mast if the mast passes through the deck (not cabin top). These beams are generally thicker than all other beams and normally about twice the width. They are constructed and installed in the same manner. The partner beams will constitute a pair and are installed in the same manner as all other beams except that they must be notched at the beam shelf in order to be in proper line with the sheer when set.

The partner beams will be interconnected with the partner timbers using mortised joints. The partner timbers will be as wide as the mast itself and half the depth of the partner frame thickness. The distance between the partner timbers should be enough to accommodate the passage of the mast plus a minimum of 2" on each side for hardwood wedges.

The partner construction will be clearly shown on your general construction drawing. (Fig. 318) (Plate 168)

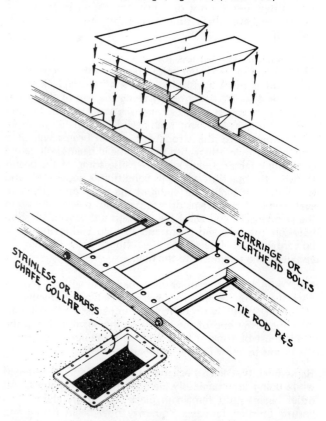

STAINLESS OR BRASS CHAFE COLLAR

CARRIAGE OR FLATHEAD BOLTS

TIE ROD P&S

Fig. 318. Installing mast partners into the deck framing system. Notice the use of longitudinal tie rods to prevent the separation of joints due to the fore and aft surging of the mast.

324

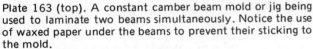

Plate 163 (top). A constant camber beam mold or jig being used to laminate two beams simultaneously. Notice the use of waxed paper under the beams to prevent their sticking to the mold.

Plate 164 (middle). The foredeck framing of a 46' **Andromeda** ketch clearly shows the fairness which can be achieved through this system. Notice the use of knees in way of the companionway opening as well as the strongbacks located at heavy hardware positions. A Bingham design.

Plate 165 (bottom). Solid strongbacks or backing blocks should be "halved" into the beam system at all heavy hardware positions. If your construction drawing does not show the strongback detail, you must refer to your Deck Plan drawing for proper locations.

Plate 166 (top). The half-beams as seen from under the deck of a 20' **Flicka**. Inspection of this photograph will clearly show the sheer clamp laminate, the beam shelf, the cabin trunk carlin, the cabin carlin facing and the deck beam through-bolts.

Plate 167 (middle). The deck framing of a 20' **Flicka** as seen from under the foredeck. Here you can see the breasthook shaped to the deck crown, the sheer clamp laminate, the beam shelf, full decking beams, half-beams in way of the hatch, the hatch carlins and the cabin trunk forward face attachment beam.

Plate 168 (bottom). The mainm'st partners of a 46' **Andromeda** ketch. These have been constructed of solid timber and are halved into the deck beams. Tie rods have not as yet been installed.

Laying the Deck

One of the basic drawbacks of the beamed deck construction is that the beams will rarely fall at the convenient intervals for using plywood sheets full size. Almost every panel to be laid must be cut to a length which will allow the plywood transverse joints to fall directly along the center of the beams. This is not necessary with the longitudinal joints, as they can be temporarily battened until the second ply layer has been fully attached.

As I have just intimated, two plywood layers should be used to cover the deck, not just one. A double layer is considerably stronger because the crown shape is laminated into it and the joints become discontinued and staggered. Just prior to setting the first layer, a strong, watertight seal must be provided. This may be done by positioning two narrow strips of 1.5 oz. fiberglass along the upper edge of hull and sheer clamps, thoroughly saturated with epoxy or polyester resin. An alternative is to trowel the sheer edge with a heavy application of epoxy grout. As the first deck layer is fastened, be sure that someone is stationed inside the hull to clean up the drippings.

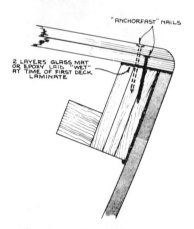

Fig. 319. Attaching the plywood decking along the sheer of the vessel over a framed deck system. Epoxy grout or saturated fiberglass mats provide for a watertight sealing.

Each layer is fastened separately, being nailed to the beams, sheer clamp and carlins with glue in between. Two-inch (2") Anchorfast nails on 5" centers (min.) should be used along the timbers with closely placed galvanized or bronze screws between. The screws should be driven as soon as the second ply layer is positioned to prevent air from being trapped between the plys. To enhance the strength of the deck, all plywood joints should be cut to a 45° (or more) scarf.

If your vessel has been constructed with a bulwark, it will be necessary to pull patterns with a tick-stick to ensure the proper fit of the ply panels along the hull. Prefit all pieces before gluing and fastening. The deck should not be finally trimmed at the cabin and hatch carlins until it has been completely installed. At the edge of the hull (flush sheers only) a ¾" radius should be fashioned for the most professional appearance, as the toe-rail will not set fully outboard. If any gaps become apparent between the hull and deck along the sheer (inside or outside), they must be thoroughly packed with an epoxy grout made of epoxy resin and talcum powder or Cab-O-Sil. Don't use polyester body putty, glazing compound or bedding compound. After

packing, however, the sheer edge may be dressed with polyester glaze and it can also be used to finish the plywood end grain. Do not fiberglass over the deck edge; it is rarely successful. (Fig. 319)

Laminated Deck Construction

After all major structures have been installed in the hull (engine, tanks, sole, bulkheads and the majority of large joinery), the deck is ready to be fabricated. This will first require the cutting of temporary deck crown forms from scrap wood. The development of deck crown patterns has been discussed in Chapter 12 (Figs. 128–130). These forms should be positioned a maximum of 18" apart and attached at each sheer clamp with small wooden cleats. If possible, arrange the forms so that the first deck layer plywood seams fall on form edges in order to support the plywood ends. These particular forms are best cut from 2" x 8"s or 2" x 12"s. All others may be cut from ¾" lumber. Needless to say, the bulkheads which reach up to the deck will also serve as forms, as they will have been prefitted with deck attachment beams.

Now lay off the cabin and/or cockpit carlin half-breadths on the bulkhead forms (if not already cut for the carlin) and the deck crown forms. Once marked, spring a batten to check fairness, then cut notches to receive this timber. The carlin, premilled to shape of clear fir or spruce, is bent and inserted into the carlin notches, halving at the end beams. Cut the fore and aft ends to proper length, then bevel the upper edge to the deck crown angle. If the carlin lies below the edge of the crown forms when inserted into the notches, drive wedges under the timber to force it upward slightly.

Prior to laying the plywood, check the fairness of your deck shape by laying some longitudinal battens across the tops of the form. Raise or lower the forms, as necessary, to evolve a fair line or plane them as the circumstances dictate. (Fig. 320)

Laying plywood onto the deck will be done one sheet at a time. Do not proceed to subsequent sheets until the previous ones have been completely glued and nailed. Now let's begin with the first sheet.

Position the plywood as economically as possible, marking with pencil along the sheer and/or deck openings. The ends of the plywood should rest onto one-half of a deck form at each end, if possible. Cut the plywood to rough shape leaving, say, ½" trim margin at the sheer and openings. Mix an adequate quantity of epoxy bonding resin and brush it liberally onto the sheer clamp. Cut two matching strips of 1.5 oz. fiberglass mat to the width of the hull edge and sheer clamp, then place in the proper position. Apply epoxy resin to the upper surface of the fiberglass mat until it becomes transparent, then add a little more. You want the mat to be quite wet, as some of it will become absorbed into the first layer of plywood. Now place your first plywood sheet in its intended position. Nail the plywood to the sheer clamp, bulkhead beams (if any) and cabin carlin (if applicable), using 1½" Anchorfast nails on 3" centers (max.). Do not use resin, glue or Anchorfast nails at the temporary crown form positions. Here you should use only very small 2d finishing nails.

Continue attaching one sheet after another, successively laying individual wet fiberglass gaskets along the edges of the permanent deck timbers and Anchorfast nailing in place. Remember that only small finishing nails are used at the crown forms. In this way you will complete the first deck laminate. The openings for that cabin, hatches and along the sheer will be left with a rough overhang. This will be trimmed later.

The cutting and positioning of the second and subsequent laminate layers will not depend upon butting together on crown forms, as the entire sheets will be supported by the first ply layer. When cutting the second (or third, etc.) layer panels, stagger their butt joints well away from those underneath. Mix an adequate amount of resorcinol or phenolic

When the laying of all deck laminate has been completed, remove all temporary deck crown forms from the inside of the hull. The protruding nails from the deck underside may be pulled through from underneath with pliers, and their resulting holes glazed over at a later time. You will find, now, that the deck without structural beams is "rock" hard, with little tendency to pant or bounce. Of course, further stiffening will occur when adding the cabin, hatch frames, cockpit coamings, etc. The appearance from the inside will be an amazing vastness of incredible headroom and uncluttered expanse.

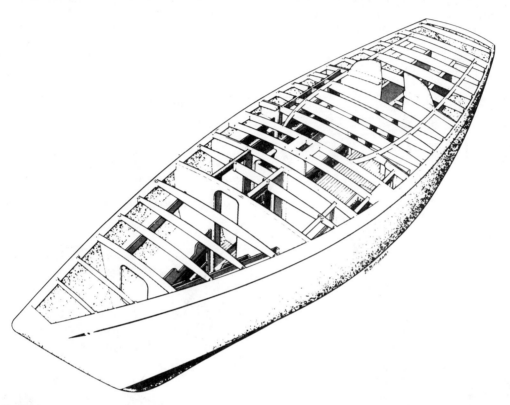

Fig. 320. **Brenda's** temporary deck camber forms in place just prior to the laying of the laminated deck covering. Several permanent deck beams may be seen in way of the partners and along the upper edges of the bulkheads. The temporary forms are completely removed upon the completion of deck lamination. A Bingham design.

glue and roll it onto the previously laid deck laminate in the area of the new layer panel. Immediately position the panel, again Anchorfast nailing along the vessel's permanent timbers, while using small finishing nails along the crown forms. Now use additional fastenings between the deck forms. These will be removed before adding successive ply layers.

The deck will still be quite flexible at this time (it will be very stiff after completion of the second laminate) and nailing will be difficult unless you position someone inside the vessel to hold a brick against the underside of the deck. This will not only aid in driving the nails but will also help in providing a very positive glue bond. Once each panel is fastened, place concrete or cinder blocks on it until the glue has had a chance to cure. Avoid placing so many blocks that the weight would cause distortion. Continue covering the deck with successive panels until completing the entire deck.

The third and all successive deck laminates are attached in almost exactly the same way as the second.

You may now carefully trim the deck along the cabin and cockpit carlins. Locate the exact hatch perimeters and cut accurately then shape the deck at the sheer with a smooth ¾" radius. Your deck is ready for fiberglassing or planking. (Fig. 321)

Fiberglass Cabin Trunks

This section is aimed specifically to my own builders because several of my designs are distinguished by "bubble" cabins, such as the **Andromeda** (Fig. 26) and **Panella** (Fig. 29). These are both basically flush deck boats upon which bubble cabins appear quite low and sleek.

The bubble cabin, however, is very difficult to build out of wood because of its drastic compound shape and accentuated rake of sides and faces. I have worked with so much custom fiberglass that I no longer think fiberglass presents problems, nor need it be avoided by amateur builders. We will not go into the actual details of fiberglassing here, as the details are available in many books in public libraries. I will describe an inexpensive male mold which is now used by many of my builders for constructing fiberglass hulls. The same basic approach may be used for building fuel and water tanks and other complicated shapes.

We are going to use a temporary male mold upon which the fiberglass shape will be fabricated. It will first be necessary to cut out and install cabin forms between the cabin carlins at, say, 2' intervals (see Cabin Forms, this chapter). When cutting the forms, you must allow for the thickness of the cabin trunk (usually about ¼") and the mold thickness (3/8"). The forms may be temporarily cleated into position. (Fig. 322—1) Now cover the forms with 1/8" x 1½" strips of Masonite. These strips do not have to fully enclose the cabin forms but must clearly define the cabin shape. If

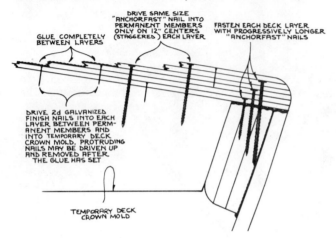

GLUE COMPLETELY BETWEEN LAYERS

DRIVE SAME SIZE "ANCHORFAST" NAIL INTO PERMANENT MEMBERS ONLY ON 12" CENTERS (STAGGERED) EACH LAYER

FASTEN EACH DECK LAYER WITH PROGRESSIVELY LONGER "ANCHORFAST" NAILS

DRIVE 2d GALVANIZED FINISH NAILS INTO EACH LAYER BETWEEN PERMANENT MEMBERS AND INTO TEMPORARY DECK CROWN MOLD. PROTRUDING NAILS MAY BE DRIVEN UP AND REMOVED AFTER THE GLUE HAS SET

TEMPORARY DECK CROWN MOLD

Fig. 321. The building up of a laminated deck covering. The small finishing nails which are driven into each ply while the glue is setting are driven upward and removed before proceeding with each successive plywood layer. Epoxy grout or saturated fiberglass mat is used along the sheer of the vessel to provide a watertight seal.

any unfairness is detected, the forms may be chipped away to reduce bumps or wedged to correct flat spots. The strips should not protrude beyond the inside edge of the carlin or onto the deck surface. (Fig. 322-2) Cover the cabin form with a layer of ordinary window screen using a stapling gun. You may fold, cut, overlap or patch the screen as necessary to create a fair shape. The screen will have nothing to do with the strength of the finished cabin. (Fig. 322-3)

Now lay strips of waxed paper over the cabin form. Tape all of the paper edges with ordinary cellophane tape. (Fig. 322-4) An alternative to waxed paper is troweling the entire form with plaster of Paris to a thickness of about 1/16". You can mix and apply the plaster by hand. Because this type of plaster hardens in only five minutes, you will be able to shape the plaster with a "Sureform" file almost immediately. If there are flat spots in the plaster, you can simply add a little more. Use your own judgment here. When you have completed the shaping, completely cover the form with waxed paper, using common cellophane tape along the paper joints. By the way, the tape won't stick unless the plaster has been wiped free of dust. No open gap should be allowed at the corner of the cabin and deck but all of the plywood must be exposed in order to create a strong joint.

Laying of the fiberglass to specified thickness and finishing proceeds in a normal manner for male molding. (Fig. 322-5) The cabin-to-deck joint, however, is created by allowing progressively wider margins of fiberglass to overlap on the

plywood deck. The width of the finished margin should result in a 2½" joint. (Fig. 322-6) This joint will be finished to a smooth radius using a polyester fairing compound before the fiberglass deck covering is laid on the deck. (Fig. 322-7) With a little patience, the outcome will be strikingly handsome and graceful.

The finishing of the fiberglass trunk will first require troweling the fiberglass with a thin polyester glaze. A creamy substance may be made by adding a small amount of styrene to any ordinary polyester body putty. (Fig. 322-8) Carefully mark and cut the opening for the windows and companionway. (Fig. 322-9) The cabin is now sanded out with carborundum paper. If any divots in the fiberglass remain, you may add a little more fairing compound. (Fig. 322-10) The painting of the fiberglass cabin will proceed in a normal manner except that you must use a base coat specially formulated for fiberglass. (Fig. 322-11)

Fiberglassing the Deck

If you do not intend to plank over your plywood decking (this should not be done unless you have reduced the sub-deck thickness accordingly), fiberglassing will provide the most permanently watertight and maintenance-free surface possible. It is easy to do; however, it would be wise to have someone with experience help you the first time you try it. The basic glassing will only take one day (even on a large

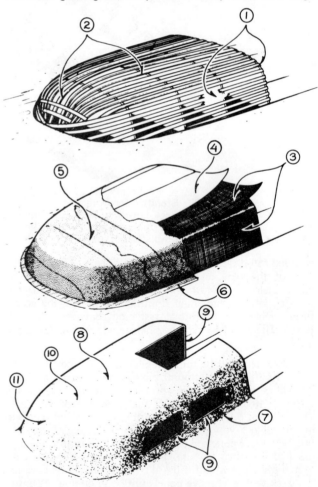

Fig. 322. A simple temporary mold system used for the construction of fiberglass "bubble" cabin trunks. The same system may be used for the construction of custom fiberglass tanks.

boat). But don't call him in until you have purchased the necessary materials and all is in readiness.

Carefully measure the deck area in terms of the length of fiberglass panels you will need. Fiberglass is available in various widths so check with your local supplier before you begin. Personally, I believe glass rolls over 36" wide incur more waste than narrower panels. Unusually wide fiberglass panels are also more difficult to apply quickly. The following glass combinations may be used regardless of the size of deck being covered: 1) one layer of 1.5 oz. fiberglass mat and one layer of 10 oz. fiberglass cloth, or 2) two layers of 1 oz. mat, or 3) two layers of 10 oz. fiberglass cloth. My first recommendation is the mat/cloth combination because of its superior adhesion to the plywood, its resistance to splitting and easily finished surface. The other combinations bear these attributes individually but not jointly.

You will need about 7.5 ounces of "boat" (structural) resin and 1.25 ounces of finishing (surfacing) resin for each square foot of deck. The catalyst will come with the resin in most cases but, if not, buy 1% of catalyst (M.E.K.) per weight of resin. You will need several paint rollers, pans, a half dozen throwaway 4" brushes, about two dozen old (but clean) coffee cans, an Exacto knife, a pair of scissors, a rubber squeegee (the kind used by silk screen printers, available at art stores and fiberglass suppliers) and several gallons of acetone for cleaning tools and yourself.

Pick a day when temperature is between 60° and 70°, if possible, but never below 60°. The warmer it is the faster the resin will "kick off," which might catch you with a wrinkled or bubbled panel. At 60°, the resin will "gel" in about 45 minutes using a 1% catalyst mixture. Your plywood deck must be absolutely dry. To ensure this, allow the sun to beat down on the deck for at least a week before glassing. At night, cover the deck with plastic sheets to prevent dew from forming and soaking in. By all means, don't let it rain directly on the deck. If it does, you **must** increase your drying time. If you short-cut these precautions, you could be fiberglassing moisture into the wood which would eventually disrupt the bond of the glass as well as promote dry rot within the deck laminate. Before starting, make sure the deck is pristine clean.

Many amateur glassers think the resin should be painted onto the surface being covered before laying down the glass. This is generally true when fiberglassing compound surfaces, but fiberglassing a large flat surface must be handled differently. I have found it is much easier to position the fiberglass directly onto the dry wood and then roll on the resin a few feet at a time, saturating as I go. I have fiberglassed at least six large decks in this manner and have found it much faster, easier to "de-wrinkle," and less conducive to bubbling.

Cut the first panel while allowing about a 4" safety margin; position it onto the deck. It may be taped or stapled to prevent it from blowing. Mix a small quantity of structural resin (no more than you can conveniently use in a half-hour). Begin rolling it onto the fiberglass at one end, working only a few feet of the panel at a time. As you proceed, have an extra man gently pull out the wrinkles from the opposite end of the panel. You'll get the knack of this very quickly.

Refrain from applying so much resin that the fiberglass becomes absolutely glossy. When properly saturated, the fiber pattern should be distinctly visible in the reflections, although the glass itself will be absolutely transparent. If the glass appears milky at any point, it may be that it needs a little more resin or that there is a nerd or bubble under the glass. If this is the case, immediately cut a small slit with your Exacto knife to allow the escapement of entrapped air or conduct minor surgery to remove the nerd particle.

When you are satisfied with the laying quality of your first panel, you may begin applying an adjacent panel. Do not overlap any of the joints between panels. The only acceptable joint is a careful butt joint but not so close together that the two panels push against each other. The second and succeeding ones should be laid in exactly the same way as the first.

Trimming the panel edges is very critical and if not done properly, the whole job could be ruined. Do not cut the panel edges with scissors; use an Exacto or other very sharp knife. The timing for cutting is also important and a small section should be tested from time to time in order to determine the proper movement. The fiberglass should have cured slightly (not hard) before putting the knife to it. The glass will not be wet or even tacky but it must be pliable. At the edges of cabin carlins and hatches, run the knife directly down the side while pulling away the excess glass carefully. If your boat has a bulwark, run the point of the knife right along the corner formed by the deck and hull. If it does not have a bulwark, make the cut down the center of the toe-rail position. In this way, the toe-rail will cover the glass edge without fear of it ever pulling away or appearing awkward.

The second layer of glass is laid in exactly the same way as the first but be careful to stagger the joints well apart. Use the same procedure to saturate the glass to the same degree and trim. When the deck has been completely covered with the second glass layer, allow it to cure fully before proceeding to the next step.

The actual surfacing of the fiberglass will be done with the surfacing resin, not the structural resin. The structural stuff will only clog sandpaper and it will always remain somewhat soft. Surfacing resin sands out beautifully and with comparable ease. So let us proceed.

Once again, mix several gallons of resin—this time the surfacing type. Roll it very heavily over the entire deck, then allow it to cure. You may give it a second coat in a few hours. The desired result will be a glossy finish which completely covers the fiberglass fiber texture. After a day or two of complete hardening, you may sand the deck to satin smoothness.

If, when sanding, you wear the resin down to the fiberglass, the deck will require another coat. Sanding should begin with an 80-grit Carborundum. Progressively reduce the coarseness of the paper until you have surfaced with a 120-grit. The deck is now ready for a plastic paint primer, epoxy undercoat, or non-skid covering.

Non-Skid Deck Surfacing

Before applying the non-skid, it is imperative that you locate and draw onto the deck every piece of hardware to be eventually attached. The areas at the bases of the hardware must not receive non-skid, otherwise you will be inviting certain leakage. There is no guarantee that the hardware bedding compound will fully work itself into the rough pattern.

Carefully mask the hardware areas and around the bases of

hatches, toe-rail, cabin trunk or anything else that is to be attached. Develop a simple, logical, consistent and attractive format when masking. Now paint the unmasked areas heavily with structural resin. Immediately throw sand onto the deck so that it is completely covered (leave no thin or bare spots). **Note:** When applying the sand, the resin must be wet, not jelled. Allow the resin to cure completely.

When the resin has hardened, gently vacuum the deck. Try to avoid dislodging the sand which has stuck to the resin. You will find that the remaining sand layer will be quite even and consistently thick. Before removing the masking tape, give the non-skid areas one coat of surfacing resin. When cured, the deck will be ready for cabin trunk, hatches, toe-rail and painting.

Cabin Trunk Types

Here, again, we confront more building options! There are basically four different types of construction which may be used interchangeably regardless of how the designer has de-

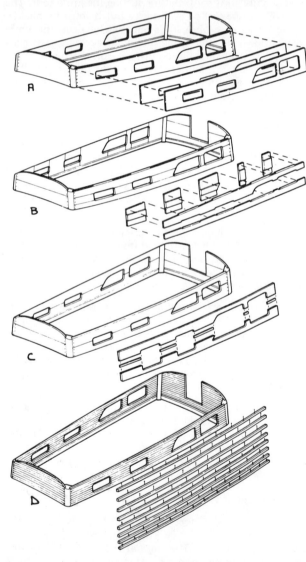

Fig. 323. A) The two-layer plywood cabin trunk; B) The "halved" solid plank cabin trunk; C) The splined or shiplapped solid plank cabin trunk; D) The strip-planked cabin trunk. Notice the use of round mitered corner posts in all of the above examples.

tailed his plans. He could not possibly show every structural variation, so the decision must be made by the builder.

The first (and most popular) is **a two-layer plywood laminate.** Each layer is precut to shape and bent and fastened to the deck over corner posts and temporary plywood forms to ensure proper shape. Between the plywood layers is, of course, a liberal film of glue. This is the easiest type of cabin trunk to build. (Fig. 323A)

Next is the **half-planked** trunk side. This construction is more difficult because of the necessity of having to bend a thicker plank to the proper shape. In essence, the cabin trunk is comprised of upper and lower horizontal planks, half-joined by vertical planks at the cabin ends and between windows. This type of trunk should be confined to the construction of trunks with many large or closely spaced windows. (Fig. 323B)

A **spline-planked** cabin trunk is the most professional but somewhat difficult to execute. It requires the building up of two or more planks to the lumber width required of the trunk by using splined edge joints. These joints are often reinforced with hardwood or brass dowels. (Fig. 323C)

An **edge-planked** or **strip-planked** cabin trunk is the most beautiful and one of the simplest to construct. The individual planks are easily bent to shape and then finished to the contour of the cabin after edge-gluing and nailing all strakes. (Fig. 323D)

On the following pages, each type of cabin construction will be discussed under a separate heading.

Temporary Cabin Forms

Many of the mysteries and complexities can be eliminated from cabin building by following this time-proven method employed by numerous fine professional custom yacht yards. I found out the hard way after simply trying to save myself a little bit of extra work!

Regardless of the type of cabin construction or deck attachment you have decided to use, you will find it very helpful if you install temporary cabin forms spanning the cabin carlins at, say, 2' intervals. These forms can be cut out of some dirty old plywood or built up from used lumber. Their shape, of course, will represent the inside of the cabin.

The cabin forms must first reflect the proper angle along the sides. The height of the cabin corner may be determined from your cabin pattern drawing, outboard profile, or sail plan (if it is a large-scale drawing). All form sides must be of equal angles. If any of the major bulkheads protrude from the cabin side, the form angles must match these bulkheads. The proper cabin crown should also be cut into the cabin forms after locating the upper cabin corner.

Once you have installed the cabin forms, check all surfaces for fairness with a wooden batten. If a bump along the batten is revealed, remove the respective form and reshape it slightly, then replace it and check for fairness again. If, on the other hand, a dip is found along the batten, simply build up the form slightly by gluing chips of wood to the form. Be sure to develop the cabin crown very carefully in this way, as it will definitely ease the problems of construction later on.

If you plan to have beams in the cabin, notches must be cut in each form at the upper corner in order to receive the beam shelf (or bearer). Be sure, at this time, that this corner line is very fair and make whatever adjustments are necessary to create a "sweet" curve.

Once the forms have been set up, you will find that it is easy to make an accurate cabin trunk template with a "tick-board" bent to the cabin side plane. This will also facilitate locating the corner post rabbets. By laying several battens along the crown, you will be able to fit or cut the fore and aft cabin faces to finished accurately, if this has not been done previously. Remember that all bulkheads which extend into the cabin trunk must be considered as the controlling factors for correcting the cabin form shapes. (Fig. 324)

Cabin Trunk Corners

Regardless of the type of cabin trunk you are building, the corner construction must be your second consideration before actually proceeding to build. Of prime importance here are watertightness, strength and appearance. The best

rounded at the outside corner. Normally the bases of the corner posts must be cut away so they will fit snugly into the carlin corners, while their upper portions lap over onto the deck to the exact thickness of the planking. If your vessel has cement decks, you will face a considerably different problem.

With cement decks, it is possible to install a carlin under the deck using epoxy and through-bolts. With this construction, the cabin corner post is installed in exactly the same way as with a wooden deck. If you intend to attach the cabin trunk by through-bolting into a cement carlin flange or by screwing into an upper carlin batten, the corner post will be cut in the same way as before but through-bolted into the flange or carlin batten. (Fig. 324)

Bulkhead Cabin Frames

Before actually beginning the construction of the cabin trunk, it will be necessary to provide for the attachment of the cabin sides to those bulkheads which protrude above

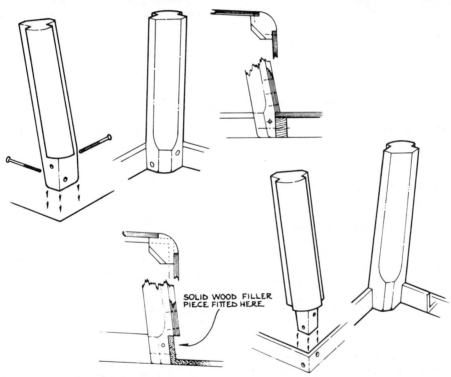

SOLID WOOD FILLER
PIECE FITTED HERE

Fig. 324. Typical cabin corner-post construction and installation.

of these corners are built with rabbeted corner posts which are installed prior to attaching any cabin side planking. They are always well glued and thoroughly bolted into position.

I heartily disapprove of cleated or battened cabin corners because they require the exposing of cabin trunk plank end grain unless an outer corner covering batten is used. Mitered cabin corners never stay tight for long and are rather weak.

The posts are cut and constructed in much the same manner as those for the interior of the vessel. They will usually be slanted in two different directions, however, which complicates their shape as well as making it necessary for them to be cut to a parallelogram (in section) in order to match the angles of the cabin side and front. The corner posts will be rabbeted to receive the cabin trunk planking, then

the main deck and touch the cabin side. This attachment will be by means of frames fastened along the edge of the upper portion of the bulkhead.

These frames (or cabin cleats) will be cut from mahogany or fir to 1½" x 1½". They will be glued to the least visible side of the bulkhead but screwed from the opposite plywood to provide for a deeper bury of the fastening. After the glue has set, the edge of the bulkhead and frame must be beveled accurately to the lay of the cabin side. Trim the bulkhead frame flush with the top of the bulkhead.

The attachment of the cabin top must also be provided in a similar manner. This will require the laminating or cutting of a beam which, again, will be fastened to the least visible side of the bulkhead. If fitted, these beams should be of the same dimensions as all other cabin beams. In some vessels the bulkhead is actually cut to bridge over doors or passageways. In this case, the bulkhead cabin beam traverses the entire width of the cabin. If, on the other hand, the bulkhead is installed as separate pieces (one port, one starboard), the beam may still traverse the full cabin width or it may be discontinued between the two bulkhead sections. Of course, if the bulkhead is located only on one side of the vessel, the bulkhead cabin beam will continue only as far as the inner edge of the bulkhead. (Fig. 325).

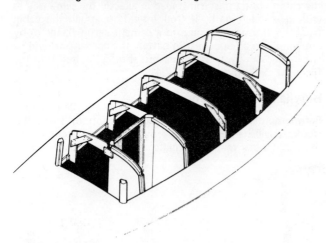

Fig. 325. Temporary cabin forms can eliminate much of the mystery of cabin trunk construction. These forms may be corrected for fairness prior to actually laying of side planks or cabin top.

Cabin Faces

Frequently the aft end of the cabin trunk is simply an upward extension of a major bulkhead. If the cabin face is vertical, no unique problems exist if the trunk is to be painted. If you are constructing the trunk with mahogany or teak, it will be necessary to veneer the cabin portion of the bulkhead or to cut the bulkhead flush with the main deck camber to allow for its continuance with the different lumber. The joint, here, may be executed with a 45° scarf reinforced with a heavy butt batten (say, 1" x 5" cut to the shape of the crown) or by building up the deck beam thickness as an attaching member (if a deck beam happens to fall at the bulkhead position).

If the aft end of the cabin is slanted forward above the deck but continues vertically downward into the hull as a bulkhead, the joint between the upper and lower sections must be made by shaping a transverse beam or butt batten so that the two angles are accommodated. If a deck beam falls along the forward side of the cabin face, a beam doubler may be attached to the beam in way of the cabin trunk to allow the outboard edges to fit snugly into the corner-post rabbets.

A beam must be formed to fit the inside upper edge of the rounded cabin face. This beam will be curved in two directions, i.e., the cabin crown and cabin face radius. To construct this beam, laminate several layers of solid wood over the cabin face radius form. The lumber used must be wide enough to accommodate the cabin camber as well as the beam depth. When the beam laminate has cured, set it inside the curved cabin face and scribe the proper crown, then draw a second line representing the beam depth. Cut the radiused beam at the bevel which corresponds to the angle of the cabin top at the face. After surfacing, the beam

may be fastened to the cabin face using glue and screws driven from the forward side of the cabin face. The outboard ends of the beams must be fashioned to provide a clean, neat joint at the corner posts.

Before attaching the corner posts or cabin face, check the accuracy of the side angles and crown by temporarily positioning the cabin face and springing the batten around the cabin forms. Make whatever corrections are necessary for fairness for the attachment of the cabin face. This doubler must be cut along its aft side to the exact angle of the slanting cabin face.

Regardless of the configuration of the aft cabin face, it must be cut so as to accommodate the cabin corner posts on each side. It will also be necessary to fit a laminated beam or cut beam along the upper edge of the cabin face to provide for fastening the cabin top. Cut this beam flush with the edge of the companionway opening, then bevel the face and beam to the exact angle of the cabin top.

The forward cabin face, if vertical or slanted slightly flat across its surface, should pose no special construction problems. The flat cabin face should employ the same type of construction as the cabin sides, i.e., plywood, spline-planked, half-planked (if large windows are being installed) or strip-planked. These methods are described separately on following pages. In any event, the template for shaping the face must be drawn to full size from the pattern shown in your Cabin Construction drawing. As with the construction, the cabin face will employ the same method of attachment as the cabin sides. The top of the cabin face must be fitted with a laminated or cut beam to provide for fastening the cabin top.

If you are constructing a curved forward cabin face, you will be required to laminate the face over a bending form, much in the same manner as in laminating deck beams. You may use three or four ¼" plywood layers for laminating the shape or a sandwich of 1/8" mahogany or teak veneers. After the glue has fully cured, the expanded cabin face template is layed on the wood for marking the finished shape. When cutting, make allowance for the required bevels.

Plywood Cabin Sides

The strongest plywood cabin sides are constructed of two "half" layers rather than only one heavy thickness. To enhance the trunk strength, the laminating of the two layers may be accomplished in position over the cabin trunk forms attached to the carlins. In this way, the strain of the bend will not be nearly as severe on the plywood. You may, however, elect to do your laminating on the floor, where it is much easier to accomplish.

A pattern must first be made by enlarging to full size the small-scale template shown in your Cabin Construction plan or by employing the tick-stick pattern method. In the latter case, the tick-board must be sprung around the temporary cabin forms in order to correctly reflect the slope of the cabin sides.

Once the pattern has been drawn on meat wrapping paper, it must be transferred to the first plywood sheets which are positioned end-to-end with 45° (min.) scarf joints. A second plywood layer is also positioned and drawn to the pattern shape but you must take care to ensure that the joints of the ply sheets are no closer than one foot from the joints of the first layer.

If you plan on laminating the cabin sides in position, the plywood joints of each layer must be temporarily held together with butt blocks (about 6" wide). Locate the first ply layer on the cabin form and clamp it in position. You may trim the ends of the plywood to fit into the corner posts at this time. You may also correct any slight fitting problems which may exist at the cabin deck joint. Now raise the second ply layer and fit it in the same way as the first. Roll glue onto the surface of the first ply layer and immediately clamp the second layer to it. While the glue is still setting, drill, countersink and drive small screws in neat rows. The screws will ensure that no air remains entrapped between the ply layers. Of course, you will plug these screws later on. Do not fasten the sides permanently to the deck or posts at this time.

When the glue and screws have been set, you may remove the curved cabin side from the forms, mark and cut the windows, and carefully bevel the lower edge of the cabin side to the proper angle of the deck. Fastening of the trunk side may take many different forms (see Fig. 326).

As mentioned earlier, the two-layer cabin sides may be laminated on the floor, if you wish. This is easier than joining on the cabin forms. You will not have to screw the layers together but place weights on the laminate in the same manner as shown in Fig. 305. When the glue has completely set, accurately spring the side around the cabin forms and fit snugly to the corner post rabbets. Correct any errors at the deck/cabin joint and bevel the edge to the crown. The fastening of the cabin to the deck is shown in Fig. 326.

Spline-planked Cabin Sides

Most of the "good old boats" have cabin trunks constructed in this way. The reason for this is that plywood was not readily available to the public until after World War II, so solid planking was the only alternative. These cabin trunks are very strong but somewhat difficult to construct because of the accuracy of the joint required between the planks and the necessity for drilling many very long, straight bolt-holes. If you intend to varnish your cabin trunk to expose the beauty of the wood and your work, the spline-planked cabin will be much more effective than a plywood or veneered one.

First, a pattern must be made by enlarging to full size the small-scale template shown in your Cabin Construction plan or by employing the tick-stick pattern method. In the latter case the tick-board must be sprung around the temporary cabin forms in order to correctly reflect the slope of the cabin sides.

The planking must be milled to the proper thickness and cut to widths which will slightly more than accommodate the cabin trunk heights when the planks are set side by side. The joint edges of the planks must be very accurately surfaced so that no gap, whatsoever, will appear when the planks are finally fastened together.

The spline is a small, rectangular section of wood which is to be inserted into slots cut into the edges of the two mating planks. The slots are cut by placing the planks on edge and running them through the table saw several times until the exact spline width has been achieved. The spline slots should be about ½" deep and ¼" wide in each plank and their positions must be absolutely dead on the center on the plank edge. The spline to fit these slots would then be 1" by ¼" and should be milled to the accuracy that will

just allow a "press" fit. The spline should not have to be driven into the slot nor should it be loose.

Carefully join the planks by applying glue to the plank edges and into the slot for the spline; insert the spline into the edge of the first plank, then immediately press the second plank onto the first. Use pipe clamps at close intervals to ensure that the planked joint has been drawn up tightly. After the glue has completely set, transfer the cabin pattern to the joined planks. Accurately cut the upper and lower edge of the cabin side but leave at least 1/8" margin at the ends to allow for possible minor corrections. Bevel the lower edge of the cabin side to the deck crown and cut out the window shapes. Now raise the cabin side to the deck and bend it around the temporary cabin forms. Mark closely and accurately the positions of the rabbets in the fore and aft corners posts, as well as any necessary modifications required along the deck joint. When this fitting is completed, your cabin side is ready for installation. Its attachment to the deck is shown in Fig. 326.

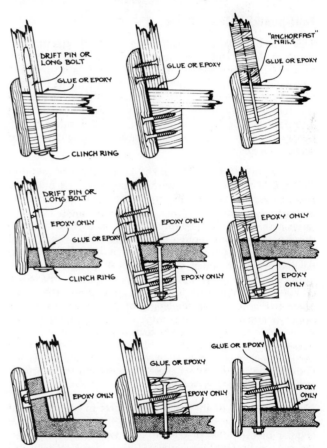

Fig. 326. Cabin side attachment variations.

A modification to the foregoing procedure is that of first cutting the planks to shape, fashioning the slot to receive the spline, then beveling the bottom edge of the lower cabin side plank. This lower plank is then accurately drilled at close intervals to receive long, bronze bolts or driftpins. The lower cabin side plank is raised and bent to the shape of the temporary cabin forms, fitting it snugly into the corner post rabbets. The plank is then clamped securely in its position to the deck. The predrilled holes are continued through the deck (and cabin carlins, if applicable). Once drilled, the plank is removed and epoxy or glue is liberally applied along the area of the deck joint. The plank is carefully repositioned onto the deck, taking care to align the bolt-holes properly. The bolts are driven down through the holes from the top of the plank and then drawn up tightly from the underside of the carlin. (Note here that the upper heads of the bolts must be counterbored into the top of the plank edge to a depth which will not interfere with the placement of the spline. These bolt heads, so driven, must be thoroughly covered with white lead or a plastic glazing compound to eliminate any possibility of rot.) The upper edge of this plank is now glued, then the spline is inserted into the slot and the upper plank immediately clamped firmly to the first.

Prior to fastening both the upper and lower cabin side planks, they must be accurately fitted to the deck, the corner post rabbets, and to each other. At the time of bolting to the deck or of clamping the planks to each other, the forward and aft ends must be glued and screwed to the corner post.

Half-planked Cabin Sides

This type of construction is the most difficult type of cabin trunk to build. The halved joints in the relatively wide timbers must be formed with hairline accuracy and the many pieces required must be carefully laid out and cut to shape prior to their joining. When properly executed, it yields a fine looking cabin trunk, particularly when varnished, but if you have any problems with fitting, the cabin side can turn out to be a complete mess, incapable of a bright finish of acceptable quality.

Construction of the half-planked cabin side first requires that the accurate full-size patterns be drawn by enlarging the small-scale template shown on the Cabin Construction plan or by employing the tick-stick pattern method. In the latter case, the tick-board must be sprung around the temporary cabin forms in order to correctly reflect the slope of the cabin sides. Once the perimeter of the cabin side has been drawn, you must superimpose all windows in their correct positions. By doing this, you will be essentially dividing the cabin side into three basic parts: the upper plank (which will span the length of the cabin between the upper window line and the edge of the cabin trunk), the lower cabin plank (which will span the length of the cabin between the lower window line and the deck), and the vertical or angular planks between the windows and at the ends of the cabin. All of these parts should be drawn directly on your pattern for clarity.

Now transfer all of the cabin plank parts to the surface of the lumber to be used, remembering that the vertical and angular pieces between the windows will completely overlap on the horizontal members. Don't forget to take this additional length into account. While drawing these patterns, also clearly indicate all edges of all half joints on all of the planks. When choosing the plank to be used for fabricating the cabin side parts, you must also take into account any radii employed at window corners.

Carefully cut each of the cabin parts and lay them into their proper positions on the cabin side pattern. At the same time, place the vertical planking in the positions they will occupy in the finished work. If there are any inaccuracies in the marking of the halved joints, they must be corrected now. Halving the cabin timbers may be accomplished on your table saw or with a router, but do not attempt to cut the entire joint depth on the first pass. If

you inadvertently overcut the depth of the joint, there will be no correcting of this error and your valuable lumber will be lost. It is advisable to cut just slightly less than the joint depth on the first pass, then test the joints for accuracy. Any corrections required, then, may be made a little at a time; this is a surer procedure.

When all of the joints have been fitted snugly, apply glue to all of the joints, then position the halved counterparts together. Use long pipe clamps across the cabin width to pull the joints tightly together while placing cinder blocks over the rest of the cabin side as an additional precaution against loose fitting. When you are sure that the glue has completely set, spring the cabin side around the temporary cabin forms and mark the trimming required for a snug fit into the cabin corner posts. Also indicate any corrections that are necessary along the deck/cabin joint. This will complete the basic construction of the half-planked cabin side. Fastening of the cabin side to the deck is shown in Fig. 326.

Strip-planked Cabin Sides

As stated earlier, I believe the strip-planked cabin side is somewhat easier for the amateur to construct than the half-planked or spline-planked cabin side trunk. Although there are many more pieces of wood to be handled, this type of construction does not require the purchasing of large timbers of the continuous length of quality that is necessary with the other cabin types. Also, the stripping up of a cabin side does not entail the critical accuracy of joints as with the others, except for the positions at the ends of the strakes where they fit into the cabin post rabbet. Another benefit to be considered is that the stripped cabin side normally employs a relatively thinner lumber than the other two planked cabin options.

The construction of the strip-planked cabin side will require drawing an accurate full-sized cabin side pattern. This may be enlarged from the small-scale template shown on your Construction Plan or by employing the tick-stick pattern method. In this case, the tick-board must be sprung around the temporary cabin forms in order to correctly reflect the slope of the cabin sides. After the perimeter has been drawn out, all windows must be superimposed in their proper position. The cabin side is not actually constructed on top of this pattern but it will be used primarily as a guide for locating windows as the cabin side is formed in its actual position. Of course, the pattern will also be used when scribing the upper cabin edge prior to final cutting.

The lumber to be used in the strip-planked cabin side must be milled to the proper thickness, then cut into long strakes of square (or slightly more) sections. These strakes do not have to extend the full length of the cabin, since it may be difficult to purchase stock of these dimensions. If the lumber is not full cabin length, you may scarf the lumber (1 to 10) to produce the necessary pieces. When your lumber is well matched in tone and color, these scarfs will be quite handsome if handled carefully.

The first strake is the most critical, as its lower edge must be beveled to reflect the angle of the cabin side at the deck.

If you have built a ferro-cement deck, the deck must be predrilled in order to receive closely spaced through-bolts. This may also be done on a wooden deck, although you will see that it is not necessary. After the deck has been drilled (ferro-cement), the strake is bent to shape, trimmed at the

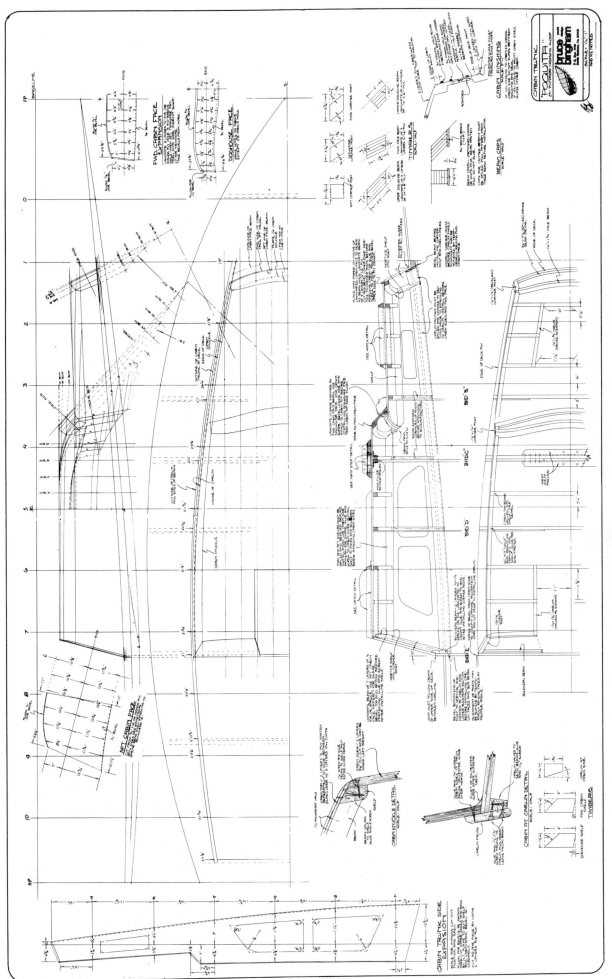

Fig. 327. **Poquita's** complete cabin trunk lines development and construction drawing shows the methods for expanding curved faces and cabin sides as well as all structural components and attachments of the cabin trunk to the deck carlins. In this drawing take care to notice the beveled beam used for constructing the knuckle by bulkhead "E" where it transcends into the slanted cabin end. A Bingham design.

ends to fit the corner post, clamped into position, then the bolt-holes are drilled into the strake from under the deck. Remove the strake; lay a film of epoxy along the joint area. Place the plank back into its proper position and bolt it firmly home.

If you have constructed a wooden deck on your vessel, the first strake laid will not require bolts unless you want them. The strake is carefully cut to fit the rabbets of the cabin corner posts. A layer of epoxy (fiberglass) or glue (over bare wood) is applied along the strake joint. The strake is then sprung into position, firmly clamped, then fastened to the deck and carlin using, say, 2¼" or 2½" Anchorfast nails. You may drill and screw the first strake but the fastening will not be any stronger.

After the first strake has been attached, all successive strakes are attached, one on top of the other, in exactly the same fashion. Obviously, it will not be necessary to bevel any of the additional strakes. Their cuts may be square on all sides. Between the successive strakes you may use glue or epoxy. You may also elect to use galvanized finishing nails instead of Anchorfast, unless you specifically order small gauge nails. Otherwise you could risk the possibility of splitting the planking. When nailing, do not be too concerned about staggering your fastening, as it will be apt to occur automatically.

Now we consider the use of the cabin side pattern. As the cabin side takes on its vertical shape, frequently superimpose the cabin pattern. This is to prevent driving fastenings into those areas which will require cutting through (as necessary) around the window positions. Your planking will continue upward a short distance past the upper edge of the cabin trunk. The last few strakes should not be nailed— simply clamped, so that you will not have to pass your saw through these fastenings. After the upper cabin edge has been cut, nails may be driven and counterset to allow for the cabin crown bevel. When the glue has completely set throughout, the windows may be marked off, cut, surfaced, and rabbeted. To achieve a smooth finish inside and out, sand the sides of the cabin trunk with a blocking board or carefully with a belt sander, using a relatively fine grit paper. Refrain from doing this if you feel there is a possibility fo gouging the cabin side with the belt sander. Your cabin trunk is now ready to receive the cabin top.

Cabin Top Framing

The bending of beams and general construction of the cabin top framing are quite similar to framing in the main deck system. There are two primary differences—perhaps three. Since the cabin top form frequently assumes a much more severe compound shape than that of the main deck, it is usually necessary for each cabin top beam to be indivually laminated to its particular curve. The cabin top camber is always more pronounced in height than that of the main deck. It is often designed as a rapidly decreasing radiused curve toward the edge of the trunk rather than as the continuous radius frequently applied in the main deck crown. It will be almost impossible to use cut beams for the cabin top framing unless its camber is almost flat, and this looks ridiculous on anything but powerboats. A third difference, in essence, hinges on the angle created at the cabin top and the cabin side, as it will affect the type of attachment used at this junction. I will discuss this attachment subsequently.

Since the designer seldom draws a Lines Plan for the cabin

trunk or top (Fig. 327), you may not be able to visualize the finished shape until you begin its construction. This is why I heartily recommend "working the bugs out" by using temporary cabin forms (Fig. 225) to allow for an accurate predetermination. Once the temporary cabin forms have been faired and corrected with battens, the individual shape of each cabin beam, then, may be derived directly from the forms themselves. Otherwise it may be necessary to literally loft a cabin top shape full size in order to derive accurate cabin beam shapes.

The cabin top beam shapes, derived in any manner you choose, are laminated exactly the same way as shown in Fig. 315. Each beam will be numbered, then set aside. A beam shelf or supporting timber must be cut to its proper section, sprung along the upper inner edge of the cabin trunk, accurately cut to length and fastened to the inner upper edge of the cabin trunk sides. When attaching the beam shelf, be careful to cut its ends so as to produce a clean, neat fit against the cabin trunk corner posts.

The beam shelf may take several different forms and will usually be detailed by your designer. If the beam shelf chosen does not extend fully to the cabin trunk edge, it is common to insert spacers between the beams on top of the shelf, after the beams have been fastened. These spacers will have the effect of widening the upper edge of the cabin trunk, thus providing a wider base for glue and nails when the cabin top is attached. The spacers will also ease the trouble of painting and cleaning the inside corner of the cabin trunk. (Fig. 328)

With the beam shelf fashioned and attached, the cabin top beams are placed in their proper positions, marked for exact length and bevel, then carefully cut. The beam attachment to the shelf is by way of through-bolts with washers and glue. I recommend carriage bolts because flatheads often split the beam end when drawn tight. The heads of the bolts must be countersunk in order to allow the cabin top ply to lie flush along the beams. It is a good practice to counterbore the lower end of the bolt-hole for the washer and nut. An exposed fastening will look rough and could possibly cause personal injury if scraped against. Before attaching the cabin top, fill the bolt heads with white lead or other compound to prevent possibility of rot. (Fig. 329)

The Cabin Top

Will the builder's options never cease? Cabin tops are no exception. The most common type of cabin top is a two-ply laminate over beams. This system may be used, however, only when the camber of the cabin top is rather flat and gentle. There are really no special tricks to know here. Although some rather flat cabin tops do, in fact, involve a compound shape, it is normally not so pronounced that dart cuts are required in either of the two-ply layers. The subtle shapes can usually be forced. When attaching the two-ply cabin top, the first layer is sprung over the deck beams and cabin edges after glue has been applied along the cabin trunk structural members; then Anchorfast nails are driven all around. The edge of the plywood may protrude slightly beyond the cabin trunk sides. This plywood layer, then, is rolled out with resorcinol or resin glue and the second ply layer is immediately sprung and Anchorfast nailed through the first ply layer and again nailed well into the cabin top structural members. After the glue has cured, the perimeter of the cabin top is very carefully trimmed and generously radiused. You must note

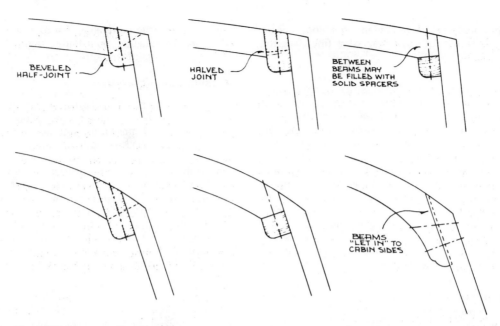

Fig. 328. Beam shelf and cabinet trunk beam attachment variations for both low and high crown shapes.

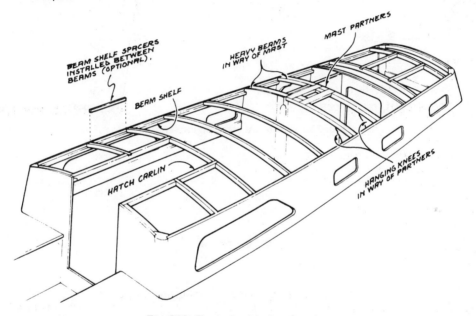

Fig. 329. Typical cabin top framing.

here that the corner radius will be dictated by the positions of the fastenings around the cabin top perimeter so it is well to think ahead here when driving the second layer of fastenings, i.e., the farther inboard you are able to place the second fastenings, the larger a radius you will be able to produce. Once so attached, the hatch openings may be cut and finished flush with their respective carlins. The cabin top is now ready for fiberglassing and trimming. (Fig. 330)

If your cabin top has been designed with a generous camber, you may not be able to get away with using the two-ply layer system but rather breaking down the required cabin top thickness into many more layers of, say, quarter-inch ply. The laminated cabin top may be constructed over a cabin top framing system in exactly the same way as for the two-ply layer system. In some designs

the cabin top beams may be eliminated completely, using only temporary camber forms much the same way as for the laminated deck system. If you choose the latter method, the first ply layer is glued and Anchorfast nailed to the cabin trunk clamp as well as the cabin faces and all existing bulkhead beams. However, only small finishing nails are used in the temporary cabin forms. These will be removed as soon as the glue has cured along the permanent structural members. The second, third and fourth layers will be laid on top of the preceding layers which are rolled with resorcinol or resin glue prior to fastening. Each layer will require Anchorfast nails in the areas of the cabin clamp, bulkhead beams and cabin faces. Once again, these successive layers will be fastened along the temporary cabin forms but they will be removed as before. Once so laid, the temporary cabin forms may be removed from the inside of the trunk, the hatch opening cut and the upper cabin corners radiused smoothly. You may now fiberglass the the cabin top (see Fig. 330).

Occasionally a cabin top is developed as a compound shape whose severity will not allow the forming of plywood sheets without locking in an excessive amount of stress. You will become aware of these stresses at the time you begin to force the plywood to the desired shape. If the plywood refuses to form itself, you will have to cut long tapered darts into the plywood layer. It is much better here to cut, say, three very narrow darts rather than attempt to form the cabin top by use of one large dart. If you are constructing a two-layer cabin top over the cabin beams, these darts should be so well staggered that there is no possibility of their occurrence within the same region. If you are constructing a laminated cabin top, you will also have to stagger the darts, although one or two of these darts

may fall within the same area. Try to avoid this if possible, but, generally speaking, your failure to do so will not result in a dangerous loss of strength.

Miscellaneous Joinery

I could easily triple the size of this chapter if I were to cover all of the various details concerning the use of tools, woodworking and marine joinery. So rather than dwelling upon these complexities and procedures verbally, I think much more will be gained by referring to the following detailed illustrations, as you proceed from one stage to the next in the finishing of your vessel. Again, I urge you to tour your local marina and personally view the examples of the various elements which you are about to construct. It is the fine joinery that will represent your philosophy as well as the very spirit and soul of your boat. Refrain from rushing your finished work and attempt to form the habit of objectively criticizing your handiwork.

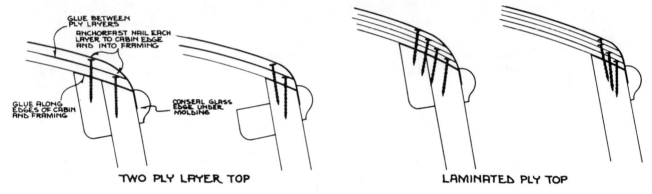

TWO PLY LAYER TOP

LAMINATED PLY TOP

Fig. 330. Cabin top attachment variations.

• • • •

Contents of illustrations on following pages:

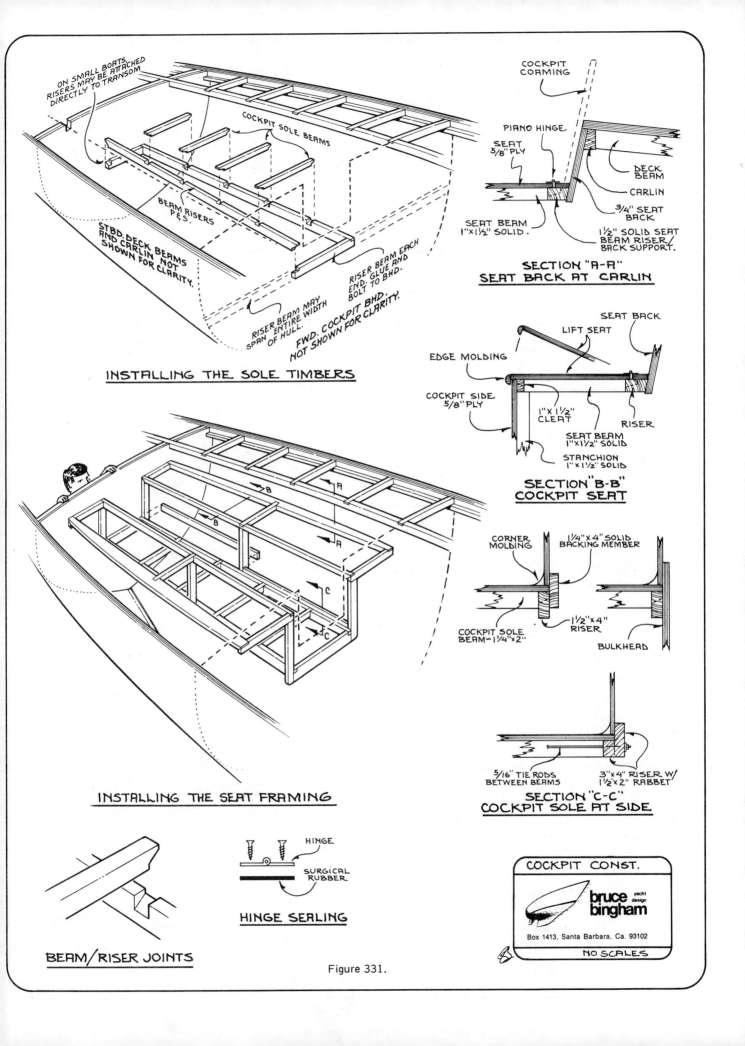

ON SMALL BOATS RISERS MAY BE ATTACHED DIRECTLY TO TRANSOM

COCKPIT SOLE BEAMS

BEAM RISERS P&S.

STBD. DECK BEAMS AND CARLIN NOT SHOWN FOR CLARITY.

RISER BEAM MAY SPAN ENTIRE WIDTH OF HULL.

RISER BEAM EACH END. GLUE AND BOLT TO BHD.

FWD. COCKPIT BHD NOT SHOWN FOR CLARITY.

INSTALLING THE SOLE TIMBERS

COCKPIT COAMING

PIANO HINGE

SEAT 5/8" PLY

DECK BEAM

CARLIN

3/4" SEAT BACK

SEAT BEAM 1"x 1½" SOLID.

1½" SOLID SEAT BEAM RISER/ BACK SUPPORT.

SECTION "A-A" SEAT BACK AT CARLIN

SEAT BACK

LIFT SEAT

EDGE MOLDING

COCKPIT SIDE 5/8" PLY

1"x 1½" CLEAT

SEAT BEAM 1"x 1½" SOLID

RISER

STANCHION 1" x 1½" SOLID

SECTION "B-B" COCKPIT SEAT

CORNER MOLDING

1¼"x 4" SOLID BACKING MEMBER

COCKPIT SOLE BEAM—1¼"x 2"

1½"x 4" RISER

BULKHEAD

INSTALLING THE SEAT FRAMING

5/16" TIE RODS BETWEEN BEAMS

3"x 4" RISER W/ 1½"x 2" RABBET

SECTION "C-C" COCKPIT SOLE AT SIDE

HINGE

SURGICAL RUBBER

HINGE SEALING

BEAM/RISER JOINTS

COCKPIT CONST.

bruce bingham yacht design

Box 1413, Santa Barbara, Ca. 93102

NO SCALES

Figure 331.

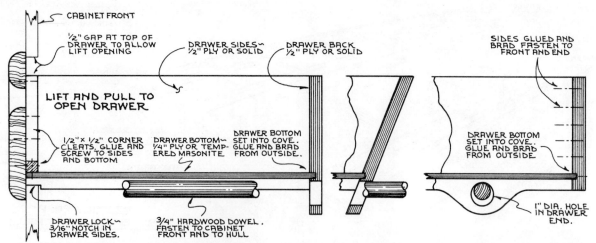

CABINET FRONT

½" GAP AT TOP OF DRAWER TO ALLOW LIFT OPENING

DRAWER SIDES~ ½" PLY OR SOLID

DRAWER BACK ½" PLY OR SOLID

SIDES GLUED AND BRAD FASTEN TO FRONT AND END

LIFT AND PULL TO OPEN DRAWER

½" X ½" CORNER CLEATS. GLUE AND SCREW TO SIDES AND BOTTOM

DRAWER BOTTOM~ ¼" PLY OR TEMPERED MASONITE

DRAWER BOTTOM SET INTO COVE. GLUE AND BRAD FROM OUTSIDE.

DRAWER BOTTOM SET INTO COVE. GLUE AND BRAD FROM OUTSIDE

DRAWER LOCK~ 3/16" NOTCH IN DRAWER SIDES.

¾" HARDWOOD DOWEL. FASTEN TO CABINET FRONT AND TO HULL

1" DIA. HOLE IN DRAWER END.

THIS DRAWER CAN'T FALL OUT, "JUMP THE TRACK", JAM, TILT, SPILL, OR OTHERWISE GO HAYWIRE... AND IT'S VERY EASY TO INSTALL.

THE BASIC SELF-LOCKING DRAWER WITH SIMPLE NON-MISALIGNING CENTER RUNNER

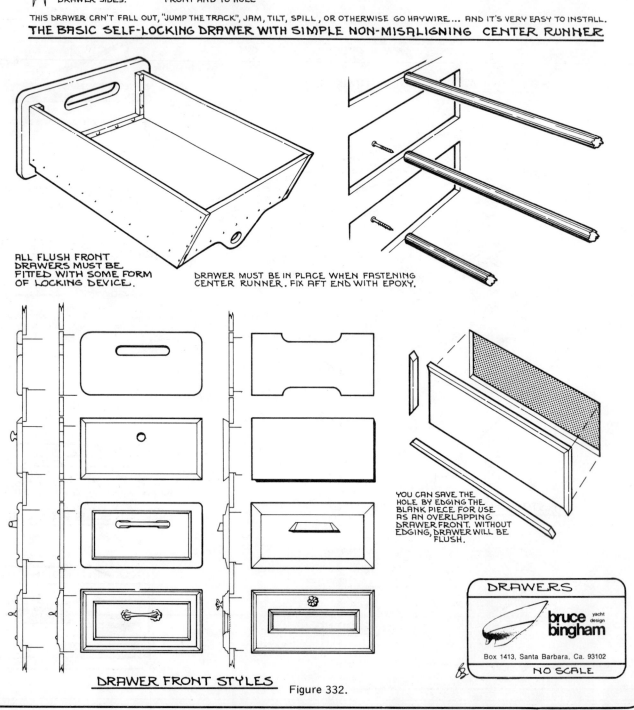

ALL FLUSH FRONT DRAWERS MUST BE FITTED WITH SOME FORM OF LOCKING DEVICE.

DRAWER MUST BE IN PLACE WHEN FASTENING CENTER RUNNER. FIX AFT END WITH EPOXY.

YOU CAN SAVE THE HOLE BY EDGING THE BLANK PIECE FOR USE AS AN OVERLAPPING DRAWER FRONT. WITHOUT EDGING, DRAWER WILL BE FLUSH.

DRAWERS

bruce bingham yacht design

Box 1413, Santa Barbara, Ca. 93102

NO SCALE

DRAWER FRONT STYLES

Figure 332.

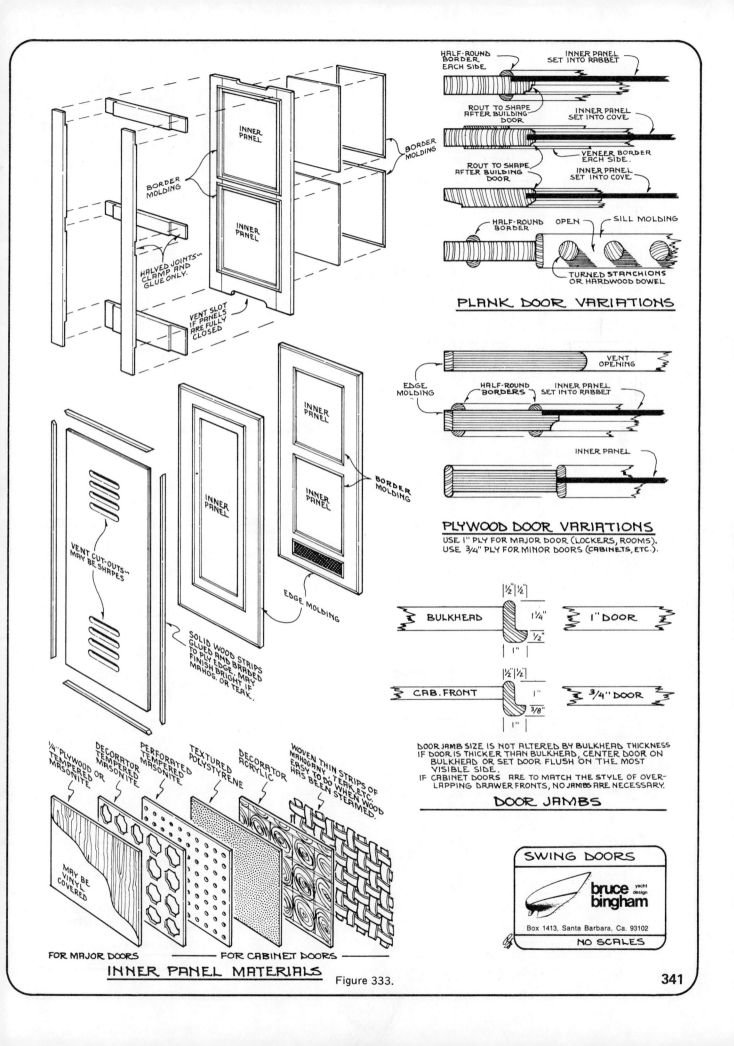

HALF-ROUND BORDER EACH SIDE

INNER PANEL SET INTO RABBET

ROUT TO SHAPE AFTER BUILDING DOOR

INNER PANEL SET INTO COVE

ROUT TO SHAPE AFTER BUILDING DOOR

VENEER BORDER EACH SIDE.

INNER PANEL SET INTO COVE

HALF-ROUND BORDER

OPEN

SILL MOLDING

TURNED STANCHIONS OR HARDWOOD DOWEL

PLANK DOOR VARIATIONS

VENT OPENING

EDGE MOLDING

HALF-ROUND BORDERS

INNER PANEL SET INTO RABBET

INNER PANEL

PLYWOOD DOOR VARIATIONS

USE 1" PLY FOR MAJOR DOOR (LOCKERS, ROOMS).
USE 3/4" PLY FOR MINOR DOORS (CABINETS, ETC.).

BULKHEAD   1/2" 1/2"   1 1/4"   1" DOOR
            1"   1/2"

CAB. FRONT 1/2" 1/2"   1"   3/4" DOOR
            1"   3/8"

DOOR JAMB SIZE IS NOT ALTERED BY BULKHEAD THICKNESS
IF DOOR IS THICKER THAN BULKHEAD, CENTER DOOR ON BULKHEAD OR SET DOOR FLUSH ON THE MOST VISIBLE SIDE.
IF CABINET DOORS ARE TO MATCH THE STYLE OF OVER-LAPPING DRAWER FRONTS, NO JAMBS ARE NECESSARY.

DOOR JAMBS

BORDER MOLDING

INNER PANEL

INNER PANEL

BORDER MOLDING

HALVED JOINTS CLAMP AND GLUE ONLY.

VENT SLOT IF PANELS ARE FULLY CLOSED

INNER PANEL

INNER PANEL

INNER PANEL

BORDER MOLDING

VENT CUT-OUTS MAY BE SHAPES

INNER PANEL

EDGE MOLDING

SOLID WOOD STRIPS GLUED AND BRADED TO PLY EDGE, MAY FINISH BRIGHT IF MAHOG. OR TEAK.

1/4" PLYWOOD TEMPERED MASONITE
MAY BE VINYL COVERED

DECORATOR TEMPERED OR MASONITE

PERFORATED TEMPERED MASONITE

TEXTURED POLYSTYRENE

DECORATOR ACRYLIC

WOVEN THIN STRIPS OF MAHOGANY, TEAK, ETC. EASY TO DO WHEN WOOD HAS BEEN STEAMED.

FOR MAJOR DOORS   FOR CABINET DOORS

INNER PANEL MATERIALS

Figure 333.

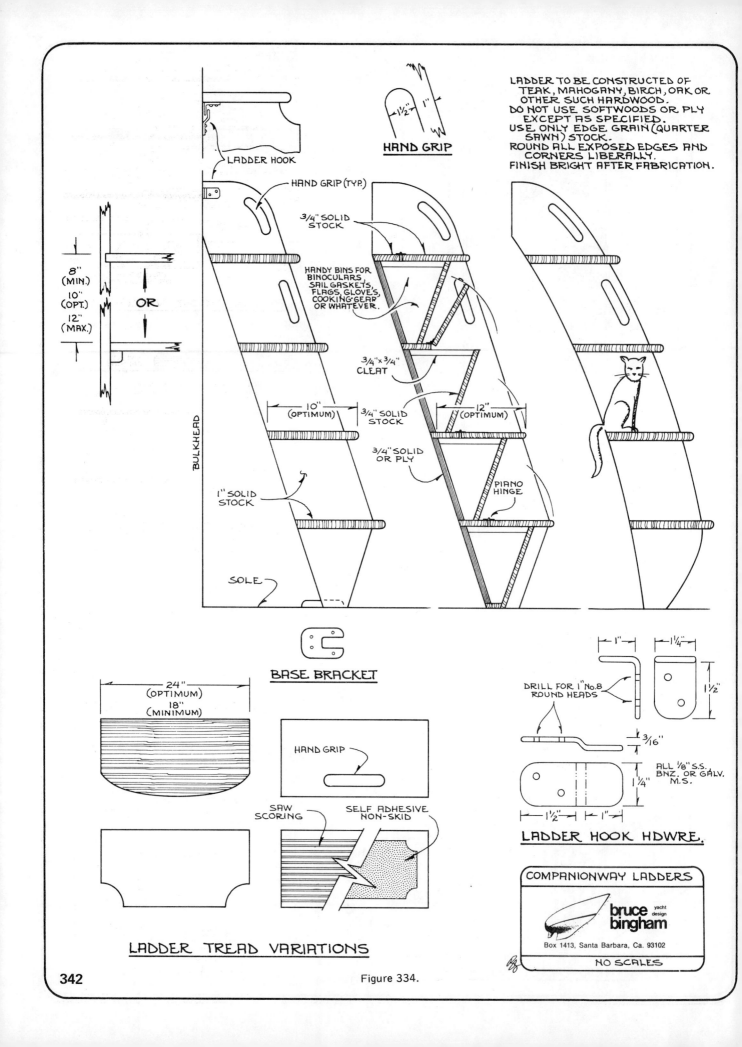

HAND GRIP

$1\frac{1}{2}$" 1"

LADDER TO BE CONSTRUCTED OF
TEAK, MAHOGANY, BIRCH, OAK OR
OTHER SUCH HARDWOOD.
DO NOT USE SOFTWOODS OR PLY
EXCEPT AS SPECIFIED.
USE ONLY EDGE GRAIN (QUARTER
SAWN) STOCK.
ROUND ALL EXPOSED EDGES AND
CORNERS LIBERALLY.
FINISH BRIGHT AFTER FABRICATION.

LADDER HOOK

HAND GRIP (TYP.)

3/4" SOLID STOCK

HANDY BINS FOR
BINOCULARS,
SAIL GASKETS,
FLAGS, GLOVES,
COOKING GEAR
OR WHATEVER.

3/4" x 3/4"
CLEAT

3/4" SOLID
OR PLY

PIANO
HINGE

8"
(MIN.)
10"
(OPT.)
12"
(MAX.)

OR

BULKHEAD

10"
(OPTIMUM)

3/4" SOLID
STOCK

12"
(OPTIMUM)

1" SOLID
STOCK

SOLE

BASE BRACKET

24"
(OPTIMUM)
18"
(MINIMUM)

HAND GRIP

SAW
SCORING

SELF ADHESIVE
NON-SKID

LADDER TREAD VARIATIONS

1" $1\frac{1}{4}$"

DRILL FOR 1" No.8
ROUND HEADS

$1\frac{1}{2}$"

3/16"

$1\frac{1}{4}$"

ALL 1/8" S.S.
BNZ. OR GALV.
M.S.

$1\frac{1}{2}$"   1"

LADDER HOOK HDWRE.

COMPANIONWAY LADDERS

bruce yacht
bingham design

Box 1413, Santa Barbara, Ca. 93102

NO SCALES

Figure 334.

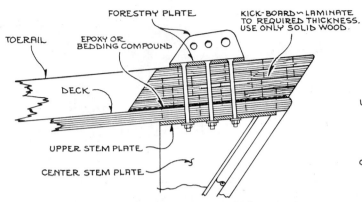

FORESTAY PLATE

EPOXY OR BEDDING COMPOUND

KICK-BOARD ~ LAMINATE TO REQUIRED THICKNESS. USE ONLY SOLID WOOD.

TOERAIL

DECK

UPPER STEM PLATE

CENTER STEM PLATE

TOERAIL KICK-BOARD FOR EXTERNAL FORESTAY PLATE

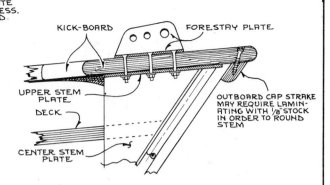

KICK-BOARD

FORESTAY PLATE

UPPER STEM PLATE

DECK

CENTER STEM PLATE

OUTBOARD CAP STRAKE MAY REQUIRE LAMIN-ATING WITH 1/8" STOCK IN ORDER TO ROUND STEM

BULWARK KICK-BOARD FOR EXTERNAL FORESTAY PLATE

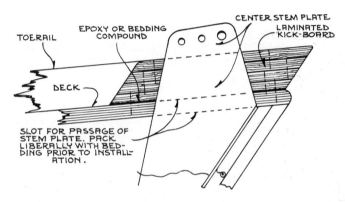

EPOXY OR BEDDING COMPOUND

CENTER STEM PLATE

LAMINATED KICK-BOARD

TOERAIL

DECK

SLOT FOR PASSAGE OF STEM PLATE. PACK LIBERALLY WITH BED-DING PRIOR TO INSTALL-ATION.

TOERAIL KICK-BOARD FOR INTERNAL FORESTAY PLATE

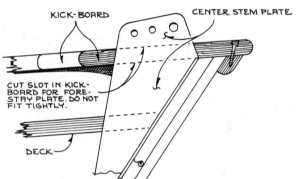

KICK-BOARD

CENTER STEM PLATE

CUT SLOT IN KICK-BOARD FOR FORE-STAY PLATE. DO NOT FIT TIGHTLY.

DECK

BULWARK KICK-BOARD FOR INTERNAL FORESTAY PLATE

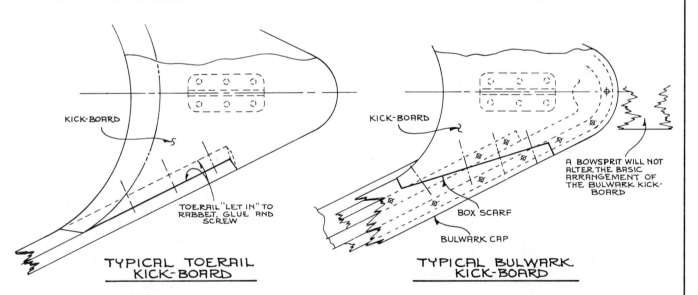

KICK-BOARD

TOERAIL "LET IN" TO RABBET. GLUE AND SCREW

TYPICAL TOERAIL KICK-BOARD

KICK-BOARD

A BOWSPRIT WILL NOT ALTER THE BASIC ARRANGEMENT OF THE BULWARK KICK-BOARD

BOX SCARF

BULWARK CAP

TYPICAL BULWARK KICK-BOARD

STEM KICK-BOARDS

bruce bingham yacht design

Box 1413, Santa Barbara, Ca. 93102

NO SCALES

Figure 335.

343

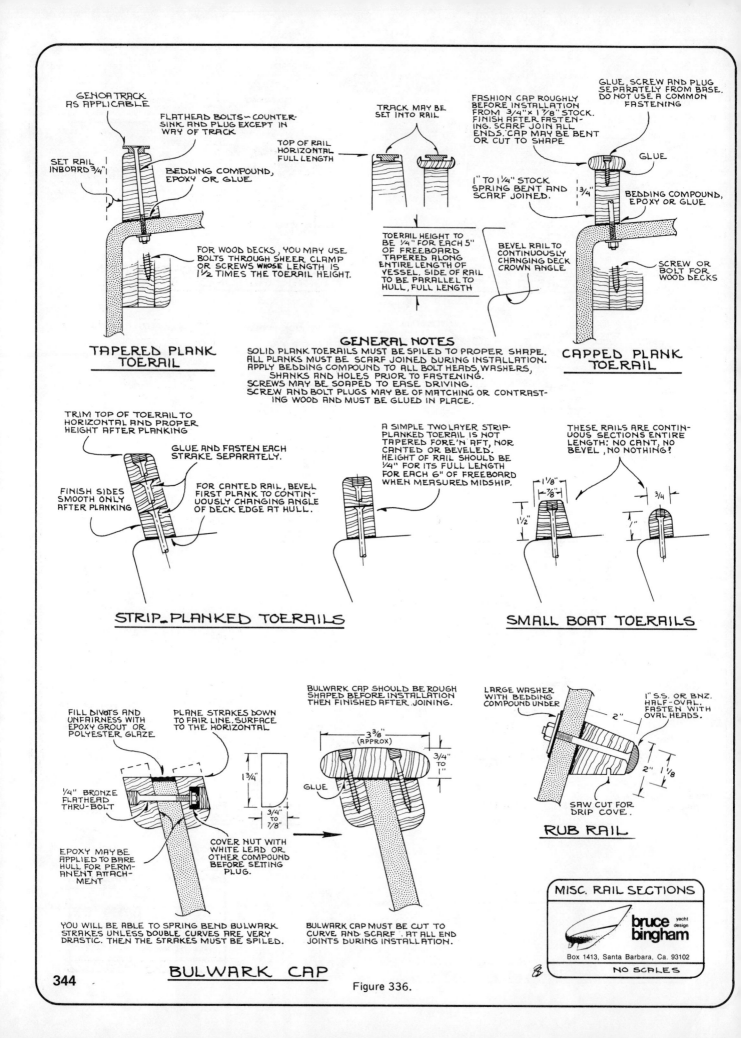

GENOA TRACK AS APPLICABLE

FLATHEAD BOLTS – COUNTERSINK AND PLUG EXCEPT IN WAY OF TRACK

SET RAIL INBOARD 3/4"

BEDDING COMPOUND, EPOXY OR GLUE

FOR WOOD DECKS, YOU MAY USE BOLTS THROUGH SHEER CLAMP OR SCREWS WHOSE LENGTH IS 1½ TIMES THE TOERAIL HEIGHT.

TRACK MAY BE SET INTO RAIL

TOP OF RAIL HORIZONTAL FULL LENGTH

TOERAIL HEIGHT TO BE ¼" FOR EACH 5" OF FREEBOARD TAPERED ALONG ENTIRE LENGTH OF VESSEL. SIDE OF RAIL TO BE PARALLEL TO HULL, FULL LENGTH

GLUE, SCREW AND PLUG SEPARATELY FROM BASE. DO NOT USE A COMMON FASTENING

FASHION CAP ROUGHLY BEFORE INSTALLATION FROM 3/4" x 1 7/8" STOCK. FINISH AFTER FASTENING. SCARF JOIN ALL ENDS. CAP MAY BE BENT OR CUT TO SHAPE

GLUE

1" TO 1¼" STOCK SPRING BENT AND SCARF JOINED.

3/4"

BEDDING COMPOUND, EPOXY OR GLUE

BEVEL RAIL TO CONTINUOUSLY CHANGING DECK CROWN ANGLE

SCREW OR BOLT FOR WOOD DECKS

TAPERED PLANK TOERAIL

GENERAL NOTES
SOLID PLANK TOERAILS MUST BE SPILED TO PROPER SHAPE. ALL PLANKS MUST BE SCARF JOINED DURING INSTALLATION. APPLY BEDDING COMPOUND TO ALL BOLT HEADS, WASHERS, SHANKS AND HOLES PRIOR TO FASTENING.
SCREWS MAY BE SOAPED TO EASE DRIVING.
SCREW AND BOLT PLUGS MAY BE OF MATCHING OR CONTRASTING WOOD AND MUST BE GLUED IN PLACE.

CAPPED PLANK TOERAIL

TRIM TOP OF TOERAIL TO HORIZONTAL AND PROPER HEIGHT AFTER PLANKING

GLUE AND FASTEN EACH STRAKE SEPARATELY.

FINISH SIDES SMOOTH ONLY AFTER PLANKING

FOR CANTED RAIL, BEVEL FIRST PLANK TO CONTINUOUSLY CHANGING ANGLE OF DECK EDGE AT HULL.

A SIMPLE TWO LAYER STRIP-PLANKED TOERAIL IS NOT TAPERED FORE'N AFT, NOR CANTED OR BEVELED. HEIGHT OF RAIL SHOULD BE ¼" FOR ITS FULL LENGTH FOR EACH 6" OF FREEBOARD WHEN MEASURED MIDSHIP.

THESE RAILS ARE CONTINUOUS SECTIONS ENTIRE LENGTH: NO CANT, NO BEVEL, NO NOTHING!

1⅛"
⅞"
1½"

¾"
1"

STRIP-PLANKED TOERAILS

SMALL BOAT TOERAILS

FILL DIVOTS AND UNFAIRNESS WITH EPOXY GROUT OR POLYESTER GLAZE

PLANE STRAKES DOWN TO FAIR LINE. SURFACE TO THE HORIZONTAL

¼" BRONZE FLATHEAD THRU-BOLT

EPOXY MAY BE APPLIED TO BARE HULL FOR PERMANENT ATTACHMENT

COVER NUT WITH WHITE LEAD OR OTHER COMPOUND BEFORE SETTING PLUG.

BULWARK CAP SHOULD BE ROUGH SHAPED BEFORE INSTALLATION THEN FINISHED AFTER JOINING.

3 3/8" (APPROX)
1¾"
¾" TO 1"
¾" TO 7/8"

GLUE

LARGE WASHER WITH BEDDING COMPOUND UNDER

1" S.S. OR BNZ. HALF-OVAL. FASTEN WITH OVAL HEADS.

2"
2" 1⅛"

SAW CUT FOR DRIP COVE.

RUB RAIL

YOU WILL BE ABLE TO SPRING BEND BULWARK STRAKES UNLESS DOUBLE CURVES ARE VERY DRASTIC. THEN THE STRAKES MUST BE SPILED.

BULWARK CAP MUST BE CUT TO CURVE AND SCARF AT ALL END JOINTS DURING INSTALLATION.

BULWARK CAP

Figure 336.

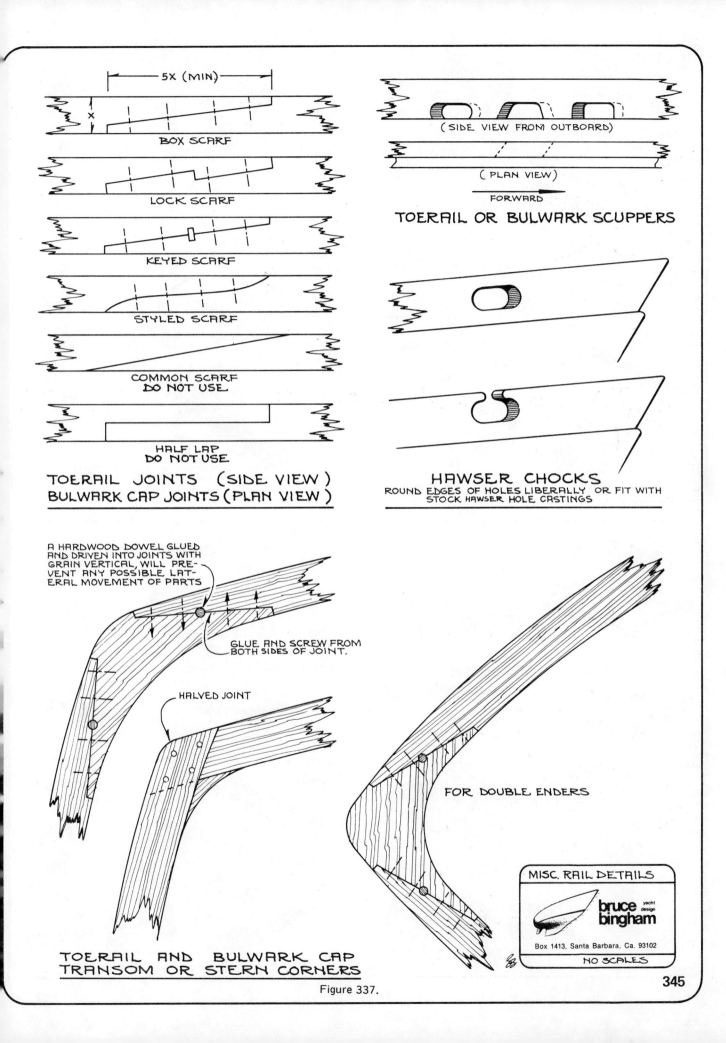

5X (MIN)

X

BOX SCARF

LOCK SCARF

KEYED SCARF

STYLED SCARF

COMMON SCARF
DO NOT USE

HALF LAP
DO NOT USE

TOERAIL JOINTS (SIDE VIEW)
BULWARK CAP JOINTS (PLAN VIEW)

(SIDE VIEW FROM OUTBOARD)

(PLAN VIEW)

FORWARD

TOERAIL OR BULWARK SCUPPERS

HAWSER CHOCKS
ROUND EDGES OF HOLES LIBERALLY OR FIT WITH
STOCK HAWSER HOLE CASTINGS

A HARDWOOD DOWEL GLUED
AND DRIVEN INTO JOINTS WITH
GRAIN VERTICAL, WILL PRE-
VENT ANY POSSIBLE LAT-
ERAL MOVEMENT OF PARTS

GLUE AND SCREW FROM
BOTH SIDES OF JOINT.

HALVED JOINT

FOR DOUBLE ENDERS

TOERAIL AND BULWARK CAP
TRANSOM OR STERN CORNERS

MISC. RAIL DETAILS

bruce bingham yacht design

Box 1413, Santa Barbara, Ca. 93102

NO SCALES

Figure 337.

345

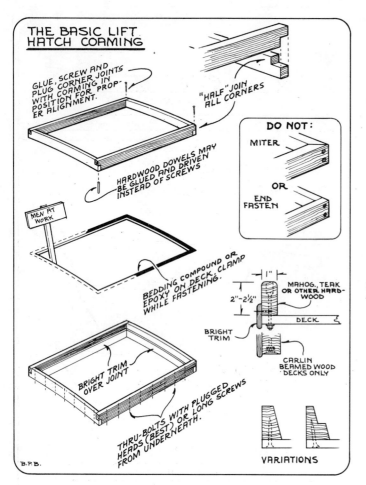

THE BASIC LIFT HATCH COAMING

GLUE, SCREW AND PLUG CORNER JOINTS WITH COAMING IN POSITION FOR PROPER ALIGNMENT.

"HALF" JOIN ALL CORNERS

HARDWOOD DOWELS MAY BE GLUED AND DRIVEN INSTEAD OF SCREWS

DO NOT:
MITER
OR
END FASTEN

MEN AT WORK

BEDDING COMPOUND OR EPOXY ON DECK. CLAMP WHILE FASTENING.

BRIGHT TRIM OVER JOINT

THRU-BOLTS WITH PLUGGED HEADS (BEST) OR LONG SCREWS FROM UNDERNEATH.

MAHOG., TEAK OR OTHER HARD-WOOD

1"
2"-2½"
DECK

BRIGHT TRIM

CARLIN BEAMED WOOD DECKS ONLY

VARIATIONS

B.P.B.

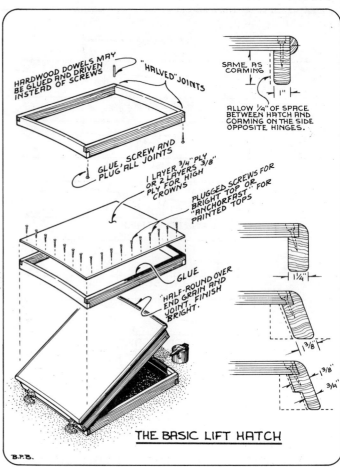

HARDWOOD DOWELS MAY BE GLUED AND DRIVEN INSTEAD OF SCREWS

"HALVED" JOINTS

SAME AS COAMING

1"

ALLOW ¼" OF SPACE BETWEEN HATCH AND COAMING ON THE SIDE OPPOSITE HINGES.

GLUE, SCREW AND PLUG ALL JOINTS

1 LAYER ¾" PLY OR 2 LAYERS ⅜" PLY FOR HIGH CROWNS

PLUGGED SCREWS FOR BRIGHT TOP OR "ANCHORFAST" FOR PAINTED TOPS

GLUE

"HALF-ROUND OVER END GRAIN AND JOINT. FINISH BRIGHT.

1¼"
⅜"
⅜"
¾"

THE BASIC LIFT HATCH

B.P.B.

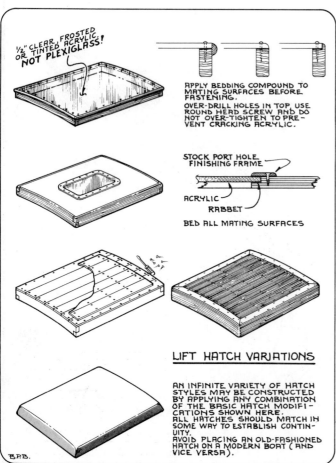

½" CLEAR, FROSTED OR TINTED ACRYLIC. NOT PLEXIGLASS!

APPLY BEDDING COMPOUND TO MATING SURFACES BEFORE FASTENING.
OVER-DRILL HOLES IN TOP, USE ROUND HEAD SCREW AND DO NOT OVER-TIGHTEN TO PREVENT CRACKING ACRYLIC.

STOCK PORT HOLE FINISHING FRAME

ACRYLIC
RABBET

BED ALL MATING SURFACES

LIFT HATCH VARIATIONS

AN INFINITE VARIETY OF HATCH STYLES MAY BE CONSTRUCTED BY APPLYING ANY COMBINATION OF THE BASIC HATCH MODIFICATIONS SHOWN HERE.
ALL HATCHES SHOULD MATCH IN SOME WAY TO ESTABLISH CONTINUITY.
AVOID PLACING AN OLD-FASHIONED HATCH ON A MODERN BOAT (AND VICE VERSA).

B.P.B.

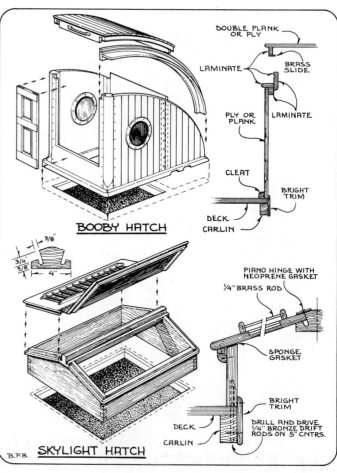

DOUBLE PLANK OR PLY

LAMINATE

BRASS SLIDE

PLY OR PLANK

LAMINATE

CLEAT

BRIGHT TRIM

DECK
CARLIN

BOOBY HATCH

⅜"
¾"
⅝"
4"

PIANO HINGE WITH NEOPRENE GASKET

¼" BRASS ROD

SPONGE GASKET

BRIGHT TRIM

DECK
CARLIN

DRILL AND DRIVE ¼" BRONZE DRIFT RODS ON 5" CNTRS.

SKYLIGHT HATCH

B.P.B.

Figure 338.

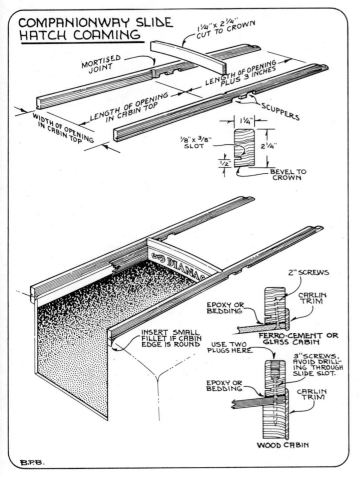

COMPANIONWAY SLIDE HATCH COAMING

1¼" × 2¼" CUT TO CROWN

MORTISED JOINT

LENGTH OF OPENING PLUS 3 INCHES

LENGTH OF OPENING IN CABIN TOP

WIDTH OF OPENING IN CABIN TOP

SCUPPERS

1¼"
⅛" × ⅜" SLOT
2¼"
½"
BEVEL TO CROWN

DIANA

INSERT SMALL FILLET IF CABIN EDGE IS ROUND

2" SCREWS
CARLIN TRIM
EPOXY OR BEDDING
USE TWO PLUGS HERE
FERRO-CEMENT OR GLASS CABIN

3" SCREWS. AVOID DRILLING THROUGH SLIDE SLOT.
EPOXY OR BEDDING
CARLIN TRIM
WOOD CABIN

B.P.B.

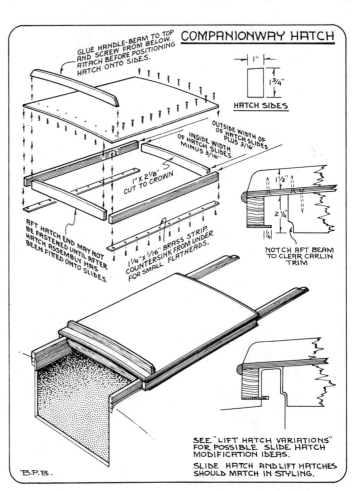

COMPANIONWAY HATCH

GLUE HANDLE-BEAM TO TOP AND SCREW FROM BELOW. ATTACH BEFORE POSITIONING HATCH ONTO SIDES.

1"
1¾"
HATCH SIDES

OUTSIDE WIDTH OF HATCH SLIDES PLUS 3/16"
INSIDE WIDTH OF HATCH SLIDES MINUS 3/16"

1" × 2⅛" CUT TO CROWN

AFT HATCH END MAY NOT BE FASTENED UNTIL AFTER HATCH ASSEMBLY HAS BEEN FITTED ONTO SLIDES.

1¼" × 1/16" BRASS STRIP. COUNTERSINK FROM UNDER FOR SMALL FLATHEADS.

1½"
2¼"
1¼"
NOTCH AFT BEAM TO CLEAR CARLIN TRIM

SEE "LIFT HATCH VARIATIONS" FOR POSSIBLE SLIDE HATCH MODIFICATION IDEAS.

SLIDE HATCH AND LIFT HATCHES SHOULD MATCH IN STYLING.

B.P.B.

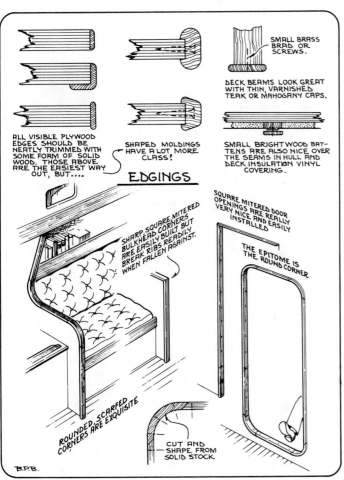

ALL VISIBLE PLYWOOD EDGES SHOULD BE NEATLY TRIMMED WITH SOME FORM OF SOLID WOOD. THOSE ABOVE ARE THE EASIEST WAY OUT, BUT....

SHAPED MOLDINGS HAVE A LOT MORE CLASS!

SMALL BRASS BRAD OR SCREWS.

DECK BEAMS LOOK GREAT WITH THIN, VARNISHED TEAK OR MAHOGANY CAPS.

SMALL BRIGHT WOOD BATTENS ARE ALSO NICE OVER THE SEAMS IN HULL AND DECK INSULATION VINYL COVERING.

EDGINGS

SHARP SQUARE MITERED BULKHEAD CORNERS ARE EASILY BUILT BUT BREAK RIBS READILY WHEN FALLEN AGAINST.

SQUARE MITERED DOOR OPENINGS ARE REALLY VERY NICE AND EASILY INSTALLED

THE EPITOME IS THE ROUND CORNER

ROUNDED SCARFED CORNERS ARE EXQUISITE

CUT AND SHAPE FROM SOLID STOCK

B.P.B.

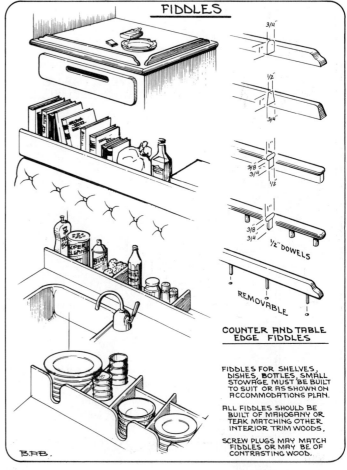

FIDDLES

3/4"
1"
½"
1"
¾"
1"
3/8"
3/4"
½"
1"
3/8"
3/4"
½" DOWELS

REMOVABLE

COUNTER AND TABLE EDGE FIDDLES

FIDDLES FOR SHELVES, DISHES, BOTTLES, SMALL STOWAGE MUST BE BUILT TO SUIT OR AS SHOWN ON ACCOMMODATIONS PLAN.

ALL FIDDLES SHOULD BE BUILT OF MAHOGANY OR TEAK MATCHING OTHER INTERIOR TRIM WOODS.

SCREW PLUGS MAY MATCH FIDDLES OR MAY BE OF CONTRASTING WOOD.

B.P.B.

Figure 339.

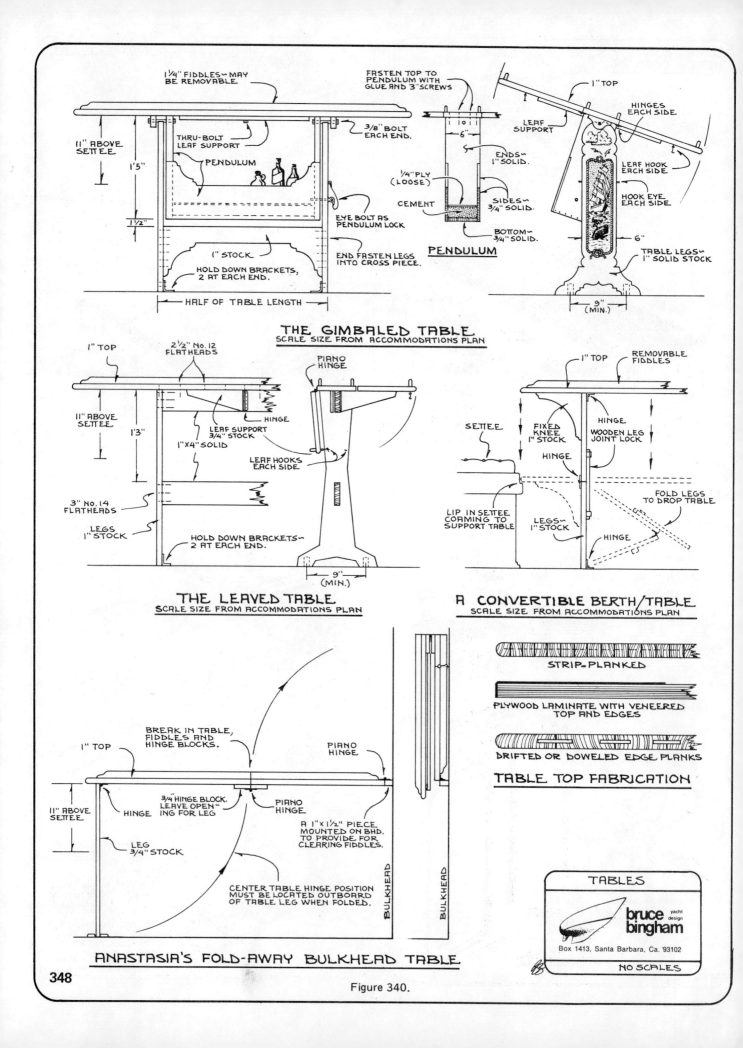

THE GIMBALED TABLE
SCALE SIZE FROM ACCOMMODATIONS PLAN

PENDULUM

THE LEAVED TABLE
SCALE SIZE FROM ACCOMMODATIONS PLAN

A CONVERTIBLE BERTH/TABLE
SCALE SIZE FROM ACCOMMODATIONS PLAN

TABLE TOP FABRICATION

ANASTASIA'S FOLD-AWAY BULKHEAD TABLE

TABLES

bruce bingham   yacht design

Box 1413, Santa Barbara, Ca. 93102

NO SCALES

348

Figure 340.

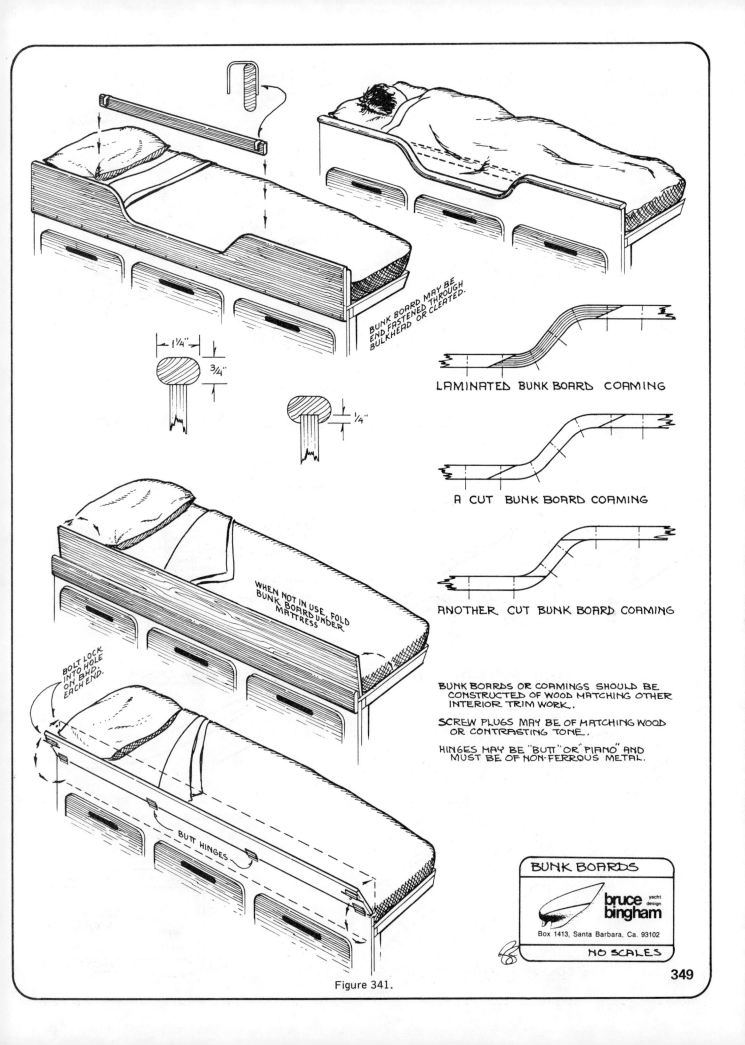

BUNK BOARD MAY BE END FASTENED THROUGH BULKHEAD OR CLEATED.

1¼"

¾"

¼"

LAMINATED BUNK BOARD COAMING

A CUT BUNK BOARD COAMING

ANOTHER CUT BUNK BOARD COAMING

WHEN NOT IN USE, FOLD BUNK BOARD UNDER MATTRESS

BOLT LOCK INTO HOLE ON BHD EACH END.

BUTT HINGES

BUNK BOARDS OR COAMINGS SHOULD BE CONSTRUCTED OF WOOD MATCHING OTHER INTERIOR TRIM WORK.

SCREW PLUGS MAY BE OF MATCHING WOOD OR CONTRASTING TONE.

HINGES MAY BE "BUTT" OR "PIANO" AND MUST BE OF NON-FERROUS METAL.

BUNK BOARDS

bruce bingham yacht design

Box 1413, Santa Barbara, Ca. 93102

NO SCALES

349

Figure 341.

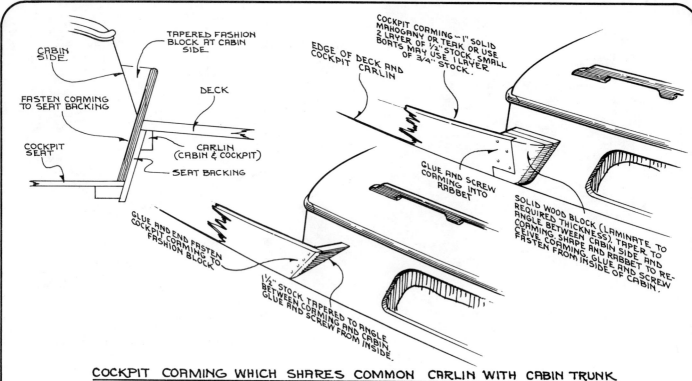

CABIN SIDE.

TAPERED FASHION BLOCK AT CABIN SIDE.

FASTEN COAMING TO SEAT BACKING

DECK

COCKPIT SEAT

CARLIN (CABIN & COCKPIT)

SEAT BACKING

COCKPIT COAMING — 1" SOLID MAHOGANY OR TEAK OR USE 2 LAYER OF ½" STOCK. SMALL BOATS MAY USE 1 LAYER OF ¾" STOCK.

EDGE OF DECK AND COCKPIT CARLIN

GLUE AND SCREW COAMING INTO RABBET

SOLID WOOD BLOCK (LAMINATE TO REQUIRED THICKNESS). TAPER TO ANGLE BETWEEN CABIN SIDE AND COAMING. SHAPE RABBET TO RECEIVE COAMING. GLUE AND SCREW FASTEN FROM INSIDE OF CABIN.

GLUE AND END FASTEN COCKPIT COAMING TO FASHION BLOCK

1½" STOCK TAPERED TO ANGLE BETWEEN COAMING AND CABIN. GLUE AND SCREW FROM INSIDE.

COCKPIT COAMING WHICH SHARES COMMON CARLIN WITH CABIN TRUNK

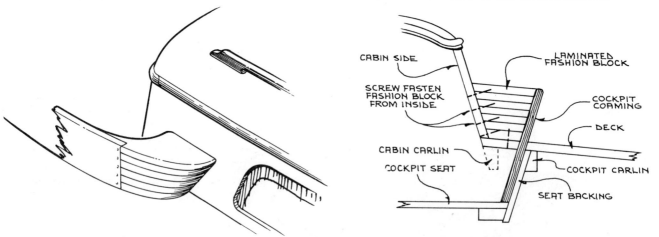

CABIN SIDE

SCREW FASTEN FASHION BLOCK FROM INSIDE.

CABIN CARLIN

COCKPIT SEAT

LAMINATED FASHION BLOCK

COCKPIT COAMING

DECK

COCKPIT CARLIN

SEAT BACKING

COCKPIT COAMING WITH INDEPENDENT CARLIN FROM CABIN

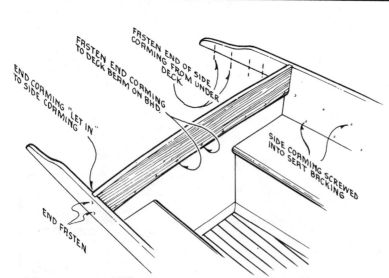

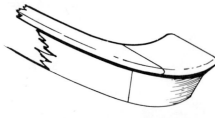

END COAMING "LET IN" TO SIDE COAMING

FASTEN END OF COAMING TO DECK BEAM ON BHD.

FASTEN END OF SIDE COAMING FROM UNDER DECK

SIDE COAMING SCREWED INTO SEAT BACKING

END FASTEN

FOR A SLIGHTLY MORE TRADITIONAL FLAVOR, ANY OF THE COAMING VARIATIONS SHOWN ABOVE MAY BE CAPPED.

COCKPIT END COAMING

Figure 342.

COCKPIT COAMINGS

bruce bingham yacht design

Box 1413, Santa Barbara, Ca. 93102

NO SCALES

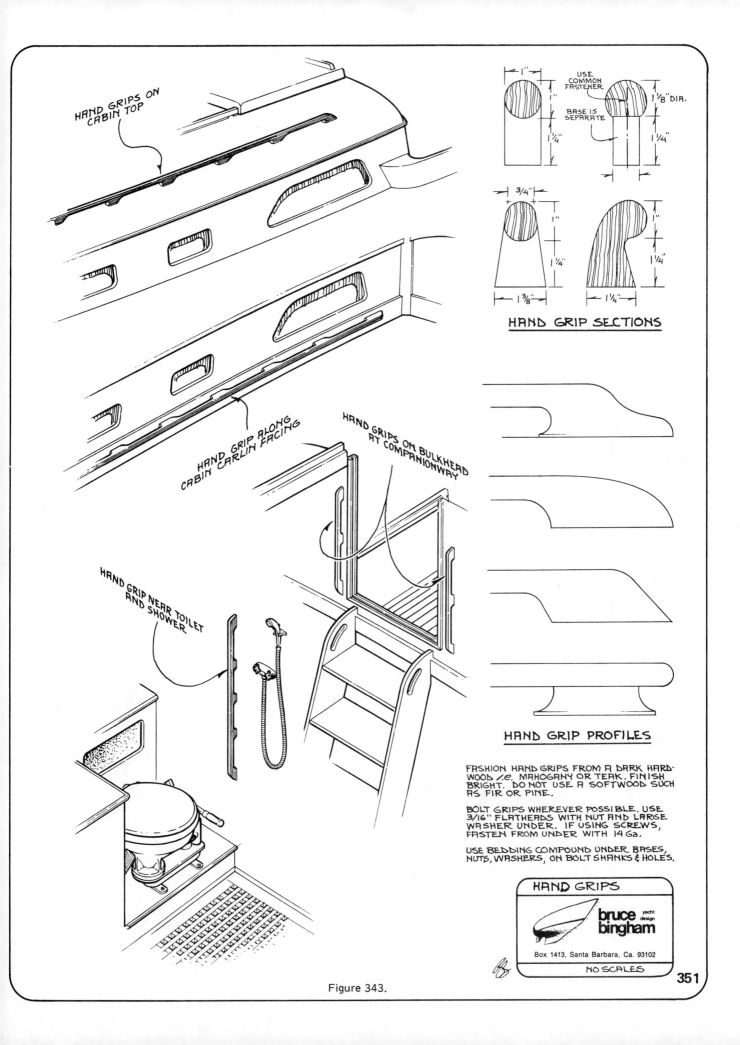

HAND GRIPS ON CABIN TOP

HAND GRIP ALONG CABIN CARLIN FACING

HAND GRIPS ON BULKHEAD AT COMPANIONWAY

HAND GRIP NEAR TOILET AND SHOWER

USE COMMON FASTENER

BASE IS SEPARATE

1"
1 1/4"
1 1/8" DIA.
1 1/4"

3/4"
1"
1 1/4"
1 3/8"
1 1/4"
1 1/4"

HAND GRIP SECTIONS

HAND GRIP PROFILES

FASHION HAND GRIPS FROM A DARK HARD-WOOD *i.e.* MAHOGANY OR TEAK. FINISH BRIGHT. DO NOT USE A SOFTWOOD SUCH AS FIR OR PINE.

BOLT GRIPS WHEREVER POSSIBLE. USE 3/16" FLATHEADS WITH NUT AND LARGE WASHER UNDER. IF USING SCREWS, FASTEN FROM UNDER WITH 14 Ga.

USE BEDDING COMPOUND UNDER BASES, NUTS, WASHERS, ON BOLT SHANKS & HOLES.

HAND GRIPS

bruce bingham
yacht design

Box 1413, Santa Barbara, Ca. 93102

NO SCALES

351

Figure 343.

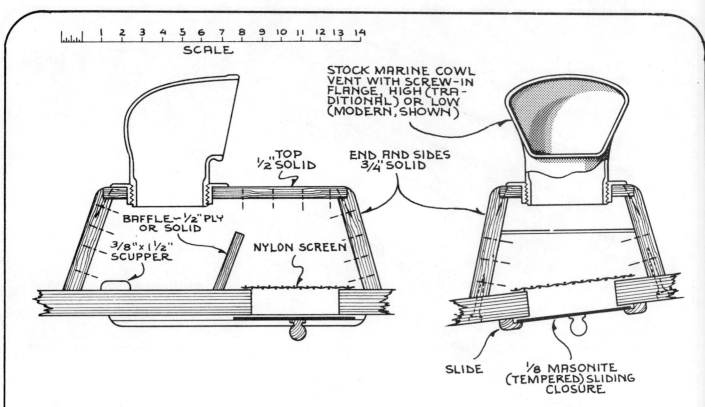

SCALE | 1 2 3 4 5 6 7 8 9 10 11 12 13 14

STOCK MARINE COWL VENT WITH SCREW-IN FLANGE, HIGH (TRADITIONAL) OR LOW (MODERN, SHOWN)

½" TOP SOLID

END AND SIDES ¾" SOLID

BAFFLE-½" PLY OR SOLID

⅜"x1½" SCUPPER

NYLON SCREEN

SLIDE

⅛ MASONITE (TEMPERED) SLIDING CLOSURE

GENERAL NOTES

DORADE BOX SHOULD BE CONSTRUCTED OF MAHOGANY OR TEAK, FINISHED BRIGHT.

JOINTS MAY BE RABBETED (SHOWN) OR LAP-JOINED.

FASTENINGS MAY BE PLUGGED, BRONZE, FLAT-HEAD SCREWS OR GALVANIZED FINISHING NAILS, COUNTERSUNK AND GLAZED OVER WITH PLASTIC WOOD.

GLUE ALL JOINTS DURING ASSEMBLY.

APPLY SEVERAL COATS OF VARNISH TO INSIDE OF BOX PRIOR TO MOUNTING ON DECK.

ROUND ALL EDGES AND CORNERS LIBERALLY PRIOR TO APPLYING EXTERIOR FINISH.

MOUNTING

AFTER SHAPING TO FIT DECK CROWN, MARK THE PERIMETER OF THE BOX ONTO THE DECK. POSITION AND CUT THE VENT HOLE INTO THE DECK. SEAL THE HOLE EDGE WITH PAINT. DRILL THE SCREW HOLES INTO THE DECK AT 3" INTERVALS AROUND

PLACE THE DORADE BOX IN ITS MARKED POSITION ON THE DECK. DRILL SCREW PILOT HOLES UPWARD FROM BELOW INTO THE BOX EDGES WHILE SOME ONE HOLDS IT FIRMLY TOPSIDE.

COAT THE BOX EDGES AND BAFFLE WITH BEDDING COMPOUND. PLACE BOX IN ITS MARKED POSITION. NOW SCREW IT HOME.

FOR FERRO-CEMENT DECKS: EPOXY MOUNT WOOD ¾"x¾" CLEATS ONTO THE DECK THEN SCREW THE DORADE ONTO THE CLEATS.

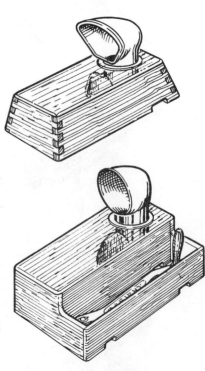

VARIATIONS

VENT DORADES

bruce bingham

NO SCALES

Figure 344.

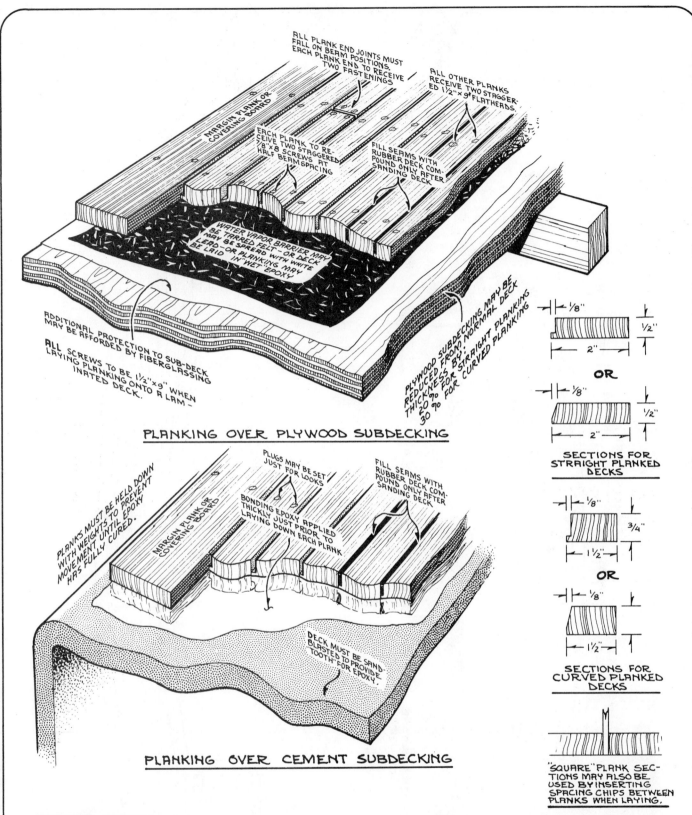

ALL PLANK END JOINTS MUST FALL ON BEAM POSITIONS. EACH PLANK END TO RECEIVE TWO FASTENINGS

ALL OTHER PLANKS RECEIVE TWO STAGGERED 1½"x9" FLATHEADS.

MARGIN PLANK OR COVERING BOARD

EACH PLANK TO RECEIVE TWO STAGGERED ⅞"x8 SCREWS AT HALF BEAM SPACING

FILL SEAMS WITH RUBBER DECK COMPOUND ONLY AFTER SANDING DECK.

WATER VAPOR BARRIER MAY BE TARRED FELT- OR DECK MAY BE SPREAD WITH WHITE LEAD- OR PLANKING MAY BE LAID IN WET EPOXY

ADDITIONAL PROTECTION TO SUB-DECK MAY BE AFFORDED BY FIBERGLASSING

ALL SCREWS TO BE 1½"x9" WHEN LAYING PLANKING ONTO A LAMINATED DECK.

PLYWOOD SUBDECKING MAY BE REDUCED FROM NORMAL DECK THICKNESS BY: 20% FOR STRAIGHT PLANKING 30% FOR CURVED PLANKING

PLANKING OVER PLYWOOD SUBDECKING

PLUGS MAY BE SET JUST FOR LOOKS

FILL SEAMS WITH RUBBER DECK COMPOUND ONLY AFTER SANDING DECK

PLANKS MUST BE HELD DOWN WITH WEIGHTS TO PREVENT MOVEMENT UNTIL EPOXY HAS FULLY CURED.

MARGIN PLANK OR COVERING BOARD

BONDING EPOXY APPLIED THICKLY JUST PRIOR TO LAYING DOWN EACH PLANK

DECK MUST BE SAND BLASTED TO PROVIDE "TOOTH" FOR EPOXY.

PLANKING OVER CEMENT SUBDECKING

⅛"
2"
½"

OR

⅛"
2"
½"

SECTIONS FOR STRAIGHT PLANKED DECKS

⅛"
1½"
¾"

OR

⅛"
1½"

SECTIONS FOR CURVED PLANKED DECKS

"SQUARE" PLANK SECTIONS MAY ALSO BE USED BY INSERTING SPACING CHIPS BETWEEN PLANKS WHEN LAYING.

GENERAL NOTES:

ALTHOUGH TEAK IS GENERALLY THE PREFERRED DECKING WOOD, IT IS EXTREMELY EXPENSIVE, PARTICULARLY WHEN PURCHASED IN THE PROPER GRADE AND CUT. IT HAS ALSO BECOME ALMOST UNAVAILABLE AS AIR DRIED. IT MAY BE PRUDENT, THEREFORE, TO CONSIDER USING QUARTER SAWN FIR, SPRUCE OR PINE. THESE WOODS ARE VERY ATTRACTIVE, EASY TO WORK, INEXPENSIVE, LIGHT WEIGHT, AND LONG LASTING WHEN OILED WITH COD OR LINSEED OR VARNISHED,

ALL DECKING LUMBER MUST BE QUARTER SAWN, AIR DRIED AND OF "A", "E", OR "SELECT" GRADE. NOTHING ELSE WILL DO AT ALL.

CHECK WITH YOUR DESIGNER FOR POSSIBLE WEIGHT PROBLEMS, FIRST OFF!

DECK OVER-PLANKING

bruce bingham yacht design

Box 1413, Santa Barbara, Ca. 93102

NO SCALES

Figure 345.

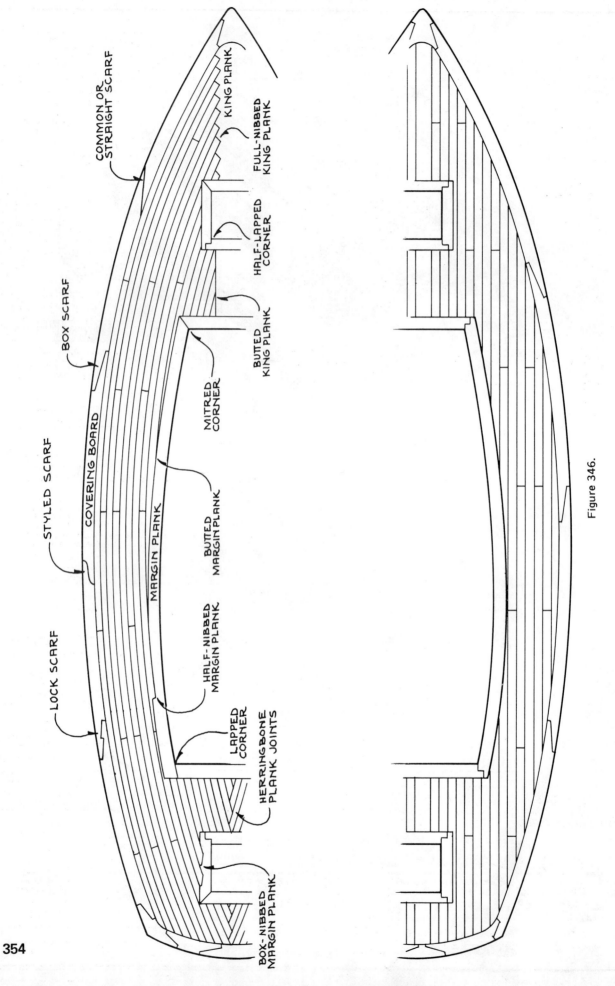

COMMON OR STRAIGHT SCARF

KING PLANK

FULL-NIBBED KING PLANK

HALF-LAPPED CORNER

BOX SCARF

BUTED KING PLANK

MITRED CORNER

COVERING BOARD

STYLED SCARF

MARGIN PLANK

BUTTED MARGIN PLANK

HALF-NIBBED MARGIN PLANK

LOCK SCARF

LAPPED CORNER

HERRINGBONE PLANK JOINTS

BOX-NIBBED MARGIN PLANK

Figure 346.

354

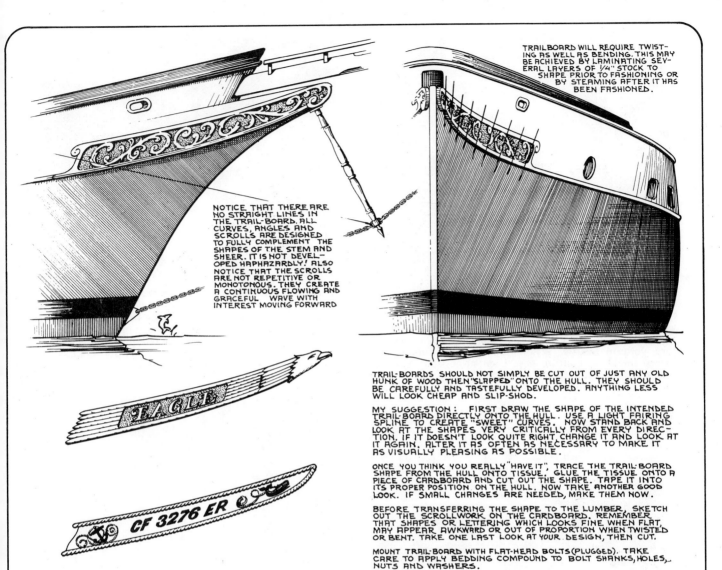

TRAILBOARD WILL REQUIRE TWISTING AS WELL AS BENDING. THIS MAY BE ACHIEVED BY LAMINATING SEVERAL LAYERS OF 1/4" STOCK TO SHAPE PRIOR TO FASHIONING OR BY STEAMING AFTER IT HAS BEEN FASHIONED.

NOTICE THAT THERE ARE NO STRAIGHT LINES IN THE TRAIL-BOARD. ALL CURVES, ANGLES AND SCROLLS ARE DESIGNED TO FULLY COMPLEMENT THE SHAPES OF THE STEM AND SHEER. IT IS NOT DEVELOPED HAPHAZARDLY! ALSO NOTICE THAT THE SCROLLS ARE NOT REPETITIVE OR MONOTONOUS. THEY CREATE A CONTINUOUS FLOWING AND GRACEFUL WAVE WITH INTEREST MOVING FORWARD

TRAIL-BOARDS SHOULD NOT SIMPLY BE CUT OUT OF JUST ANY OLD HUNK OF WOOD THEN "SLAPPED" ONTO THE HULL. THEY SHOULD BE CAREFULLY AND TASTEFULLY DEVELOPED. ANYTHING LESS WILL LOOK CHEAP AND SLIP-SHOD.

MY SUGGESTION: FIRST DRAW THE SHAPE OF THE INTENDED TRAIL-BOARD DIRECTLY ONTO THE HULL. USE A LIGHT FAIRING SPLINE TO CREATE "SWEET" CURVES. NOW STAND BACK AND LOOK AT THE SHAPES VERY CRITICALLY FROM EVERY DIRECTION. IF IT DOESN'T LOOK QUITE RIGHT CHANGE IT AND LOOK AT IT AGAIN. ALTER IT AS OFTEN AS NECESSARY TO MAKE IT AS VISUALLY PLEASING AS POSSIBLE.

ONCE YOU THINK YOU REALLY "HAVE IT", TRACE THE TRAIL-BOARD SHAPE FROM THE HULL ONTO TISSUE. GLUE THE TISSUE ONTO A PIECE OF CARDBOARD AND CUT OUT THE SHAPE. TAPE IT INTO ITS PROPER POSITION ON THE HULL. NOW TAKE ANOTHER GOOD LOOK. IF SMALL CHANGES ARE NEEDED, MAKE THEM NOW.

BEFORE TRANSFERRING THE SHAPE TO THE LUMBER, SKETCH OUT THE SCROLLWORK ON THE CARDBOARD. REMEMBER THAT SHAPES OR LETTERING WHICH LOOKS FINE WHEN FLAT, MAY APPEAR AWKWARD OR OUT OF PROPORTION WHEN TWISTED OR BENT. TAKE ONE LAST LOOK AT YOUR DESIGN, THEN CUT.

MOUNT TRAIL-BOARD WITH FLAT-HEAD BOLTS (PLUGGED). TAKE CARE TO APPLY BEDDING COMPOUND TO BOLT SHANKS, HOLES, NUTS AND WASHERS.

TRAIL-BOARDS, THE VERY SOUL OF THE TRADITIONAL BOAT

DEVELOP QUARTER-BOARD WITH A TRACING AND CARDBOARD PATTERN IN EXACTLY THE SAME MANNER DESCRIBED FOR DESIGNING THE TRAIL-BOARDS.

QUARTER-BOARD MAY REQUIRE STEAMING OR LAMINATING TO FINISHED THICKNESS IF THE RADIUS OF TRANSOM IS SEVERE.

QUARTER-BOARD MAY BE CARVED, ROUTED OR CAREFULLY PAINTED.

MOUNT QUARTER-BOARD WITH FLATHEAD BOLTS. TAKE CARE TO APPLY BEDDING COMPOUND TO BOLT SHANKS, HOLES, NUTS AND WASHERS.

TRAIL & QUARTER BOARDS

bruce bingham yacht design

Box 1413, Santa Barbara, Ca. 93102

NO SCALES

QUARTER-BOARDS

Figure 347.

Plate 169. The finished interior of Jay Benford's 17' cat-boat, **Puffin,** is the epitome of tradition and rugged, but tasteful, finishing. Her "open" bulkhead layout adds to the vessel's spaciousness while dark brightwoods lend warmth and personality. While this vessel was executed at the Benford yard in Seattle, Washington, literally hundreds of sisters are being built or have been successfully commissioned. (A Roy Montgomery Photo; courtesy, Jay R. Benford and Assoc., Inc.)

Plate 171. The joinery typified in the galley of the Monk-designed 65' schooner, **Wayward Wind,** is the most exemplary work I have ever seen from the hands of an amateur builder. J.W. McWilliams, a onetime cabinet maker, devoted nine patient years to the creation of this lovely vessel—a monument to fortitude for which all builders should strive. (Courtesy, Mr. and Mrs. Walter J. Kiefer)

Plate 170. Looking into 20' **Flicka's** interior from the companionway hatch betrays an astounding spaciousness for a boat of her length. Teak trimming has been used exclusively throughout even to that of sealing exposed portions of the hull. At a glance you would think she was a wooden vessel. This boat was executed by Ferro-Boat Builders of Edgewater, Maryland, and is now a production model. Three settee berths may be clearly seen as well as her 6'3" galley which is convertible to a fourth berth. Shelving and stowage binds are ample and, again, the "open" bulkhead arrangement provides elbow space in her compact layout. (A Bingham design. Courtesy, Ferro-Boat Builders)

Plate 172. Another impeccable example of J.W. McWilliams' handiwork is seen in **Wayward Wind's** master head. Here, as elsewhere throughout the vessel, he has avoided becoming overly nautical while maintaining a practical and comfortable approach. (Courtesy, Mr. and Mrs. Walter J. Kiefer)

Plate 173. The starboard side of the main salon of Tom and Jean Carroll's **Windrift.** This is the most tasteful and warm interior I have ever encountered aboard a ferro-cement boat. This Hartley-designed 47' hull was constructed at the yard of Ferro-Craft in Auckland, New Zealand, while her finishing was in the hands of Jim Young Marine. The owner, an American civil architect, designed her own accommodations and, at this writing, has successfully completed an extensive tour of the South Pacific and a crossing from Auckland to the coast of California. (A Hartley design. Courtesy, Tom and Jean Carroll)

Plate 175. **Windrift's** galley is open and airy yet arranged for security in heavy seas. Lifelines can be seen hanging from the left side cabinet front. The planked pine sole gives excellent footing while eliminating most difficult cleaning problems. Notice here that the fiddles along the counter top have been discontinued at their corners as this also aids cleaning. (A Hartley design. Courtesy, Tom and Jean Carroll)

Plate 174. **Windrift's** salon dinette is convertible to a double berth by way of folding table legs. The quality of construction shown in this photo is typical throughout her interior. (A Hartley design. Courtesy, Tom and Jean Carroll)

Plate 176. A detail view of **Windrift's** salon credenza and salon bookshelf shows the end result of patience and fortitude, the common key to any quality yacht. (A Hartley design. Courtesy, Tom and Jean Carroll)

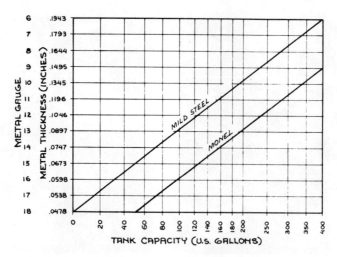

Fig. 348. Recommended metal thicknesses to be used for
black iron/mild steel or Monel fuel and water tanks.

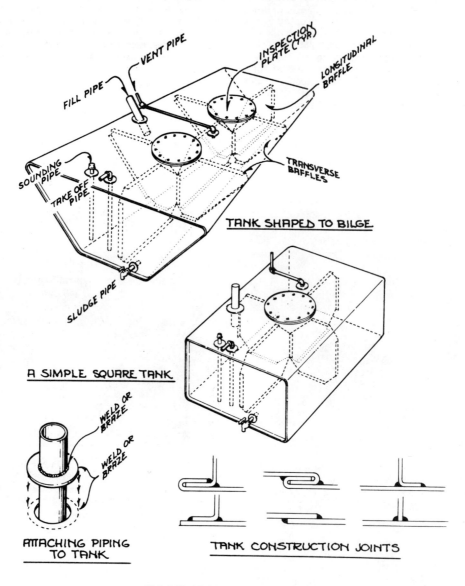

Fig. 349. Metal tank construction.

25

miscellaneous installations

Unfortunately, there are very few set standards upon which to rely when it comes down to the specific installations of machinery, electrical equipment, plumbing and tankage. Until just recently, almost all vessels were put together by men who simply had a "feel" for what they were doing. Generally, the designer stayed out of everyone's way when it got down to this stage of construction lest he find himself bombarded by dirty looks and words from the men who knew their trades better than he. Even Lloyd's and A.B.S. have very little to say when it comes to installations. The most outspoken and helpful of all the regulatory groups is the American Boat and Yacht Council Inc., 15 E. 26th St., New York, N.Y. 10010. It would be a wise idea to send $10 for their book, *Safety Standards for Small Craft.* Fortunately, your engine will be delivered with an extensive service manual which will include installation and wiring details and much of your other equipment (galley range, toilet, bilge pump, electronic gear, etc.) will be supplied with concise diagrams. Unfortunately, every piece of equipment from different manufacturers has its own individual quirks so it will be difficult for me to pin down hard-and-fast step-by-step procedures which can be followed to the letter. You are really going to have to feel your way along as you go. Once again, I must advise you, first, to call upon professional experts in their respective fields to help you stay out of trouble. A marine electrician is worth his weight in gold as well as a marine refrigeration specialist. Do not discount putting out a little money for their services, as the possible cost of damages suffered by hit-and-miss work may cover their expenses many times over.

Metal Fuel Tanks

As previously stated, I must discourage ferro-cement fuel tanks. I've heard report after report of leaking, cracking, bleeding and fouling of lines. My preference is for the use of metal or fiberglass tanks set into the hull, spaced away from the structure by means of wooden cradles or wood or styrofoam packing blocks.

I think the best tanks are constructed of Monel sheet or plate even though they are expensive. They will never pit or rust and are absolutely immune to corrosion. My second preference would be tanks of black iron or mild steel. In this case, the water tanks should be galvanized but not those for fuel.

When designing tanks, they should be provided with both longitudinal and transverse baffles, not only to prevent excessive sloshing, but to stiffen the tank as well. If the tanks are provided with curved surfaces (to the general shape of the hull), they will also be considerably stronger.

Working upon the assumption, then, that no expanse of tank surface shall exceed 18", the accompanying metal gauges should be chosen for construction of various tank sizes, regardless of shape. (Fig. 348)

All tanks should be fitted with a fill pipe which is fully enclosed from the deck fitting to within, say, 2" from the tank bottom. The fill pipe should be discontinued at some point for the insertion of a length of special flexible fuel hose. This hose section will help to prevent strain on tank and pipe fittings and joints caused by vibration and shifting. The gap between the metal fill pipe parts should be bridged by a braided ground wire.

Each tank must be provided with an air or overflow vent pipe. This pipe is attached to the tank top and led topside or overboard, its open end fitted with a metal gauze flame arrester or metal cloth screen. Each tank should be fitted with a sludge drain pipe with a shutoff valve in an accessible position. This pipe must be located at the lowest part of the tank and is used for purging or flushing the tank clean of foreign elements. Each tank is to be fitted with a "take off pipe" entering the top of the tank and reaching downward to within, say, 1½" of the tank bottom. Any closer will possibly cause the picking up of solid particles which inevitably find their way into the tank.

Each tank, if not fitted with a float gauge, should be fitted with a sounding pipe used for determining tank content. If a tank is being designed for diesel fuel, it may be necessary to provide a fuel overflow return inlet. You must check your engine installations manual to see if this provision is required.

I believe the best tanks are constructed using folded seams, sealed with a soft solder bead. This may not be practical with very large tanks using heavy gauge metal. In this case, welded joints (both sides) should be used wherever the tank material cannot be bent to form a corner. Baffle plates must be welded to the tank sides and bottom but may be riveted at their inner sections, if desired. When welding baffles, it is not necessary to use a continuous bead. I recommend the use of only small tanks welds on, say, 2½" centers, alternating from side to side. The swash holes cut into the baffle plates must be positioned to provide for the passage of fuel at any degree of heel or roll. The total area of swash hole openings in any given baffle need not exceed 20 square inches. (Fig. 349)

Fiberglass Tanks

These tanks are relatively inexpensive, long lasting and quite easy to fabricate. Their one main drawback is that they lend their plastic taste to the drinking water. In this regard, it should be pointed out that most fresh water outside of your native country will taste very foreign anyway, so I do not think this is a valid complaint. The fiberglass taste problem can still be licked by either steaming the completed tank for 18–24 hours or by coating its interior with Neoprene prior to closing off the top.

The construction of a fiberglass tank will first require the building of a temporary male mold. This is really no big thing at all. First determine the shapes of ends of the tank, allowing for proper space around the tank for blocking. A pattern for the tank ends may be made by using the "tick-stick" or by cutting a piece of cardboard to fit. Transfer the end patterns to ¾" plywood. Now erect the plywood tank ends, top down, onto a flat surface. Align the ends at their proper angles and distance end to end, and square to a center line. Once done, place 1/8" Masonite sheets over the bottom and sides of the tank forms. It should be noted that the corners and edges of the tank must be well rounded. It is advisable to install a "chine timber" of 1½" x 1½" stock along these corners. These timbers will not only support the Masonite edges but may be shaped to a generous radius. The lower edges of Masonite (at the ground or supporting surface) must be provided with some type of flange forming device. This device may be the supporting surface itself, and should be covered with waxed paper. Now, cover the finished tank form with waxed paper, using cellophane tape for attachment. Overlap all paper edges so that no wood or Masonite is left exposed. The tank body mold is now ready for glassing. (Fig. 350)

I will not go into the actual mechanics of fiberglassing, as these procedures are fully covered in many other good

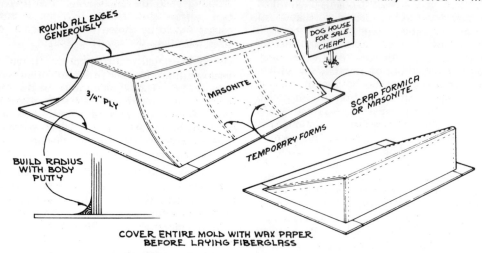

TYPICAL FIBERGLASS TANK MOLDS

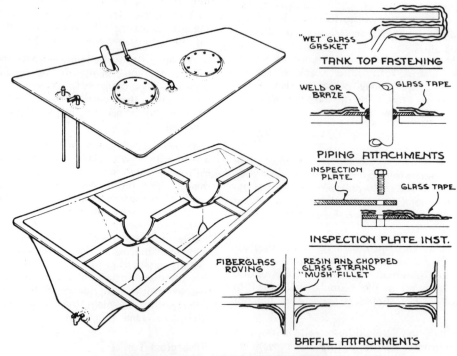

FIBERGLASS TANK ASSEMBLY

Fig. 350. Fiberglass tank construction.

books. The formula for achieving the required tank thickness, however, is shown in the accompanying table. The tank body, top and baffles should follow this schedule. (Fig. 351)

When all fiberglass layers have been applied to the tank mold and have been fully cured, the tank must be inverted, then the mold removed from within. The tank body is ready to receive the baffles and top. These units are cut from preformed fiberglass sheets. Fabrication of such sheets is accomplished by covering a piece of Masonite or plywood with waxed paper and simply building up the required fiberglass layers upon it. When fully cured, the fiberglass panel is lifted off and cut to the required shapes.

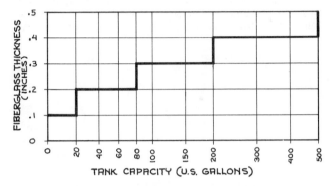

Fig. 351. Fiberglass thicknesses to be applied to tanks of varying capacities. Each .1" equals a laminate consisting of two layers of 1.5 ounce fiberglass mat and one layer of 24-ounce fiberglass roving.

The baffles are fitted into the tank and attached to the tank interior through the use of a generous polyester putty fillet, then covered with two layers of 24 oz. fiberglass roving tapes out to 6" widths. These tapes must be applied to each side of all baffle joints. A top attachment flange must be provided along the upper edges of the baffles. This may be provided by cutting 3" strips of prelaminated panel and fiberglassing them to the baffle edges. The tank should now receive the sludge drain pipe.

This pipe fitting is made in the same way as shown in Fig. 349. Their attachment to the tank, however, is provided by glassing over the pipe flange once the fitting has been placed into its hole. The pipe fittings being placed into the tank top will be attached in an identical manner.

The tank top, being accurately cut and fitted with all necessary piping connectors, may now be permanently attached to the tank. First cut several lengths of 1½ oz. fiberglass mat to 2" widths. Place these strips along the tank and baffle edge flanges. Now place a second layer of strips on top of the first. Mix a quart or so of slow curing bonding polyester and completely saturate the fiberglass mat strips. It will be impossible to use too much resin here. Once done, immediately position the top onto the tank. Place concrete block weights onto the tank top along the positions of the baffles to force the top to make a positive contact. Quickly place "C" clamps around the tank top flange at as close of intervals as possible. You will be able to see the nature of the joint as you apply pressure to the clamps. Do not clamp so tightly that you force all of the resin out of the mat gasket. Allow the gasket to cure fully. Once completed, wrap the tank top flange with 2 layers of 10 oz. fiberglass cloth and polyester resin for a final seal. The tank is now ready for testing and installation. (Fig. 350)

I must point out that the shower sump tank and waste holding tank may be fabricated in exactly the same manner as those for fuel and water. The piping is somewhat different, however. For these arrangements, refer to Fig. 352.

Tank Installations

Installing the tank, of course, requires that the tank cannot shift during heavy rolling or pitching. Heavy wooden chocks or frames should be fashioned to fit both tank and hull and are attached to the hull using ferro-cement webs or brackets or through the use of an epoxy grout fillet shown in Fig. 308. When positioning tanks, be sure to leave enough space under the sole beams to allow the running of piping. At least 2" of space should also be provided between the tank and the hull to allow for bilge ventilation and the avoidance of tank damage in case the hull is fractured. All lower surfaces of the tank should be as evenly supported as possible. If the tank support framing does not provide this adequately (although it should), additional packing blocks must be inserted between the tank and hull. These blocks may be fashioned out of varying thicknesses of wood or styrofoam and are simply epoxied to the hull. They should not be attached to the tank. Shaping of these blocks, unfortunately, must be accomplished on a trial and error basis and you may have to place and remove the tank a dozen times until a snug fit is finally achieved. Once the tank is in place, it must be well secured with metal straps or wooden chocks drawn tight with threaded rod.

The piping to and from tanks should be copper or brass tubing or Neoprene hose. Connections to couplings may be of the flared or threaded type. All couplings should be well coated with flexible gasket cement before drawing tight.

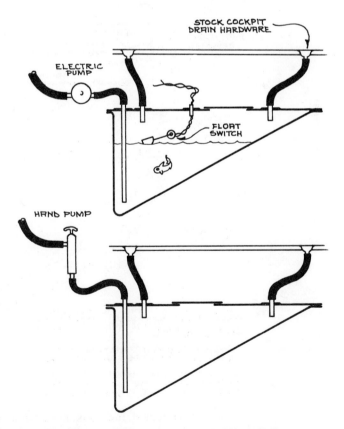

Fig. 352. Typical shower sump tank installation.

Water and fuel takeoffs should be fitted with bronze valves at the tank. Once again, insulation from vibration must be provided by inserting lengths of flexible Neoprene or special fuel hose, well clamped to the tubing or pipe ends. (Fig. 353)

Propeller Shafting Assembly

Before describing this installation, I must first set forth a word of warning. The metals generally developed for shafting are very specialized. The formulas have been determined to produce the highest torsional strength while reducing corrosion and electrolytic pitting. Many amateur builders simply purchase from their local metal suppliers a piece of mild steel, bronze or stainless steel bar stock of the proper diameter and label it "shaft." This will not do!

Shafting must be purchased from those companies who specialize in marine manufacturing and fabrication. When choosing the proper shaft, be very clear to explain the construction around the deadwood and the metals used in this vital area. Your shafting and propeller dealer, then, can more readily ascertain the proper metal formulation. If possible, take your Aft Installations drawing with you when you go shopping.

Your shafting and propeller supplier should also be given the liberty of suggesting the proper stuffing boxes, stern bearings and shaft flange couplings. In this way, the entire shaft assembly may be prefitted in his shop, which will insure the correct critical tolerances necessary for smooth and troublefree operation. Do not even think of turning this job over to an amateur machinist. He may not be fully versed in the S.A.E. standards necessary in this vital area. The taper of the shaft end, the propeller hub, the keyways

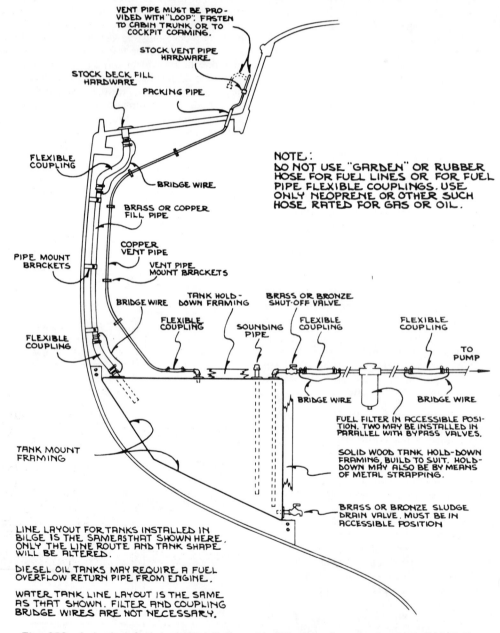

Fig. 353. A typical fuel tank installation. Modification for a water tank installation requires only the removal of the fuel filter and the elimination of bridge wires between flexible couplings.

at the prop and engine couplings must be extremely accurate so do not let just anyone tamper here.

The shaft train is composed of the shaft, the shaft tube or log, the tail or stern bearing, the stuffing box, the intermediate support bearings (if the shaft is very long), the shaft-to-engine flange coupling, the propeller, the propeller fastening nut and the propeller lock nut.

The shaft tube will have been previously constructed into the deadwood assembly. The stuffing box is first fastened to the inboard end of the shaft tube and slackened to allow the insertion of the shaft. The stern bearing is then placed in a bucket of ice to cause its slight shrinkage. When thoroughly cold, the stern bearing is inserted into the aft end of the shaft tube. Do not attempt to insert the shaft until the stern bearing has been allowed to warm slowly back to ambient temperature. Now, wet the bearing lining with soapy water and carefully slide the prop shaft into the tube and fully through the stuffing box. You may now affix the propeller if you wish. There are no special tricks here. Place the prop key into the slot provided, insert the propeller, place and tighten the first retaining nut, then the second hard against the first and finish it by inserting the cotter pin into the hole at the very end of the shaft. (Fig. 354)

Preliminary Engine Alignment

Prior to permanently installing the prop shaft, the engine must be lowered onto the engine beds. The engine mounting jack screws should be preset at mid-height (if fitted) and the shaft coupling flange attached to its counterpart on the gear box. Now sight through the prop shaft tube. You should be able to see the engine shaft coupling. Adjust the position of the engine transversely until it appears to be on the center line. Then raise or lower the engine mounts so as to accurately represent the shaft angle. The chances are that the engine coupling will still be below the sighted shaft line. While sighting through the tube, have another person place a tape measure over the shaft coupling. Read the distance between the shaft and the coupling flange. This distance will represent the thickness of the bed logs or hardwood shims which must be placed under the engine mounts to bring it up to its proper position. Remove the engine, fashion the wooden shims and place them onto the engine beds. Replace the engine.

Now the fun begins! Once again, sight through the shaft tube toward the shaft coupling. You will actually be able to tell whether the engine is angled too much or too little or whether the engine is too high, low, or off center. In a way, it is much like sighting down the barrel of a gun. The nature of the target (the shaft coupling) at the end of the sight

(the shaft tube) will indicate in what directions the engine must be moved. (Fig. 355)

When the engine alignment looks right to you, you may insert the prop shaft. Do not attempt to force the shaft into the coupling if it seems to bind. This will only bend the shaft or distort the shaft bearings. Go slowly. Adjust the engine jack screws until the shaft slips easily into the coupling.

Final Engine Alignment

First, loosen the shaft flange bolts then tighten the shaft coupling set screw. Now, tighten the shaft coupling bolts slightly so as to be certain that the flange bases are in contact with each other. There will be a small gap apparent between the flanges. Insert feeler gauges into this gap to ascertain the differences in the alignment angle. Adjust the engine jack screws upward or downward very slightly or insert thin galvanized sheet steel shims under the engine mounts until successive readings with the feeler gauges show an equal gap around the entire perimeter of the coupling flange. Rotate the shaft slowly to detect any possible binding. (Fig. 356)

If all seems in order, tighten the bolts or compression hex ring of the stuffing box. Now you are in business. Some

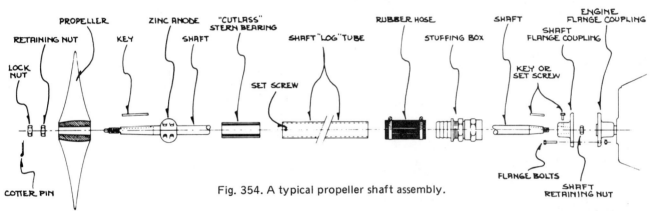

Fig. 354. A typical propeller shaft assembly.

realignment may be necessary after the vessel has been in commission for, say, twenty engine hours because there is always a slight amount of settling due to vibration or what have you. Do not be negligent in this regard. A perfectly aligned engine today may be slightly out of alignment within its first week of service.

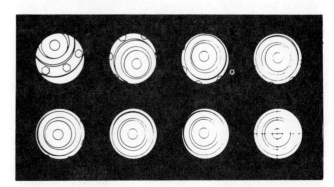

Fig. 355. Sighting up the propeller shaft tube reveals the position and orientation of the shaft coupling flange at the engine. It is quite easy to see sequence of engine movements from view to view as the engine is moved up, down, sideways and tilted about its axis. The final sighting (indicated by cross hairs) shows proper engine positioning.

363

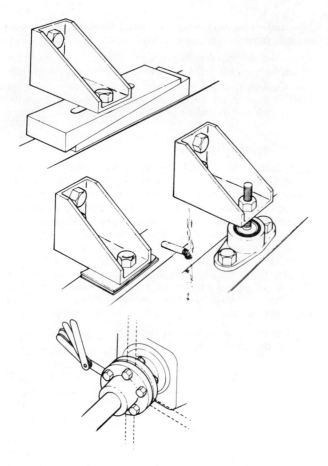

Fig. 356. Final engine alignment. Here are shown three variations of engine mount adjustments including double wedges, steel shims, and adjustable jack screws. Checking the engine's final alignment requires the use of a feeler gauge inserted between the coupling flanges from all sides in order to ascertain "parallel."

Throttle and Gearbox Controls

On all but the very largest boats, both throttle and gearbox controls are actuated through the use of push-pull cables. Such cables have been specially designed for manual as well as hydraulic transmissions. The push-pull cables must be made up in proper lengths by the manufacturer or distributor and measurements for such cables should be made with a flexible piece of wire passed along the control cable route. When making these measurements, be sure to allow for very generous radii to prevent crimping or binding of the control cable as well as neat and convenient mounting along bulkheads and/or beams. The control cables should not be pulled up tightly so be sure to allow a few inches of slack when taking measurements.

The dealers or distributors of push-pull cable control units will undoubtedly be familiar with the mounting requirements for your engine unless you have an "off brand" plant. Even in the latter event, the choice of proper clamp brackets and cable connections should not be difficult if you are able to supply pictures and installation drawings of the engine.

The typical push-pull control will consist of the following, working toward the engine.
1. Hand lever connected to a rocker arm.
2. Cable core pivot or eye terminal connected to a rocker arm.

3. Cable jacket anchor bracket at control lever.
4. Push-pull cable assembly to engine.
5. Cable jacket anchor bracket at engine.
6. Cable core pivot, jaw, ball joint or terminal eye on throttle or gearbox control lever.

To install the push-pull control unit, first locate and affix the helmsman's lever assembly at the desired position within easy reach of the wheel or tiller. Route the cable assembly from the helmsman's control lever to the throttle or gearbox lever on the engine. Attach the cable core terminals at both ends to the respective levers. Now fasten the cable anchor bracket to the engine or gearbox according to your engine or cable installation instructions. Affix the cable control jacket sleeve to the anchor bracket so that the sleeve cannot move backward or forward. Now place the gearbox and helmsman's control levers in the neutral positions or the throttle levers in lowest speed positions. Affix the cable jacket sleeve to the anchor bracket at the helmsman's end of the cable then fasten the anchor bracket to the bulkhead or backside of the control panel. Test the control. If any adjustments are necessary, they are made by altering the position of the cable anchors at the helmsman's control end of the cable or by resetting the cable core terminal into a different hole in the helmsman's control lever rocker arm. Fasten the cable to bulkheads or beams along the cable route using cable "U" clamps.

It is possible to set up multiple station control units but the details of such installations are much too complicated to be described here. Very large engines may require chain/sprocket control units but, because they are so specialized and variable, I will simply suggest that you consult with your local control specialists. (Fig. 357)

Choke or Fuel Shutoff Control

These units will use push-pull cable controls in much the same manner as for throttle and gearbox controls. Choke or fuel shutoff controls, however, do not require lever action but rather a simple push-pull knob or "T" handle. The measurement and installation of the choke or fuel cutoff control is much the same as the throttle and gearbox control except that the upper end of the control cable does not require an anchor bracket to fix the position of the cable jacket sleeve. Such fastening is accomplished by simply screwing the control panel retaining nut onto the cable jacket threaded sleeve.

Exhaust System Prerequisites

There are many variables in the design of an exhaust system and these variables can take many different forms. All types of exhaust systems should have several common factors:
1. The system must be absolutely gastight from engine manifold to hull outlet. This is particularly important in the event that the ship must be run with closed ventilation such as might be the case during heavy weather.
2. There must be no restrictions to the free flow of exhaust, as this otherwise may cause back pressure which could damage the engine. Restrictions may take the form of tight bends in the exhaust route, kinks in the flexible rubber hose couplings, too small of an exhaust line, a silencer of insufficient size or solid water in the line.
3. Solid water must not be allowed to enter the system in quantities which could cause back pressure or which could find its way into the manifold or engine. This precaution is most usually satisfied by constructing a loop in the exhaust route well above the stern or quarter wave crest height.

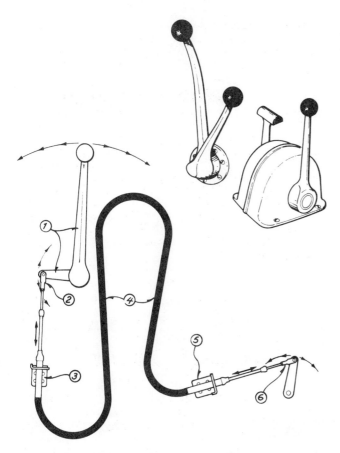

Fig. 357. "Push-pull" table controls. They are most commonly used for clutch, reversing gear shift, throttle, choke, diesel fuel shutoff, and many other applications. Many variations of the push-pull control are available from a variety of manufacturers specifically designed for a given purpose.

Transom exhaust flanges are also available with metal flappers which prevent large amounts of water from entering the pipe. They are a good bet. Some builders suggest installing a seacock within the exhaust line, but this may cause an engine explosion, particularly if starting the engine with the seacock closed. Condensation and excess cooling water may also collect in the exhaust system thus causing back pressure upon starting. A "sump loop" must be provided within the exhaust route which will prevent this water from entering the engine or causing restriction. Such a loop should be fitted with a drain cock for purging the exhaust system when necessary.

4. Exhaust cooling or insulation is a must. Most pleasure boats employ a "wet" exhaust system whereby the engine cooling water is injected into the hot exhaust gas, thus reducing the line temperature to a safe degree. Dry exhaust systems do not draw upon any cooling method and, consequently, the exhaust piping temperatures can cause the risk of fire, excessive heat inside the boat, or the severe drying of structural timbers within the vicinity of the line. With the dry system, insulation in the form of fiberglass padding covered with asbestos should encircle all piping.

5. The exhaust piping must be insulated against shifting and vibration. Such elements can cause the loosening or failure of joints, excessive noise within the vessel, possible damage to the engine exhaust manifold, the breaking down of silencer components or the effectiveness of the watertight seal at the piping through the hull.

6. Vibration insulation is most often achieved by installing lengths of flexible exhaust hose between the engine manifold and exhaust pipe, at each end of the silencer, and the connection between the piping and thru-hull fitting. Solid portions of the exhaust line (pipe sections) are generally secured to the hull, sole and deck beams or bulkheads through the use of chock blocks covered with metal straps. Of course, a well aligned shaft and finely tuned engine can eliminate most severe vibration from the beginning.

Exhaust Line Material

Although most exhaust detailing will show galvanized or black iron pipe for the consideration of cost, it is rarely the preferred alternative. Copper or bronze pipe is far superior because it is long lasting although very expensive. Never allow different types of metals to touch. Provide (except for diesel installations) rubber exhaust hose "bridges" between different metal types. More and more often we see professional builders constructing the entire exhaust line using flexible exhaust hose with the exception of the manifold "takeoff" section assembly, the water injection pipe and sump loop. Exhaust hose is often simply termed "steam" hose. Of course, this material is very inexpensive and will allow you to commission your vessel at a considerable savings . . . but the use of hose should be restricted to those areas where the exhaust route is conveniently accessible for periodic inspection and replacement. It definitely should not be employed where it will be out of sight. Exhaust hose does become brittle and will rot after some length of service, much in the same way as the cooling pipeline on your car.

Each and every joint along the exhaust route should be coated with a flexible hot gasket cement prior to assembly and each end of all hose sections should be well fastened with **two** stainless steel hose clamps.

Exhaust System Types

Wet line. This has become the most popular among both sail and power boat owners because it provides the coolest piping and it does not require fancy silencer types or complicated weldments. Its primary drawback is that raw salt water is constantly attacking the inside of the exhaust line which, of course, causes the deterioration of the pipe. This is accelerated through wetting and drying. Silencers are readily available for the wet line system but, here again, its replacement should be anticipated at more regular intervals than with a dry exhaust system.

Your engine cooling system will be equipped with a raw water outlet coupling. A small diameter pipe is connected to this coupling, then bridged with a flexible hose for vibration insulation, thence coupled to the exhaust manifold takeoff pipe. Your engine manufacturer will provide drawings showing the minimum drop and distance from the manifold to the takeoff pipe. The turns of the pipe manifold should be constructed with a series of 45° "els" in order to "soften" the severity of the bend in the pipe line. The exhaust pipe should run some distance downhill where it will be fitted with the sump loop drain cock. This drain cock must be the lowest point in the exhaust system. Now, the line will proceed upward toward the "water back-up" loop then downward again to the thru-hull flange. The silencer may be placed at any convenient position from the sump loop aft. (Fig. 358)

Wet standpipe line. This system cools the exhaust line in much the same manner as the wet injection line, but rather

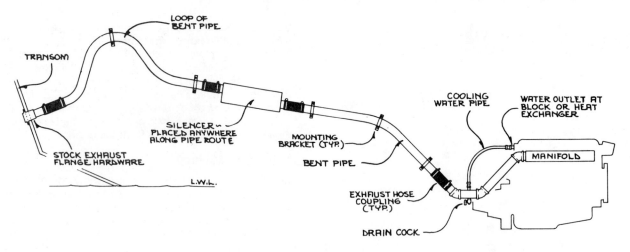

Fig. 358. A wet line exhaust system.

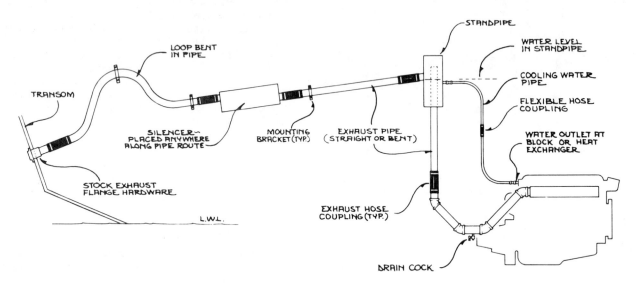

Fig. 359. A wet standpipe exhaust system.

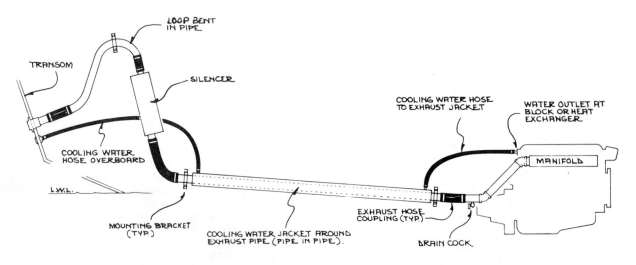

Fig. 360. A wet jacketed dry line exhaust system.

than allowing the cooling water to be directed immediately into the exhaust system, it is collected in an expansion tank to a given level, then "spilled" into the exhaust line. This system has the one benefit of preventing raw water from entering the exhaust manifold from the aft end. The expansion tank is a somewhat complicated iron weldment (galvanized after fabrication) which is located as near the engine as possible as well as above the waterline. This latter requirement usually creates difficult space problems. (Fig. 359)

Dry line with wet jacket. This system is generally only used on power workboats or yachts provided with exhaust funnels. The exhaust line rarely leads to the transom or anywhere near the waterline but rather is routed through the cabin or superstructure top. This is often necessary when it is impractical to pass the exhaust line through fish holds, cargo spaces, or when the exhaust line would otherwise be extremely long. The water pressure build-up due to the exhaust length or height under these circumstances, with a wet line system, could not be overcome by the exhaust pressure and thus would find itself back into the manifold.

The dry line with wet jacket system requires that the exhaust takeoff pipe be fitted into the manifold where it first leads downward to a sump loop and drain cock, thence upward to the exhaust funnel. A welded, galvanized jacket then encloses the greatest possible portion of the exhaust pipe. The cooling water is piped from the engine to the base of the jacket where it fills the jacket with water, thus cooling the exhaust pipe. The water is then piped from the top of the jacket downward and over the side via a thru-hull fitting. (Fig. 360)

Dry line. This system is used only when a short exhaust route is encountered which passes through large, well-ventilated spaces not affected by high heat. The dry line is most usually found on power workboats with short vertical stacks or on small power launches and tenders. This system is simply a direct piping of hot exhaust. No cooling is provided, but it is equipped with a silencer as with all other systems. I cannot recommend the dry line for use on sailboats. (Fig. 361)

North Sea exhaust line. This system may be wet injected dry line with wet jacket, wet standpipe, or dry. Its principal characteristic is that it is designed with two exhaust outlets, one on each side of the vessel. The attributes of the North Sea line are: (1) it shortens what otherwise would be a long route to the transom or funnel, (2) the exhaust route eliminates excessive heat build-up under compartment flooring, (3) it absolutely eliminates any possibility of exhaust back pressure caused by water entering the exhaust line. Essentially, the make-up of this system only requires that a pipe "T" be installed in the line any place aft of the water back-up loop, where separate straight pipes are directed to the sides of the hull a few inches above the waterline. As with other systems, it will be fitted with a sump-loop, silencer, hose bridges for vibration insulation, and water back-up loop. It is primarily only a modification of the other foregoing systems.

Auxiliary Generators

Your main engine will be equipped with a small generator or alternator when delivered by the dealer. This generator (or alternator) will usually be of a capacity which will provide recharging of batteries within one hour of running after, say, twelve engine starts. The generators so supplied

are not designed to meet the normal electrical draw of the ship's "house" circuitry under live-aboard conditions unless you are prepared to operate your engine for many uneconomical hours every day. The standard generator equipped engine may be sufficient for occasional weekend sailing but hardly adequate for extended cruising. It is, therefore, quite normal to order an optional "heavy duty" engine mounted generator (or alternator) in addition to the smaller standard generator. Usually each generator, then, will be attached to opposite sides of the engine but powered with a common "V" belt. If provisions for a second generator have not been allowed by your engine manufacturer, you will have to build a bulkhead mount or weldment for engine mounting which will be placed in such a position so as to align the belt pulley of the extra generator with the "takeoff" pulley on the main engine. With such an installation, your small standard generator, say 40 amp, would be used for charging the engine starting batteries and the larger generator, say 80 amp, would be patched into the "house" circuit batteries. If your daily "house" draw is calculated to be beyond the reasonable output of an 80 amp generator, you should consider the installation of a large capacity unit equipped with its own engine.

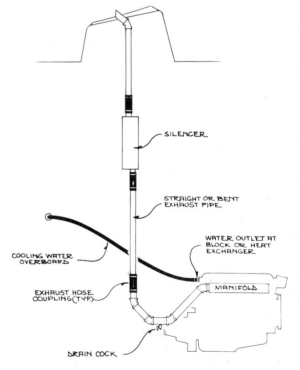

Fig. 361. A dry standpipe exhaust system.

Fortunately, almost all large generators are available as power unit/dynamo assemblies. Rarely will a builder require the coupling of unmatched units. Virtually all available generators are supplied on vibration insulated pads bolted directly to steel mounts. You need only provide simple, horizontal bedding to which to bolt the generator unit. Many small generators are equipped with silencers and heat exchangers so their installation boils down to leading the fuel line with filter, raw waterlines (which can be taken off of other internal salt waterlines to avoid extra thru-hull fittings) and routing an exhaust line.

The prerequisites of the generator exhaust system are identical to those of the main engine so no more need be said here. Of course, there is no gearbox control and the power unit throttle, if remote, is a simple push-pull cable exactly like your main engine choke or fuel shutoff control. All but the smallest generators are fitted with a governor and the speed of the power unit is adjusted so as to just actuate the governor when the generator is under a load. A change in the electrical load, then, normally requires a change in the power unit throttle.

I have seen 60 amp generators powered by tiny single cylinder engines no larger than a loaf of bread. These are great for charging ships' batteries if your requirements are not heavy because the generator can be installed almost anywhere within the vessel without cramping otherwise valuable space. They may be attached to engine compartment lockers, inside the cockpit lockers, lazarette, forepeak bulkheads, etc. When you get into larger generators, however, you will have to depend upon the equipment installation manual for specific instructions. Whichever type of generator you decide to install, the patching in of current to the ship's batteries will be essentially the same and will be shown schematically under **Electrical Systems.**

Basically, the requirements of a generator engine are the same as your main engine. In this regard, special attention should be paid to accessibility for servicing, ventilation, compartment sound-proofing and fire protection. Do not discount these factors even though the unit may be small. The installation, of course, will be subject to Coast Guard inspection.

Rudders

A rudder's sole purpose is to alter the hydrodynamics of the hull, thus creating transverse lifting forces which cause the vessel to change its direction. The proper application of the rudder in the hands of the helmsman will depend upon his interpretation of movement of the vessel and the forces playing against the rudder. The feel of the rudder, therefore, must be purely hydrodynamic. All other forces which may otherwise influence the helmsman's interpretation must be eliminated as completely as possible. The rolling of the vessel, for instance, should not cause a tendency for the rudder to swing back and forth nor should the heeling of the vessel under sail cause the rudder to fall to leeward or float to windward. It is imperative, therefore, that the rudder be designed and constructed in such a way so as to cause it to be neutrally buoyant, i.e., it should neither float nor sink, thus its weight will have no effect whatsoever on the vessel, the wheel or the tiller.

Over the past several years I have seen a dangerous trend toward the construction of steel plate or ferro-cement unbalanced rudders, as most often found on sailboats. Frankly, I can think of no other factors that could more readily affect the safety or handling characteristics of a yacht. It has been a time proven and accepted premise that a well-balanced sailboat will always head into the wind so long as her helm is unattended. This is a particularly necessary tendency when a boat is being driven under severe wind and sea conditions. Should a vessel under such conditions be knocked down momentarily by heavy gusts, she would automatically "head up," thus relieving the heeling forces on her. When a vessel is headed downwind and rolling from side to side, the rudder should be angled away from the direction of the roll, not toward it. Such a rudder angle would be directly opposed to the inherent tendency of a negatively buoyant blade. With a steel blade or ferrocement rudder, then, the helmsman would continually find himself having to fight against the weight of the rudder,

which would be acting in the exact opposite direction of that desired. A negatively or positively buoyant rudder will have a severe (if not overwhelming) influence on the self-steering characteristics of any boat, as well as placing unreasonable loads on an automatic pilot, should one be installed.

So what's the answer? Many of today's very finest production racing boats and cruisers are being equipped with rigid foam core fiberglass covered rudders. In themselves, they are positively buoyant but this is usually offset through the weight of the rudder stock and stock extension plates. Such rudders are very efficient and are easily built, even in the hands of amateurs. Efficient rudder blades may also be constructed by laminating several layers of marine plywood together in alternating directions. Marine glue is used for laminating and the blade is fiberglass covered after shaping. Here again, the positive buoyancy of such a blade is offset by the rudder stock and stock extension plates.

I have also seen some very fine strip planked rudders fashioned by edge gluing and nailing square fir or mahogany planks. The buoyancy of these blades is normally counteracted through the insertion of a lead plate or block. The attachment of the blade to the rudder stock is through the use of bronze, Monel, or stainless steel rods.

Drifted planked rudders, most commonly found on boats constructed prior to 1950, are still considered efficient and lasting, especially when fiberglass covered. Neutral buoyancy is provided here by the weight of the drift pins and stock connecting rods.

The types of rudder construction briefly described above may also be used on power boats as well as sailboats. (Fig. 362) Most power boats, however, employ a semi-balanced rudder which tends to offset some of the buoyant forces on the blade. The weight of a power boat rudder, therefore, is rarely as critical as that to be expected on a sailboat.

The hydrodynamic forces required of a rudder to steer the boat depend solely upon the transverse lifting forces which the rudder is able to create. The size of the blade is important, but size alone does not dictate the hydrodynamic efficiency. The size of a rudder may often be reduced in area simply by constructing a more hydrodynamically efficient shape while maintaining the same lifting force. Flat plate is one of the least efficient rudder sections known and thus should be avoided if at all possible. On a sailboat with the rudder attached to a skeg or keel, the rudder blade should be shaped and faired in such a way so as to become as much of an integral hydrodynamic foil with the keel or skeg as possible. In the case of a free-standing rudder, however, it should take the shape of a wing foil. Such shaping described can be fashioned through any of the construction methods previously mentioned. Of course, there are many variations which may be applied to these construction methods and you must use your own discretion if your architect has not detailed this installation. (Fig. 363)

Cable Steering

The components for cable steering from the wheel to rudder stock are generally as follows: steering wheel locking nut, wheel (available in a half dozen different styles), the hub/chain sprocket unit, sprocket chain, flexible stainless cables (port and starboard), central double idler sheaves (directly below the chain sprocket), cable fairlead sheaves (port and starboard; several pairs may be required), quadrant, cable tension adjustment eyes (fitted to quadrant), and quadrant-to-shaft locking key.

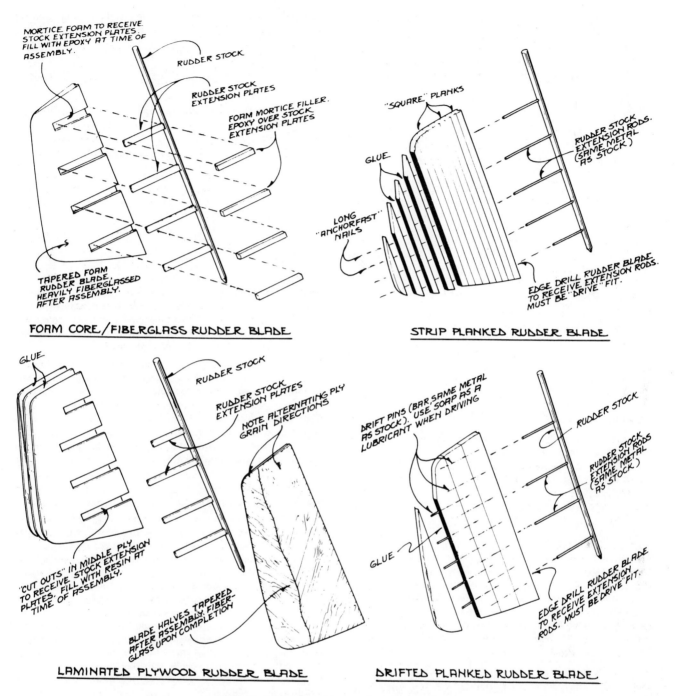

MORTICE FORM TO RECEIVE STOCK EXTENSION PLATES. FILL WITH EPOXY AT TIME OF ASSEMBLY.

RUDDER STOCK

RUDDER STOCK EXTENSION PLATES

FORM MORTICE FILLER. EPOXY OVER STOCK EXTENSION PLATES

TAPERED FORM RUDDER BLADE HEAVILY FIBERGLASSED AFTER ASSEMBLY.

FORM CORE/FIBERGLASS RUDDER BLADE

"SQUARE" PLANKS

RUDDER STOCK EXTENSION RODS. (SAME METAL AS STOCK.)

GLUE

LONG "ANCHORFAST" NAILS

EDGE DRILL RUDDER BLADE TO RECEIVE EXTENSION RODS. MUST BE "DRIVE" FIT.

STRIP PLANKED RUDDER BLADE

GLUE

RUDDER STOCK

RUDDER STOCK EXTENSION PLATES

NOTE ALTERNATING PLY GRAIN DIRECTIONS

"CUT OUTS" IN MIDDLE PLY TO RECEIVE STOCK EXTENSION PLATES. FILL WITH RESIN AT TIME OF ASSEMBLY.

BLADE HALVES TAPERED AFTER ASSEMBLY. FIBER-GLASS UPON COMPLETION

LAMINATED PLYWOOD RUDDER BLADE

DRIFT PINS (BAR, SAME METAL AS STOCK). USE SOAP AS A LUBRICANT WHEN DRIVING

RUDDER STOCK

RUDDER STOCK EXTENSION RODS (SAME METAL AS STOCK)

GLUE

EDGE DRILL RUDDER BLADE TO RECEIVE EXTENSION RODS. MUST BE DRIVE FIT.

DRIFTED PLANKED RUDDER BLADE

Fig. 362. Sailboat rudder construction variations.

The steering compass, wheel sprocket unit and even engine controls may be housed in a commercially available or custom-built pedestal. If you construct your own pedestal, say out of mahogany or teak, then you must specify a "bulkhead" type of sprocket unit. This unit, then, is placed within the pedestal and through-bolted to its front. If you choose a commercially-built pedestal, order your sprocket unit from the same manufacturer, as he has engineered these items for complete compatibility. When ordering your sprocket unit, I strongly suggest that you also order a steering hub brake assembly, as this will allow you to lock your wheel in a given position from time to time. This is very handy when on watch alone at night, as it will give you the liberty of trimming sheets or warming your hands.

Some bulkhead mounting hub/sprocket units are available with integral brakes and I highly recommend this unit type as your first choice.

The idler unit is a weldment or casting supporting two steering cable sheaves. The purpose of the idler is to maintain parallel cable alignment downward through the pedestal thence directing the cable to the first pair of fairlead sheaves. The idlers are available in several styles meant specifically for different idler-to-fairlead angles. As this angle is very critical and adjustment of great benefit to reduce the risk of the cable from jumping the sheaves or possible wear on the cable, I think your choice should be toward the swivel type idler. This type, however, may not be used if

369

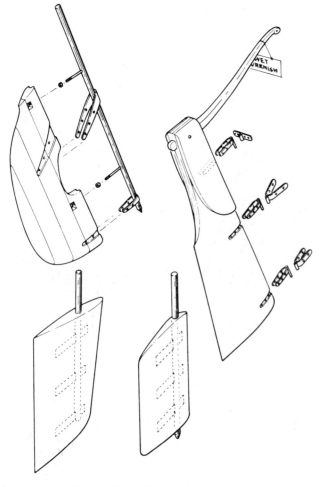

Fig. 363. Rudder design variations.

the run of the steering cable is parallel to the center line (which rarely occurs).

The installation of the idler is just a little tricky because the forward edges of the sheaves must be directly below the edges of the chain sprocket within the steering pedestal. I have found it a lot simpler to install the idler **before** the pedestal (and I've installed a few steering units). First, generally locate the pedestal on deck or cockpit sole and mark its position. Now place the idler upside down on the deck so that it corresponds to its installed position. Mark and cut the hole (or holes) for the passage of steering cable, then drill holes for the idler mounting bolts. Now place the idler into its installed position from the underside of the deck, bolting it firmly home. Once done, place the steering pedestal in its general position on the deck or cockpit sole. With the top of the pedestal removed, drop a small plumb bob downward from the sprocket edges. Adjust the position of the pedestal until the plumb bob aligns perfectly with the idler sheaves. Bed and bolt the pedestal into its permanent position. Now place the sprocket chain onto the sprocket gear, dropping the steering cables down into their respective idler sheaves.

Aligning and attaching the fairlead sheaves may also be a little tricky as, once again, their alignment is very critical. It will be first necessary to place the quadrant onto the rudder stock. Now run string along the steering cable route, i.e., around the cable slots of the quadrant, outward to the hull at the aft fairlead general positions, then to the forward

fairlead positions (if they are to be installed), thence to the idler sheaves. You may be able to hold the string at the hull with tape but do not pull the string too tightly, otherwise it will become dislodged. Adjust the vertical position of the string representing the aft fairleads so that they are in a direct alignment with the quadrant plane. Place a straight line onto the hull corresponding to the angle of the string leading forward from the fairlead position. At the same time, place a mark where the string changes its angle.

Procedure for establishing the angle of intermediate fairlead sheaves will be identical to that just described.

The attachment of the fairlead or intermediate sheaves to the hull should not pose any special problems except that it will require a considerable amount of drilling through the hull. The sheaves must be first aligned with the wire lead line previously established with the string. Carefully mark and drill holes into the hull which correspond to the holes in the sheave base. These holes should be countersunk for large, flat head bolts (stove bolts). The head of the bolt need only be set, say, 1/16" below the surface of the concrete. The fairlead is best installed against a hardwood block which has been fashioned to fit the shape of the hull snugly. This block must also be accurately drilled to receive bolts and should be well bedded to the hull. The sheaves may now be bolted permanently into position.

The steering cable is passed through the fairlead sheaves, into the coves of the quadrant and the cable clamp fastened to the tension adjusting eyes. These eyes should be tightened just enough to remove the slack from the cable. The steering cable should not become banjo string tight, as it will tend to bind up the steering system and wear out the bearings. Now test your steering system for alignment of cables, feel, rudder angle and binding. (Fig. 364)

Hydraulic Steering

The installation of hydraulic steering is relatively simple and I much prefer its use over that of cable steering because it does not use up the valuable interior space of the boat by virtue of the fact that the hydraulic lines are small, they can be bent to relatively sharp angles and the lead of the hydraulic lines can take any route desired. The helm hydraulic pump is no larger than the helm sprocket unit found on cable steering systems and is easily installed in almost any type of steering pedestal or may be bulkhead mounted. Generally only two holes are required to fasten the helm pump to the bulkhead or inner side of the steering pedestal. No special alignment is necessary at this juncture. The helm pumps are usually constructed of a non-ferrous metal so you need not be concerned about its effect on compass accuracy.

The piping or tubing used with the hydraulic systems may be of either metal tubing or pipe providing that it is rated for a minimum of 1,000 p.s.i. working pressure. The inside diameter of the piping or tubing will be specified by your manufacturer upon delivery of the equipment. Most hydraulic systems will require the installation of a by-pass cross valve. This will allow "free swinging" the rudder for the alignment of the steering wheel reference or when doing maintenance work on the rudder, etc. Hydraulic hose must not be used other than for the installation of short pieces which interconnect the tubing with the steering ram. Such short hose sections will provide an adequate measure of vibration insulation. Most hydraulic steering pump units are also equipped with a helm lock which is a handy feature when single handling, as it will allow you to tend sheets or halyards while the vessel continues on its way.

Vent lines will be required when more than one helm pump is installed and these vent lines should lead into an expansion tank. This expansion tank, partially filled with oil, will prevent air from becoming entrapped in the hydraulic lines and will assure a continual and adequate supply of oil.

The oil most generally used in hydraulic steering systems is automatic transmission fluid type "A" or type "B" or a non-detergent S.A.E. 10 engine oil designated for A.P.I. service "M." If two steering stations are being installed, the lowest steering station reservoir must be filled first, turning the wheel in one direction until the helm is solid then re-

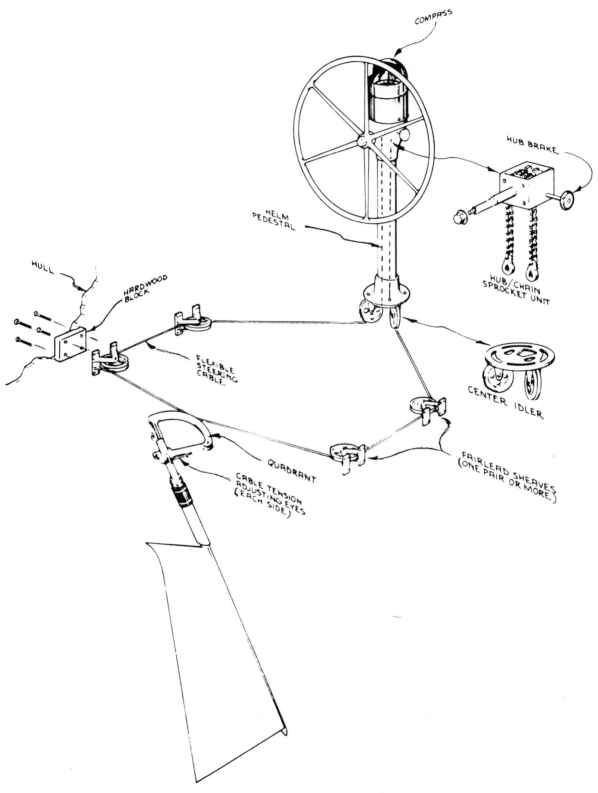

Fig. 364. A typical cable steering installation.

versing the helm direction until it is once again solid. As each steering station reservoir is filled, the helm is turned hard over each way to exert pressure on the line for about half a minute. This will allow any remaining air in the system to pass through the lock valve. After the system has been installed, it may be slightly stiff for the first few days of operation. This will usually be due to small amounts of air remaining in the system, but will eventually work its way to the pump and at such time the oil level will become stable.

Most manufacturers of hydraulic steering systems provide all necessary parts for the complete installation. When ordering the rudder stock tiller, you must specify the necessary diameter of bore. When machining the tiller to your specifications, the manufacturer will provide the proper keyway. Your rudder stock must be machined to match this keyway.

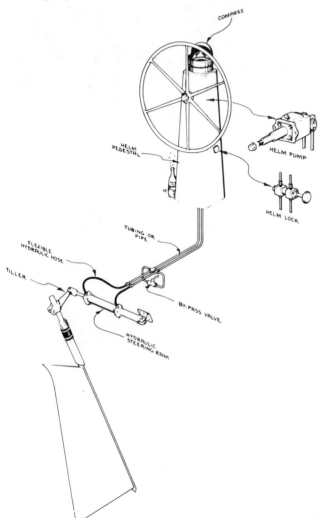

Fig. 365. A typical hydraulic steering installation.

It is impossible to cover all of the variables involved in the installation of hydraulic steering. I have, therefore, included a hydraulic steering diagram that will help you understand the system and its components. Of course, I must advise your reading your hydraulic installation manual thoroughly before proceeding. (Fig. 365)

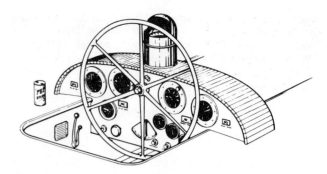

Fig. 366. The steering station or pedestal may take many different forms. Here is **Andromeda's** instrument, wheel, and binnacle mount which may be moved forward or aft according to the individual helmsman's taste. Flexible hydraulic steering lines allow such movement. The instruments visible in this drawing are speed log, depth sounder, wind direction indicator, wind speed indicator, engine tachometer, engine oil pressure ammeter, engine water temperature, instrument panel light switch, steering lock valve, inclinometer, heat start switch, fuel shutoff switch. Throttle and gear levers are mounted on the cockpit side next to the leeward bilge blower outlet. The opposite side of this pedestal (removable for instrument servicing) houses the navigational light switches as well as bins for binoculars and winch handles. A Bingham design.

Gravity Water Pressure System

At the outset, most yachtsmen generally assume that to have water pressure is to install an electric pump of some type. This, of course, does not come cheaply. The pump, itself, costs quite a bit of money as well as the valuable amperage it will use. The answer, then, is to consider a gravity pressure system.

Basically, the gravity system requires a normal water storage tank with all of the essential piping. The takeoff pipe, however, is not led to the pump at the sink but rather to a centrifugal or reciprocating transfer pump. This pump may be located anywhere in the vessel and its purpose is that of filling a small gravity storage tank (say 5 or 10 gallons) positioned high above the fresh water outlets. The gravity tank may be located in a box on deck or in the hanging locker (provided there is no shower) or somewhere in the bow of the vessel. The higher the better.

You may still provide hot water for the shower with the gravity system. The only difference between this and mechanical pressure devices is that the water pressure is natural. Of course, the gravity system does require a little manpower occasionally to keep the gravity tank full or the transfer pump may be electric. I believe the latter destroys the prime benefit of the system. (Fig. 367)

Motorized Water Pressure Systems

The most common layout for motorized water pressure is to locate one pressure pump in a convenient and accessible place within the vessel, piping water from the tank takeoff to the pump inlet, thence to a manifold. The manifold may be an assembly of pipe "tees" or it may be a series of "Y" couplings installed within the waterlines. The water pump manufacturers often recommend a larger or smaller pump according to the number of water outlets but I have never understood this rationale. The rarity of more than one outlet open at one time does not seem to justify the purchasing of larger than necessary equipment.

There are several basic ways of actuating the water pressure pump. The simplest and most popular method is by installing one water pressure switch anywhere within the water piping layout. This water pressure switch will instantaneously sense an open faucet and close the electrical circuit to the pump. Water runs from the faucet immediately.

Many water pressure pumps are available with integral pressure switches. I think these are the best. There are also individual pressure pump switches which are mounted at the base of each water faucet. This type of faucet is quite expensive and requires additional wiring. I do not think they are necessary. The water pressure pump may also be actuated by installing a toggle switch into the cabinet front at each faucet location. Here, again, the electrical circuitry will be unnecessarily complicated by such an installation.

There are two types of pumps most normally found on pleasure boats, i.e., reciprocating plunger and centrifugal. There is little doubt that the plunger type is superior as it generally provides more water pressure per amp than does the centrifugal pump. The centrifugal pump often utilizes a rubber impeller which softens when in contact with hot water, thus losing efficiency and possibly causing permanent damage. (Fig. 368)

Thru-Hull Systems

I, personally, like to install a seacock at the hull for every hose or piping line which leads outside of the vessel below the waterline. This may not be absolutely necessary if a particular thru-hull system leads above or terminates above the waterline inside of the vessel. A seacock at each thru-hull fitting, however, will provide the convenience of being able to shut the line off in order to remove the hose or pipe for cleaning and servicing. I have already discussed the provisions for the thru-hull holes and fittings in Chapter 18, Integral Structures. In the same chapter appears considerable detail as to the benefits of nylon thru-hull fittings and seacocks. I should add here, though, a few more words about completing the system.

If you decide to install thru-hull fittings other than nylon, your choice should be for bronze only. Do not even consider using brass gate valves, which are available from your local hardware store, as they will freeze and corrode readily through electrolysis. Several well-known American stock yachts have even gone to the bottom in recent years through the use of such brass fittings which had completely turned to powder through electrolytic corrosion. It is quite possible, however, to use P.V.C. (polyvinyl chloride) gate

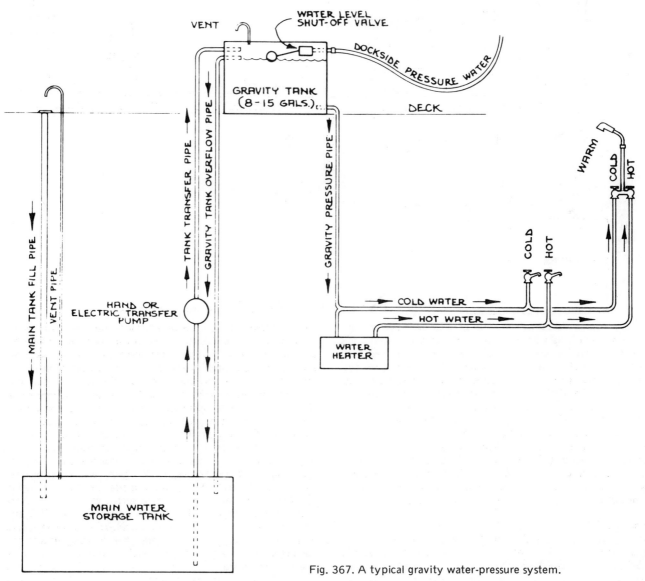

Fig. 367. A typical gravity water-pressure system.

valves and these may be purchased from your local plumbing supply store. When installing the thru-hull hardware or any piping or hose within the system, be sure to seal the connections by applying a coating of a flexible piping gasket cement. The inner flanges of the thru-hull fittings should be shimmed away from the hull using a hardwood block carefully shaped to fit the contour of the vessel. Bedding compound should be applied to both surfaces of this block as well as to the flanges of the hardware and to the inside of the hole through the hull. All intake thru-hulls should be fitted with a strainer screen to keep out foreign particles. This is particularly important in the case of main engine and generator cooling water intakes.

reduce the number of holes through the hull, you will find that it is not necessary to provide two thru-hull fittings for two or more cockpit drains, as they may be interconnected with "Y" couplings and discharged through the hull with one fitting only. It is also possible to interconnect the sink drains with the cockpit thru-hull system in order to eliminate another hole through the hull. Other similar interconnections are also possible. The galley salt water pump may be connected to the engine salt water cooling line. The shower sump drain line may be coupled to a "Y" gate at the bilge pump, thus discharging both systems through the same seacock. "Y" couplings, "Y" gate valves, reducing couplings, etc., are all available in P.V.C. material. I would

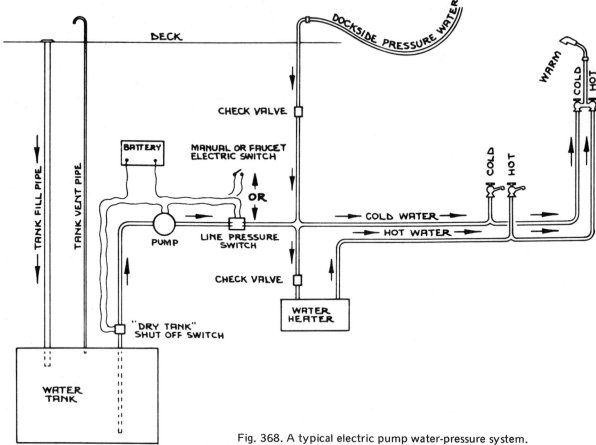

Fig. 368. A typical electric pump water-pressure system.

I have always discouraged the use of rigid piping within any type of vessel, as vibration can cause fatigue at joints and corners as well as creating additional electrolytic problems within the piping itself. Flexible synthetic tubing is my first choice. I can never recommend rubber as it has a tendency to rot in the dark, warm, moist areas of the bilge, etc. Rubber hosing can also disintegrate completely if it comes into contact with oil, gasoline, or other caustic bilge elements. Clear Neoprene hose is by far the best as instantaneous inspection along the length of any line will readily show any causes for clogging. This type of hose is almost immune to any type of chemical disintegration. It is available in a variety of useful diameters and may be purchased from any auto supply or marine hardware store.

The routing of your thru-hull lines should be as direct and as simple as possible. All thru-hull tubing should be well supported every two feet or so with metal "U" brackets. Two stainless steel hose clamps should be used at all connections of tubing and coupling hardware. In order to

choose these over the use of any metal hardware. Under no circumstances should you consider using a smaller diameter of tubing or hose than that recommended by the manufacturer of your plumbing equipment. If you are planning to install bronze thru-hull hardware, all of these fittings should be interconnected with a ground wire led to a zinc sacrificial anode. When installing a toilet, the discharge hose should be provided with a loop leading well above the waterline. This will prevent the possibility of back flow and flooding of the system in the event of heavy heeling or rolling. If a holding tank is being provided in the vessel for retaining the toilet refuse, a "Y" gate valve may be provided in order to allow the refuse to pass overboard when at sea. I have not been able to locate such a valve for 1½" hose or tubing manufactured in nylon or P.V.C., but it is my understanding that such valves are available in bronze and may be purchased through most fire fighting equipment suppliers. If an electric bilge pump unit is being installed as the primary purging device, the vessel should also carry a manual bilge pump, rigged, ready and accessible from top-

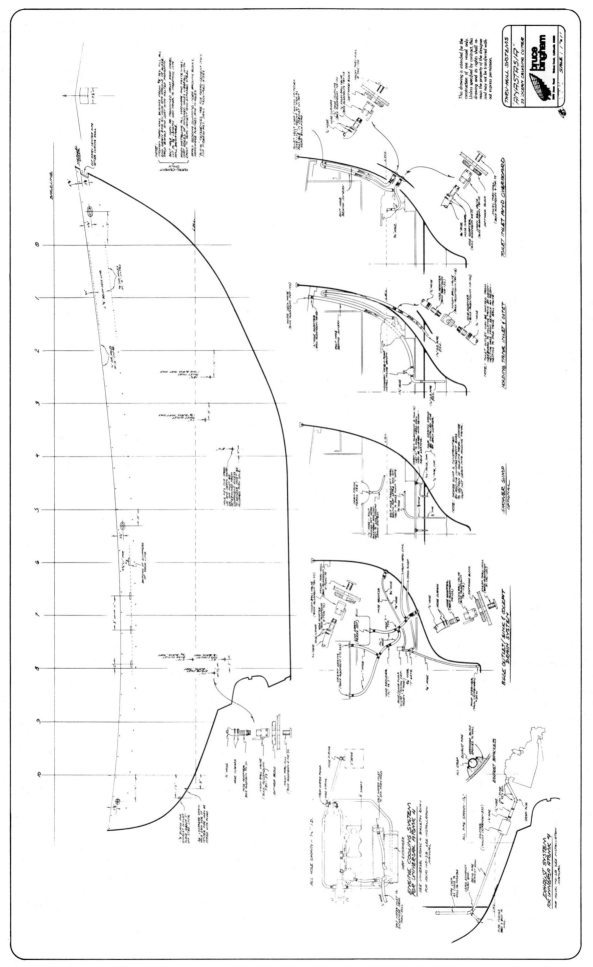

Fig. 369. Anastasia's Thru-hull System Drawing showing all intakes, outlets, couplings, connections and pipe routes.

side. I might also point out here that one of the most efficient emergency bilge pump units will be installed aboard your boat at the time you fit your engine. In dire emergencies, provisions should be made to disconnect the engine salt water cooling hose from the salt water pump in order to reconnect a hose leading from the bilge. Of course, the bilge end of this hose should be fitted with a strainer. By simply running the engine as necessary, vast amounts of bilge water may be removed through the engine cooling system. You should be cautioned, however, against allowing the bilge to run dry while the engine is still running.

Batteries

The type and size of batteries required on your vessel cannot be specifically pinned down by your designer as he cannot be fully acquainted with your daily draw of electricity under variable circumstances. Your choice, however, should not be arbitrary.

Determining your daily draw of electrical power will require your estimating as closely as possible the number of minutes or hours that each electric appliance is to be used. This estimate should include, for instance, the duration of use of each light bulb, bilge pump, water pressure pumps, navigation lights, refrigeration, etc., over a 24-hour period. When doing so, you should establish three separate criteria, i.e., dockside, light cruising (or weekending), and extended cruising (long periods of independent sailing). Once you have established durations of use, refer to your electrical schematic to determine the amperages of each of the appliances by multiplying the number of hours (or percentages thereafter) by the given amperages. You will be able to determine the daily "draw" of each appliance. By adding the daily draws of these appliances in each of the three criteria columns, you will derive the battery amperage capacity necessary to operate the vessel's "house" system for one day without recharging. It is now up to you to decide how long you intend to conveniently operate the vessel without having to recharge the house batteries. Of course, this will have a lot to do with your choice of generator or alternator. If, for instance, you discover that your daily draw under extended cruising conditions would normally be 120 amps, then a 60 amp alternator mounted on your engine would require two hours of running each day to provide the required electric daily current. Considering the same daily draw, a 120 amp alternator or generator would only require one hour of running. An 80 amp alternator or generator would require 1.6 hours of operation daily to maintain the house circuits. The choice is yours.

You will have noticed by now that I have not included the engine circuitry when determining the daily house draw. I much prefer to maintain the engine circuit on a separate battery and generator (or alternator) entirely. In this way if there is a severe electrical breakdown in the house line, you will not be left without a means of starting the engine and at the same time you will be able to switch the house circuit onto the engine starting battery thus maintaining primary lighting and navigation units. I believe this arrangement to be an absolute necessity on a boat intended for long-range cruising. If, on the other hand, you only intend to sail your vessel locally, then you will be able to get away with only one large battery or two smaller ones, provided with a master circuit switch capable of directing the current to either the engine starter or the house circuit. This is the setup that is usually provided by stock yacht manufacturers.

The type of batteries most often installed in vessels is not unlike those found in your car. They are of the lead plate-acid electrolyte type. They are the most serviceable and the most readily available. They do, however, produce volatile gases, particularly if they become overcharged. For this reason they should not be stored within the body of the vessel if other practical areas are available.

Nickel cadmium batteries have also become quite popular as they are generally of higher amperage capacities. I believe the nickel cadmium battery should be restricted to uses where large current draws are required for short periods of time such as engine starting or when the lighting load is relatively low.

When installing the batteries, be sure that they are placed onto a tray or other receptacle lined with a non-acid corrosive material. If you intend to use alkaline batteries, however, such a receptacle may be constructed of steel. The battery compartment must be well ventilated. This may be natural or mechanical and it is best to locate the vent inlet level at the lowest portion of the battery compartment. An outlet vent should also be provided, but this vent should not project deeply into the battery compartment. If at all possible, the battery compartment should be completely independent from other spaces of the yacht and avoid installing other electrical equipment in the same space. Of course, make absolutely sure that your battery cannot shift during heavy rolling or heeling of the vessel and provide for easy access and removal. (Fig. 370)

Electrical Systems

This is a subject that has always been abstract and complicated to me. I have never fully understood it. I have known very few people who do. One thing I can tell you, however, is that when things go wrong in electrical circuitry, they really go wrong and that no other element aboard the boat can be as much a headache if improperly executed (except for a persistent leak above your bunk). I have had to call upon Frank Beumer to give me a hand in this area. He is a delightful man who runs a marine electrical shop just down the road from my own office and he talks in terms of circuits and grounds and diodes and tubes and regulators as ordinarily as you and I would talk about the daily weather. I am still not quite sure that I understand all of Frank's definitions and descriptions but I will pass on to you as much of what made sense to me as possible.

I have always been confused between generators and alternators. To me, they were always things that looked like motors with a few extra wires at each end which essentially did exactly the same thing. Frank, however, tells me that there is a very definite difference. He explains that a generator produces only direct current and that the faster the generator turns, the more amperage and voltage it will produce. The r.p.m. required of a generator to produce a serviceable current begins at about 800 to 1,000 r.p.m. Therefore, if you are in dire need of rapidly charging your batteries with a considerable amperage, it would be necessary to rev your engine up quite highly. An alternator, on the other hand, produces only alternating current. This current is converted to direct current through the use of an integral rectifier. It has been explained to me that the prime advantage of an alternator is that it is capable of producing peak amperage and voltage at a lower engine r.p.m. than a DC generator.

Both charger and alternator will start charging at about the same engine r.p.m. but the alternator will reach its full output with less r.p.m. than that of the DC generator and will require a little less horsepower from the engine. The alternator, therefore, will charge the batteries quicker than a DC generator.

Very few stock American engines are delivered with generators. Both the marine and automobile industry have settled almost exclusively upon the benefits of alternators. They are capable of producing more amperage and voltage at a lower r.p.m. and with considerably less loss of horsepower from the engine.

Thus, an alternator should be your choice if purchasing generating equipment separately from the power plant.

The voltage produced by an alternator or generator is dependent upon the speed of the alternator or generator as well as the state of charge of the battery. If the battery is in a state of low charge, the amperage and voltage output of the generator or alternator will be high. As the condition of charge of the battery rises, the amperage and voltage of the generator or alternator will become lower. Because of this voltage and amperage variation, a voltage regulator must be installed between the generating equipment and the battery. Frank tells me that engine manufacturers normally supply the proper voltage regulator with the equipment. This is also true when purchasing an alternator or generator separately, so I will not go into any details concerning the choice of voltage regulating equipment or their types. It is important to note, however, that the voltage regulator which you install must be compatible with the generating

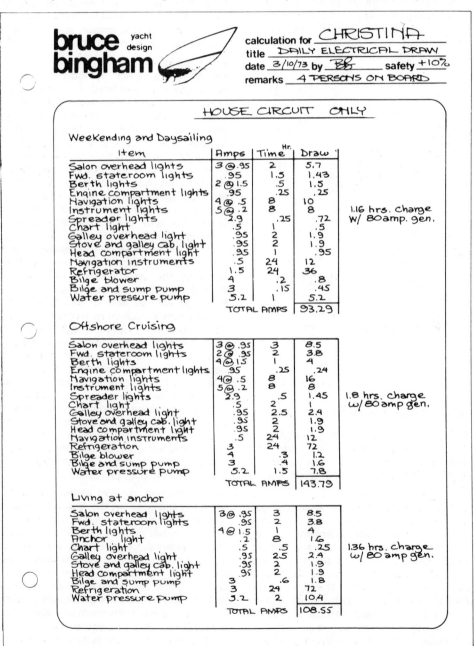

Fig. 370. A typical electrical requirements calculation for the 49' ketch, **Christina** under different service conditions. Generator size was chosen specifically to avoid more than two hours of daily charge.

equipment and should not be changed or modified even if tempted by a "good deal." Unfortunately, there is no stock schematic which can be applied to the wiring of all voltage regulators. However, such comprehensive instructions will be included with your equipment.

Some designers and builders have recommended that two alternators be supplied with the engine. A small alternator, say 40 amps, will be wired to serve the charging of only the engine starting battery. The larger alternator, say 65 or 80 amps, would then be wired only to serve the charging of the house circuit. Each alternator must be equipped with its own voltage regulator and each circuit with separate amp meters. This seems to have been the most common setup aboard pleasure boats in the past. A master circuit switch is usually installed in this kind of system which will allow you to transfer the current from the engine starting battery onto the house circuit as well as the house battery current to the engine starting circuit or from both batteries together onto either circuit at the same time.

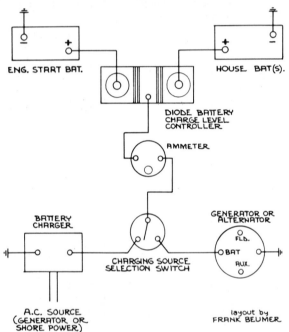

Fig. 371. Diode battery charge level control schematics.

Such flexibility in varying the circuitry through a switching system is very good as it allows the rerouting of power in the event of either battery's failure. Such a switching system is in complete compliance with all Coast Guard safety requirements.

Frank suggests that the separate alternator system is not necessary on small boats which are not intended for extensive cruising. He believes that one large alternator is more than sufficient to supply charging to both the engine start system and the house circuitry. With this setup only one voltage regulator is employed and the current is diverted to whichever particular battery is in need of charge and such diversion is accomplished through a network of diodes capable of sensing variations in voltage. Through the use of diodes, therefore, the current produced by the generator or alternator will be directed immediately to the engine start battery at the time the engine is being started and will continue to charge the engine start battery until the voltage of that battery reaches exactly the same level as all of the other batteries aboard the vessel. If you were to turn on the

bilge pump, the diode system would then direct the charging current to that battery being drawn up for the bilge pump, again for such a period of time required to bring that battery up to the same level as all other batteries aboard the vessel. This single generating system sounds quite complicated to me. Frank assures me that it is much simpler to install and maintain than that of using separate generators and that it is most definitely worth consideration for the builder sensitive to simplicity. (Fig. 371)

Wiring

The type of wire used aboard a boat should never be of the solid core type; only twisted or braided copper should be used, as its flexibility will reduce fatiguing through vibration. While rubber and Neoprene wire insulation is very commonly found aboard many boats, vinyl insulation should be your first choice, as it cannot be cut easily and may sustain abrasion far better.

Vinyl insulation will not "flame up" as a result of overheated wiring caused by short circuits or overloaded circuits. The vinyl insulated wire, therefore, should be considered as a viable safety feature. The abrasion resistance of vinyl insulated wire is also worth considering because the wire aboard a boat will be in almost constant motion through rolling, pitching and engine vibration. This motion over a long period of time could wear the insulation off of other types of wire. Needless to say, all of the wiring aboard your vessel should be clearly color coded. Such color coded insulations are available in literally dozens of combinations and sizes. As each circuit is run through the vessel, the wire color code should be marked on your schematic for future reference. All wiring aboard the boat must be kept well above the highest conceivable bilge water slosh level. You should even take into account the possible event of the vessel's taking on vast amounts of water through hull damage. Keep your wiring "high and dry" so that you will be able to maintain your bilge pump and lighting during such an emergency. The ship's wiring should not simply run loosely through holes in the bulkhead and behind joinery. It should be as completely enclosed as possible in some form of conduit for as great a distance as possible. This will also help to keep your wiring neat and organized. P.V.C. or P.V.A. pipe makes an excellent conduit which can be bent easily to the shape of the hull. When assembling the wiring conduit, provisions should be made to eliminate all possible entries of water but you must also provide for its disassembly in case of necessary wiring repairs. Obviously, all wiring conduit should be completely accessible. Never run wiring through the bilge, except, of course, for the bilge pump circuit.

The connecting of wires to the control bus boards and to the poles provided on the appliances should never be done by simply twisting an eye into the stripped wire. Stock electrical terminals must be employed throughout. Avoid using "jaw" terminals, as they can work themselves loose and dislodge through vibration. It is imperative, therefore, that only "eye" terminals should be used through the electrical system. Upon the completion and testing of the electrical system, all wire connections should be sprayed with silicon to protect them from corrosion due to the moist marine atmosphere. This one precaution alone can increase the trouble free life of your system by many, many years. Your choice of wire sizes should not be arbitrary or "eyeballed." In my own work, I do provide the recommended wire sizes on the Electric Schematic Drawing based upon reasonable and calculated appliance amperages. Realizing, however, that some builders will be providing for more appliances or increasing the watts of those shown on the Electric Schematic Drawing, I have included a chart to aid the builder's choice of wire sizes. (Fig. 372)

LENGTH OF ROUTE FROM POWER SOURCE TO FARTHEST APPLIANCE

6 VOLT SYSTEM

TOTAL AMPS ON CIRCUIT	10	15	20	25	30	35	40	45	50	55	60
5	14	14	14	12	12	12	10	10	10	10	8
10	14	12	10	10	8	8	8	8	6	6	6
15	12	10	8	8	8	6	6	6	4	4	4
20	10	8	8	6	6	6	4	4	4	4	3
25	10	8	6	6	4	4	4	4	3	3	2

12 VOLT SYSTEM

TOTAL AMPS ON CIRCUIT	10	15	20	25	30	35	40	45	50	55	60
5	14	14	14	14	14	14	14	14	12	12	12
10	14	14	14	12	12	12	10	10	10	10	8
15	14	14	12	10	10	10	8	8	8	8	8
20	12	12	10	10	8	8	8	8	6	6	6
25	10	10	10	8	8	8	6	6	6	6	4

24 VOLT SYSTEM

TOTAL AMPS ON CIRCUIT	10	15	20	25	30	35	40	45	50	55	60
5	16	16	16	16	14	14	14	14	14	14	14
10	16	16	14	14	14	12	12	12	12	12	10
15	14	14	14	14	12	12	12	10	10	10	8
20	14	14	12	12	10	10	10	10	8	8	8
25	12	12	12	10	10	10	10	8	8	8	6

32 VOLT SYSTEM

TOTAL AMPS ON CIRCUIT	10	15	20	25	30	35	40	45	50	55	60
5	16	16	16	16	16	16	14	14	14	14	14
10	16	16	16	16	14	14	14	14	14	14	14
15	16	14	14	14	14	14	14	12	12	12	12
20	14	14	14	14	12	12	12	12	10	10	10
25	14	14	14	12	12	12	10	10	10	8	8

110 A.c. VOLT SYSTEM

TOTAL AMPS ON CIRCUIT	10	15	20	25	30	35	40	45	50	55	60
5	14	14	14	14	14	12	12	12	12	12	12
10	14	14	12	12	12	12	12	12	10	10	10
15	14	12	12	12	12	12	10	10	10	10	10
20	12	12	12	12	10	10	10	10	8	8	8
25	12	12	10	10	10	10	8	8	8	6	6

A.W.G. SIZES SHOWN
LENGTHS OF WIRES IN FEET
CHARTS BASED UPON DOUBLE LEAD TWISTED COPPER

Fig. 372. To determine the proper wire size to be used for any sub circuit enter the proper voltage chart with the total number of amps on the given circuit. Reading from left to right choose the wire size which complies with the length of the circuit to the most distant appliance on that circuit. The length of the circuit must be measured from the battery to the most distant appliance, even though the sub circuit may begin at the control box.

The control box is the real center of the electrical system. It is essentially a switchboard through which the common power is supplied and rerouted. The more separate circuits provided aboard the vessel, the more individual switches, the more circuit breakers, then the safer your system will be from possible failure and even fire. Granted, it may be a pain in the neck to have to turn a switch at the control panel, in the compartment and at the light bulb just to turn on the berth light in the forward cabin, but such multiple switches do provide for a much finer isolation of individual circuits should a problem develop. Of course, with the multiple switching system, it will not really be necessary to turn all of the switches on and off at each appliance every time. The major circuit switches may be left on at all times except during periods of repair, so it is really not as bad as

it sounds. When going aboard American stock manufactured boats, I am continually amazed to see an electrical control panel provided with only two or three electrical circuit breakers, a master house switch, an engine circuit start switch, an interior lighting switch and a navigational lighting switch. While certainly such a setup is far easier to design and install, it also means that should the spreader light overload, its dependent navigational light circuit will not only lose the spreader light but the light at the compass, engine instruments, wind indicator and everything else being led into the same bus board. I would much rather see a main house circuit switch, a navigational circuit switch, an instrument light switch, running light switch and, then, separate spreader light switch in progression. Not only will the master house circuit have an overload circuit breaker but so will the entire navigational circuit as well as individual circuit breakers to cover each one of the appliances on that circuit. Granted, this means developing an electrical control panel with many more breakers and switches, but it could also mean preventing the burning up of a large portion of the ship's wiring and even a major fire on board. (Fig. 373)

Two large wires will be run from the opposing poles at the battery to separate bus boards within the control box. These wires should be of a very heavy flexible No. 4 to No. 2 battery cable. Of course, one of these wires will be considered "ground." Naturally, two wires are required to activate any given appliance. The appliance switch or circuit switch need only be provided within one of these wires in order to disconnect the circuit.

It would be far too difficult to describe all of the house circuit wiring verbally. You should, therefore, refer to the accompanying schematic, which may be used as a general guide for almost all moderate sized cruising boats. (Fig. 374)

Engine Wiring

Once again, it is virtually impossible to describe verbally the engine circuitry system and all of its many variations. Every engine and each model will have its individual quirks and requirements. Some engines, for instance, will require different types of instrument "takeoff" systems. Some engines will be equipped with fuel pre-heat elements. The type of generating equipment operating with each engine will demand variations in the basic wiring layout so follow your engine installation manual faithfully. Realizing, however, that many amateur builders have located and plan to install used engines, I have found it necessary to supply general wiring diagrams for them, as no other graphic material may be available. (Figs. 375-376)

If it is your intention to use your boat for long-range cruising or for live-aboard for extensive periods of time, the previously shown basic electric schematics may be modified slightly so as to include heavier duty battery and charging equipment than that normally provided aboard the "day sailer" or "weekender." (Fig. 377)

Electrolysis Protection

Every vessel is plagued with some degree of "loose" electrical current flowing through the hull, metal parts and electronic equipment, most particularly when the boat is at dockside and being energized with 110 volt current. Such loose current sources are often mysteriously undetectable, even with all circuit switches turned off. Such loose current, if allowed to flow through a metal part which is in contact with sea water, can cause the severe electrolytic

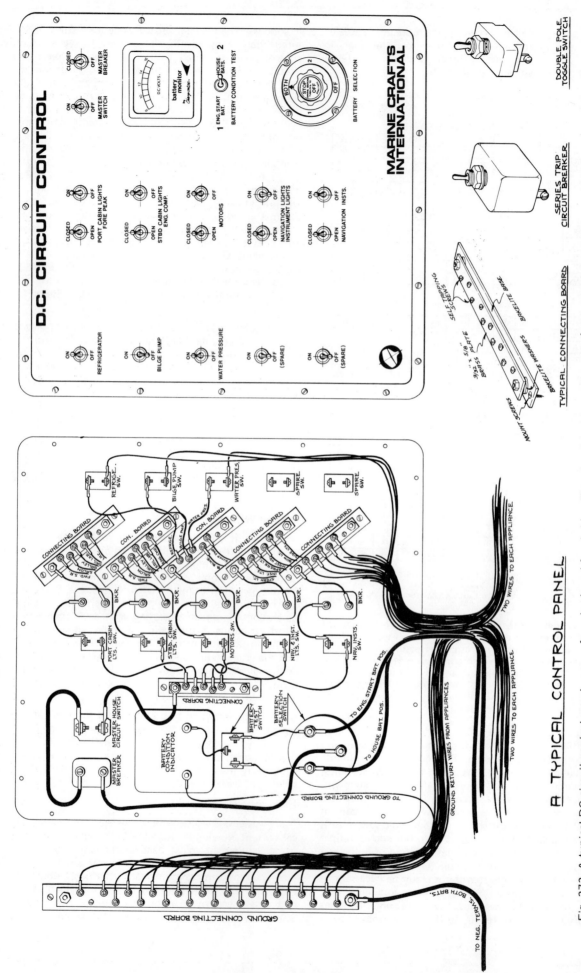

Fig. 373. A typical DC circuit control panel as seen from back and front sides. The circuit breaker capacities have not been listed here as they must be determined by the particular loads being carried by the individual circuits on your vessel. The components shown in this control panel may be purchased at virtually any electronics or electrical supply store. The actual schematic for the control panel may be clearly seen in Fig. 374.

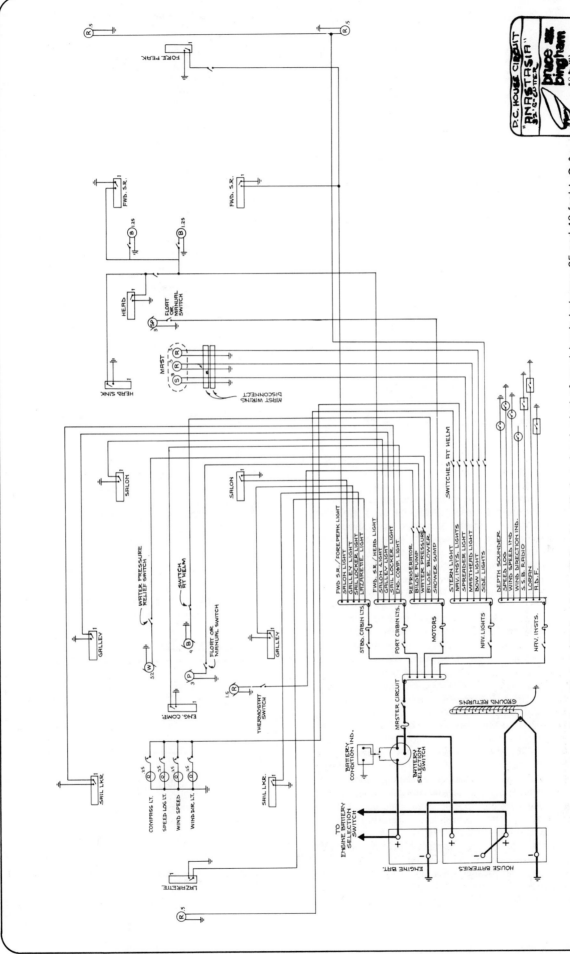

Fig. 374. A typical house circuit schematic for the 32' Anastasia. This schematic is not untypical of most boats between 25 and 40 feet L.O.A. Wire sizes and breaker capacities have not been indicated as they must be individually based upon the specific loads of each circuit on your own vessel. The numbers shown at each appliance indicate the recommended amperages. A Bingham design.

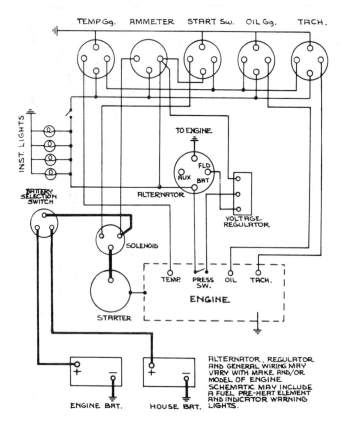

TYPICAL DIESEL ENGINE WIRING
by Frank Beumer

Figure 375.

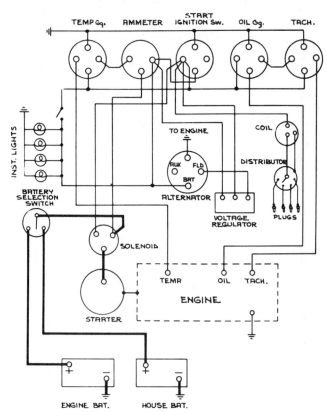

TYPICAL GAS ENGINE WIRING
by Frank Beumer

Figure 376.

disintegration of that part or others in its vicinity. It is absolutely imperative, therefore, that certain precautions be taken to avoid such disintegration. The most common method is by grounding all metal units to a sacrificial zinc anode below the load waterline. Grounded units should include the engine, generator, all separate pump and motor chassis, all electronic equipment chassis, toilets, fuel tank, sump tanks, water tanks, chain plates, aluminum mast (if one is to be installed), etc. Frank recommends such grounding in the following way: a continuous 18 gauge x 1½″ copper strap is mounted longitudinally along the full length of the hull. This grounding strap should extend from stem to stern and may be mounted to the hull by lag bolting it to wooden blocks epoxied to the hull's shell. The strap should be positioned well above the bilge water slosh level. This strap should not come into direct contact with the hull, as it would then be subject to moisture condensation and possible direct water flow from deck leaks and the like. The unit chassis ground wires are attached at the unit by using eye terminals fitted to any convenient bolt providing that such bolt is not also used as a terminal for AC or DC current flow. The ground wire from the unit chassis to the copper strap may be, say, 8 or 10 gauge twisted insulated copper. This ground wire is attached to the copper strap by means of an eye terminal on the wire being fastened with a self-tapping screw into the copper strap. Because there is always a possibility of the tapping screw vibrating loose or becoming loose through service, the eye terminal should receive a drop or two of noncorrosive flux. This will also insure a positive electrical contact. Along the length of the copper strap may be located literally a dozen or such chassis groundings. The copper strap, in turn, is wired to the zinc anode on the outside of the hull. This wire should

be of a 10 gauge twisted insulated copper bolted to the strap and attached to the zinc anode by means of an eye terminal passed over the anode through bolts. Such zinc anodes should be checked and all replaced at least every six months. Sacrificial anodes should also be fitted to the rudder stock, propeller shaft and any other metal part below the waterline.

The method of grounding the propeller shaft to the copper ground strap is by means of a pair of copper brushes. Such an arrangement is necessary to allow for the turning of the shaft while still providing a positive electrical contact. (Fig. 378)

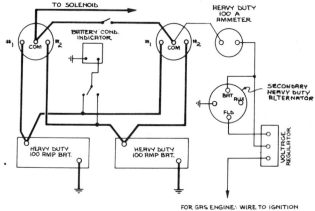

SECONDARY HEAVY DUTY ALTERNATOR WIRING SCHEMATIC

Figure 377.

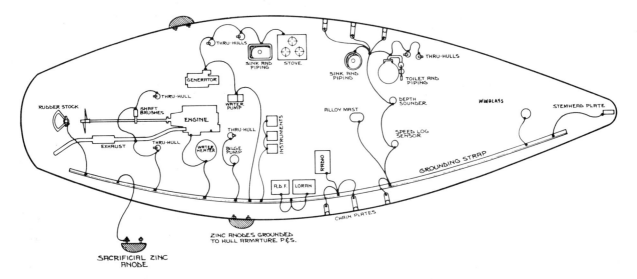

Fig. 378. A simplified electrolysis grounding schematic shows the method for connecting all metal masses aboard the vessel to the sacrificial zinc anode. The elements shown in this schematic are only a sampling of the many possibilities of diverting stray electronic currents. The wiring to the ground strap must not be connected to "live" appliance wires but only to the chassis of the appliances. Attention to these details may add measurably to the longevity of your vessel.

Alternating Current Circuitry

Needless to say, AC appliances most definitely add to the quality of life aboard any boat, although it is very rare to operate AC appliances when at sea except aboard very large vessels which have been equipped with heavy AC generating equipment. Livability can most definitely be improved through the convenience of a toaster, vacuum cleaner, television, hi-fi, air conditioning, electric iron, razor, electric frying pan and the like. From the standpoint of maintaining the vessel, it is also very handy (and almost necessary nowadays) to be able to operate an electric drill, sanders, and jig saw. The wiring for the AC circuit and its schematic will be completely separate from the DC system. The exception to the separation of these systems may be through the coupling of an AC to DC battery charger. Such a charger may derive its current either from the dockside power or from an AC generator.

The wiring sizes to be used throughout your AC circuit will be derived in exactly the same manner as for the wire sizes chosen for your DC system. You may refer to Fig. 372 as a guide for ascertaining such wire sizes. It is important to note, however, that while amperages are used for the wire size selection, almost all AC appliances are rated in watts. You will be required, therefore, to convert the appliance wattage to amps in order to enter the wire size table properly. To do this simply divide the appliance wattage by the voltage. Example: $\frac{1000\ W}{110\ V} = 9$ amps. For your convenience, I have listed the approximate anticipated wattage of some of the most commonly used appliances as a general guide. Of course, each particular appliance may vary somewhat with model so be sure to check the specific plate attached to your equipment before going ahead. (Fig. 379)

The use of transistorized electronics, for instance, will certainly reduce the wattage listed very significantly.

In a ferro-cement boat, the polarity of the current may have a very severe electrolytic influence on the hull armature underwater and metal parts. For this reason, I highly recommend the installation of an isolation transformer with single phase, ungrounded secondary. Such an isolation

transformer is shown in the accompanying schematic. It is also advisable to install a visual or auditory ground-detection device as an additional precaution against electrolysis. If you do not intend to install a transformer, your best protection against an inverted polarity is through the use of two-pronged plugs and sockets which allow the insertion of a socket in one direction only. When installing such sockets, you must be very sure that the positive wire from the sockets is connected to the positive wire of the shore connection plug. Once again, the installation of an isolation transformer will alleviate this necessity.

Your AC wiring should be of the double lead type. Your switches, sockets, line receptacles, etc., should be rated for no less than 20 amp service, although I would recommend 30 amp as being the safest. Once again, your wiring should be plastic or vinyl insulated of stranded copper type TW or THW. Your outlet sockets should be of Bakelite, not metal.

Additional protection against electrolysis should be provided by grounding all of the AC appliance chassis to your

APPLIANCE	WATTS
ELECTRIC RANGE	1500-2000
HOT WATER HEATER	800-1000
ELECTRIC IRON	1200
ELECTRIC SKILLET	1100-1200
MICRO-WAVE OVEN	1200-1350
VACUUM CLEANER	800-900
ELECTRIC COFFEE POT	400-600
3 cu. ft. REFRIGERATOR	150-200
AM-FM RADIO	50-80
16" TELEVISION (COLOR)	110
19" TELEVISION (COLOR)	270
DRILL & JIG SAW	250-400
CIRCULAR SAW	600
ELECTRIC RAZOR	12-15
ELECTRIC CLOCK	3-4
AIR CONDITIONER	600-800

Fig. 379. Typical power requirements for common AC appliances.

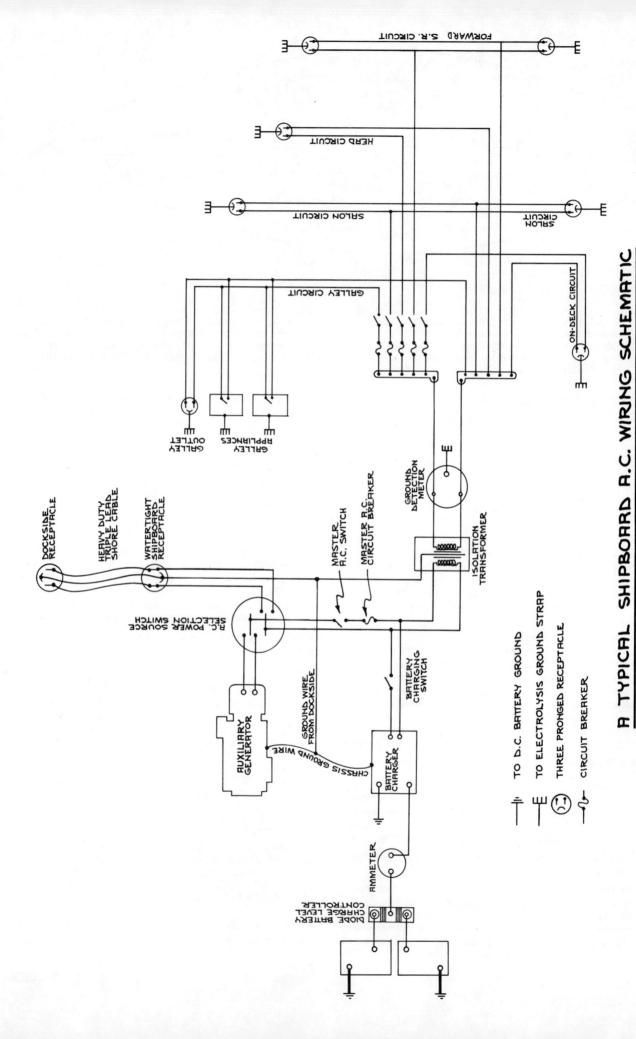

A TYPICAL SHIPBOARD A.C. WIRING SCHEMATIC

Figure 380.

copper electrolysis ground strap. No circuit carrying wires should be grounded to this strap.

Most marinas are equipped with three-pronged electrical receptacles, but occasionally you will still find two-prong outlets. It is important, therefore, that you make allowances for either system through the use of a two-pronged plug and socket adapter. The ground wire from the dockside receptacle should be routed to the ground terminal or chassis at all shipboard plug receptacles. When docking at a marina equipped with two-pronged dockside receptacles, the grounding of the shipboard sockets may be accomplished through the use of an alligator clip attached to the plug adapter ground wire, thence to the dockside receptacle metal chassis.

As with the DC electrical system, your AC system will require a control box for the purpose of switching and routing current to the various sections of the boat or for carrying individual appliances. You may, for instance, provide your vessel with, say, five or six major circuits within the AC system.

1. Galley appliance outlets
2. On-deck tools
3. Main salon appliance outlets
4. Water heater filament
5. Head appliances
6. Aft cabin appliances, etc.

Each circuit should be supplied with a circuit breaker at the control box which must be rated for the maximum anticipated load. All on-deck sockets, including the dockside connection socket, should be of an approved waterproof covered type which may be found in almost any marine hardware catalog. Your shore connecting cord must be of a size capable of carrying the maximum load of the entire AC circuit and such wire size must be maintained from the dockside to the bus boards of your control box.

Once again, I cannot overemphasize the value of having your AC system installed and/or completely checked over by a professional marine electrician. (Fig. 380)

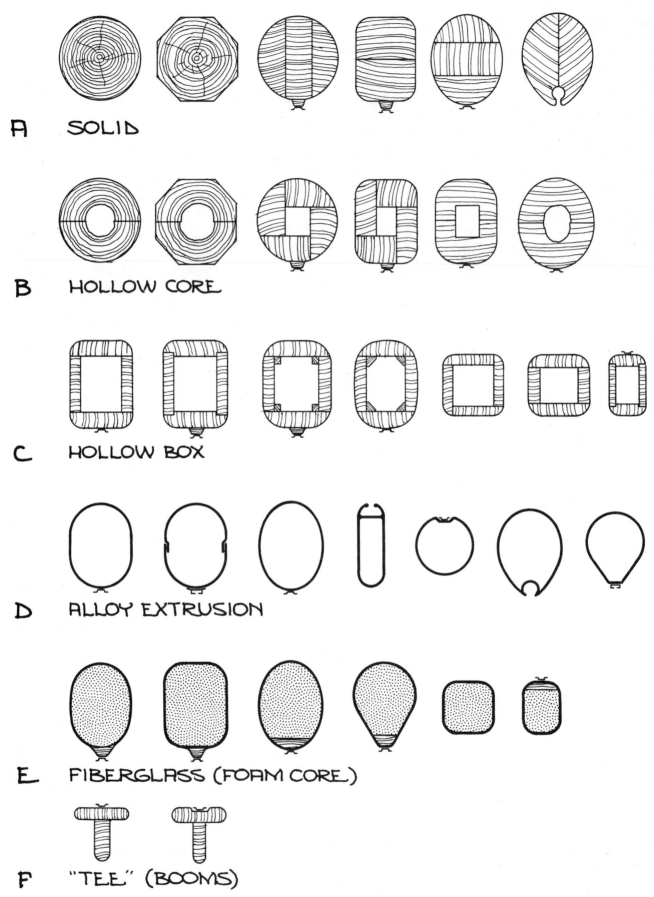

A SOLID

B HOLLOW CORE

C HOLLOW BOX

D ALLOY EXTRUSION

E FIBERGLASS (FOAM CORE)

F "TEE" (BOOMS)

Fig. 381. Mast construction types.

spars and rigging

To the neophyte standing on the dock looking up, a mast may seem to be simply a pole of sorts held in place by a network of wires. Even to many seasoned sailors a mast is defined as something to hang your sails on every weekend, varnish twice a year and take down to clear low bridges! From shore, one can hardly make out the spars on a boat more than a quarter mile away and few people have ever had the opportunity to see one built or even know what goes on in the inside of a mast. Well, no wonder so little real thought is given to them! It is rare to see a vessel with a broken mast nowadays and I hardly ever hear of one being replaced. "Just keep it straight" says the sailmaker, "and everything will fit just fine!" Hmmmm!

Masting and rigging has always been grossly oversimplified by the boat owner. Its complexities have been left to the designer and builder. Unfortunately, oversimplification has given birth to statements like: "all you have to do is. . . ." No other phrase in this business scares me more than, "I want mine extra strong!" Between these two philosophies I do not know which has caused more drastic failures. Proper masting and rigging is not quite so arbitrary as a lot of people think nor is it as scientific as one might suppose. There is a tremendous amount of recorded experience from which your designer draws upon to establish safe, workable sizes and materials but he also liberally applies his "feel" for sound design as well. You are going to have to trust his rationale completely here, otherwise you could get into a lot of trouble by juggling his numbers.

I cannot overemphasize that the performance of your boat will depend upon the propriety of her rig, its workmanship and the quality of material. It is going to cost a lot of money to put a proper mast into your boat and to purchase proper hardware. There are few other areas in yacht construction where it is more tempting to scrimp and save, but there is no quicker way to sacrifice safety, workability and performance. Beating the established system here could be absolutely disastrous.

Mast Types

Solid. Until forty years ago, almost all spars (masts, booms, clubs, gaffs, sprits, yards) were of solid wood. The very large spars may have been constructed of a single piece of wood or built of several layers to the desired diameter, or scarfed to a given length, or even constructed of pie shaped sections and banded together. Nevertheless, they were solid. Sometimes the solid sections were round, octagonal or square but seldom elliptical. Old schooners rarely had anything else but round spars and this became a part of their character. Unfortunately for them, their personalities have

changed very little since the beginning of this century and round spars still look best on them. It is very difficult to build a hollow round spar as it takes an exceptional amount of experience and equipment. The use of solid spars persisted until the nineteen forties because there were no glues available of the kind that could withstand the sheering and tension loads of mast stresses, particularly in the continually humid marine environment. Certainly, there were hollow glued masts at that time, but their use was still restricted mostly to high performance racing machines and, even then, their longevity was somewhat questionable.

Today, however, very few solid masts are constructed because they are very heavy for a given strength and large "perfect" lumber stock is almost impossible to get as air-dried stuff. It is true that a solid spar is usually of a smaller diameter than a hollow one but it is also comprised of considerably more material, too. Solid spars fashioned from a single piece of wood also tend to check and twist over the years.

While it seems as if I am completely negative regarding solid spars, I really am not. . . altogether. I think their use should be restricted to spars of small diameter and to those which would be impractical to build as hollow ones, i.e., jib clubs, booms, gaffs and bowsprits. Even in these units I think it far better to avoid solid construction if at all possible in order to keep lofty weights to a minimum. (Fig. 381A)

Hollow Core. Masts constructed to this fashion are neither considered hollow nor solid as they are somewhat of a cross between the two. In essence, a hollow core spar begins as a solid or planked member with a small portion of its center removed. This construction does not truly fall into the hollow spar category because the wall thickness is much thicker and the weight considerably heavier but still they are lighter than solid spars.

I much prefer a hollow core mast to a solid one, although they are considerably more difficult to construct. This type of construction may be used for masts, booms, gaffs, bowsprits and small diameter stays'l clubs. If your vessel has been designed for solid masts, it would be wise to consider hollow core spars instead, as an increase in the vessel would be very marked without a sacrifice of strength and safety. (Fig. 381B)

Hollow Box. Almost every wooden mast built in the last three decades has been of this type and there is little wonder why. They may be constructed of more readily available materials than with any other type of spar and they are extremely light for a specific strength, thus enhancing the vessel's stability. It is well known that a square or rectangular shape withstands bending forces far more readily than a round or oval shape which is why almost all wooden booms, clubs and spinnaker poles are square. Although the rectangular shape sacrifices some aerodynamic efficiency, it is the most easily constructed and strongest of all hollow spar types. This should be your first choice of mast construction type. (Fig. 381C)

Aluminum. A tour of any marina will attest to the popularity of alloy spars. There must be good reasons. Although aluminum spars are almost twice as expensive as home-built wooden spars, they are virtually maintenance free, extremely light for a given strength, are available in many aerodynamic shapes and may be purchased "off the shelf" with a minimum of hassle and involvement. A broken or kinked alloy mast is very rare but easily repaired while being visually in vogue aboard a modern boat. I, personally, would hesitate rigging a character schooner or ketch with an aluminum mast, simply because of an aesthetic disruption but I still agree that it would be extremely practical.

Aside from the warmer appearance of a wooden mast, I do dislike aluminum from one other major standpoint. . . noise. There is nothing worse when living on a boat than continually hearing the "kling klang kling" of slapping halyards on the mast. This can be partially solved by cinching the halyards to the shrouds when not in use as well as liberally applying halyard fairleads to their hauling ends. But as every sailor knows, there are always one or two halyards high aloft that refused to be silenced. They are relatively hard to live with over a period of time. Most aluminum mast manufacturers will fill the mast extrusion with polyurethane foam as a means of deadening the sound transmission but this is rarely completely effective.

If it is your desire to equip your vessel with aluminum spars, you can be sure that the manufacturer is eminently qualified to supply the proper rig based upon your existing Sail Plan and Wooden Mast Drawing. Assuming that your vessel is not of an extreme beam/length ratio or of an exotic rig or hull type or of an unusual sail area/displacement proportion, the vessel's metacenter will normally fall approximately within 3% to 3.5% of the extreme waterline beam width above the L.W.L. With this assurance, conversion from wood to aluminum masting is not a complicated matter. Your aluminum spar maker will use your drawings as a basis for determining tang design, hardware location, spreader lengths and wire sizes. If I were to order an aluminum mast tomorrow, I would leave the whole affair to the mast builder. I would ask for a completely equipped unit with everything attached and ready to drop into the boat, i.e., mast, masthead, spreaders, tangs, winch bases, outhaul, sheet bales, gooseneck track, sail track, sheaves, blocks. . . the whole works. In this way I would know that all parts would be completely compatible, properly insulated against corrosion and strongly attached. There is very little money to be saved by purchasing a blank extrusion and trying to "fit it out" yourself. The mast builder will take no responsibility for failure or repair if you attempt the finishing work yourself, so give it some reflective thought, even though you might come out a few dollars ahead otherwise. (Fig. 381D)

Hollow Fiberglass Section. Fiberglass spars are still a rarity but primarily because of the lack of quantity production technology, not because of inferior strength or weight. Unfortunately, the few numbers of glass masts leaves us with little technical information to draw upon in terms of per-

formance and longevity. The information I have seen, though, is so promising that I can see no reason why boat builders should not consider this construction. The procedures for the production of a "one-off" spar is not complicated and can be mastered by the amateur with little trouble.

One-off fiberglass masts normally require a "core" mandril over which the fiberglass is laid and wrapped. This core is usually a polyurethane "plug" fashioned slightly under the size of the finished spar but of the same desired shape. Because the foam will not be available in the required long lengths, it will be necessary to "glue" short pieces end to end, "plank up" or to cast the foam in an inexpensive Masonite mold. The foam core is easily shaped to the required sections with a spokeshave, plane, belt sander, or what have you. This fashioning is very quick and easy.

Certain areas along the length of the foam core must be replaced with solid fir or spruce as additional reinforcement such as at the bury (foot), partners, gooseneck and winch hardware, spreaders, and masthead. This will be dictated by your designer. It is also possible to preinstall the masthead sheave cheeks and other selected hardware prior to glassing.

Fiberglass is then layered over the foam core. The type of glass used is very critical as the major layers must be linear in nature, i.e., more heavy filaments running lengthwise than transverse. Such linear fiberglass roving must be special ordered. The lay-up of the spar requires a combination of linear roving running the length of the mast and fiberglass cloth wrapped spirally around the mast. The specific formulation of roving and cloth must be detailed by your architect. Only structural resin is used during the major lay-up, but "casting" or "finishing" resin is applied to the surface to aid sanding to fairness and smoothness. Certain portions of the mast may receive an additional patch of glass in order to more adequately support hardware fastenings. Winch bases and other bosses may be sculptured to shape using "mush" (a combination of resin and ¼" chopped fiberglass strand).

It is imperative that no screws be used in a fiberglass mast unless a wood core has been built-in beforehand. Mast track and plate hardware must be screw attached to wood cores or feathers while cleats, winches, tangs, etc., should be supplemented with bolts. Bolt nuts may be impossible to reach in many places, however, which will assure the necessity of the wood core. With care, internal halyards may be routed within a fiberglass mast, either by inserting aluminum tubing through the foam, coating tubing directly into the foam core or by partially "digging" the foam out of the spar after the fiberglass has fully cured. (Fig. 381E)

Tee Section. This type of construction is extremely simple and requires little skill to execute but it is restricted to the use as booms and stays'l clubs which are not fitted for roller reefing. Tee spar construction is both light and strong but not as aesthetically pleasing as a round, elliptical or square member. Aerodynamically, the tee boom is quite superior to all other types, as it forms a "vortice plate" or "vortice shelf" along the foot of the sail, as the upper tee flange is usually designed as a wide plank. (Fig. 381F)

Mast Material

For years I have been told that Sitka Spruce was the first and only choice for building a fine wooden mast. Sitka Spruce is a native timber of Alaska and, unfortunately, never was plentiful. Due to the wood's specialized use, little concern was given to its cultivating and programmed re-planting as it was being hauled away in years past. The inevitable result is that it has become almost impossible to find Sitka Spruce, either on the mountainside or at your boat lumberyard. Green Sitka is hard enough to come by, let alone dry stock of the grade necessary for spars. Even the regular professional customers have a hard time getting it and more quality builders are turning to fir and alternative spruce grades.

There are slightly lesser qualities of spruce which may be used as a reasonable substitute for Sitka. Such a spruce is often referred to as "aircraft" grade. Its primary use is that of wing spar timbers and ribbing in small airplanes, so you can imagine that its quality is very high. It is usually available from specialized lumberyards on special order in random widths and lengths, and is worth its cost and waiting for.

Douglas fir has always been considered a very fine wood for spars, although it is a little heavier than spruce. A fir spar is also about 7% larger in section than one built of spruce. A major manufacturer of some of the finest sailboats in the world now uses fir exclusively for mast making because of the assurance of consistency of grade and availability. In this age of shortages on every front, it should not be considered as degrading to have to settle with a fir mast.

Whatever lumber you choose to pursue, it should be hand picked. All timbers should be quartersawn with its annual rings equally spaced and with its grain as straight as an arrow. The grain should leave the timber as seldom as possible, as it will definitely reduce the advent of splitting and absorbing moisture. Your lumber must be absolutely knot free, absent of checks, shakes and sap pockets. Purchase lumber in as long of length as possible but do not buy an inferior piece of wood on this basis. Have your yard mill the lumber to the required finished thicknesses because your light-duty planer may leave shallow, almost imperceptible corrogations along the surface which could affect the glue bond. Of course, your lumber must be air-dried prior to milling or mast fabrication, as excessive moisture will be detrimental to the strength of the glue and the shrinking of green lumber can easliy part a fine joint or cause splitting on an otherwise well-built mast.

The glue used in mast work is either casein or phenolic resin. Nothing else will do. What about epoxies? I do not really know how it would work, as there is little, if any, information available concerning this particular application. I must suggest continuing with the glue that has proven itself over the years.

Taper

Regardless of the spar construction types specified for your vessel, chances are that no one member will be of a continuous section throughout. Most usually at least one end is tapered, if not both. The reason for this is that the compression and bending loads on a spar are not constant throughout the full length and the section of the spar at any given point is engineered to withstand its specific load while keeping weight to a practical minimum. Masts are never designed to simply support the maximum anticipated load under an adverse condition, as ample safety factors are always taken into consideration, so you need not worry about tapering having weakening effects upon the structure. Tapering of a spar is also very pleasing to the eye as it adds a delicate touch noticeable at a glance. Tapering of an aluminum mast is accomplished by the manufacturer upon special order, which adds to the cost markedly without

appreciably saving mast weight. But there is nothing more cumbersome looking to me than an untapered aluminum mast, which is the way most stock yachts are delivered.

If you are constructing a solid spar, your designer will most likely have shown tapers by simply indicating the various diameters on the Sail Plan. Taper development is not critical here because you will not be required to shape planks to a fine tolerance before assembly. The taper proportions for solid spars are quite standard and I have included an easy reference chart for your use. (Fig. 382) When using this chart it is assumed that your spars will be of no other shape than round or octagonal and no allowance has been made for the reduction in sections caused by the "letting in" of hardware.

Unless you are constructing a very high performance racing boat with a "bendy" rig, your sailmaker will automatically assume that your main luff and foot are to be straight. He will cut your sails accordingly unless told to do otherwise. It will be entirely up to you to construct your mast and boom so that the sides of these spars at the sail are absolutely straight when rigged and tuned. If these sides are not, they will cause irremovable wrinkles or "soft" or "hard" spots in an otherwise fine sail. Because the power of your sails, the windward performance of your boat, even her heeling characteristics depend upon a proper sail "foil," the straightness of luff and foot cannot be overemphasized.

Assuming, then, that the trailing edges of all masts are to be straight, these spars will be tapered on the forward edge and both sides. Booms, also with straight upper edges, will be tapered on each side and bottom. Gaffs will be constructed with a straight lower edge while being tapered on each side and top edge. Jib clubs, which are not fitted with a continuous length of sail track nor with the sail laced to them, will be tapered equally on all four sides (no one side being straight). Spinnaker poles and whisker poles are also tapered equally on all four sides. Bowsprits may take on different shapes in profile. Some bowsprits (the best looking ones) seem to "hog" downward as they lead forward. Such bowsprits are tapered along their upper edges and sides while being straight along the bottom. Occasionally a bowsprit will be tapered on all four sides but it is rare. No good looking bowsprit curves upward with a straight upper edge, as it makes the vessel look very "snooty."

If you are constructing hollow box or fiberglass spars, your designer will have provided a section development drawing along with the spar construction details. The section development is a strange looking configuration at the outset because it is usually drawn to a small vertical scale and a large horizontal scale. This change of scale, in essence, shortens the vertical proportion of the spar thus accentuating the taper curves for additional accuracy. It will be up to you to faithfully scale a full size upon your lumber (or foam core) the resultant true curves as developed from the "compressed" spar views. (Fig. 383) Your curves should be carefully drawn with the aid of a fairing batten as any lumps or hollows in these lines will surely affect the contact and bond of your glue joints when building wooden spars. After cutting your lumber, the stave edges must be planed free of all saw scores and irregularities. Check the fairness of your cuts by sighting along the edges, placing a fairing batten onto the edges and by running your finger tips rapidly up and down the edges of the staves. One caution which must be faithfully observed when planing is that your plane **must** be held absolutely perpendicular to the sides of the stave at all times. Nothing can weaken an otherwise good spar more severely than an inadvertently beveled

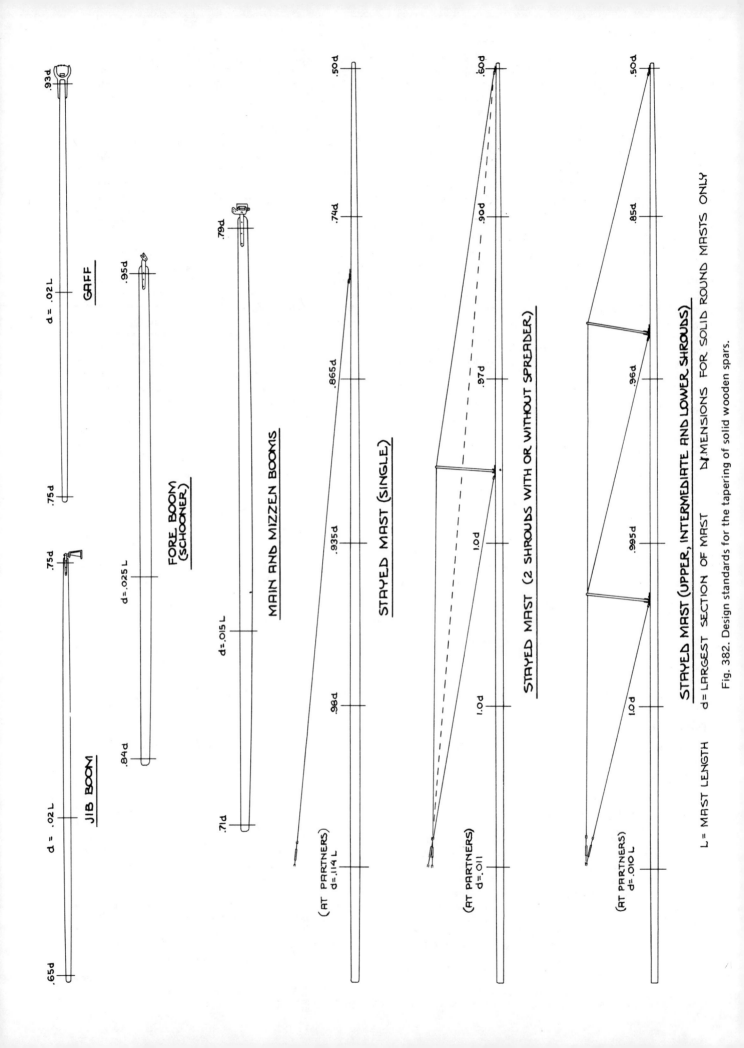

.93d

GAFF

d = .02 L

.75d

JIB BOOM

d = .02 L

.65d

.75d

.95d

FORE BOOM
(SCHOONER)

d = .025 L

.84d

.79d

MAIN AND MIZZEN BOOMS

d = .015 L

.71d

.50d

.74d

.865d

STAYED MAST (SINGLE)

.935d

.98d

(AT PARTNERS)
d = .114 L

.60d

.90d

.97d

STAYED MAST (2 SHROUDS WITH OR WITHOUT SPREADER)

1.0d

1.0d

(AT PARTNERS)
d = .011

.50d

.85d

.96d

STAYED MAST (UPPER, INTERMEDIATE AND LOWER SHROUDS)

d = LARGEST SECTION OF MAST   DIMENSIONS FOR SOLID ROUND MASTS ONLY

.995d

1.0d

(AT PARTNERS)
d = .010 L

L = MAST LENGTH

Fig. 382. Design standards for the tapering of solid wooden spars.

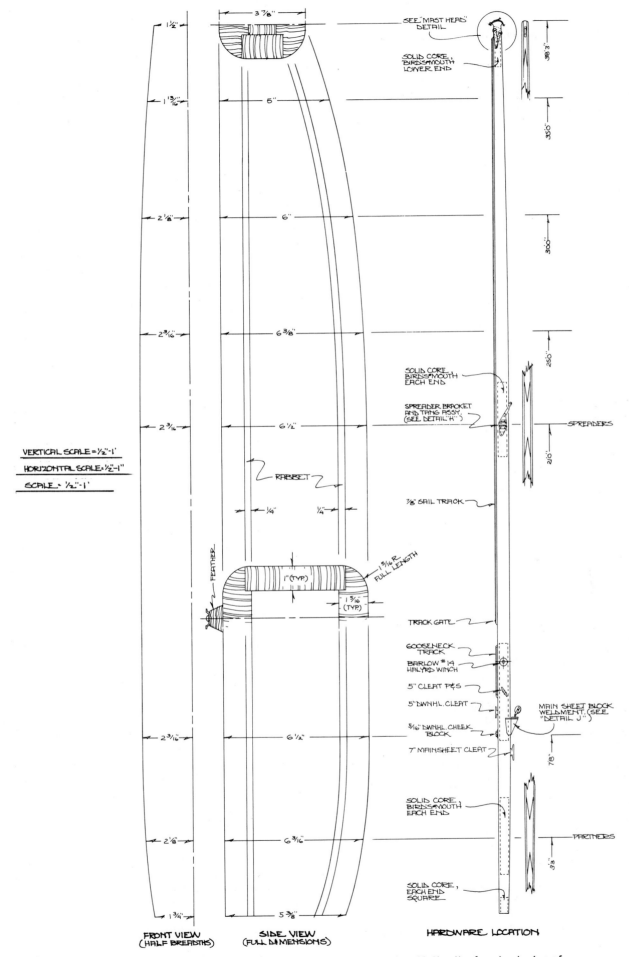

Fig. 383. **Doreana's** mainmast taper and construction drawing. Notice the foreshortening of the vertical scale in order to exaggerate the mast shape. A Bingham design.

glue joint. My suggestion here is that you construct a jig out of plywood to fit onto your plane to help insure this perpendicular position. (Fig. 384) When tapering the gluing edges or sides of a plank or stave, never use a belt or disk sander. In fact, never sand these surfaces in anyway, as it will cause "hair" which detrimentally increases the glue thickness.

Most often the tapers of booms and clubs will not require a "compressed" taper development drawing as in the case of masts. The primary reason for the more simplified boom drawing is that these members are never subjected to the severity of compression loads as found with masts. The tolerance for error, therefore, is considerably more liberal. In any event, the tapers for booms should be faired full size upon the lumber in the same manner as that of laying out mast timbers. (Fig. 385)

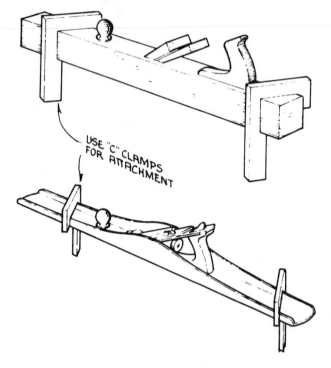

Fig. 384. Jigs clamped to spar plane to ensure perpendicular alignment.

Solid Spar Construction

If your spars are very short, say under 20' stick length, you may be able to locate solid timbers of suitable quality to accommodate the full sizes of your spar requirements. Keep in mind, though, that dried lumber will most often be checked at its ends so you will have to purchase length a few feet longer than the finished spar dimensions. If you are not able to purchase the continuous lengths necessary, do not be concerned. A properly scarfed timber will be as strong as its continuous solid counterpart, although the joint may be visually disruptive.

One of the strongest methods for joining two solid spar sections is through the use of "vee" scarfs. The accuracy of the cuts for such a scarf, however, is extremely difficult to produce without using sophisticated equipment and jigs. If such a scarf is to be attempted by an amateur builder, it should be practiced several times using scrap lumber in order to get a feel for the work. The length of such a scarf should be no less than 12 times the thickness of the lumber. Scarf lengths are usually referred to as ratios, i.e., 12 x 1.

A common scarf may also be used to join spar timbers. It will not be quite as attractive as the "vee" scarf, but will be far easier to execute. (Fig. 386)

If you are not able to find the timbers necessary to encompass the spar diameters, you will have to laminate several planks to the required dimensions. These planks should be professionally planed to a "perfect" finish at the yard. Do not try it yourself. If anyone or all of the planks are too short to accommodate the full spar length, they will have to be accurately scarfed, but not joined together until actually being laminated to an adjacent plank. A common scarf is quite suitable here but again, the scarf ratio must be very long, say at least 12 x 1 or more. Because these scarfs must be cut perpendicular to the edge of the plank, they should not be attempted with a saw as the required accuracy will be impossible to achieve. Such scarfs must be cut only with a plane using a scarfing jig, as shown in Fig. 387, to insure the perfect matching of both joining planks. When laminating planks to the given timber thickness, it is not necessary to join all of the layers at one time. Because casein or phenolic glues produce the strongest bond when used within 15 minutes of mixing, you would be better off by laminating one adjacent layer per session, as this will allow you to slow down a bit in order to do your work well. When gluing up such a spar, it is best to use many "C" clamps, say, one every 10" or so. Do not use too much pressure when clamping, however. Your clamps should be turned just until the glue begins to ooze from the joints. (Fig. 388)

Solid Spars, Shape Layout

Whether your solid spars are to be perfectly round in section, octagonal or elliptical, the method of layout before hewing is essentially the same. The mast timber is first marked off for taper on each applicable side. The taper marks are joined and drawn as a curved line using a batten for fairness then the timber is cut to its inclusive square or rectangular section using the taper lines as guides. Once done, plane the sides of the mast timber to fairness and free of saw scores.

Even if your mast is not to be finished as an octagon, its shaping to round or oval will begin as such. Each side of the mast must be lined off as three sectors. The method for establishing the sector widths is shown in Fig. 389. A simple gauging tool may be used for drawing consistently spaced sectors onto the mast regardless of the width or thickness at any given point. The gauge will automatically adjust itself properly to the mast width as it is drawn along the timber. Once the octagonal sectors have been drawn onto the timber, the sector corners may be planed away to produce the basic octagon. If you are producing a round or elliptical mast section, careful planing away of the remaining corners will rough in the desired shape. Once so done, it is best to shift to the use of a blocking board then to a sanding strap. (Fig. 390) When the mast has been finely shaped, sculptural detailing may proceed, such as the fashioning of the truck and morticing for flush hardware. You are now ready for painting or varnishing and the attachment of fittings.

There are an infinite number of variations to the solid mast construction shape and procedure. You may, for instance, attach a track feather to the mast trailing edge or maintain a square or octagonal section below the gooseneck while rounding the mast above. The choice here must be based upon your own experience so look at plenty of boats before committing your final direction.

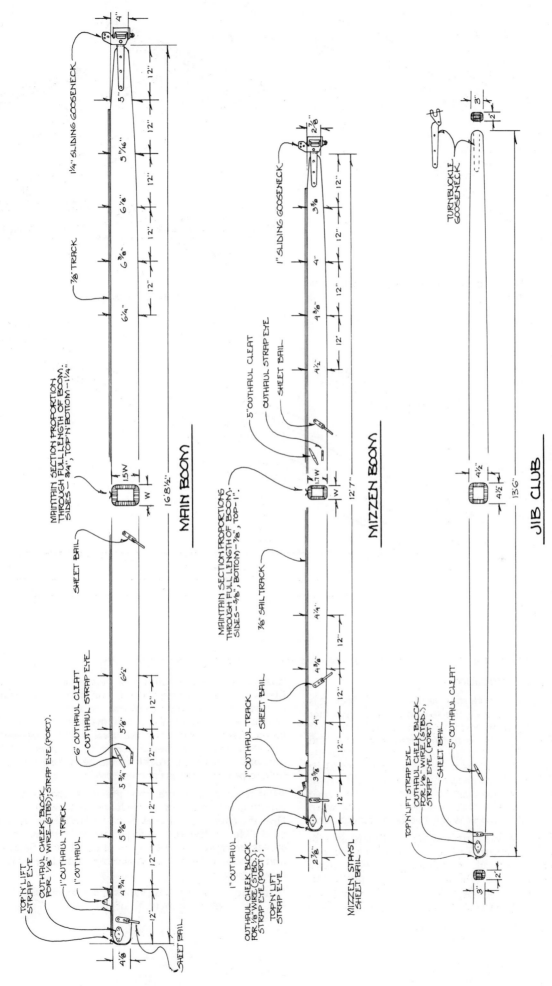

Fig. 385. Boom taper drawing for the **Doreana** cruising ketch showing the construction as well as the placement of all critical hardware. A Bingham design.

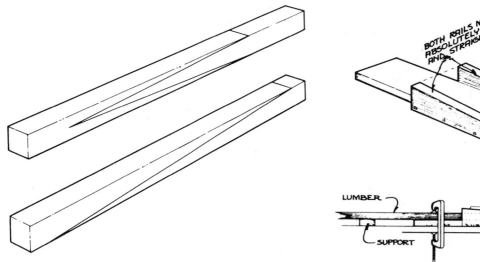

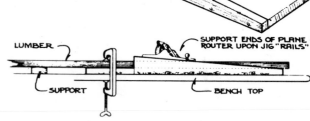

Fig. 386. The most commonly used scarfs as applied to solid mast construction.

Fig. 387. A scarfing jig.

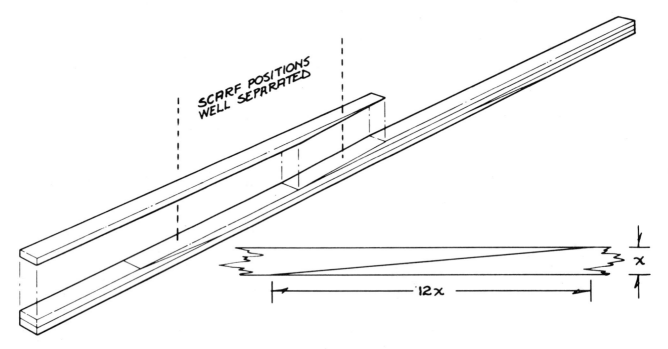

Fig. 388. Building up a planked solid mast showing the proper use of scarfs.

Hollow Core Construction

The most basic hollow core spar begins as a solid spar. It is built or fashioned in exactly the same manner as any other normal solid spar. It is hewn to shape and finished off smoothly. The positions of the partners, gooseneck and spreaders are clearly marked for future reference. Now the spar is carefully cut down the middle along the fore-and-aft center line. It is just like slicing a banana for a banana split. The cut edges of the two mast halves should not be planed or sanded at all because they will match perfectly when rejoined even if your saw cut is not a perfectly straight line.

Removing the core of the mast is usually accomplished with a blow torch and a gouge. To prevent the joining edges of the spar from being burnt, they should be covered with a metal adhesive tape available at most auto supply stores. This tape will provide ample insulation. When coring, the wood is burnt to a shallow char, then it is gouged out. Do a small area at a time until you have a feel for the work. Check the wall thickness of the spar with calipers to insure consistency. The wall thickness of a hollow core spar should be approximately 1/3 of the spar diameter. When coring, keep in mind that the areas of the foot, partners, spreaders and truck should remain solid. These solid core portions (except the truck) should be drilled with a small hole to provide drainage and ventilation.

After removing the core, sand the inside spar sections with progressively finer grits of sandpaper until velvety smooth. Give the core a liberal coat or two of Cuprinol, allowing it to dry for a few days, then apply at least two coats of shellac or varnish to seal the inside of the spar. Be sure that you do not shellac or varnish any mating surface which is to receive glue.

Rejoining of the two spar halves is a relatively easy matter. Mix and apply a thin layer of glue to all of the mating

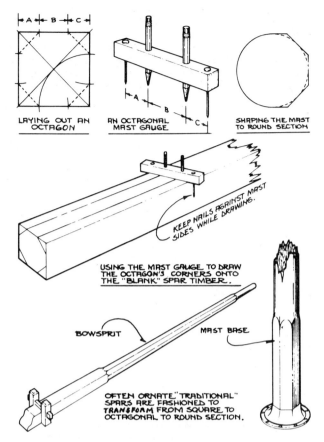

Fig. 389. Laying out a round spar section beginning with the basic octagon.

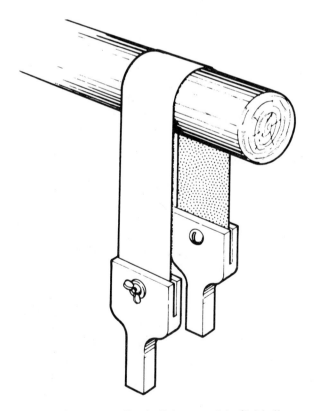

Fig. 390. Using a sanding belt to smoothly finish the spar section.

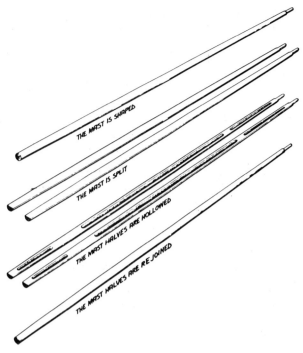

Fig. 391. Construction stages for the fabrication of a hollow core mast derived from a solid stick.

surfaces. Place the spar halves together, aligning carefully. Then clamp or band the spar halves liberally. Do not use so much clamp pressure that you squeeze out all of the glue, as this will do as much harm as using no pressure at all. You actually only need about 15 pounds of torque on the clamp handle to do the job. Once the glue has completely set and cured, remove the clamps, give the spar a final sanding and finish it out. (Fig. 391)

The second type of hollow core mast to be discussed is that which is constructed of "spirally" arranged wood staves. All of these staves are of equal thickness and quartersawn so that the grain will be arranged radially in the finished spar. Once again, the wall or plank thickness of the hollow core spar is usually 1/3 of its diameter. Each of the staves may require a 12 x 1 scarf joint to achieve the desired length but be sure that the scarfs are well staggered from each other and do not occur in the areas of the spreaders, partners or gooseneck.

The real key to the successful staved hollow core mast construction is the accurate laying of the spar taper on each of the four staves prior to their gluing. The taper of each stave may be saw cut first then carefully planed to finish. When planing, it is imperative that the plank edge remain absolutely perpendicular, otherwise the strength of the glue joint will be lost. I must recommend using a jig attached to the plane in order to prevent quivering when drawing it back and forth. (Fig. 384) The taper layout will have been drawn out by your designer so there is little more to be said here.

Once the individual staves of the hollow core spar have been shaped, they should be given several coats of Cuprinol wood preservative. Keep the Cuprinol away from the gluing surfaces if possible.

The staves of the hollow core spar are rarely glued up all at once because this will require rotating the spar while the glue is wet. This could reopen joints and cause twisting and stave misalignment. It is better to glue one stave at a time.

The first stave to be laid down will be the aft side (mast), upper side (boom) or lower side (gaff) because these sides will always be straight. In the case of a club or spinnaker pole it will make no difference because all sides will be tapered. The first stave must be clamped or wedged firmly to the bench rack or trough to prevent any possible springing, twisting or other distortion. The glue is now mixed and thinly applied to the mating edge of the second stave. The second stave is placed, edge up, onto the first stave, making sure that the second stave aligns with the taper lines marked on the first stave. Clamp or wedge the lines marked on the first stave. Clamp or wedge the staves together liberally at regular intervals, say, 9" to 14". Do not remove these clamps or wedges or move the staves in any way until the glue has completely cured.

The spar section is now rotated 90° and replaced onto the bench, rack or trough so the aft side of the spar is once again maintained as a straight line. The second stave, however, will not be straight because it has been sprung to the taper and, therefore, it must be wedged along its length in order to support it at its predetermined shape. With this curve so supported, the edge of the third stave is glued and positioned onto the second stave. Again, take care to align the third stave along the taper line marked upon the second. Change or wedge the assembly as before, allowing it to cure fully before movement.

At this point you will have formed a "U" section within the three staves. It will now be necessary to shape and insert the solid wood cores at the foot and truck of the spar (or at the ends of booms, gaffs or clubs). The fittings of these cores must be very accurate and snug but they should not be force fitted, as this can strain the spar severely. Often spar builders will fashion the cores with "bird's-mouths" in order to relieve strains on the spar at the ends of the cores. This is an excellent idea and is easily accomplished by cutting and removing long "vee" sections from the ends of the cores. These bird's-mouthed solid cores may be clearly seen in Fig. 383. The axis of the "vee" should orient fore-and-aft. Once the cores have been installed, they must be planed flush with the edge of the first stave. Give the inside of the spar several coats of shellac or varnish as a precaution against rot.

Now the fourth and last stave may be placed. Once again, be sure that the aft side of the spar is maintained as a straight line while wedging as necessary to provide the proper support for the taper of the spar sides. Apply glue to the exposed surfaces of the spar cores, the upper edge of the third stave and the mating edge of the fourth stave. Position the fourth stave onto the spar assembly and clamp or wedge until cured. Do not move the spar for a day or two in order to be sure that it has completely set.

The finished spar assembly will have four protruding stave edges. These must be planed flush with the spar sides before final shaping. The shaping itself will follow exactly the same procedure as described under **Solid Spars, Shape Layout.** (Fig. 392)

Hollow Box Construction

This type of spar construction is the real "meat" of the mast builder's trade. Although there are other types of spar construction which are even more complicated, it is this method which comprises the majority of all spars built even though it is quite difficult. Although categorized under one heading, hollow box spars may take many different forms. Each form has its own purpose (namely, the allowance for

finished shape) but generally, no one form may be considered superior to the other. Even though your architect may have detailed the simplest of these box forms, you may elect to choose one of the others, taking into account your own capabilities and equipment. Such variations have been shown in Fig. 381. Essentially, the conversion from

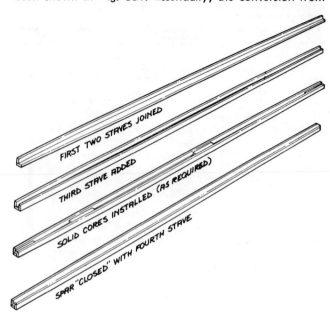

Fig. 392. Construction stages for the fabrication of a staved hollow core mast. The insertion of solid cores at the foot, spreaders and stock may be eliminated on some designs and will be dictated by your architect.

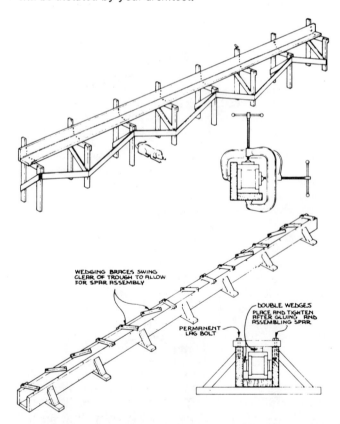

Fig. 393. Spar bench and trough showing their use in fabricating a hollow box spar section.

one box form to another requires that you do not alter the spar section modulus (surface area of the spar section) nor the outside overall dimension of the section. It must be pointed out here that when choosing a hollow box type, you must keep in mind that complex corner joints do not weaken a spar when properly built, most especially if the joint provides a larger gluing surface. More often than not the glue joints will be stronger than the wood surrounding them. Rarely will a mast fall at a joint but rather in the wood proper.

As with all other spar types, hollow box construction will require a substantial spar bench, rack or trough in order to prevent twisting, distortion or bending while gluing. (Fig. 393) Prior to the assembly of the four staves, they must be scarfed to full length taking care not to locate the scarfs near the partners, spreaders or gooseneck. The scarfs must be well distributed from each other so that there will not be a concentration of joints in the same areas. The scarf lengths must be at least 12 times the thickness of the lumber. The lumber thicknesses will be determined by your architect.

The hollow box construction procedure requires first that each continuous length stave be accurately drawn out for taper using a flexible fairing batten. The staves then are accurately cut and edges planed for "perfect" smoothness. To insure that the planed edge is absolutely perpendicular to the sides of the staves, it is best to attach a jig to the plane as shown in Fig. 384, as this will prevent the tool from rocking as you pull it back and forth. If your staves are to be rabbeted, they should be cleanly cut with a router at this time.

Each stave should be treated with Cuprinol preservative then given several coats of shellac or varnish as a precaution against rot. Do not apply any finishing to gluing surfaces, however, as it will destroy the joint strength.

The solid wood cores are drawn out and shaped prior to gluing any of the staves. They are most usually located at the foot, partners, gooseneck, spreaders and truck. These cores must be very accurately shaped and prefitted (but not fastened) to the various staves. If the core is even so much as a paper thickness too large on any side, it will prevent the proper mating of the staves, thus causing inevitable spar failure. If, on the other hand, the cores are slightly too small, they will not benefit from the strength of glue contact either, thus serving no purpose whatsoever. In order to relieve the concentrated bending strains at the top and/or bottom of the solid wood cores, they may be fashioned to a bird's-mouth with the axis of the "vee" oriented fore-and-aft. Such bird's-mouths may be clearly seen in Fig. 383. Once so fashioned and accurately prefit, the cores may be glued and prefitted in their proper positions atop the aft mast stave, bottom gaff stave, top boom stave or otherwise first stave to be laid down onto the rack.

If your hollow box spars are to be very generously radiused along the corners, the exceptional loss of wood thus incurred may require the installation of corner fillets or gussets. These fillets are simply small continuous pieces of wood located along the inner spar corners as additional reinforcement. They are sometimes cut as square sections but more often (and properly) fashioned as triangular shapes. These fillets must be glued to the side staves prior to gluing the spar assembly. When attaching the fillets, care must be exercised to see that their surfaces are absolutely flush with the stave edges. The fillets will not continue through the solid cores but must be cut so as to snugly butt

against them. Allow the fillets' glue to cure fully before proceeding. Such fillet variations may be seen in Fig. 381.

Before gluing the spar assembly, "walk" through a practice assembly. Preset your "C" clamps at approximate openings so that you will not waste time on gluing day by having to turn down your clamps. Remember that every minute will count once your glue has been mixed. Make sure to have enough help on hand so that each person will not have to concentrate on a longer spar section than, say, 7' or 8'. Premeasure your glue proportions ahead of time to gain valuable seconds later and make sure that all gluing surfaces are dust- and oil-free. Even the oils from your hands can interrupt a glue bond, therefore, some spar builders wear cotton gloves when handling their lumber.

Gluing time has arrived. All four sides will be assembled at the same time. Some amateurs believe that it is easier to glue only the side staves on one day then to close up the spar with the last stave later on. The fallacy here is that it is quite possible that the side staves may be set up slightly out of "square" then must be forced to proper angles when attaching the last stave. This will strain the glue severely and must be avoided at all costs.

I would suggest mixing two separate pots of glue. The first pot will be used for attaching the side staves to the bottom while the second pot, mixed a little later, will be used for setting the upper stave. The staggered mix will insure that only the freshest possible glue will be used at any given time.

Glue and set up the side staves first. You may use masking tape to prevent them from falling over while you prepare the last and upper stave for positioning. Now carefully place the upper stave. An occasional finishing nail may be driven here and there to hold the mast together until you are able to place your "C" clamps or wedges. When clamping, you need only to apply about 15 pounds of torque to the handles as additional pressure will only serve to drive out all of the glue from the joint. Check the mast carefully for "square" as you go, driving small correcting wedges as necessary. Once set up, take the day off as any tampering or movement of the spar now will disturb the glue bond severely. The longer the mast remains without movement, the better, so give it a few days to rest.

Once set, the spar corners may be radiused, the masthead fashioned and the surfaces sanded and finished out. These details will be covered later in this chapter.

Hollow Fiberglass Construction

Make no mistake about it. Fiberglass spar construction is messy work requiring the ultimate of skill in the glassing art. There is no room for error. One bad batch of resin (too hot or too mild), improper resin impregnation of the fiberglass (too rich or too lean), or bubbles and voids can completely destroy the effectiveness of a fiberglass spar. But when the operation has been pulled off successfully, the results are superb and few other spar types can match the light weight and high strength of fiberglass.

One of the tricks of fiberglass spar construction is building a proper spar bench or rack. The key is to support the mast during curing with as little physical contact as possible to prevent sticking during curing. It is also imperative that the spar is able to be rolled back and forth during the application of the glass. I think the best spar supports are 2' lengths of steel or aluminum angle coated with wax. These angles can be easily dropped out of the way or unmounted at times when they get in the way of laying glass, then repositioned once the work has passed by. Fiberglassing of small spars may also be accomplished by rolling the fabric onto the core in much the same way as a window shade. (Fig. 394)

A fiberglass mast, although designed to be as strong as a wooden one in terms of compression, will generally be more flexible. Actually this is a built-in safety feature of a fiberglass spar, as they have far less tendency to break than a wooden one. The fiberglass spar will bend like a noodle if a shroud or stay becomes dislodged and the spar can most often be saved with a minimum of damage if the helmsman

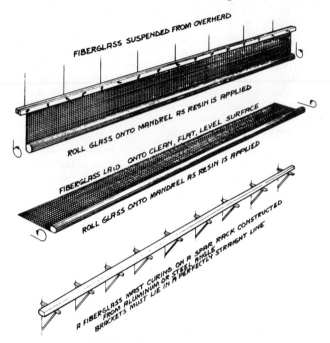

Fig. 394. Methods of laying linear fiberglass around a foam mandrel.

is able to round up into the wind swiftly. Of course, there is a practical limit to acceptable flexibility. You cannot have a mast as limber as a vault pole! Some type of core is necessary, not only to add stiffness but also to serve as a construction mandrel. Additional stiffness is also provided by employing larger corner radii than normally found on a wooden spar. It is even better to construct a fiberglass spar with all sides rounded slightly.

The most popular spar core is rigid polyurethane foam. This material is available in long block timbers and planks as well as a two-part liquid which can be mixed and molded to any desired shape. I will not describe the molding process for polyurethane foam, as it is a procedure far more practical in professional hands. If you choose to use foam block timbers as a core for your mast, the blocks may be plastic glued end to end to achieve the required length for fashioning. If you intend to use flat foam planks, they may be laid atop one another to the required thickness, again using a plastic glue. When building up the plank thickness, avoid butt joining two or more plank layers in the same area.

Before shaping the spar core, the areas of the foot, partners, winches, gooseneck, spreaders and truck must be located and marked. These are areas of high stress as well as attachment points straining hardware. Because fiberglass does not hold a metal fastening well, these areas must be reinforced

with solid wood sections. The wood sections should be scarfed into the foam core by using a "bird's-mouth" cut. (Fig. 395) Such a scarf joint will release concentrated strains on the spar walls. Once the spar has been fiberglassed, tangs, cleats, winch bases, etc., may be safely screwed or bolted into or through the solid wood cores.

With the wood cores scarfed into the foam core, shaping may now begin. Carefully lay out the proper tapers onto the core as detailed by your architect. The shaping of the foam may be easily accomplished with a spokeshave, Sureform file and sandpaper. The solid wood sections must be carefully faired with a plane or belt sander. No bumps or hollows may remain at the foam to wood joints as this may create weak spots along the spar length. If the spar is to receive sail track, the best attachment is by means of screws driven into a wooden feather laminated onto the aft side of the mast. The wood grain of the feather should run perpendicular to the screw axis. The feather section should be shaped to provide a smooth transition into the foam core as well as developing a more aerodynamically efficient air flow at the sail luff. Fastening the feather to the spar may be accomplished through the use of epoxy.

Developing the scantlings for fiberglass spars is one of essentially matching the structural characteristics of the glass laminate to the calculated radii of gyration and compression loads of each particular vessel. There are no simple thumb rules here and the formulation of the spar laminates should be left strictly to your designer. As a general comparison, however, a 30' tapered hollow box spruce mast, stayed with upper and lower shrouds and one set of spreaders would measure approximately 3½" x 5" at the maximum section with end and side walls of 1¼" and ¾" respectively. Such a mast would weigh 37 to 44 pounds unrigged. A fiberglass mast engineered to the same strength requirements would be of the same outside dimensions with

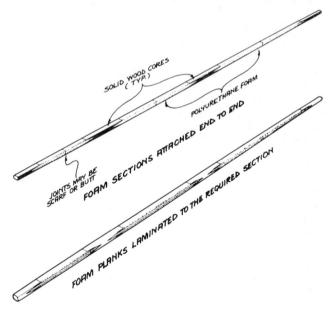

Fig. 395. Fiberglass foam mandrel variations clearly showing the installation of bird's-mouth solid wood cores.

more rounded corners and sides. The fiberglass wall thickness, however, would average only 3/16" while the weight of the mast would be reduced to 28 to 33 pounds. I am assuming, in this comparison, that the fiberglass mast would be constructed of 20% (by weight) boat cloth and 80% (by weight) of linear glass roving. This comparison should not be interpreted as a thumb rule, as every mast of every rig will have different requirements. Ask your architect.

The lay-up of the fiberglass onto the spar core should proceed slowly and at an ambient temperature between 60°–75°. Only attempt to apply as many layers of glass as you know you can handle conveniently and within about 35 minutes. All of the fiberglass layers must be very tightly wrapped around the core; no looseness may be tolerated. Be very sure that your resin is accurately catalyzed. Do not guess at it! Too much catalyst will produce a brittle laminate while too little catalyst will result in a lay-up which is too flexible. The actual fiberglassing procedure is far too complicated to be described fully here so I would suggest referring to many other fine books available on the subject as well as the use of experienced helpers during the lay-up operation.

Upon the completion of the major lay-up, your spar will inevitably display some bumps and hollows. Very sharp and severe bumps may be removed with a belt or disc sander, but I must caution you against gouging the spar or removing too much of the fiberglass. Hollows may be faired in with a mixture of polyester resin and micro-balloons (available at your fiberglass supply store). Such a mixture should be of the consistency of a creamy glaze. The fairing of the spar is most properly accomplished prior to applying the very last layer of fiberglass cloth. In this way the fairing compound will actually become a part of the laminate rather than simply a cosmetic surfacing which could peel or chip away with time. Once so faired, finish wrapping the spar with boat cloth (not roving) using only casting or sanding resin. When fully cured, additional coats of resin should be applied, sanding between each to obtain the desired smooth finish. The spar is now ready for painting and the attachment of hardware.

"Tee" Boom Construction

This is absolutely the easiest type of boom to build and they are very strong and efficient while being relatively light. The scantlings for such booms will be called out by your architect as well as his providing section and taper drawings. I must caution you against using anything but full length lumber, however. Even the longest scarf joint can lead to failure if not "perfectly" made.

The tee boom construction requires only two pieces of wood, both accurately tapered before assembly, i.e., the horizontal "base" and the vertical "flange." The flange is glued and fastened with long screws along the center of the base, taking care to insure that the base is absolutely straight while gluing. Often a "feather" is attached to the upper surface of the base plank to support the sail track, but I do not think this is necessary. The sail track should be attached with separate screws from those of the boom structure. There is little more to it other than finishing and attaching hardware. (Fig. 396)

Hardware in General

When I was a young and budding draftsman under Charles Morgan, Jr., I frequently glanced at a sign which hung at one end of the long drawing room which read "MAKE IT LIGHTER BUT STRONGER." This sign was a constant reminder to the design staff that additional pounds of material which served no structural function would only impair a vessel's stability and, hence, performance. Charlie was absolutely fanatical on this premise, but the incredible

racing records of such notable yachts as **Paper Tiger, Sabre** and **Maredia** went to prove that he was right. Of course, a great deal of time and expense went into the engineering and fabrication of special, custom hardware for the purpose of serving the required function with the least cost of material weight. This is particularly important with the design or choice of mast hardware which may be swinging around as much as 60' or 70' above the center of buoyancy.

Jack Corey and I worked together on the engineering of a masthead weldment to be fitted to an aluminum mast for a 60-footer. We were very proud of the fashionable outcome and delivered the working drawings to the machine shop with great confidence. The finished piece was beautiful and sparkling in its newness but imposing with inefficiency. It was truly a striking piece of hardware. But . . . enter Charlie Morgan . . . ; end of our momentary elation. With a black grease pencil he proceeded to mark the fitting as an indication of lightening holes and the removal of unnecessary

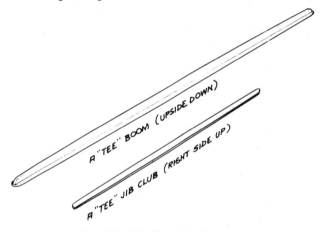

Fig. 396. "Tee" section booms.

metal. Before sending the part back to the machine shop, the masthead weldment was weighed at 13 pounds. When the revised part was placed on the scale, however, it topped out at only 10½ pounds.

Although the slight difference of only two and a half pounds does not sound very significant, it would have been located 84' above the waterline! In terms of moments, the same weight would have been translated into an astounding 205 foot-pounds! That would have taken exactly 30 pounds of ballast to offset its effect on stability. After this realization, Charlie laughingly threatened to "dock" our paychecks for every unnecessary pound of rigging weight detected after engineering. What a lesson!

So often I receive requests to design a rig which is "extra strong." Many owners want extra strong shrouds and stays, extra strong tangs, extra strong spreaders, extra strong halyards, super shackles, double weight sail material, an increase in the dimensions or wall thickness of the normal mast section and then add baggy wrinkles, ratlines, crow's nests and other paraphernalia to boot! Well, on a 40' boat, the inventory above could easily weigh 2,000 foot-pounds! The sad fact is that very few boat owners realize that most yacht designers automatically inject a 400 percent safety factor into all rigging calculations over and above that which would be anticipated under the severest conditions. Frankly, I think the most significant step that can be taken by any boatbuilder toward developing an extra strong rig is to "do it right" all the way through and to loosen the strings on his purse.

Only a mile from my own office there is a ferro-cement boat under construction whose owners chose to build a rolled steel plate mast step support and bow sprit. The mast itself must certainly weigh many hundreds of pounds and

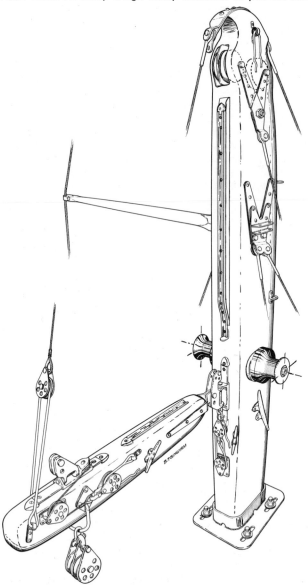

Fig. 397. Basic mast and boom hardware. While there are many variations of outhauls, downhauls, sheet leads, topping lifts and tangs, the accompanying illustration portrays the most common applications that are found on a masthead rigged vessel. The prospective in this drawing has been foreshortened for clarity so try to visualize the mast and boom lengthened to their proper proportions.

yet the boat is only a 33-footer. To add to this incredible weight factor a galvanized pipe ladder was installed up the forward side of the mast. This is one of the most flagrant abuses of weight I have ever seen and, like so many other construction errors, it was done in the name of strength.

I must grant that purchasing proper hardware and rigging is an expensive proposition but, that is yachting! Cutting back on the quality of castings, plate material, or commercial fittings here and substituting propriety for the scrounging of makeshift gear can reduce your rig's safety factor to naught.

There is an infinite variety of ways to rig every element of your boat. There must be a dozen methods alone for setting up downhauls, halyards, mainsheets and another dozen methods for rigging tangs, reefing and runners. I cannot detail every system, nor can I possibly recommend the individual pieces of hardware which make up the systems. The best education regarding rigging and sail handling comes from thousands of miles of sailing and years of personal offshore experience. If you do not have that behind you, then start sailing every chance you have. And read everything you can get your hands on. While you are at it, study the marine hardware catalogs profusely and walk the dock to see "how it is done" on the other fine boats. Rigging a yacht could be a volume in itself. In fact, check your local library; you may find it! Figure 397 will give you just an inkling of the profusion and variety of fine commercial fittings which are readily available, as well as an excellent indication of the most common rigging methods and location of gear.

Plate Rigging Hardware

Tangs. It would be useless to design a set of tangs or even chain plates or turnbuckles that would exceed the strength rigging wire to which they were attached. Of course it could be disastrous to lessen the strength of the hardware to under that of the wire. In theory, then, the configuration of a shroud should cause a failure under the same tension as its accompanying shroud. All additional metal over that required to meet the given tensile strength is nothing more than unnecessary weight. A safety factor of 1.25 is often added to the tensile requirements to account for possible

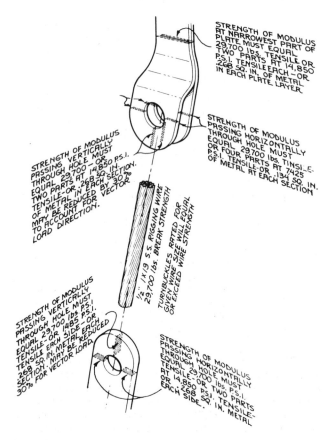

Fig. 398. The basic engineering premise for the design of chain plates and tangs. Notice here that all portions of the rigging chain equal the same strength.

fabrication inaccuracies, saw scores or stresses in the material caused by bending and heating. The tensile strength must be accounted for at all portions of the tang as well as the shearing strength of the tang attachment. (Fig. 398) Granted, simplicity of fabrication is a necessary consideration when designing for amateur construction so the examples shown should be viewed as somewhat ideal.

tangs. Their tension vectors are directly in line with the shrouds or stays attached to them. As you can imagine, screw fastenings alone could not possibly withstand the thousands of hours of tremendous loads without eventually loosening, fatiguing or stripping the wood surrounding them. Through-bolts, therefore, should always be considered as the primary attachment, but they must not be

FOR "EYE" TYPE PLATES & TANGS

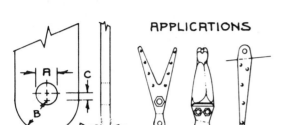

APPLICATIONS

FOR "JAW" TYPE TANGS

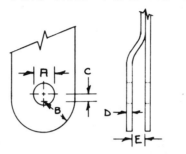

APPLICATIONS

WIRE DIA.	TENS. STRENGTH	A	B	C	D	TURN-BUCKLE
1/8	2,100	1/4	3/8	1/16	1/8	1/4
5/32	3,300	5/16	7/16	3/32	3/16	5/16
3/16	4,700	3/8	1/2	3/32	3/16	3/8
7/32	6,300	7/16	9/16	1/8	1/4	7/16
1/4	8,200	1/2	11/16	1/8	1/4	1/2
9/32	10,300	9/16	3/4	5/32	5/16	1/2 A
5/16	12,500	5/8	13/16	3/16	5/16	5/8
3/8	17,600	11/16	7/8	3/16	7/16	5/8
7/16	23,400	3/4	1	7/32	1/2	3/4
1/2	29,700	7/8	13/16	1/4	1/2	7/8
9/16	37,000	7/8	1 1/4	9/32	5/8	7/8
5/8	46,800	1	1 3/8	5/16	11/16	1
3/4	59,700	1 1/4	1 5/8	5/16	3/4	1 1/4

WIRE DIA.	TENS. STRENGTH	A	B	C	D	E	TURN-BUCKLE
1/8	2,100	1/4	3/8	1/16	1/16	7/32	1/4
5/32	3,300	5/16	7/16	3/32	3/32	7/32	5/16
3/16	4,700	3/8	1/2	3/32	3/32	9/32	3/8
7/32	6,300	7/16	9/16	1/8	1/8	5/16	7/16
1/4	8,200	1/2	11/16	1/8	1/8	5/16	1/2
9/32	10,300	9/16	3/4	5/32	3/16	11/32	1/2 A
5/16	12,500	5/8	13/16	3/16	3/16	3/8	5/8
3/8	17,600	11/16	7/8	3/16	3/16	13/32	5/8
7/16	23,400	3/4	1	7/32	1/4	13/32	3/4
1/2	29,700	7/8	1 3/16	1/4	1/4	1/2	7/8
9/16	37,000	7/8	1 1/4	9/32	5/16	19/32	7/8
5/8	46,800	1	1 3/8	5/16	3/8	11/16	1
3/4	59,700	1 1/4	1 5/8	5/16	3/8	23/32	1 1/4

USE JAW TERMINALS ONLY   USE EYE TERMINALS ONLY

TO INSTALL RIGGING TOGGLES, USE SAME SIZE AS TURNBUCKLES

Fig. 399. Chain plate and mast tang design standards based upon the use of No. 316 stainless steel.

The most common materials presently used for tang fabrication are No. 316 stainless steel or silicon bronze. The accompanying drawings and tables have been engineered for stainless. You will notice the distinct avoidance of sharp corners and angles except at the spreader base brackets as well as a profusion of screw fastenings in order to distribute the rigging load to as large a spar surface as possible. The most critical part of any tang or chain plate is the area of the clevis pin hole, as this must be positioned properly, taking into account the thickness of the metal and the width of the material between the hole and the plate edge. Figure 399 may be used as a general guide for the location of such clevis holes and design of the clevis areas of chain plates and tangs. When finishing your plate hardware, be sure to round all plate edges including the perimeters of bolt and clevis holes. Be absolutely sure that no burrs remain, as they can snag sails and fiber line as well as hands.

Compression Tubes

Of course, the greatest strains on any rig are at the mast

arbitrarily placed. The purpose of the through-bolt is to transfer the rigging load to the opposite side of the mast where the load will become compressive in nature rather than in tension. The tightening of the tang through-bolts, however, must not crush the wood of the mast or place a strain on an aluminum or fiberglass hollow section. To prevent this, the bolt should be inserted through a steel or bronze pipe which is cut flush with the outer surfaces of the spar. When the tang bolts are tightened, their forces will be on the pipe or compression tube rather than on the mast itself.

The tang through-bolts need not necessarily be bolts, as it is a very common practice to use stainless steel pipe which has been threaded at both ends. The prime benefit of the use of threaded pipe instead of bolts is that the bearing surface of the fastening is much greater, while not appreciably increasing weight. (Fig. 400)

Spreaders

If a single set of spreaders is used, it will be located at half the distance between the partners and the upper shroud tang. If a double set of spreaders is to be used, it will be located at "third" distances between the partners and upper shroud tang. The spreader lengths and section shapes will be dictated by your designer, but normally they are tapered toward the outboard ends and present an aerodynamic foil.

Of course, spruce is the best wood for spreaders because of its high compressive strength/weight ratio and resistance to bend. If spruce is unavailable, fir is an acceptable substitute. In either case, the lumber should be "perfect," dry and straight grained. Hewing should be per plan, but often specific detail is not shown such as their outer tips or the configuration at their bases. You will notice in the accompanying illustrations that metal strap is used to cover the

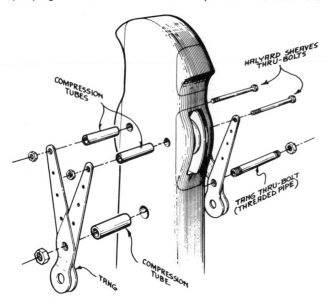

Fig. 400. The use of compression tubes.

shroud recess to prevent splitting the member under severe loads. Also notice the two small holes drilled in the spreader tips for the purpose of passing binding wire. Such wire binding around the shroud should be covered with a leather "glove" or tape to prevent damage to sails. You will also notice the expansive "footing" of the spreader base and a common spreader bracket and lower shroud tang weldment. Regarding this hardware: because of its lofty position, it is imperative that it be strong but as light as possible. One last point which should be considered of the greatest importance is that of the spreader angle. It is very critical and can result in catastrophe if not proper. If the spreaders' angle is too great, they will have a tendency to move upward then collapse under a severe load. If they are fastened horizontally, they will tend to drop downward, which can also lead to total failure and possible loss of the rig. (Fig. 401)

The Masthead

This is a study in itself! There are no two alike (except for stock yachts), not even on the same boat. Once again, there are no rules of thumb which can be applied other than "make it light . . . make it stronger" (Charles Morgan).

The most basic approach for rigging halyards is that of shackling blocks to bails or eyebolts at the truck of the mast. Unfortunately, they are heavy and noisy in a seaway while multiplying the number of parts which could possibly fail. The finer yachts with wooden spars have been con-

structed with their halyard sheaves inserted into a slot passing entirely through the masthead. With such an arrangement it is imperative to keep the slot to a minimum size. One large sheave inserted into the mast for each halyard usually requires a very large hole, thus it is most common to install two smaller sheaves for each halyard. The fore-and-aft perimeters of the sheaves should overhang the faces of the mast just enough to eliminate chafing of the lines. Each pair of halyard sheaves must be separated from each other with an aluminum plate to prevent any possibility of override.

The sheaves used must be specifically engineered for mastheads. Common pulley sheaves will not work. The groove of the sheaves must be fashioned to accommodate both the wire halyard as well as the rope tail. Such sheaves are available from most sailboat hardware manufacturers and are shown in Fig. 402.

In order to provide a suitable clearance between the sail leech or headboard from the backstay, a shaped wooden boss is usually glued to the aft side of the mast. To create the proper space between the forestay and halyard, a wooden boss is often attached to the forward side of the mast also. Wooden bosses may be fashioned around the halyard sheave apertures as well but their actual function is questionable. These bosses can be seen in Figs. 397, 400, and 402.

The Feather and Track

Several times in this chapter I have made references to the "feather" on a mast, but I have refrained from defining it until now because it more properly belongs to the spar finishing category rather than that of actual construction. The feather is nothing more than a length of wood placed onto the aft edge of a mast along the route of the sail track. On a wooden mast the purpose of the feather is twofold: (1) it lessens the number of screws to be driven into the mast proper in order to hold the sail track. While the feather is, indeed, screwed to the mast, far fewer are used than would be the case if the sail track were fastened directly; (2) the feather, tapered in section, enhances the aerodynamics around the mast.

When attaching 5/8" track, the feather is fabricated from 1/2" x 1 1/4" stock. It is then tapered to 5/8" along its trailing edge. For 7/8" sail track, begin with 3/4" x 1-5/8" stock, tapering the trailing edge to 7/8". So shaped, the feather is scarfed to full length then glued and screwed to the mast. The fastenings need not be closer together than 9" but no more than 12". (Fig. 403)

When constructing a fiberglass mast, the feather becomes much more structurally important than in the case of a wooden mast. The feather is shaped in the same manner then it is fiberglassed onto the mast's trailing edge. First a layer of, say, 10 oz. cloth is used over the feather, overlapping the cloth 1" onto the mast surface. Then a polyester glazing compound is troweled into the corners, thus fairing the feather into the mast shape. When the glazing compound has cured, it is sanded smooth. In the interest of saving unnecessary weight, the compound should be made with microballoons. Now, two additional layers of cloth are placed over the feather to complete the fastening and finishing. These cloth layers should overlap onto the mast surface progressively more than the first. Because a fiberglass laminate does not hold screw fastening worth a hoot, you may now see the feather's wood core is an absolute necessity for attaching sail track to such a structure. The additional layers of fiberglass over the feather also stiffen the mast slightly. (Fig. 403)

Rigging Wire

There are many types of cable on the market today and each has been designed and manufactured for a specific purpose. Today it is most common to use stainless steel and I will not enter a discussion of plow steel in this text. 1 x 19 is the most common type of cable used aboard sailboats. 1 x 19 means 19 individual wire threads wound into one single strand. It is an extremely stiff wire and, therefore, should not be used where flexibility is a prerequisite. 1 x 19 wire, however, is the strongest of all wire types and, hence, it is the first choice for standing rigging, i.e., shrouds and stays. It should never be used for halyards or sheets. The tremendous strength of 1 x 19 cable can be seen in the chart shown in Fig. 399. When ordering this wire, be absolutely sure that you specify marine wire, not aircraft wire. No other wire should be used for standing rigging, regardless of the price.

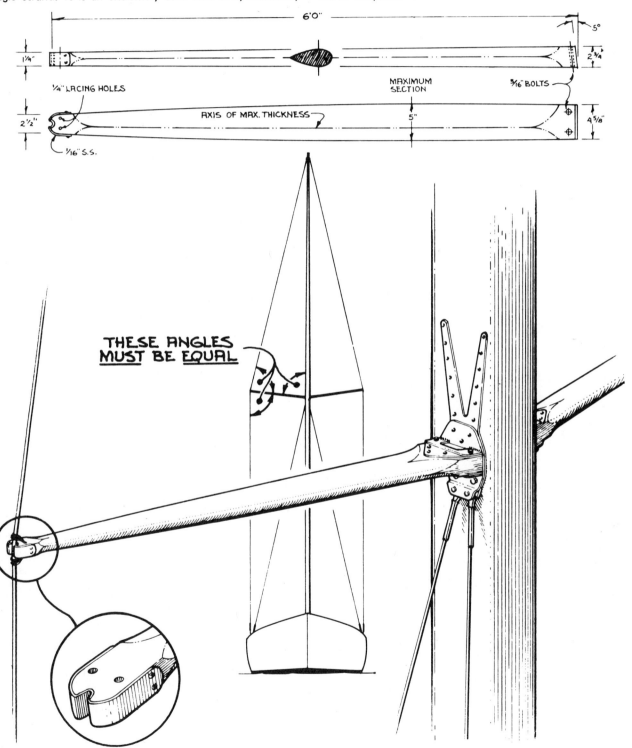

Fig. 401. Spreader design and construction.

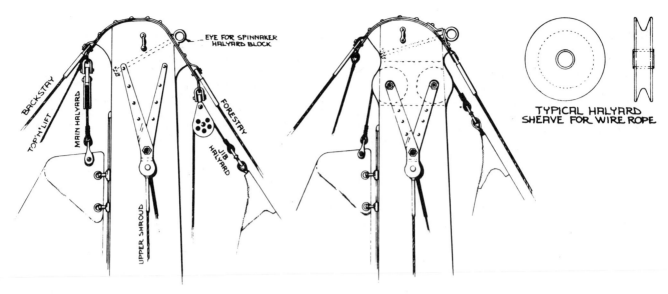

TYPICAL HALYARD
SHEAVE FOR WIRE ROPE

Fig. 402. Halyard block masthead and internal halyard sheave masthead.

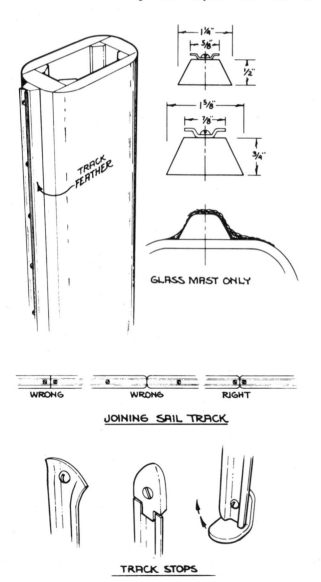

GLASS MAST ONLY

JOINING SAIL TRACK

TRACK STOPS

Fig. 403. Male sail track and feather mounting.

Most halyards and sometimes even the sheets on very large boats are made up of wire rope (a more proper term for cable) because these wires require passage through sheaves or blocks; it is imperative that they be very flexible. In wire rope, this flexibility is achieved by adding to the number of threads which make up the rope. These threads, in turn, are spun into strands. When looking through wire rope catalogs, you will see numbers like 6 x 42, 7 x 7, 7 x 19, etc. The first number refers to the number of strands which make up the cable. The second number refers to the number of threads which make up each strand. By multiplying the two numbers together, you will derive the total number of threads encased within the cable, i.e., 6 x 42 would equal 252 threads; 7 x 7 equals 49 threads; 7 x 19 would have 133 threads. In essence, the more threads within the cable, the more flexible the cable will be and, hence be less vulnerable to wear and fatigue through its movement and will handle easier. It is also important to point out here, though, that the more threads within the wire, the more expensive the wire will be. (Fig. 404) You may deduce, then, that 6 x 42 wire would make the best halyards and would ride more easily over the small halyard sheaves. This type of wire should be your first choice but many suppliers do not carry it as a stock item. Your second option, then, should be 7 x 19. One other factor that should be considered is the fact that the more flexible wire will be far easier to bend into an eye around rigging thimbles and will not be nearly as strained as would be a stiffer wire. When making up wire rope halyards, it is most common and prudent to fasten fiber or synthetic line on the pulling side of the halyard. This is done to prevent possible hand injury to crewmen, to aid in the coiling of halyards and to provide for the passing of the halyards onto a cleat. You could not do this with wire rope. Needless to say, once a halyard is drawn up on the winch, its strain can equal as much as 1,000 pounds. Under this kind of strain, synthetic fiber halyards would stretch uncontrollably.

Splicing of the fiber halyard "tail" to the wire rope is extremely difficult and is usually done on order by your rigging supplier. Do not attempt to do it yourself. It is more common today, however, to fabricate an eye in the fiber line which is passed through an eye in the wire rope. Such eyes are accomplished through the use of Nycopress sleeves and steel thimbles.

Wire Rope Terminals

In essence, a wire rope terminal is simply a fancy way of describing the hardware at the ends of the cables. There are two varieties most commonly encountered, i.e., "jaw" and "eye." Jaws are always used when attaching a wire rope to a single steel plate and to rigging toggles. Eye terminals are used when connecting a wire rope to a turnbuckle, double plate tangs, to the jaw terminals of other wire ropes or to the jaw sides of rigging toggles. As long as I have mentioned rigging toggles, I should explain that they are used to increase the flexibility of the wire connection at the wire rope terminal. Such flexibility will relieve or eliminate any possible bending strains which may be created around the neck of the wire terminal. Toggles should be used at both ends of all shrouds and stays and at the upper end of your topping-lift. You may question the additional cost of this hardware but you should not discount their effects on the longevity of your rigging. A couple extra dollars here and there can easily lengthen the replacement time of your shrouds, thus paying for themselves over a number of years.

 1 x 19 STIFF - MAXIMUM STRENGTH
USE FOR STANDING RIGGING ONLY

 7 x 7 MODERATE FLEXIBILITY
BEST FOR RUNNING BACKSTAYS

 7 x 19 VERY FLEXIBLE
USE FOR HALYARDS, LIFELINES, OUTHAULS (THROUGH LARGE SHEAVES) AND CENTERBOARD PENDANTS

 6 x 42 MAXIMUM FLEXIBILITY
USE FOR HALYARDS AND OUTHAULS (SMALL SHEAVES), SPINNAKER SHEETS, AND SAIL PENDANTS. THIS IS THE BEST WIRE FOR STEERING GEAR

Fig. 404. Wire rope types.

The most common type of wire terminal seen aboard sailboats is the roller swaged terminal. These are fabricated out of stainless steel and are available in two varieties, i.e., "marine" and "aircraft." The aircraft terminals are of much lighter construction than that of marine rigging terminals and, therefore, should not be used, even though you may be able to save a few dollars. Unfortunately, you will not be able to simply pick roller swaging terminals up at your local hardware store. They must be installed by the manufacturer of the wire rope who uses very specialized equipment for this. The measurements which you give to your wire rope supplier, therefore, must be extremely accurate. Once the terminals are fastened to the wire rope, they cannot be removed.

There are very definite plusses and minuses regarding the use of roller swaged terminals. They are extremely handsome, streamlined, and lightweight. They always make for the cleanest and most professional looking rig. However, because they cannot be removed from the cable, it is impos-

sible to inspect their connections at any time after fabrication nor may they be replaced. Roller swaged terminals have an infinity for developing cracks at the neck of the terminals through which the wire rope is passed. Once such cracks develop, the security of the terminal should be questioned. Regardless of what most people think, stainless steel will corrode. Such corrosion may occur within a roller swaged terminal where salt water may collect undetected. This corrosion could "make or break" your whole rigging system.

Norseman terminals are becoming more and more popular among cruising people and for many good reasons. Although they are somewhat heavier and bulkier than roller swaged terminals, it is possible to remove them at any time after their connection to the wire, thus allowing for inspection and replacement. The Norseman terminals, when properly attached to the wire rope, are considered to be just as strong as the roller swaged terminals. I, personally, believe that this is the way to go if extended sailing is what you have in mind. With a couple of hundred of extra feet of various types of wire rope in your bilge and a variety of extra Norseman terminals stashed away in a remote drawer, you will be able to replace virtually any wire aboard your vessel in a minimum amount of time whether at sea or at dockside. The Norseman terminals are attached to the wire rope in only a few short minutes. Therefore, it is even possible to make their connections to the lee side shrouds while the vessel is actually under sail. The benefits of Norseman terminals should be considered as a prerequisite to the cruising man's safety. They may be used virtually anywhere on the boat.

At the boat shows one often sees a third type of terminal made by bending a wire around a steel thimble then fastening the two sides of the wire with a Nycopress sleeve. When properly done with a special tool, the connection of the wire rope within the sleeve is very strong. Unfortunately, you can never be absolutely sure as to its ultimate security. Personally, I have never had any trouble with them. Fabricating an eye terminal in such a way produces a very light unit. They are, however, somewhat bulky and unprofessional looking. If any stray wires are allowed to protrude beyond the upper neck of the Nycopress sleeve, they will invariably tear sails, snag lines, and possibly injure one of your crew members with a nasty gash. These three occurrences are very common. I could never recommend using Nycopress sleeves and steel thimbles for fabricating eyes in 1 x 19 rigging cable. While the security of the Nycopress sleeve is not necessarily in doubt, they create severe stresses on the apex of the bend in the cable. The outer radius of the cable is inherently drawn up very tightly. Thus these strands and threads will be placed under extreme tension when the cable is under a load. Those threads and strands to the inside of the radius become subjected to overwhelming crushing loads. You can see, then, that the load on the cable is not uniformly distributed throughout the diameter of the wire as it passes around the rigging thimble. Less than half of the wire rope will actually be carrying the load of the rig and that is the case against Nycopress terminals. They should not be used when fabricating the standing rigging wire. Nycopress sleeves definitely have a place aboard your boat. Their use should be restricted to forming eyes in flexible halyards wire rope, outhaul wires, pendants, etc. These units, normally, are made up of 7 x 19 or 6 x 42 wires. Even though you may be swayed by their very low cost of fabrication, do not consider them anywhere else. (Fig. 405)

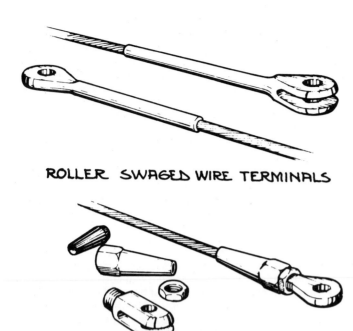

ROLLER SWAGED WIRE TERMINALS

NORSEMAN SWAGELESS TERMINALS

NICOPRESS SLEEVES AND RIGGING THIMBLE

Fig. 405. Common wire rope jaw and eye terminals.

Cleats

So you ask, "What can be said about cleats? All you do is wrap lines around them, do you not?" Well, there is a little more to it than that. Sure! Basically cleats are used for holding fiber line fast, but cleats come in many different forms specifically designed for different applications. There are four hole open base cleats, two hole open base cleats, spar cleats, jam cleats, cam cleats, "Vee" jam cleats, bits and bollards. Cleats are fabricated out of wood, aluminum, bronze, stainless steel, molded nylon, iron and fabric-reinforced plastic. They range from 2-5/8" in length to the largest normally listed in your marine catalog of 24". In my opinion, the best overall cleat is of the four hole open base type. They are quite a bit more expensive than two hole cleats or spar cleats but their security depends upon four fastenings and a very wide footing. With a four hole open base cleat, it is almost impossible to bend the mounting bolts, as often occurs with two hole cleats. The hole provided in the open base can be used as a line stopper by simply running the halyard or sheet through the base and knotting it on the other side. It is impossible to lose a line aloft through this simple precaution. The four hole open base cleats should be your first choice for securing halyards, sheets and certainly dock lines. They are the strongest of all.

In my opinion, two hole open base cleats should be restricted to those instances where you might find it impossible or impractical to set four fastenings. Frankly, I see no other advantage to them. Spar cleats are also secured with two fastenings. I think their use should be limited strictly to flag halyards, outhauls and downhauls. As I have mentioned previously, the two hole cleats are notorious for bending their securing bolts. Jam cleats may be used for securing line which is known to need continual readjustment as this type of cleat does not require the figure eight wrapping as with the other cleats previously mentioned. Unfortunately, jam cleats are also of the two hole type and are rarely available for a line size larger than 3/8". On boats under 32', though, they can be of great benefit to the single handed sailor. "Vee" jam cleats presently on the market, have been designed for use with very small line, say, 5/16" and smaller. They should not be installed for securing any line which may be subjected to strains in excess of 20 pounds. On a cruising boat they may be considered for flag halyards but that is about all. They are primarily meant for small, lightweight racing machines and sailing dinghies.

Cam cleats are now available for line sizes up to ½". The larger sizes have been engineered for three and four bolt mounting but the actual strength of the cleat is still limited to the two pivot screws. They should not be used for any line which does not require continual readjustment or where a line at the cleat will be under severe strains. They are very nice as secondary cleats on vessels intended for single handling. They are excellent for mainsheets and for jib sheets on small weekend cruisers. Do not use cam cleats for halyards or in any area where the heeling of the vessel could dislodge the line or where it would be subjected to tripping the crew members.

Cleats should be bolt fastened whenever possible. Unfortunately, this is rarely practical when mounting cleats onto the mast. In all other areas, however, bolting should be a prerequisite. Every cleat fastening nut should be backed by as large a washer as possible. On many fine custom built boats you will even find that the builder has gone so far as to make his own large washers out of stainless steel or bronze, which are as much as twice the size of the largest washer found in your local hardware store. Those cleats which are mounted on the deck should be backed up with a steel or hardwood strongback. The larger the strongback, the better. When mounting mooring cleats to a beamed wooden deck, the cleat strongback should span a pair of deck beams, thus distributing the strain over a larger portion of the structure. When positioning your cleats, the axis of the horns should not be in a direct line with the strain. Angle your cleats about 15 degrees so that the first contact of the line on the cleat is around the base facing away from the strain.

When you begin costing out your hardware list, you will find that many hundreds of dollars will be invested in your cleat hardware alone, particularly if you are constructing a sizeable vessel. Under no circumstances should you be tempted in purchasing cleats rated smaller than the line size they are to be used for. Those line sizes which are listed in the marine catalogs are considered maximum. Unfortunately, it seems that cleats can never be too big and more often much too small. It is a wise practice, therefore, to purchase cleats for a line listed one size larger. In the case of mooring cleats, bear in mind that your heavy duty anchor rods will be of a greater diameter than your dock lines and that there may be circumstances under which it will be desirable to moor your boat from astern. Also keep in mind that there will be times when you will be required to secure more than one line at a time to these cleats. I would suggest purchasing mooring cleats at least two sizes larger than that required to accommodate your docking lines alone. If, at this point in your construction, you find your commissioning budget being severely squeezed, it is possible for you to make your own cleats out of ash, oak, birch, teak, or even mahogany. I have included cleat templates with an accompanying cleat and size bolt chart for this purpose. Because these cleats are of a two hole type, they should be considered only as substitutes for their eventual replacement with metal four hole cleats. While at the outset wooden cleats are very handsome, they are an absolute headache to maintain and in time will wear away from continual line chafing. In order to deter the possible bending of the fastenings, I would strongly suggest using only stainless steel bolts or screws. If you are in question as to the use of construction of other types of cleats, consult your architect directly or talk to people along the dock who have the experience to give you "straight scoop." (Fig. 406)

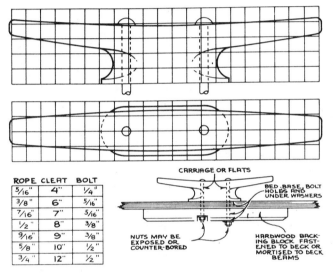

ROPE	CLEAT	BOLT
5/16"	4"	1/4"
3/8"	6"	5/16"
7/16"	7"	5/16"
1/2"	8"	3/8"
9/16"	9"	3/8"
5/8"	10"	1/2"
3/4"	12"	1/2"

CARRIAGE OR FLATS

BED, BASE, BOLT HOLES AND UNDER WASHERS

NUTS MAY BE EXPOSED OR COUNTER-BORED

HARDWOOD BACKING BLOCK FASTENED TO DECK OR MORTISED TO DECK BEAMS

Fig. 406. Cleats may be homemade using the templates shown above. Whether homemade or not, all cleats should correspond to the minimum size.

Winches

There are basically two types of winches that you will be most concerned with, i.e., single action and double action. The drum of a single action winch will turn clockwise only when the handle is turned in the same direction. Upon rotating the winch handle counter-clockwise, the winch position will remain stationary. The drum of a double action winch, however, will continue to turn clockwise regardless of the direction of the movement of the handle. Single action winches are most properly confined to use as halyard winches only on cruisers and as light sheet winches on small boats. They are far less expensive than the double action winches but they can also handicap a sailor and his crew severely. Double action winches are far more useful because the clockwise rotation of the handle actuates the drum in a high gear while the counter-clockwise actuates the winch drum in a low gear. In heavy air or when handling very large sails, then, the counter-clockwise rotation of the handle will produce far more power than a clockwise rotation of the handle. As you can see, most of your trimming in light air can be handled speedily when less power is required.

Most boat owners have become accustomed to referring to the winch size by number, i.e., No. 8 winch, No. 16 winch, No. 22 winch, etc. Upon investigation, however, you will find that different manufacturers use a different winch size number even though the power ratios are exactly the same and it is the power ratio with which you should be most concerned. In essence, the power ratio of a winch defines the number of pounds which a line from the winch can pull for every pound of pressure which you are able to place against the handle. In other words, the power ratio is the mechanical advantage of a particular winch. This mechanical advantage is achieved in two ways: (1) by the size or by the diameter of the drum and (2) the gear ratio within the winch. An 8 to 1 gear ratio in a 3½" drum will produce an entirely different power ratio than an 8 to 1 gear in a 4" winch drum. By themselves, the gear ratio and the drum size tell you very little about the power of the winch.

There are no clear-cut standards for the establishment of power ratios required on different boats of a given size. The winch requirement depends a great deal upon the lead of the sheets and their purchases, the efficiency and number of men in the crew, the degree to which a skipper pushes his boat, the normal wind conditions in the local vicinity of the vessel and the degree of speed in trimming required by the skipper. Winches of a high power ratio will be slower trimming than those of a low power ratio in light airs. In heavy airs, however, the low power winch may not be able to be turned at all. The skipper will have to weigh many factors carefully when making his final decision regarding the purchasing of proper equipment. Hopefully, he will have a considerable amount of offshore sailing experience to draw upon. In order to help you make your decision safely, I am including a chart of winch power ratio recommendations for the handling of different sails aboard normal cruising vessels. The ratios and sail areas shown in the chart do not take into account a racing capability nor should it be assumed that you will be required to purchase separate winches for each sail shown. Some experienced skippers may disagree with my recommendations but, then, it is impossible to speak for the individual. (Fig. 407)

WINCH POWER RATIOS				
10:1	20:1	30:1	40:1	50:1

MAXIMUM SAIL AREAS

	10:1	20:1	30:1	40:1	50:1
MAIN/MIZZEN/FORES'L/ SELF TRIMMING HEADS'L SHEETS (BASE ON A 4 TO 1 BLOCK PURCHASE)	TO 200	TO 400	TO 650	TO 1,000	
GENOA/JIB/WORKING STAYS'L SHEETS	TO 100	TO 200	TO 300	TO 400	TO 600
LIGHT STAYS'L SHEETS (SECONDARY COCKPIT)	TO 150	TO 275	TO 400	TO 600	TO 900
SPINNAKER SHEETS	TO 200	TO 400	TO 900	TO 2,000	TO 4,000
MAIN/MIZZEN/FORES'L HALYARDS	TO 200	TO 300	TO 400	TO 600	TO 1,000
GENOA/JIB/WORKING STAYS'L HALYARDS	TO 200	TO 275	TO 350	TO 500	TO 800
SPINNAKER HALYARD	TO 300	TO 600	TO 1,500	TO 3,500	TO 5,000
LIGHT STAYS'L HALYARDS	TO 300	TO 400	TO 500	TO 700	TO 1,000

Fig. 407. Recommended winch sizes.

Other Spar Details

Every hollow mast, whatever the type of construction, must be provided with a drain hole at its base to allow for the run-off of condensation. There is no exception to this rule. When constructing a hollow mast with solid wood

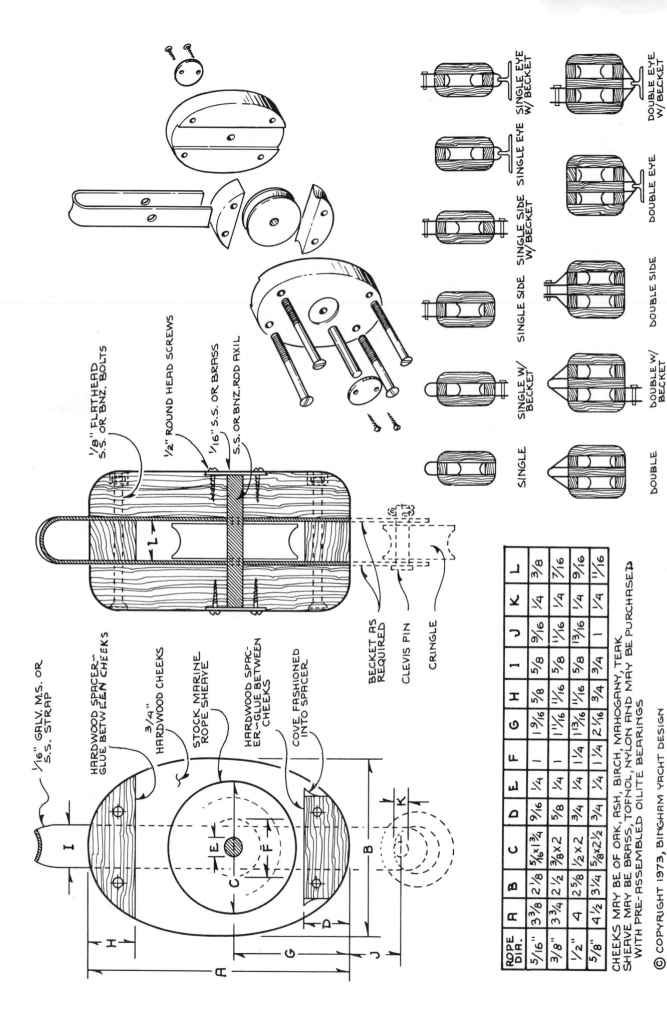

1/8" FLATHEAD S.S. OR BNZ. BOLTS

1/2" ROUND HEAD SCREWS

1/16" S.S. OR BRASS

S.S. OR BNZ. ROD AXIL

SINGLE EYE W/ BECKET

SINGLE EYE

SINGLE SIDE W/ BECKET

SINGLE SIDE

SINGLE W/ BECKET

SINGLE

DOUBLE EYE W/ BECKET

DOUBLE EYE

DOUBLE SIDE

DOUBLE W/ BECKET

DOUBLE

BECKET AS REQUIRED

CLEVIS PIN

CRINGLE

1/16" GALV. M.S. OR S.S. STRAP

HARDWOOD SPACER- GLUE BETWEEN CHEEKS

3/4" HARDWOOD CHEEKS

STOCK MARINE ROPE SHEAVE

HARDWOOD SPAC- ER-GLUE BETWEEN CHEEKS

COVE FASHIONED INTO SPACER

Fig. 408. Homemade blocks.

ROPE DIA.	A	B	C	D	E	F	G	H	I	J	K	L
5/16"	3 3/8	2 1/8	5/16 x 1 3/4	9/16	1/4	1	1 9/16	5/8	5/8	9/16	1/4	3/8
3/8"	3 3/4	2 1/2	3/8 x 2	5/8	1/4	1	1 11/16	11/16	5/8	11/16	1/4	7/16
1/2"	4	2 5/8	1/2 x 2	3/4	1/4	1 1/4	1 13/16	11/16	5/8	13/16	1/4	9/16
5/8"	4 1/2	3 1/4	5/8 x 2 1/2	3/4	1/4	1 1/4	2 1/16	3/4	3/4	1	1/4	11/16

CHEEKS MAY BE OF OAK, ASH, BIRCH, MAHOGANY, TEAK. SHEAVE MAY BE BRASS, TOFNOL, NYLON AND MAY BE PURCHASED WITH PRE-ASSEMBLED OILITE BEARINGS

© COPYRIGHT 1973, BINGHAM YACHT DESIGN

cores, the cores should be drilled to provide drainage and ventilation between the spar sections. Occasional small vent holes may be drilled through the mast walls to provide for air circulation but this is not necessary if there are no bolts passing through the hollow portions of the spar. Such vent holes when drilled should angle upward to discourage the entrance of water. If your mast is to be constructed with mains'l and heads'l halyard sheaves housed within the mast itself, the housing slot cut for such an installation must not open into the hollow portions of the mast, as this would cause water to enter the spar. When bolting into a hollow portion of any spar, it is imperative that the bolt, the hole, nuts, washers and hardware be well bedded for a watertight seal. End grain should not be left exposed at the ends of booms, clubs, gaffs or at the truck of the mast. Such end grain should be covered with a wooden cap whenever practical, as this will retard severe drying and possible checking.

When storing your spars before their installation or at any other time, be sure that they are well supported at close intervals and that they are not allowed to sag or twist. Keep them dry and out of the direct sunlight as it will take time for their moisture content to stabilize after fabrication. Sudden changes in humidity or temperature may affect the young glue bond as well as the spar shape. When shaping the individual staves for any type of spar construction, never sand the plank edges, as this will surely lead to rounding of the critical gluing surface which must be "perfectly" flat and regular. Remember to preassemble your spars prior to gluing as such a "dry run" will betray any fitting problems and it will give you an opportunity to see the stave joint regularity. Keep in mind that a mast is much, much more than just a simple pole. Your vessel's performance and even your life will depend upon your patience and diligence in this work.

There are a great many hardware items that can be fabricated in the amateur's garage, provided the equipment and experience are readily available. Other units, such as homemade goosenecks, just do not make it and when it comes to longevity or reliability, I have seen quite a few very flashy engineering drawings for such units but, unfortunately, they rarely come out as nice looking as they do on paper. Do not avoid purchasing commercial hardware, particularly if the required piece is comprised of many parts. Such hardware has been created to last many decades of faithful service under extreme conditions.

Mast Fastening Precautions

Once the rig is into the boat, the opportunity for going aloft for inspection while under sail will be rare indeed. You will rely upon faith alone to know that your hardware will remain securely fastened. As an added safety factor, however, it is a very prudent measure to crush all exposed bolt threads on the outer sides of each and every nut, however critical it might seem. In this way, you can be sure that the rigging fastenings cannot vibrate or work themselves loose during some stormy night. After placing any bolt, do not allow more than 1/16" of the shank to protrude beyond the nut as it will surely snag sails or lines or even cause personal injury. Cut all long bolts to just the size required and grind their ends free of burrs. Very large nuts and bolts may be drilled for cotter pins. This will serve the same purpose as crushing the bolt threads. When screwing fastenings into a spar, I have found that Loctite compound spread over the screw threads provides two important benefits: (1) it prevents the screw from turning out and (2) it seals the wood around the screw, thus preventing

the wood from softening from moisture absorption. As long as I am talking about screws, never use brass as they fatigue and crystallize readily in a marine environment, and never drive a screw without drilling an adequate pilot hole.

Blocks and Tackles

A block is used to either change the direction of a line, as in the case of halyard blocks and quarter sheet blocks, or to increase the mechanical advantage placed upon the line. You will have to rely upon the detailing provided by your architect regarding the specific block arrangements on your own vessel, but I will give you a few basics here to help guide the rigging of your boat.

Sheeting power is directly related to the number of running parts in a tackle, i.e., a 4-part tackle gives a 4 to 1 pound advantage, less the losses for friction in the blocks. Three-part tackles are ample on small craft whose individual sail areas do not exceed 150 to 175 square feet. For larger sails, say up to 500 square feet, a 4-part tackle is most common on the main boom, jib and mizzen sheeting. What this means is that when pulling in 4' of line, the boom will only move 1'. You may have already noticed on the winch chart (Fig. 407) that I have assumed such a mechanical advantage under the first category of winches. You may also add block purchases to your heads'l sheets. A 2 to 1 purchase on your Genoa would reduce your winch power requirement by 50%. Of course, the same would be true in the case of halyards. The problem here, however, is that your sheets would then require twice the amount of line which inevitably would pile up in the cockpit or at the base of the mast. This, in itself, can be quite a headache when handling a boat in close quarters or during night sailing.

Good blocks are expensive. A glance at any marine hardware catalog will bear this out. There are disassembled block kits available, however, which can reduce this expense significantly while still producing a strong, lasting, serviceable product. One advantage to such block kits is that their components are interchangeable from one type of block to the other, providing the components are rated for the same line size. By keeping a drawer full of extra block kit components, you will be able to fabricate any type of block on the spot while under way as circumstances dictate.

Many amateur builders have expressed a desire to construct their own blocks from scratch. Such a project can make an enjoyable, satisfying day's work. Standard block sheaves are available at most marine hardware outlets. The accompanying working drawing may be used if you decide to undertake the block fabrication on your own. (Fig. 408)

Commissioning the Boat

There are literally thousands of details which I have not been able to cover: anchors, dock lines, bumpers, safety equipment, dinghies, light sails, launching, tuning, etc. Most of these subjects are handled extremely well in such fine books as *Encyclopedia of Sailing* (Harper & Row), *The Proper Yacht* by Beiser (Adlard Coles Limited), *Sea Sense* by Henderson (International Marine Publishing Company), *Yachtman's Omnibus* by Calahan (MacMillan) and one of the best books I have ever seen on commissioning detail is *The Ocean Sailing Yacht* by Street (W.W. Norton & Company, Inc.).

I cannot overemphasize the importance of not being in a rush to launch and get under way. Many ferro-cement builders have put their boats in the water almost immedi-

Loctite is a registered trademark (May, 1956) of Loctite, 705 N. Mountain Road, Newington, Ct., 06111.

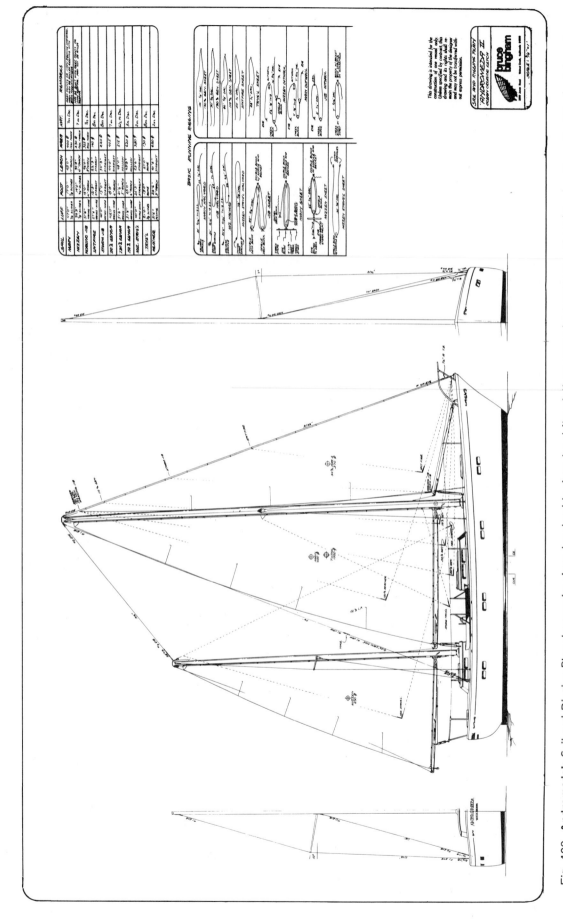

Fig. 409. **Andromeda's** Sail and Rigging Plan shows wire sizes, turnbuckle sizes, heads'l variations, spreader locations and angles, as well as all of the basic running rigging including blocks, shackles, pad eyes, track, and line lengths. As you can see, the Sail Plan is far more than just a pretty picture but a true working drawing. It also includes all of the necessary information required by the sailmaker to accurately reproduce the proper sail shapes with the recommended cloth weights. A Bingham design.

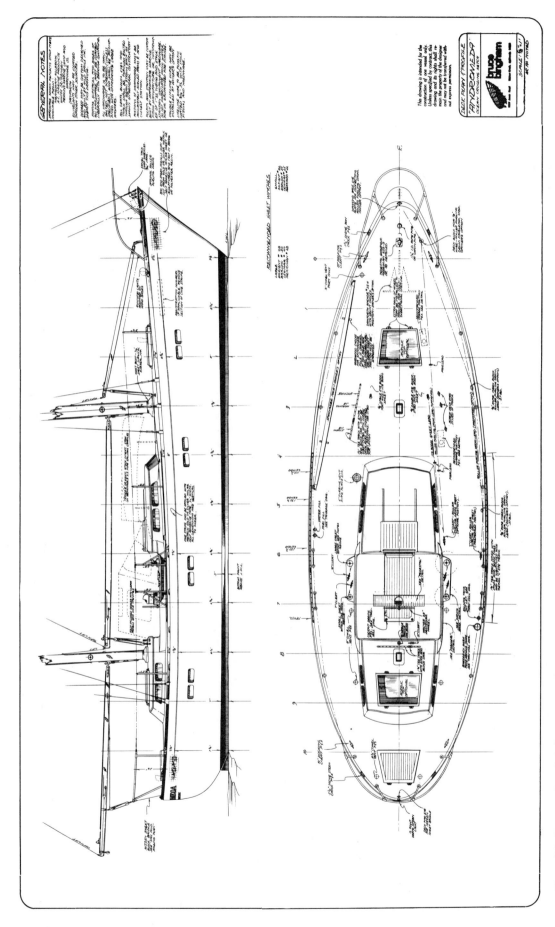

Fig. 410. Because the Sail Plan is generally of a very small scale, many designers do a separate drawing entirely to show the locations of the on-deck hardware. In this drawing you may clearly see the lifeline, stanchion and pulpit positions, pad eyes, sheet tracks, winches, cleats, ground tackle, roller furling sheet leads, chocks, spinnaker poles, water and fuel fills, hatches, dinghy location, and even the emergency bilge pump deck iron. Absolutely nothing has been left to chance. Here, the designers have even gone so far as to dimension the waterline taper, cove stripe position, registration numbers location, and bow and stern scrolls in order that the builder may faithfully execute the designed appearance of the vessel. A Bingham design.

ately after plastering and invariably have caused themselves horrendous problems, even down to the point where they are completely unable to use a carpenter's level or plumb bob. In my experience, docking charges usually exceed the rental fees for boat building property, so take your time.

Unfortunately, most stock plan packages go into very little specific detail regarding the deck layout. If you are unable to boast of extensive sailing experience you are going to be extremely surprised at the profusion of odds and ends which must be attended to before you get under way.

Your pulpits should be fabricated from at least 7/8" stainless steel tubing (not galvanized pipe or pipe fittings). Every marine catalog carries a complete line of stanchion and life line hardware. Stock pulpit kits are also available. There are few other things that can betray an amateur built boat faster than makeshift life line hardware.

There are very few thumb rules upon which you can depend for the placement of specific hardware, as every experienced skipper has a completely different idea about where things should go. The positioning of your sheet blocks, however, can be pinned down very closely. Measure the luff length of any given sail on your Sail Plan. Multiply that length by .40. Now measure the resulting distance upward from the tack of the sail along the luff and place a mark. Draw a line through the clew of the sail to the mark placed on the luff. This line, when extended to the deck, will establish a very close approximation of the fore-and-aft position of the sheet lead for that sail. This method of locating sheet leads is not always infallible, as the cut of the sails may affect it significantly, but at least it is a starting point. Only by sailing the boat will you be able to pin it down more accurately. The distance of the sheet leads from the center line and the cut of the sails varies from boat to boat.

The positioning of cleats, fairleads, ventilators, mooring cleats and other assorted paraphernalia should be kept out of walkways and positions which could cause the possible snagging of vital lines. If your designers have not so detailed this equipment, plan it out very carefully to scale before you start drilling holes. Figures 409 and 410, will give you an idea of the care and forethought which must be exercised to achieve a functional and good-looking deck layout.

All of your deck hardware must be carefully bedded to prevent leaking. Apply bedding compound to the base of the hardware, the bolt shanks, even the washers. Bed the undersides of hand rails and stanchion bases.

Test every piece of equipment before leaving the dock. Be absolutely sure that all of your rigging cotter pins are turned back in both directions. Wrap your turnbuckle pins with tape as well as anything else aboard the boat which could possibly snag sails and lines.

Run your engine at least 12 hours well below cruising speed in order to help the engine seat itself. After leaving the dock, do not operate at high speeds, as some engine misalignment will be inevitable. It will undoubtedly need some readjustment upon your return from your first day out.

Do not strain the vessel in anyway while she is new. Go at it slowly at first. Look for problems such as line chafe and improper deck, cockpit or sink drainage. Survey the boat as completely as possible when under sail so as to locate loose bolts or screws, separating joints, leaks and control problems.

At the outset your rigging should not be set up drum tight. When under sail it is common and normal for the lee shrouds to be somewhat sloppy, so do not overtighten your shrouds or stays simply to eliminate this slack. If the vessel has windward helm (tends to round up in the wind drastically), your mast may require a slight movement forward. If the vessel has a tendency to fall off (veer away from the wind), your mast should be moved aft. So do not apply a mast boot at the partners until such time as the vessel feels absolutely right.

Above all, be sure that your initial sail is done with many experienced hands on board, including your surveyor. While it will be difficult to control your enthusiasm and pride, avoid loading the boat up with land lubbers and relatives who may not be able to understand or handle an emergency situation. Do not simply head out into open water until such time as you are totally confident that you know exactly what you are doing and that both you and your boat are totally capable of such an undertaking.

Do not add your trimming ballast to the vessel to bring her down to her proper flotation line until such time as you have completely stowed her with fuel, food, navigational equipment, personal belongings, dinghy, ground tackle and all sails and hardware. And do not forget to have your compass adjusted by a professional before setting out on your first cruise.

Your patience and frustration over these past building years will be rewarded with a justified pride only if you go at it slowly and with a true sense of humility in the face of the massive power of the sea.

Plate 177. Bill Babcock's delightful little 18' canoe yawl from the board of Jay Benford has been impeccably built and rigged. Notice the use of deadeyes in lieu of rigging turnbuckles as well as the boom and mast lacing. Character boats such as this are Benford's forte, and their execution will add distinction to any waterfront.

Plate 179. Jay Benford relaxes aboard his own designed **Ragnar** on one of his rare days off. She is the epitome of both tradition and construction, and is remarkably accommodating as an excellent coastal cruiser. (A Roy Montgomery Photo; courtesy Jay Benford)

Plate 178. An **Endurance 35,** constructed by Windboats of Wroxham, England, rolls gently under a fresh quarter breeze. This vessel, from the drawing board of Peter Ibold, won the I.A.B.B.S. cruising design competition in 1973. Hundreds of these vessels are now under construction by both professionals and amateurs worldwide. (A Stuart Rodgers Photo)

Plate 180. The powerful **Limmershin,** designed by Steve Seaton and constructed by Featherstone Marine of Tampa, Florida, heads for elbowroom during her builder's trials. Few other vessels will go down in ferro-cement history as markably as this successful yacht.

Plate 181. Jay Benford's 17' cruising cat, **Puffin,** has completely disproved the claims that a successful ferro-cement boat could not be built under 25'. This photo not only shows that she is successful, but a delight, as Jay leans into a close reach. (A Roy Montgomery Photo; courtesy Jay Benford)

Plate 182. **Jefria III,** an Endurance 40 ketch, boils to weather with a "bone in her teeth." She is the product of Windboats of Wroxham, England, and the proud creation of Peter Ibold, N.A. (A Jenkins Photo)

designs and designers

The following directory of yacht designers is given only as an aid to the builder looking for those involved in ferro-cement engineering. This listing cannot be all-inclusive nor should it be assumed that it is an endorsement of these designers' works. I am certain that there are many more ferro-cement designers around the world who are eminently qualified to fulfill your needs and your own local knowledge may lead you to them.

Each designer's offerings are different in some respect. Some may offer stock plan packages while others are in-volved only in custom work. Many of the listed designers offer their consultation services for reasonable fees and some may also be qualified surveyors.

It is strongly suggested that your requests for study plans or brochures be accompanied with some recompense (particularly if a design catalog price is listed in their advertising material), as it will help the designer to defray the expenses of communicating with you. His reply (and respect) to your inquiry may be remarkably expedited if tantalized by a few dollars of consideration.

Aladdin Products
R.F.D. 2
Wiscasset, Me. 04578

Charles Bell
175 Placerville Drive
Placerville, Ca. 95667

A. L. Bellerive
20012 Wharf Street
Maple Ridge, B.C.
Canada

Jay R. Benford & Assoc., Inc.
1101 N. Northlake Way
Seattle, Wa. 98103

Bruce Bingham Yacht Design
P.O. Box 1413
Santa Barbara, Ca. 93101

Edward S. Brewer & Assoc., Inc.
Box 87
Brooklin, Me. 04616

Ronald S. Carter
528 Melcher Ave.
Akron, Oh. 44319

Coast Engineering Co.
711 W. 21st Street
Norfolk, Va. 23517

Thomas E. Colvin
Miles P.O. (Mathews County),
Va. 23114

W.I.B. Crealock
657-J W. 19th Street
Costa Mesa, Ca. 92627

Andrew Davidhazy
2822 N.W. 92nd
Seattle, Wa. 98107

John H. Davies
Box 286
Huntington, N.Y. 11746

Eldredge-McInnis Inc.
57 Water Street
Hingham, Ma. 02043

Miles D. Fitch
360 Greely Road
Cumberland Center, Me. 04021

Ken Hankinson
Box 756
Bellflower, Ca. 90706

Robert B. Harris
199 W. Shore Road
Great Neck, N.Y. 11024

Hartley's Boat Plans Ltd.
Box 30094
Takapuna North
Auckland, New Zealand

Donald G. Hogue
4107 Francis Avenue
N. Seattle, Wa. 98103

Holland Marine Design
3510 Geary Blvd.
San Francisco, Ca. 94118

Peter A. Ibold
69, Rue Galande
75-Paris (50), France

James R. Kerr
R.D. No. I, Janice Drive
Sussey, N.J. 07461

James S. Krogen & Co., Inc.
1460 Breckell Ave.
Miami, Fl. 33131

J. W. Lawson
13205 Ovalstone Lane
Bowie, Md. 20715

Marinecraft
Box 161
Brighton, Ma. 02135

Marine Crafts International
Marine Center Bldg.
The Breakwater
Santa Barbara, Ca. 93109

David P. Martin
306 23rd Street
Brigantine, N.J. 08203

Robert L. McMurray
427 Marin Avenue
Mill Valley, Ca. 94941

James McPherson
260 California Street
San Francisco, Ca. 94111

George E. Meese
194 Acton Road
Annapolis, Md. 21403

Robert Miller
96 Holly Tree Lane
Toms River, N.J. 08753

Chester A. Nedwidek
229-A Heritage Village
Southbury, Ct. 06488

Arthur D. Nelson
Daniels Street
Franklin, Ma. 02038

Ben Ostlund
515 Signal Road
Newport Beach, Ca. 92660

Robert H. Perry
2406 2nd Avenue N.
Seattle, Wa. 98109

William Preston
2899 Santiago
Florissant, Ms. 63033

Romack Marine
Box 20481
Long Beach, Ca. 90813

Jack Rouse & Assoc.
900 W. Pacific Coast Hwy.
Long Beach, Ca. 90608

Samson Marine Design Ent.
833 River Road
Richmond, B.C.
Canada

Jack E. Satterfield
12 Oak Lane
Gretna, La. 70053

Stephen R. Seaton, N.A.
2506 Cortez Road
Bradenton, Fl. 33507

Blaine Seeley Associates
1525 Superior Avenue
Newport Beach, Ca. 92627

John B. Seward
105 Edison
Corte Madera, Ca. 94925

Donald H. Smith
890 Country Club Road
Crystal Lake, Il. 60014

Eliot Spalding
Middle Street
South Freeport, Me. 04078

Victor Tchetchet
80 Knightsbridge Gardens
Great Neck, N.Y. 11023

Alexander W. Vetter
Box 376
Millbrae, Ca. 94030

Lauren Williams Yacht Design
Box 137
Mill Valley, Ca. 94941

Charles W. Wittholz
315 Lexington Drive
Silver Spring, Md. 20901

Arthur R. Wycoff
23244 West Road
Cleveland, Oh. 44138

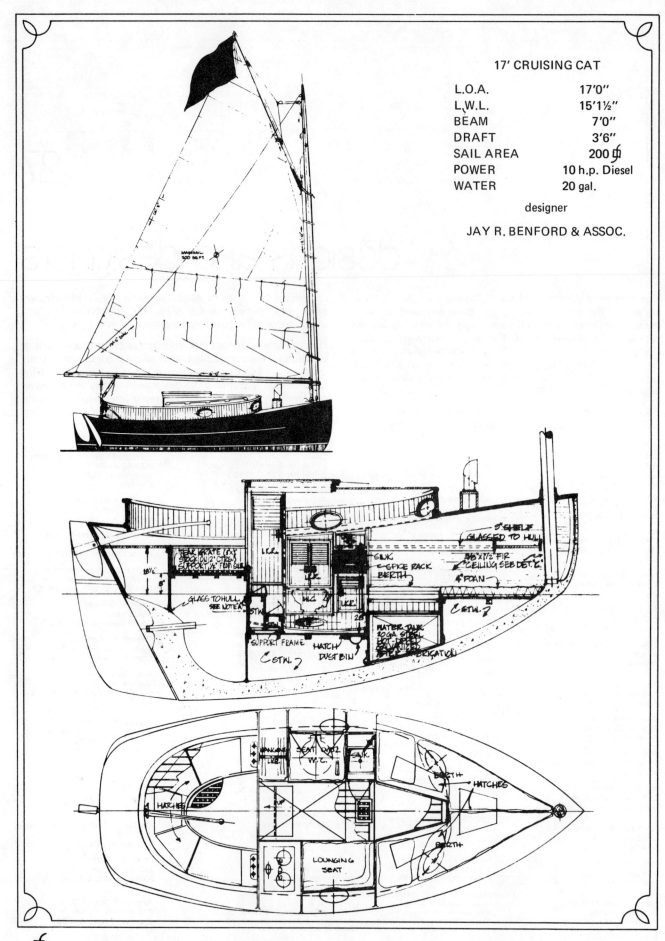

17' CRUISING CAT

L.O.A.	17'0"
L.W.L.	15'1½"
BEAM	7'0"
DRAFT	3'6"
SAIL AREA	200 ⌀
POWER	10 h.p. Diesel
WATER	20 gal.

designer

JAY R. BENFORD & ASSOC.

⌀ is the symbol for designating sail area in square feet.

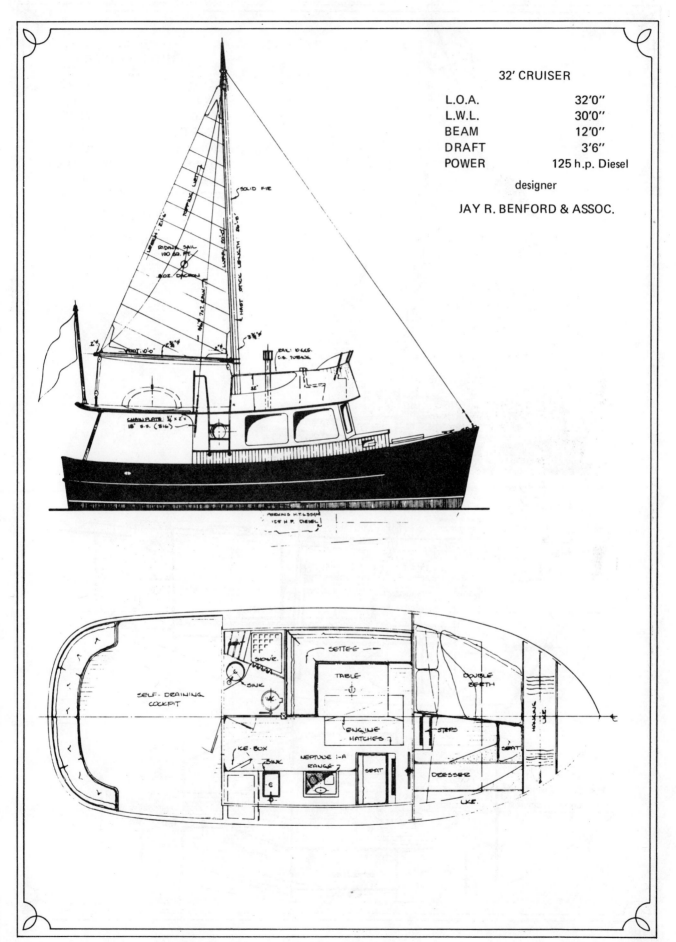

32' CRUISER

L.O.A.	32'0"
L.W.L.	30'0"
BEAM	12'0"
DRAFT	3'6"
POWER	125 h.p. Diesel

designer

JAY R. BENFORD & ASSOC.

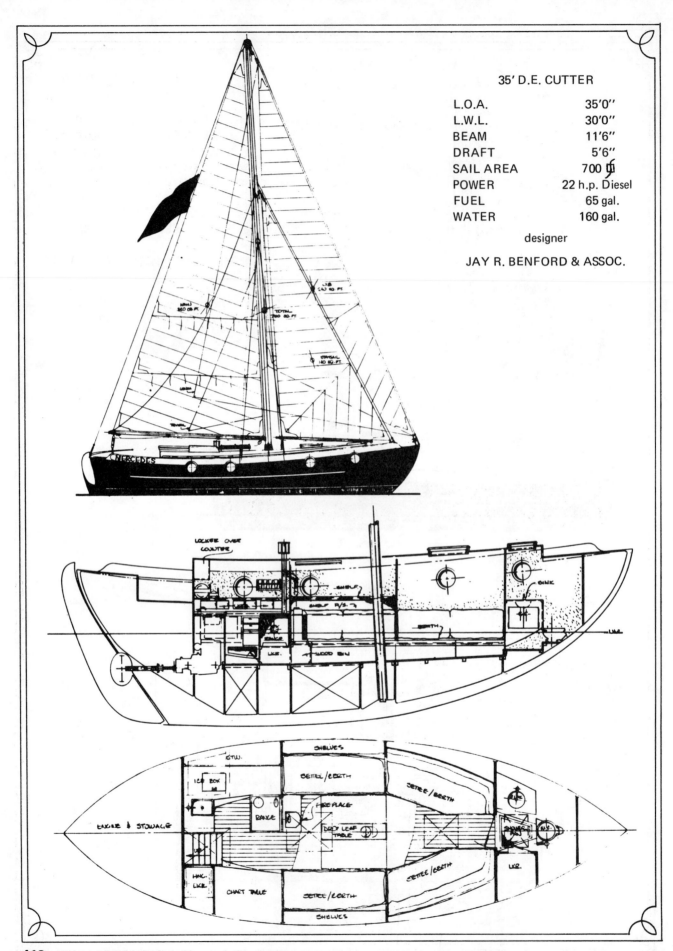

35' D.E. CUTTER

L.O.A.	35'0"
L.W.L.	30'0"
BEAM	11'6"
DRAFT	5'6"
SAIL AREA	700 sq. ft.
POWER	22 h.p. Diesel
FUEL	65 gal.
WATER	160 gal.

designer

JAY R. BENFORD & ASSOC.

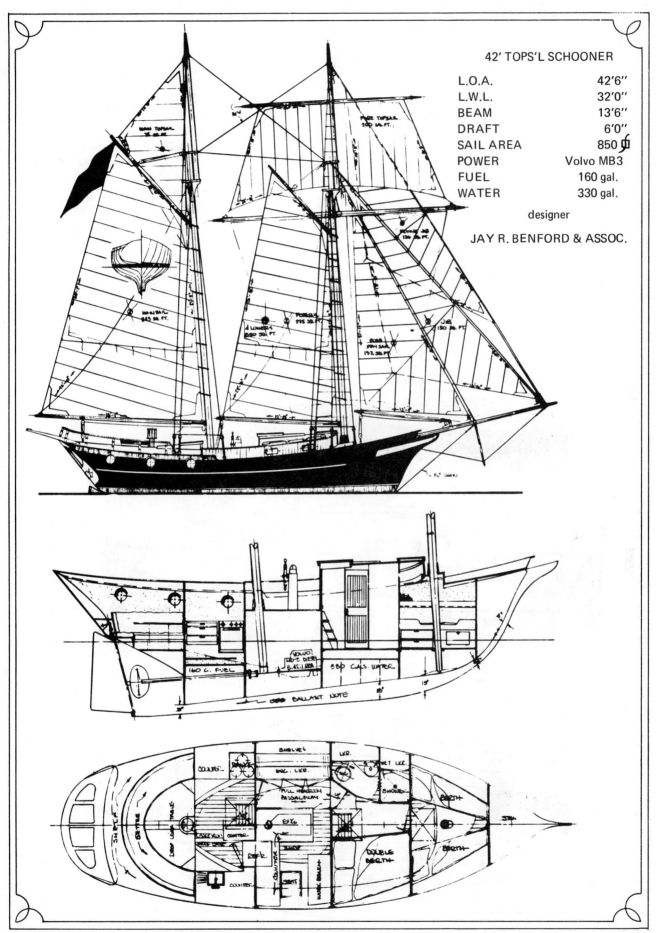

42' TOPS'L SCHOONER

L.O.A.	42'6''
L.W.L.	32'0''
BEAM	13'6''
DRAFT	6'0''
SAIL AREA	850
POWER	Volvo MB3
FUEL	160 gal.
WATER	330 gal.

designer

JAY R. BENFORD & ASSOC.

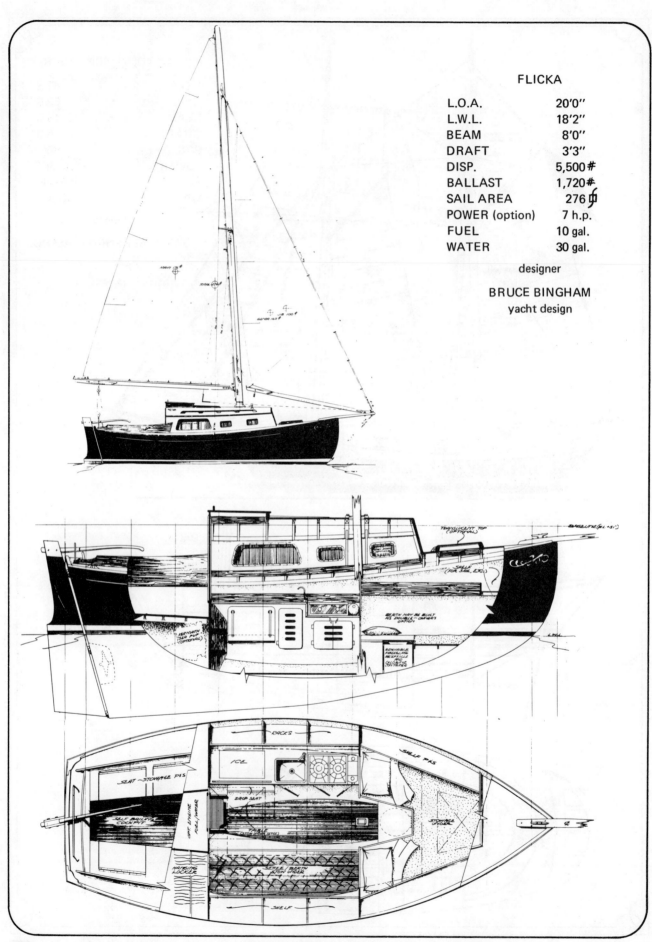

FLICKA

L.O.A.	20'0"
L.W.L.	18'2"
BEAM	8'0"
DRAFT	3'3"
DISP.	5,500 #
BALLAST	1,720 #
SAIL AREA	276 ⌀
POWER (option)	7 h.p.
FUEL	10 gal.
WATER	30 gal.

designer

BRUCE BINGHAM
yacht design

ANASTASIA

L.O.A.	32'0''
L.W.L.	25'6''
BEAM	11'0''
DRAFT	4'7''
DISP.	19,000#
BALLAST	6,500#
SAIL AREA	585 ☐
POWER	Perkins 4-107
FUEL	50 gal.
WATER	160 gal.

designer

BRUCE BINGHAM
yacht design

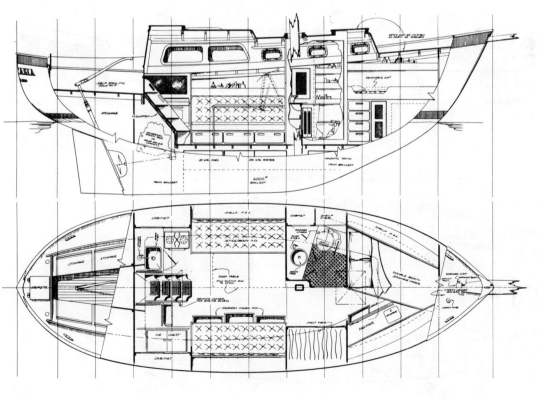

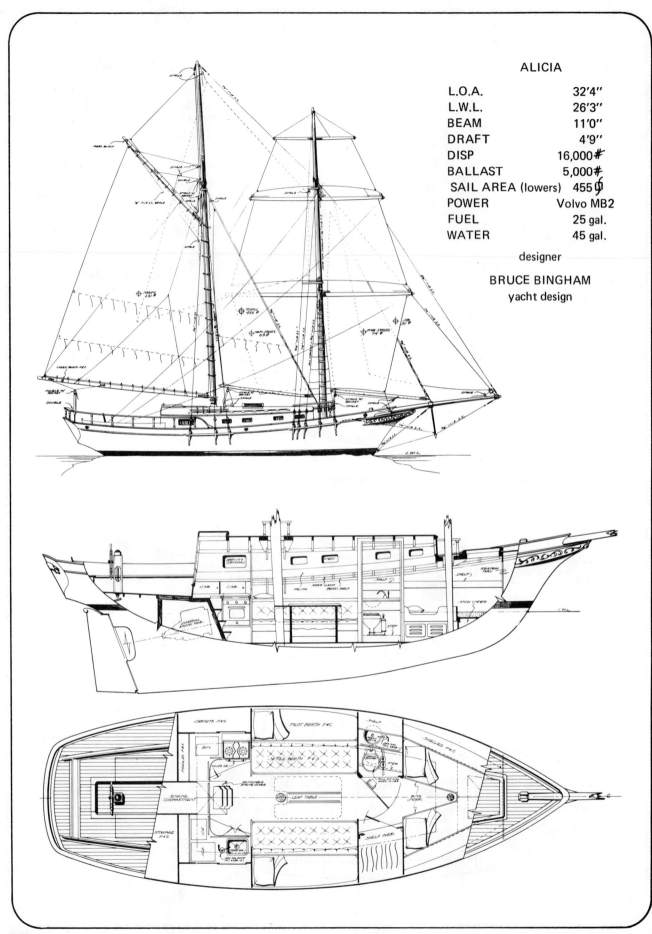

ALICIA

L.O.A.	32'4''
L.W.L.	26'3''
BEAM	11'0''
DRAFT	4'9''
DISP	16,000#
BALLAST	5,000#
SAIL AREA (lowers)	455 ☐
POWER	Volvo MB2
FUEL	25 gal.
WATER	45 gal.

designer

BRUCE BINGHAM
yacht design

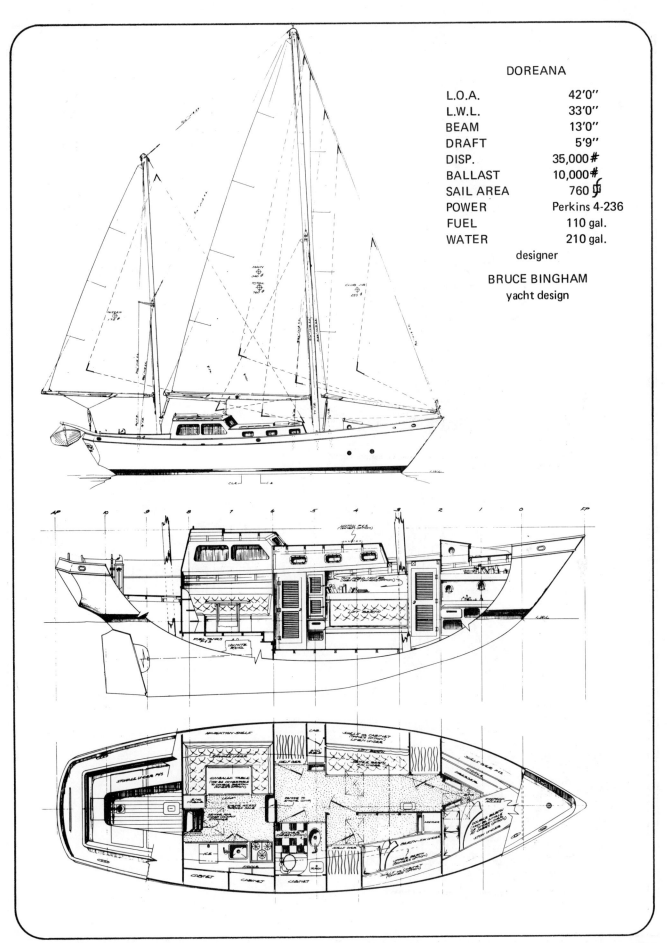

DOREANA

L.O.A.	42'0"
L.W.L.	33'0"
BEAM	13'0"
DRAFT	5'9"
DISP.	35,000#
BALLAST	10,000#
SAIL AREA	760 🔲
POWER	Perkins 4-236
FUEL	110 gal.
WATER	210 gal.

designer

BRUCE BINGHAM
yacht design

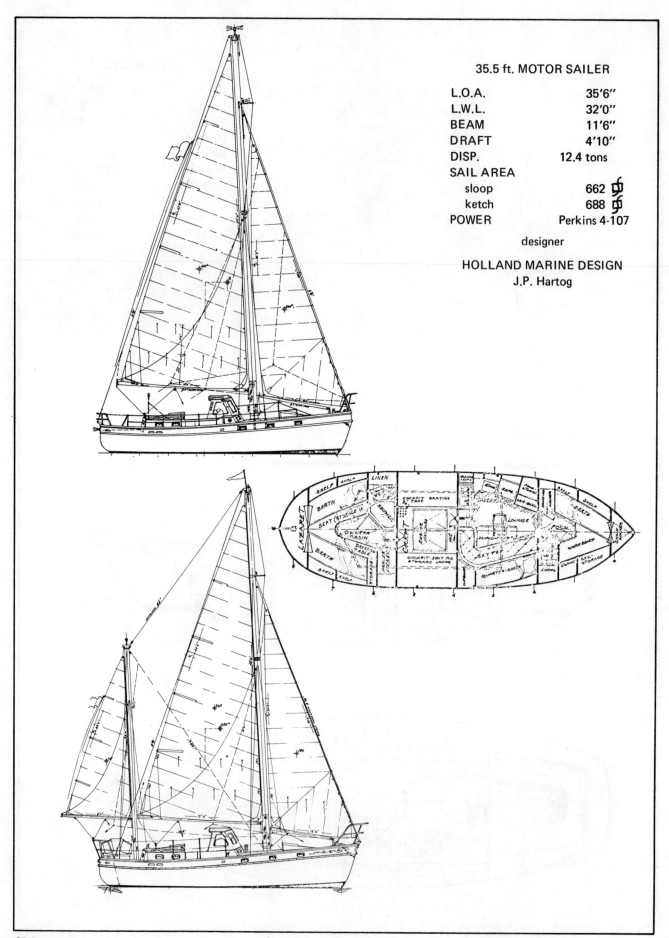

35.5 ft. MOTOR SAILER

L.O.A.	35'6''
L.W.L.	32'0''
BEAM	11'6''
DRAFT	4'10''
DISP.	12.4 tons
SAIL AREA	
sloop	662 ☐
ketch	688 ☐
POWER	Perkins 4-107

designer

HOLLAND MARINE DESIGN
J.P. Hartog

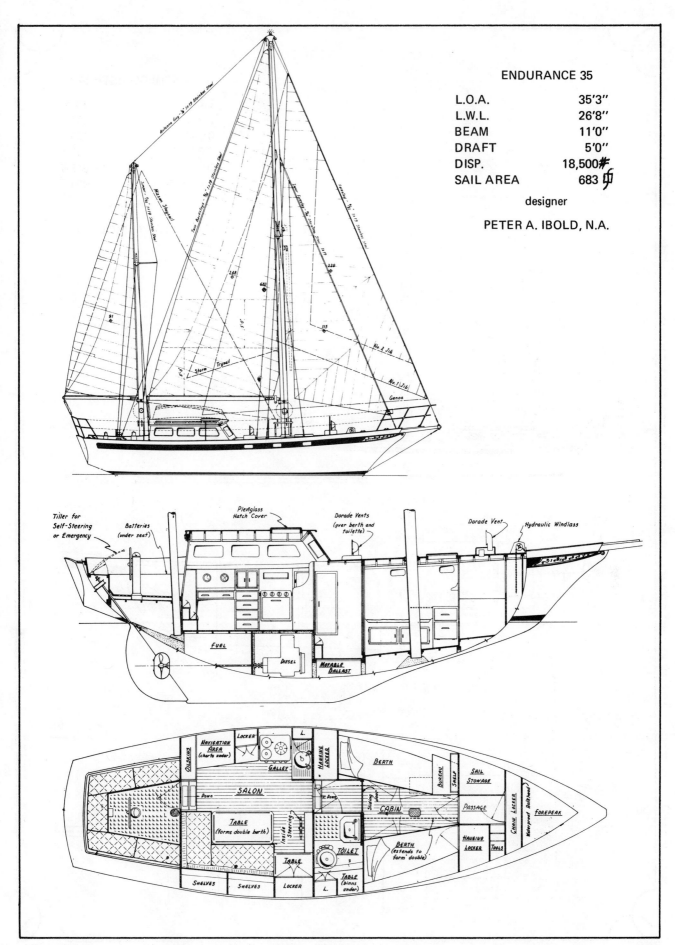

ENDURANCE 35

L.O.A.	35'3"
L.W.L.	26'8"
BEAM	11'0"
DRAFT	5'0"
DISP.	18,500#
SAIL AREA	683 ☐

designer

PETER A. IBOLD, N.A.

Tiller for
Self-Steering
or Emergency

Batteries
(under seat)

Plexiglass
Hatch Cover

Dorade Vents
(over berth and
toilette)

Dorade Vent

Hydraulic Windlass

FUEL

Diesel

MOVABLE
BALLAST

OILSKINS

Navigation
Area
(charts under)

Locker

L.

HANGING LOCKER

BERTH

BUREAU

SHELF

SAIL
STOWAGE

GALLEY

SALON

Down

Down

Sliding door

CABIN

PASSAGE

CHAIN LOCKER

Waterproof Bulkhead

FOREPEAK

TABLE
(forms double berth)

Inside
Steering

TOILET

BERTH
(extends to
form double)

HANGING
LOCKER

Tools

TABLE

SHELVES

SHELVES

LOCKER

L.

TABLE
(bins under)

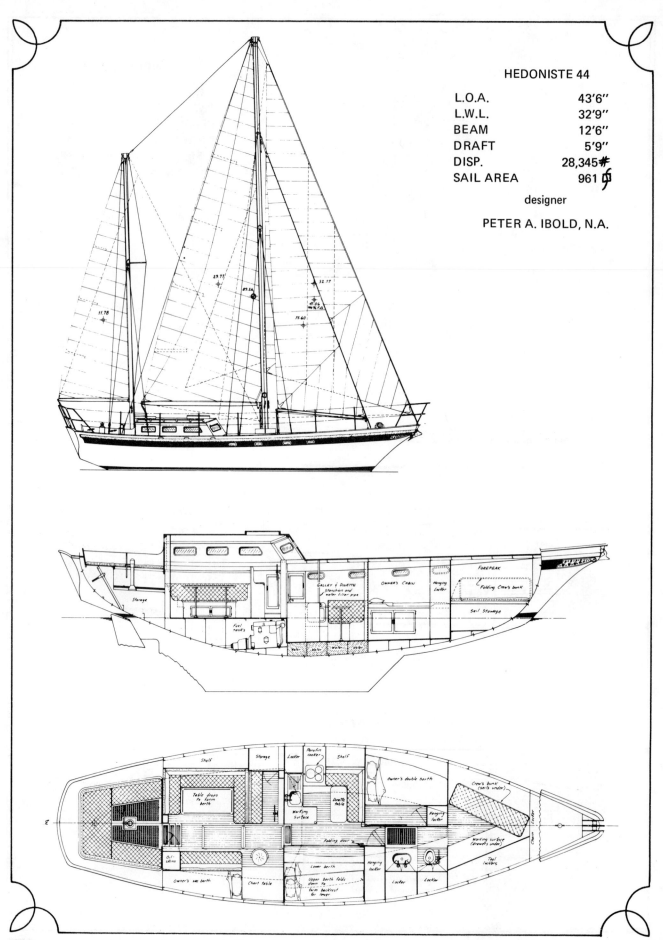

HEDONISTE 44

L.O.A.	43′6″
L.W.L.	32′9″
BEAM	12′6″
DRAFT	5′9″
DISP.	28,345#
SAIL AREA	961 ⊡

designer

PETER A. IBOLD, N.A.

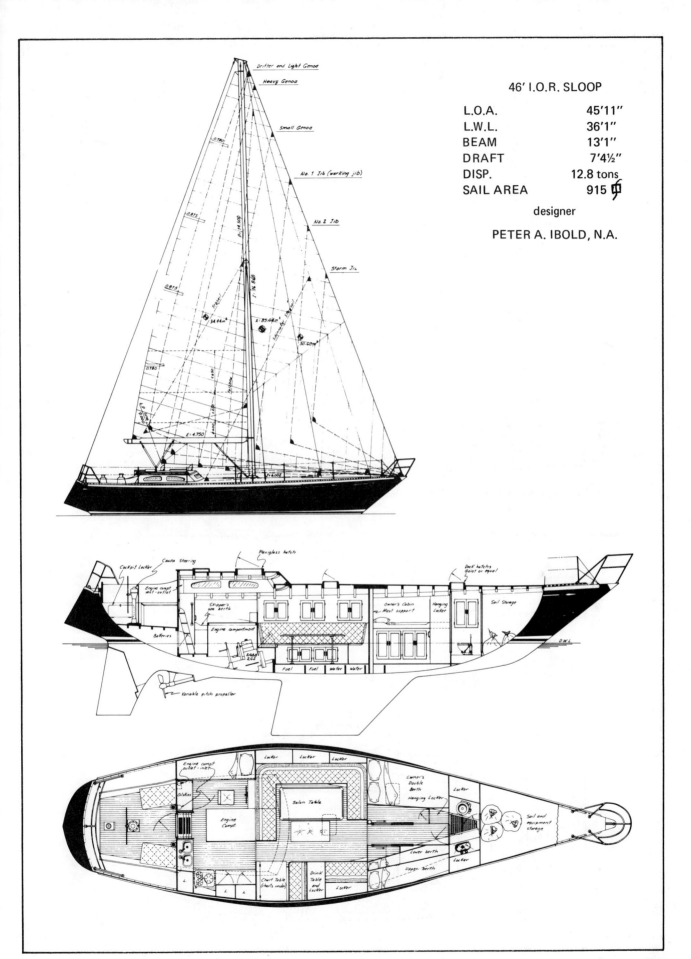

46' I.O.R. SLOOP

L.O.A.	45'11"
L.W.L.	36'1"
BEAM	13'1"
DRAFT	7'4½"
DISP.	12.8 tons
SAIL AREA	915

designer

PETER A. IBOLD, N.A.

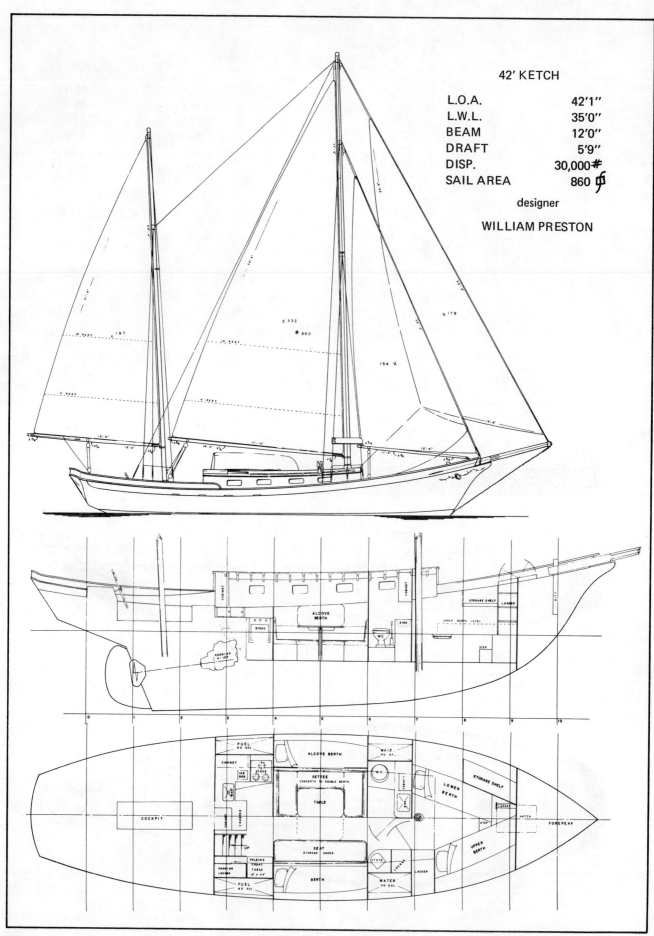

42' KETCH

L.O.A.	42'1"
L.W.L.	35'0"
BEAM	12'0"
DRAFT	5'9"
DISP.	30,000#
SAIL AREA	860 ☐'

designer

WILLIAM PRESTON

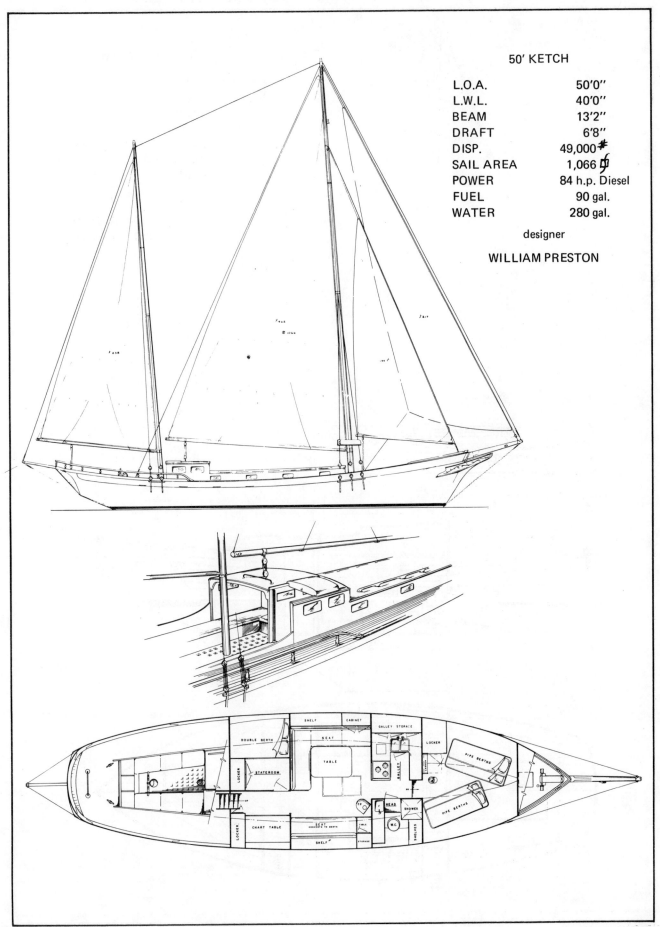

50' KETCH

L.O.A.	50'0"
L.W.L.	40'0"
BEAM	13'2"
DRAFT	6'8"
DISP.	49,000#
SAIL AREA	1,066 ☐
POWER	84 h.p. Diesel
FUEL	90 gal.
WATER	280 gal.

designer

WILLIAM PRESTON

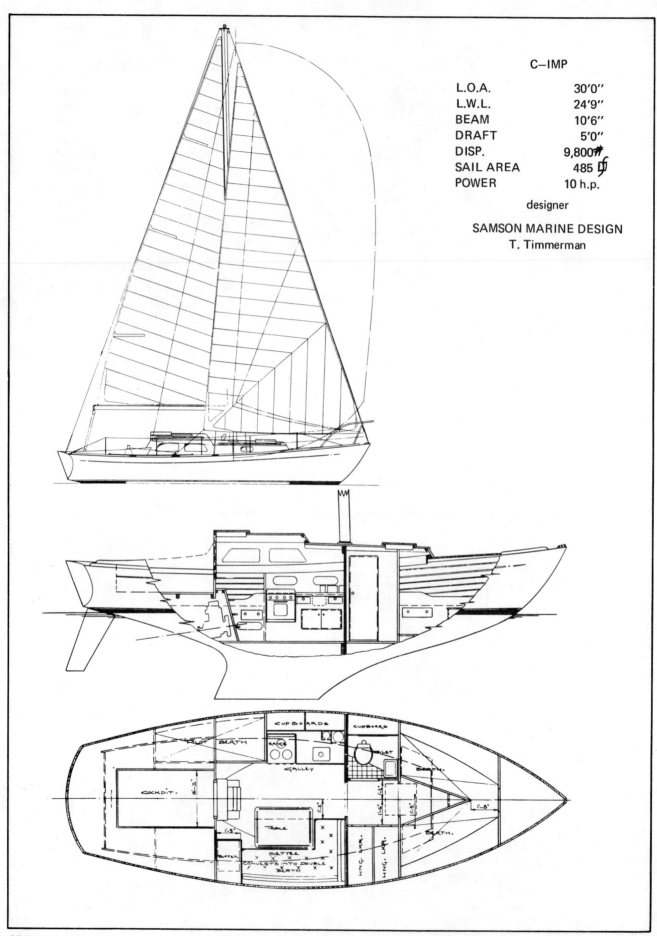

C—IMP

L.O.A.	30'0''
L.W.L.	24'9''
BEAM	10'6''
DRAFT	5'0''
DISP.	9,800#
SAIL AREA	485 ⬚
POWER	10 h.p.

designer

SAMSON MARINE DESIGN
T. Timmerman

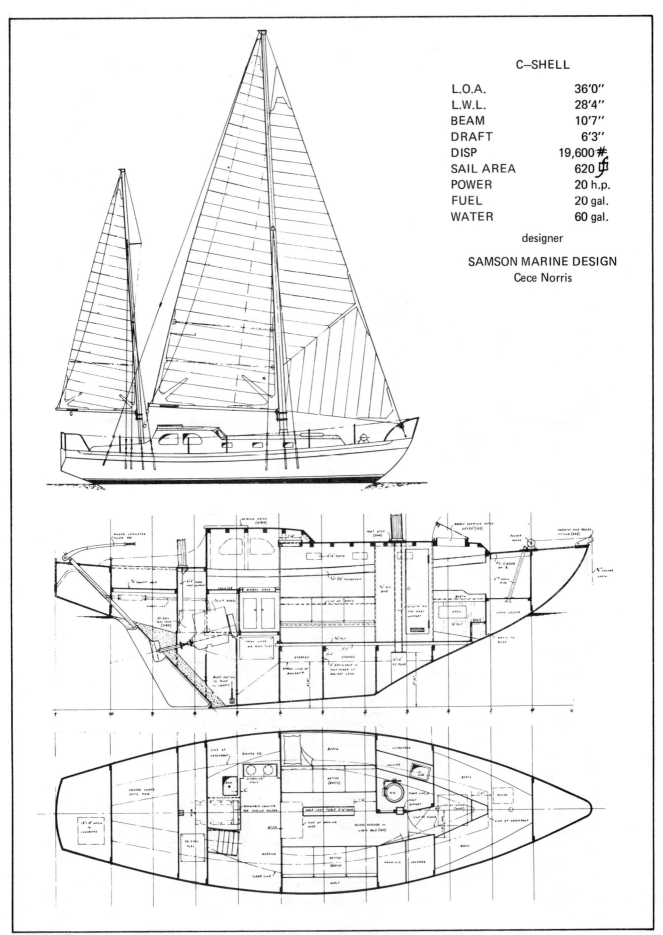

C–SHELL

L.O.A.	36'0"
L.W.L.	28'4"
BEAM	10'7"
DRAFT	6'3"
DISP	19,600#
SAIL AREA	620 ☐
POWER	20 h.p.
FUEL	20 gal.
WATER	60 gal.

designer

SAMSON MARINE DESIGN
Cece Norris

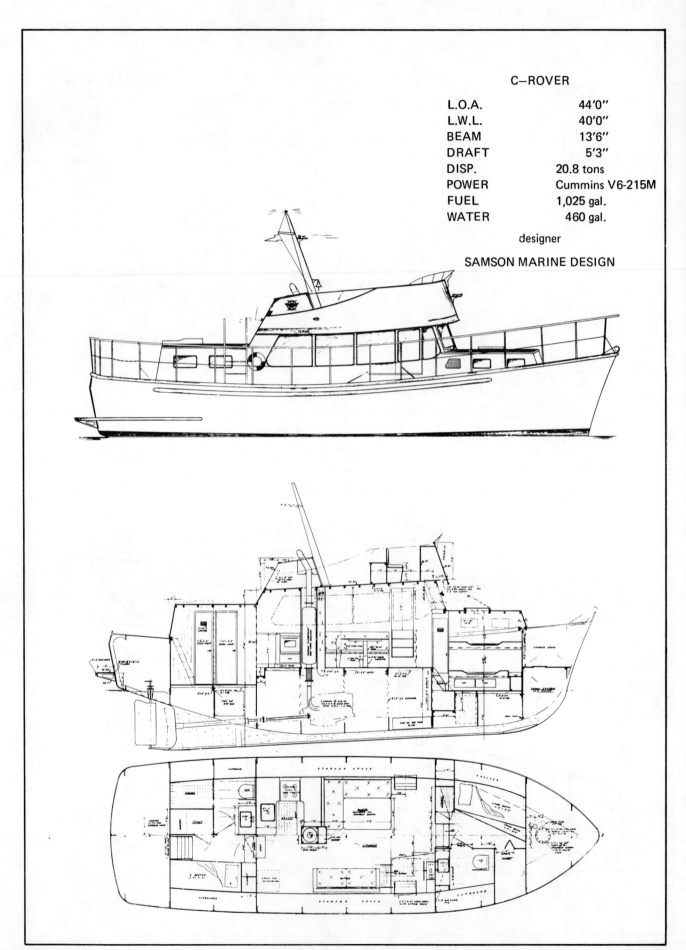

C—ROVER

L.O.A.	44'0"
L.W.L.	40'0"
BEAM	13'6"
DRAFT	5'3"
DISP.	20.8 tons
POWER	Cummins V6-215M
FUEL	1,025 gal.
WATER	460 gal.

designer

SAMSON MARINE DESIGN

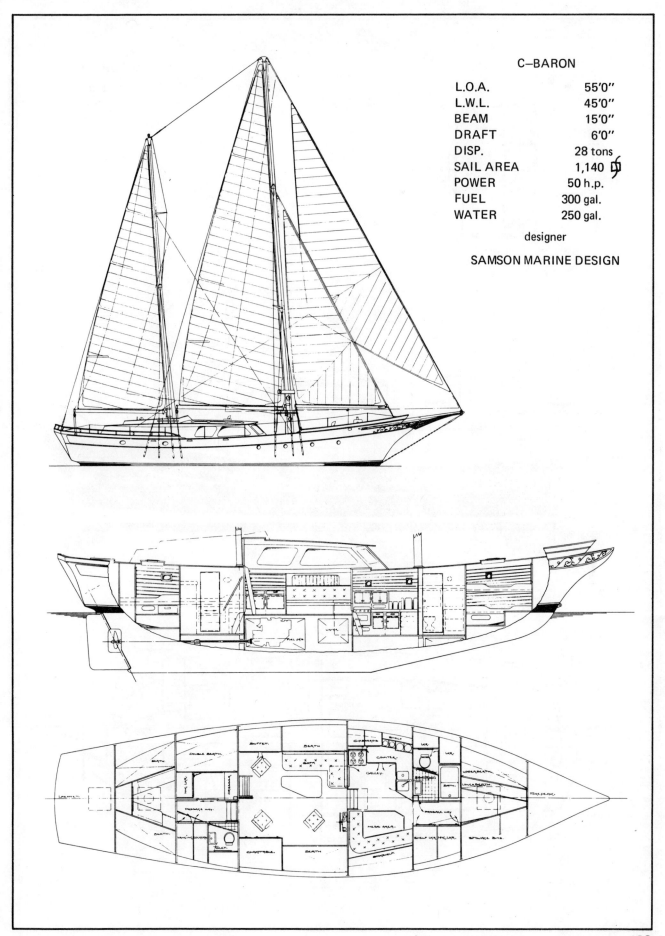

C—BARON

L.O.A.	55'0"
L.W.L.	45'0"
BEAM	15'0"
DRAFT	6'0"
DISP.	28 tons
SAIL AREA	1,140
POWER	50 h.p.
FUEL	300 gal.
WATER	250 gal.

designer

SAMSON MARINE DESIGN

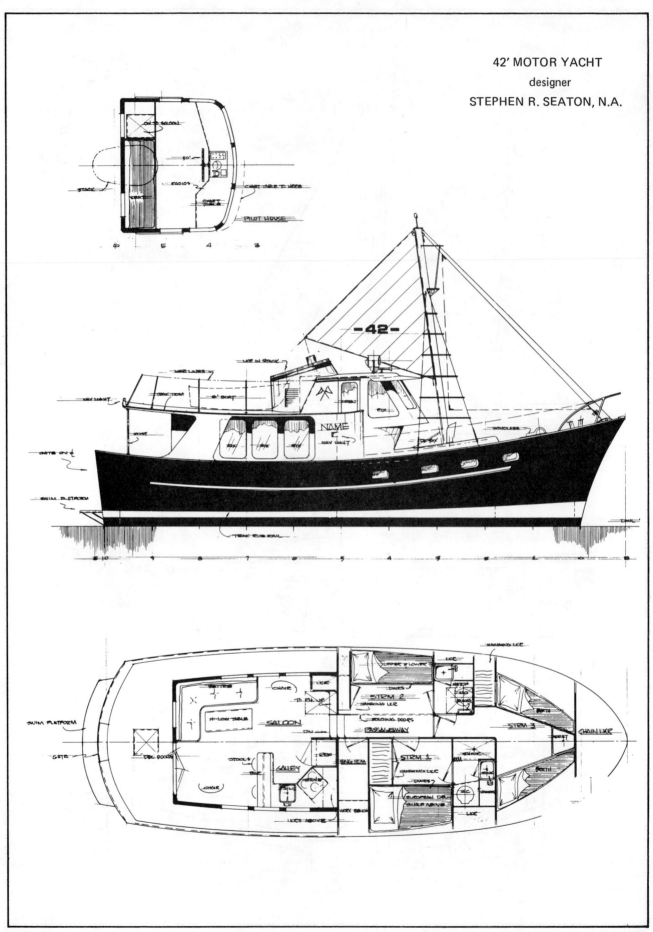

42' MOTOR YACHT
designer
STEPHEN R. SEATON, N.A.

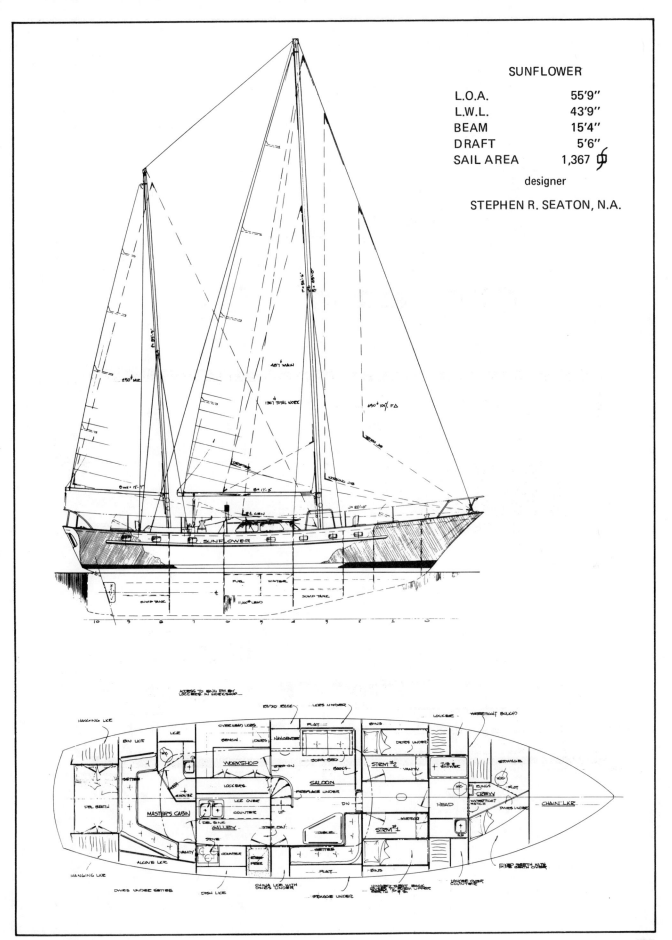

SUNFLOWER

L.O.A.	55'9"
L.W.L.	43'9"
BEAM	15'4"
DRAFT	5'6"
SAIL AREA	1,367 ☐

designer

STEPHEN R. SEATON, N.A.

435

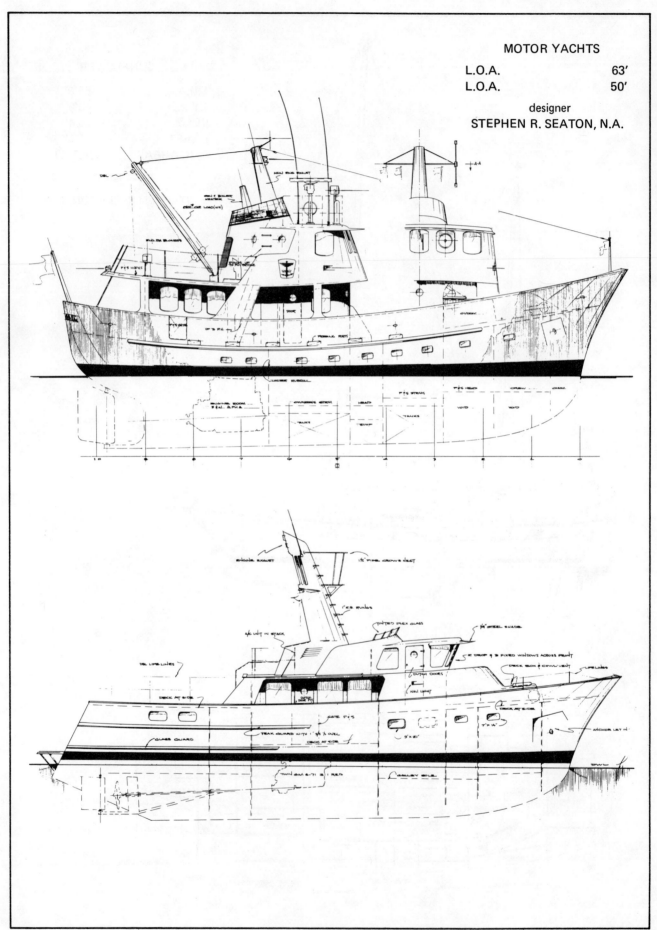

index

index